Organophosphorus Pesticides:
Organic and Biological Chemistry

Organophosphorus Pesticides: Organic and Biological Chemistry

Author

Morifusa Eto

Department of Agricultural Chemistry
Kyushu University
Fukuoka, Japan

Editor-in-Chief
CRC Pesticide Series

Gunter Zweig

U.S. Environmental Protection Agency
Washington, D.C.

CRC PRESS, INC.
Boca Raton, Florida 33431

Library of Congress Cataloging in Publication Data

Eto, Morifusa, 1930-
    Organophosphorus pesticides.

    (Pesticide chemistry series)
    Bibliography:
    1. Organophosphorus compounds. 2. Pesticides.
1. Title. [DNLM: 1. Chemistry, Agricultural.
2. Pesticides. 3. Phosphates. SB952.P5 E85o]
SB952.P5E86 632'.95'0154707   73-90239   ISBN
0-8493-5021-2

© 1979 by CRC Press, Inc.

Second Printing, January 1976
Third Printing, January 1977
Fourth Printing, June 1979

International Standard Book Number 0-8493-5021-2
Former International Standard Book Number 0-87819-024-4

Library of Congress Card Number 73-90239
Printed in the United States

Dedicated to
Professors Y. Oshima and J. E. Casida

FOREWORD

This book by Professor Eto on the organophosphorus pesticides is the second book in CRC's treatise on PESTICIDES. The first book in this series, on the chlorinated pesticides, by G. T. Brooks, was published earlier this year in two volumes. It discusses in great detail the various facets of the chlorinated pesticides, many of which are slowly being phased out, although environmental concentrations will be with us for decades to come. Professor Eto's treatment of the organophosphorus pesticides, many of which have replaced the more persistent chlorinated hydrocarbons, appears, therefore, at a very opportune time, when the use of these compounds is increasing at a rapid rate. Professor Eto's detailed discussions on the synthesis, the mode of action, and the biochemical and physical breakdown of these compounds will be of great use to scientists and decision makers. The subjects are brought up to date in a skillful fashion, encompassing the literature through the middle of 1974. The Editor has endeavored to leave the text by Professor Eto in the original language and only to edit slightly, where necessary, in order to avoid ambiguity.

Future treatises of this important series will cover the carbamate pesticides, herbicides, and fungicides, and I anticipate that the series, once it is published completely, will represent the most comprehensive treatment of the subject of PESTICIDES.

Washington, D.C.
Summer, 1974

Gunter Zweig, Ph.D.
Editor
Pesticide Series

PREFACE

Recent development in the pure and applied chemistry of organophosphorus compounds is amazing. Even though the organic derivatives of phosphoric acid occupy only a part of the family of organophosphorus compounds, the variation of the four groups attached to the phosphorus atom is almost infinite, and they display a great variety of significant biological activities. Many partial esters play essential roles in normal biological systems, whereas many neutral derivatives display biocidal activities. Since the finding of insecticidal activity in the latter part of the 1930's by Schrader, a number of neutral phosphate derivatives have been developed into practical insecticides. Owing to their high activity and biodegradability, their application to agriculture, public health, and related fields has been growing rapidly. Moreover, the pesticidal activities of organophosphorus compounds, which are not restricted to phosphate derivatives, also include acaricidal, nematocidal, anthelmintic, insect sterilizing, fungicidal, herbicidal, and rodenticidal activities.

In order to get a general concept of organophosphorus pesticides with such a variety in structure and biological activities, consideration of each aspect of chemistry, biochemistry, and the applied sciences is necessary. This book consists of these three main parts. After the presentation of the background of phosphorus chemistry in Chapter I, stress was put on the chemical and biochemical reactions of organophosphorus pesticides, including synthesis, analysis, metabolism mode of action, and other interesting aspects in Chapters II to IV, and on the structure-pesticidal activity relationship in Chapter V. Almost all commercialized and important experimental organophosphorus pesticides are mentioned. Unlike insecticides, the mode of action of other pesticides is not well established, and the scattered knowledge about them is difficult to generalize. They are described in the corresponding sections of Chapter V, "Individual Pesticides." Although this book is designed as a readable text, it has reference value also. It is the author's pleasure that this book will not only be useful to students and beginners, but will also assist specialists dealing with pesticides.

The author wishes to express his sincere thanks to Dr. G. Zweig for editing this book, to Dr. W. H. Harned for reading the manuscript, and to the staff of CRC Press for their cooperation. Figures were drawn by Miss S. Ishida. Thanks are also due to many colleagues in Kyushu University and the University of California for their encouragement and assistance. The writing of the manuscript was started in Fukuoka and completed in Berkeley.

M. Eto
Berkeley, August 1974

THE AUTHOR

Morifusa Eto graduated from Kyushu University, Fukuoka, 1952. He studied further on agricultural chemistry as a graduate student at the same university, and then was engaged in research activity at the Department of Pharmaceutical Sciences. In 1957, he was appointed an instructor at the pesticide chemistry laboratory of Professor Y. Oshima, in Kyushu University. From 1960 to 1961, he joined the research group of Professor J. E. Casida at the University of Wisconsin, Madison, to investigate the mode of action of tri-o-tolyl phosphate. His doctoral degree was obtained in 1962 from Kyushu University. In 1963, the Agricultural Chemical Society of Japan awarded him an academic prize for his chemical and biochemical studies on saligenin cyclic phosphorus esters. This work resulted in the finding of the new insecticide, Salithion. In the same year, he was appointed the Associate Professor of Pesticide Chemistry in Kyushu University In 1973, he was invited to the University of California, Berkeley, as a Visiting Associate Professor for one year. He is still working on the chemistry and biochemistry of organophosphorus compounds.

TABLE OF CONTENTS

Chapter III
Chemical Reactions

Chapter IV
Biochemistry

Chapter I

INTRODUCTION

A. HISTORY

Organic compounds of phosphorus are the essential constituent of protoplasm and play important roles for maintenance of life, for example, as nucleic acids, nucleotide coenzymes, metabolic intermediates, and phosphatides. On the other hand, many organophosphorus compounds are artificially produced for the practical uses of lubricants, oil additives, plasticizers, and pesticides.[1] Organophosphorus pesticides include not only insecticides, but also fungicides, herbicides, and others. It is surprising to know that such great varieties in chemical, physical, and biological properties are governed by the selection of groups attached on the phosphorus atom.

The research in the field of the organic chemistry of phosphorus was first undertaken by Lassaigne, in 1820, to prepare phosphate esters.[2] The chemistry of organophosphorus compounds was developed extensively by Michaelis in Germany, during the late 19th century and the beginning of this century.[3] He performed many works and gave a foundation for this field, particularly on the chemistry of compounds containing the P-N bond.[4] Overlapping the latter stages of Michaelis, a Russian chemist, A. E. Arbuzov, conducted extensive research, especially on the chemistry of trivalent phosphorus compounds, including the famous Michaelis-Arbuzov reaction to form the P-C linkage.[5] His work has been continued by his son, B. A. Arbuzov.

On the other hand, since Harden and Young disclosed, in 1905, the importance of inorganic phosphate on alcoholic fermentation and discovered fructose diphosphate as a metabolic intermediate,[6] many interesting organic phosphate esters have been found from biological sources. Since 1945, a systematic investigation on phosphorylation reactions has been carried out by the school of Sir Todd, and then by Cramer, in order to synthesize naturally occurring phosphate esters.[7,8]

Physiologically abnormal effects of organophosphorus compounds were first observed in dialkyl phosphorofluoridates by Lange and Krueger in 1932.[9] They intended to find new types of organic pesticides at that time.[10] During the Second War, Saunders in England and Schrader in Germany worked on toxic phosphorus compounds. Saunders synthesized nerve poisons, including diisopropyl phosphorofluoridate (DFP).[11] Schrader and his co-workers found, in 1937, a contact insecticidal activity in some organophosphorus compounds of the general formula:

where R^1, R^2, and R^3 are alkyl groups, and "acyl" is an inorganic or organic acid radical such as Cl, F, SCN, and CH_3COO.[12] Since this point, a variety of fruitful results have been brought forth. Schrader et al. found, in 1941, a systemic insecticide, octamethylpyrophosphoramide (OMPA), which was later named schradan after its discoverer, and also discovered a number of insecticidal organophosphorus esters, including the first practical insecticide named "Bladan," which contained tetraethyl pyrophosphate (TEPP) and was marketed in Germany in 1944. The synthesis of tetraethyl pyrophosphate was first performed by Moschnine, and then by De Clemont in 1854, and was repeated by several authors, including Nylén (1930) and Arbuzov (1938), without their taking notice of its toxicity. De Clemont tasted TEPP, but did not realize its toxicity.[10]

The great advancements in agricultural practice and scientific knowledge on the structure-activity relationship of organophosphorus insecticides were achieved by the discovery of compound No. 605, named parathion, diethyl p-nitrophenyl phosphorothionate, by Schrader in 1944. Although parathion itself is extremely toxic to mammals as well as to insects, many less toxic insecticides have been developed by slight structural modifications; for instance, chlorthion,[13] fenthion,[14] and fenitrothion[15] were discovered in 1952, 1958, and 1959, respectively.

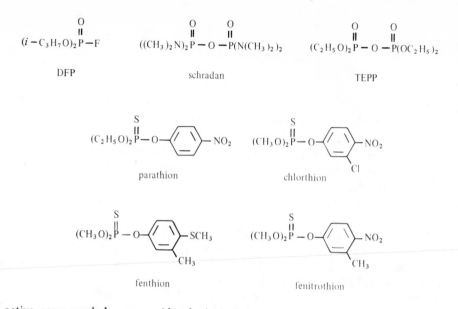

DFP   schradan   TEPP

parathion   chlorthion

fenthion   fenitrothion

All these active compounds have an acid anhydride linkage and a general formula for biologically active organophosphorus compounds first proposed by Schrader:[12]

where, R^1 and R^2 = alkyl, alkoxy or amino groups, and acyl = any acid residue.

Another important compound with low mammalian toxicity, malathion, discovered by American Cyanamid Co. in 1950, has the carboxy ester group. Demeton, found in 1951 by Bayer AG, and its related compounds are important insecticides of another class, which have a thioether group and systemic insecticidal activity.

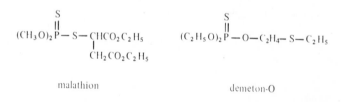

malathion   demeton-O

In 1952, the Perkow reaction was discovered,[16] and many important vinyl phosphate esters have been recently introduced as practical insecticides (see Section II.A.4).

The inhibitory activity of organophosphorus esters against cholinesterases was first found, in 1941, by Adrian and his co-workers.[17] It was found by Balls, in 1949, that the inhibition was attributable to the phosphorylation of the esteratic site. (This was initially demonstrated by the reaction of DFP with chymotrypsin.)[18] Actual thiono type insecticides do not inhibit the esterases in vitro, but their activation to potent anticholinesterases occurs in vivo. Gage demonstrated, in 1953, that the inhibitor produced from parathion in vivo was the oxo analog, paraoxon.[19] The mode of action, metabolism, and selective toxicity of organophosphorus insecticides have been actively investigated, particularly in the United States.

In addition to insecticidal and acaricidal activities, a variety of biological activities of organophosphorus compounds was discovered. Some compounds are useful as nematocides, others as anthelmintic agents. Since the fungicide, Kitazin, was discovered in 1963 in Japan,[20] a number of organophosphorus fungicides have been developed. Many of them are phosphorothiolate esters. Furthermore, herbicidal phosphorus amidoesters were also recently marketed. Plant growth regulators have also been developed, containing phosphonic acid, phosphorotrithiolate, phosphorotrithioite, and phosphonium salt. Rodenticides were recently developed, and many aziridine derivatives of phosphoric acid are actively studied for insect chemosterilants.[21] Finally, certain organophosphorus compounds are active as antitumor agents; cyclophosphamide (Endoxan) is a typical example.[22]

Because of their relatively low persistency and high effectiveness, organophosphorus pesticides are now used widely in the world. Thus, about 140 phosphorus compounds are or were used as practical pesticides (including plant growth regulators) in the world. More than 60,000 tons a year of organophosphorus pesticides are produced in the United States alone.

For the chemistry and biological chemistry of organophosphorus compounds and pesticides, many excellent books have been published.[2,3,11,13,23-36]

Kitazin   cyclophosphamide

B. NOMENCLATURE

The nomenclature of the organic phosphorus compounds is confusing because different countries use their own system. Some examples of different nomenclature systems adopted by Chemical Abstracts, Beilstein, Scandinavia, Kosolapoff's book,[3] and others have been shown by Fest.[2] In this book, the rules[37] established by the agreement of the British Chemical Society and American Chemical Society in 1952 are adopted.

Organophosphorus compounds are named as derivatives of the corresponding parent compounds: acids or hydrides. Some of the parents are listed below.

Hydrides	
Phosphine	H_3P
Phosphine oxide	H_3PO
Phosphine sulfide	H_3PS
Phosphorane	H_5P
Trivalent acids	
Phosphorous acid	$(HO)_3P$
Phosphonous acid	$(HO)_2PH$
Phosphinous acid	$HOPH_2$
Pentavalent acids	
Phosphoric acid	$(HO)_3PO$
Phosphonic acid	$(HO)_2HPO$
Phosphinic acid	$(HO)H_2PO$

As shown in the following examples, structures containing C-P bond(s) are named by prefixing the group name to the parent name.

Examples:

2-Chloroethylphosphonic acid	$ClCH_2CH_2PO(OH)_2$
Triethylphosphine oxide	$(C_2H_5)_3PO$

For the derivatives of acids, structures formed by the operation of replacing are named by insertion of the appropriate affix just preceding the valency suffix (-ic or -ous for acids and -ate or -ite for esters) in the parent name as shown in Table 1.

Esters are named by replacing the ending -ic or -ous acid with -ate or -ite, respectively. The esters of phosphoric acid and phosphorous acid are not specially named as phosphorate and phosphorite, but as

TABLE 1

List of Affixes

Affix	Operation	Name and structure	
-amid(o)-	OH → NH$_2$	Phosphoramidic acid	(HO)$_2$P(O)NH$_2$
-chlorid(o)-	OH → Cl	Phosphorochloridic acid	(HO)$_2$P(O)Cl
-imid(o)-	=O → =NH	Phosphorimidic acid	(HO)$_3$P=NH
-thio-	OH or =O → SH or =S	Phosphorothioic acid	H$_3$PO$_3$S
-thiolo-	OH → SH	Phosphorothiolic acid	(HO)$_2$P(O)SH
-thiono-	=O → =S	Phosphorothionic acid	(HO)$_3$P=S

phosphate and phosphite, respectively. Partial esters are named by inserting the term "hydrogen" between the ester radical name and the parent name.

Chemical names of some pesticides are presented as in the examples. Common and/or trade names are also presented.

Examples:

2,2-Dichlorovinyl dimethyl phosphate, dichlorvos, DDVP

(CH$_3$O)$_2$P(O)-OCH=CCl$_2$

Dimethyl *p*-nitrophenyl phosphorothionate,* parathion-methyl

(CH$_3$O)$_2$P(S)—O——NO$_2$

Diethyl *S*-(ethoxycarbonylmethyl) phosphorothiolothionate,* acethion

(C$_2$H$_5$O)$_2$P(S)SCH$_2$CO$_2$C$_2$H$_5$

S-Benzyl diethyl phosphorothiolate, Kitazin®

(C$_2$H$_5$O)$_2$P(O)SCH$_2$——

Ethyl *p*-nitrophenyl phenylphosphonothionate, EPN

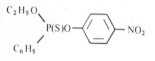

2,4,5-Trichlorophenyl diethylphosphinothionate, Agvitor

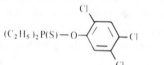

2-Chloro-4-*tert*-butylphenyl methyl *N*-methylphosphoramidate, crufomate, Ruelene®

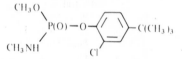

O,S-Dimethyl phosphoramidothiolate, methamidophos, Monitor®

CH$_3$O
        \
         P(O)—NH$_2$
        /
CH$_3$S

Diisopropyl phosphorofluoridate, DFP**

(*i*-C$_3$H$_7$O)$_2$P(O)F

Tributyl phosphorotrithioite, merphos, Folex®

(C$_4$H$_9$S)$_3$P

The dialkyl esters of phosphorous acid exist as the esters of tautomeric phosphonic acid. Thus, they may be named phosphonates. However, they are traditionally named dialkyl phosphites.

*Many authors prefer to name parathion-methyl *O,O*-dimethyl *O-p*-nitrophenyl phosphorothioate. "Thioate" does not distinguish between thionate (P=S) and thiolate (P-S-). Similarly, acethion may be named as *O,O*-diethyl *S*-ethoxycarbonyl-methyl phosphorodithioate.

**This is not a pesticide, but a nerve gas.

For example: dibenzyl phosphite

$$(C_6H_5CH_2O)_2P-OH \quad \rightleftharpoons \quad (C_6H_5CH_2O)_2\overset{\overset{\displaystyle O}{\|}}{P}H$$

Fully operated structures no longer having acid or ester functions are named according to the Chemical Abstracts order of precedence of functions. For example:

N,N,N′,N′-Tetramethylphosphorodiamidic fluoride, dimefox $((CH_3)_2N)_2P(O)F$

For compounds having more than two phosphorus atoms, no rule in the nomenclature is established. Pyrophosphoric acid may be used as a parent, as in tetraethyl pyrophosphate, TEPP.

$$(C_2H_5O)_2P(O)-O-P(O)(OC_2H_5)_2$$

They may be named also as anhydrides, as bis-*N,N,N′,N′*-tetramethylphosphorodiamidic anhydride, schradan.

$$((CH_3)_2N)_2P(O)-O-P(O)(N(CH_3)_2)_2$$

Group names including phosphorus atom are sometimes used for naming of complicated compounds. The group names may be derived from

phosphinyl	$H_2P(O)-$
phosphinothioyl	$H_2P(S)-$
phosphinimyl	$H_2P(NH)-$

as

dimethoxyphosphinothioyloxy	$(CH_3O)_2P(S)O-$
aminomethylphosphinylthio	$H_2N(CH_3)P(O)S-$

Actual pesticides are given as examples;

2-(Diethoxyphosphinylimino)-1,3-dithiolane, Cyolane®

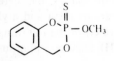

N-(Diethoxyphosphinothioyl)phthalimide, Dowco 199

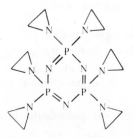

A few pesticides have a ring system containing phosphorus atom. Names of such rings are listed in Table 2.

Example:

2-Methoxy-4*H*-1,3,2-benzodioxaphosphorin-2-sulfide, Salithion®

2,2,4,4,6,6-Hexakis(1-aziridinyl)-2,2,4,4,6,6-hexahydro-1,3,5,2,4,6-triazatriphosphorine, Apholate

TABLE 2

Name of Ring Systems Involving Phosphorus

Ring Size	Ring contains nitrogen		Ring contains no nitrogen	
	Unsaturated	Saturated	Unsaturated	Saturated
5	Phosphole	Phospholidine	Phosphole	Pholpholane
6	Phosphorine	—	Phosphorin	Phosphorinane

The names of commercial pesticides are more confusing because of the variety of common and proprietary names. A proprietary name is always used with a capital letter and a registered trade name is followed by an ®. Common names are recommended by national or international committees. Unfortunately, different countries adopt their own common names for certain pesticides. For example, dimethyl 3-methyl-4-nitrophenyl phosphorothionate is named fenitrothion by the International Organization for Standard and British Standards Institution, while it is named metilnitrofos in the USSR, and MEP in Japan. No common name is recommended yet for this compound in the USA, where the trade name, Sumithion®, is used.

C. STRUCTURE AND SOME PROPERTIES OF PHOSPHORUS COMPOUNDS

1. Electronic Structure

Phosphorus is placed on the second row of Group V elements in the periodic table. Its properties are most suitably comparable with nitrogen. They make trivalent compounds and onium salts. The greatest difference is that phosphorus yields pentavalent compounds, as do arsenic, antimony, and bismuth.

The electronic configuration of the neutral phosphorus atom is denoted as $1s^2 2s^2 2p^6 3s^2 3p^3$. This means that the phosphorus atom has 5 valence electrons in the M-shell, 1 electron pair in the $3s$-orbital, and 3 electrons in the $3p$-orbitals. The activation energy to promote electron from $3s$ to $3p$ and $3p$ to $3d$ is relatively low, 7.5 and 9 eV, respectively.[25] In the case of nitrogen, having a similar electronic configuration, the corresponding activation energy from $2s$ to $2p$ and $2p$ to $3d$ is 10.9 and 12 eV, respectively. Therefore, in phosphorus compounds the d-orbital can be used much more easily for binding hybrid orbitals. This is the reason why phosphorus makes penta- or higher valent compounds.

2. Trivalent Phosphorus Compounds

Phosphorus compounds may be classified by the number of σ-bonds connected to phosphorus, as shown in Table 3.[24] Besides the listed classes of compounds which have bond numbers of 3,4,5, or 6, some singly connected, unstable phosphorus compounds, such as PN, are also known. The compounds of trivalent phosphorus exhibit the pyramidal symmetry consisting of p^3-hybridization and a certain degree of sp^3-character (Figure 1a).[24] Thus, their bond angles range between 93.5° for phosphine PH_3, and 100° for phosphorus trichloride PCl_3, instead of the theoretical 90° of a pure p^3-hybrid. These are smaller than the bond angles of corresponding nitrogen compounds, which are near the theoretical 109.5° of a pure sp^3-hybrid. This is due to the bigger radius of phosphorus (1.9 Å versus 1.5 Å of nitrogen), and to a smaller steric interaction among each substituent, because of a more extended lobe in the $3p$-orbital of phosphorus than that in the $2p$-orbital of nitrogen.

Phosphine PH_3 may be regarded as the parent compound of trivalent phosphorus compounds, and there are many organic phosphines derived by the replacement of the hydrogens with various alkyl and aryl groups. As another type of important organic trivalent phosphorus compounds, the esters of the following three oxyacids should be offered: phosphinous acid H_2POH, phosphonous acid $HP(OH)_2$, and phosphorous acid $P(OH)_3$.

The free acids of trivalent phosphorus undergo isomerization into tautomeric pentavalent acids

TABLE 3

Classification of Phosphorus Compounds

Number of σ-bonds	3	4		5	6
σ-Bond hybridization*	p^3 with sp^3 character	sp^3		sp^3d	sp^3d^2
Number of π-bonds	0	0	1	0	0
Number of lone pair	1	0	0	0	0
Directional characteristics*	Trigonal pyramidal	Tetrahedral		Trigonal bipyramidal	Octahedral
Examples of compounds	PCl_3 $P(OR)_3$	R_4P^+	$(RO)_3PO$	PCl_5	PCl_6^-

*See Figure 1.

p^3-bonding

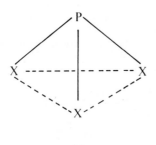

(a)

sp^3-bonding

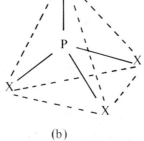

(b)

sp^3d-bonding sp^3d^2-bonding

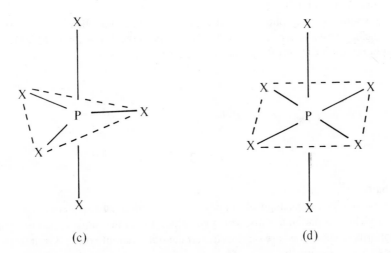

(c)   (d)

FIGURE 1. Geometry of the various types of bonding found in phosphorus compounds.

(phosphoryl form). The equilibrium is strongly towards the latter, owing to the stable phosphoryl bond (P=O).

This is also the case in partial acids such as dialkyl phosphites. They exist almost entirely in the phosphonate form, in a free state or neutral solution. As there is no lone pair electrons on phosphorus in the phosphonate form, dialkyl phosphites have few properties of trivalent phosphorus compounds, i.e., they show only very low nucleophilic reactivity.

They make salts, however, which behave as trivalent phosphorus compounds.

$$RO\diagdown P-O^- \, Na^+ \; (K^+, NH_4^+, R_3NH^+, etc.)$$

Moreover, the equilibrium shifts towards hydroxy form in a solvent (B) having the property of Lewis base, such as dioxane.[38]

On the other hand, tri-substituted derivatives in which the tautomerism is restricted are stable.

Di- and tri-alkyl phosphites and phosphorus trichloride are very important as starting materials or intermediates for the preparation of more complicated organophosphorus compounds. Trialkyl phosphites are highly nucleophilic reagents. Although dialkyl phosphites are much less reactive, their salts are active enough as nucleophiles. Many examples of reactions applied for the preparation of organophosphorus pesticides are presented in Chapter II.

However, only a small number of phosphites, themselves, have been applied for agricultural use; dibenzyl phosphite has fungicidal activity, tris-[β-(2,4-dichlorophenoxy)ethyl] phosphite (2,4-DEP) is a herbicide, and tributyl phosphorotrithioite (merphos) is useful as a cotton defoliant.

$$\cdot \, (Cl-\underset{Cl}{\diagup\!\!\!\!\diagdown}-OCH_2CH_2O)_3P \qquad\qquad (C_4H_9S)_3P$$

2,4-DEP   merphos

3. Phosphonium Salts

As would be expected by the analogy of amines, the compounds of triply connected phosphorus act as electron donors to yield phosphonium salts, which correspond to ammonium salts. The resulting quadruply connected phosphorus compounds are arranged tetrahedrally about the phosphorus atom, which is sp^3-hybridized as shown in Figure 1b.

The conversion of nonsubstituted phosphine PH_3, which is essentially p^3-hybridized (bond angle 93.5°), into sp^3-hybridized phosphonium (bond angle 109°) is energetically unfavorable. Therefore, the basicity of nonsubstituted phosphine is too weak to measure the pK_b value, which was indirectly estimated at 22—28 (compare to the value 4.75 of ammonia). Thus, nonsubstituted phosphonium salts are so unstable that they decompose in water into the corresponding components.

$$H_4 P^+ X^- \xrightarrow{\text{H}_2\text{O}} PH_3 + HX$$

However, tertiary phosphines are much stronger bases, comparable with corresponding amines; for example the pK_b value of trimethylphosphine is 5.35, and that of the corresponding amine is 4.20. There is apparently a considerable contribution of sp^3-hybridization in tertiary phosphines, since their bond angles range around 100°, instead of the theoretical 90° for a pure p^3-hybrid. Thus, quaternary phosphonium salts are as stable as ammonium salts.

Moreover, even tetraphenyl derivatives, which do not exist as ammonium salts, are known to form phosphonium salts. This is attributed to the diminished basicity of triphenylamine, due to the resonance effect between an unshared pair of electrons on nitrogen, and the π-electrons of benzene rings, and to the steric hindrance between benzene rings. On the other hand, in the phosphorus compounds little steric hindrance is expected, owing to the larger bonding radius and negligible resonance effect due to poor $p\pi$-$p\pi$ overlapping. The poor $p\pi$-$p\pi$ overlapping is due to the more extended $3p$-orbital, as compared with the more squatty $2p$-orbital as shown in Figure 2.

Certain phosphonium salts show interesting biological activities; Phosphon (tributyl-2,4-dichlorobenzyl-phosphonium chloride) is a plant growth regulator, and some other phosphonium salts show fungicidal or herbicidal activity.

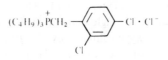

Phosphon

4. Pentavalent Phosphorus Compounds

Pentavalent phosphorus compounds may be divided into two classes by the number of σ-bonds connected to phosphorus. The majority of the compounds have four σ-bonds and one π-bond on the average for detail, see Reference. 24, Vol I, Chapter 2). A few compounds have five σ-bonds (Table 3).

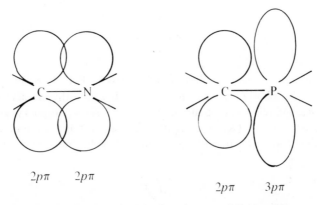

FIGURE 2. Distribution of $p\pi$ electrons of C, N, and P.

a. Compounds with Four σ-Bonds
i. pπ-dπ Bonding

All the naturally occurring phosphorus compounds have a characteristic "phosphoryl" bond shown as P = O or P → O. With only a few exceptions, all organophosphorus pesticides also have the phosphoryl bond or the analogous thiophosphoryl bond (P=S). They may be called phosphoryl compounds.

The bond angles O-P-O in phosphoric acid (PO_4^{3-}), and Cl-P-Cl in phosphoric trichloride ($POCl_3$), are 109° and 106°, respectively.[24] This indicates that phosphorus exhibits the tetrahedral symmetry corresponding to sp^3-hybridization. If only s and p orbitals are taken into account, the phosphoryl compounds will have a structure as in nitrogen oxide.

However, the phosphorus atom has vacant 3d-orbitals, the angular properties of which enable the orbitals to participate in π-bond formation. The shapes and directions of d-orbitals are shown in Figure 3. The π-bonding in a phosphoryl group arises from the donation of the nonbonding 2p-electrons of the negatively charged substituent, the oxygen atom, into a vacant 3d-orbital of phosphorus, as shown in Figure 4. This means that the polarized electrons in P^+-O^- return to phosphorus through the d-orbital. Thus, the phosphoryl group P=O includes partially pπ-dπ bonding, with a σ-bond.

Much evidence for such d-orbital bonding has been reported.[25] For example, the dissociation energy of P=O bonding (125—156 kcal/mol) is much higher than in the corresponding N→O bond in amine N-oxides (50—70 kcal/mol) and in the P-O single bond (86 kcal/mol). The observed bond length of P=O (1.44—1.55 Å) is shorter than the calculated value for P-O single bond (1.71 Å), being due to the pπ-dπ bonding.

The stabilization of the phosphorus-oxygen linkage by the pπ-dπ overlap causes the high affinity of phosphorus compounds to the oxygen atom (see Chapter II), and the formation of the stable phosphoryl bond is responsible for the motive force in various reactions, such as the Michaelis-Arbuzov reaction, which is useful for the preparation of alkyl phosphonates (see Section II.H.3).

$$(RO)_3P + R'X \longrightarrow (RO)_3P^+R' \cdot X^- \longrightarrow (RO)_2\overset{\overset{\displaystyle O}{\|}}{P}R' + RX$$

Thus, the great majority of phosphorus compounds have a form such as:

Similar pπ-dπ bonding appears to contribute more or less in P=C, P=N, P=S, and P=Se bondings in alkylidenephosphoranes, phosphorimidic acid derivatives, phosphorothionic acid derivatives, phosphoroselenonoic acid derivatives, and related compounds. Moreover, besides phosphoryl group in POX_3, unshared electron pairs on any heteroatoms (X) connected to phosphorus can overlap with the 3d-orbitals of phosphorus, even though not so very efficiently, as schematically illustrated by a phosphate ester.

Thus, all these ester bondings have a character as double bonding, in part, and are stabilized by the pπ-dπ contribution.

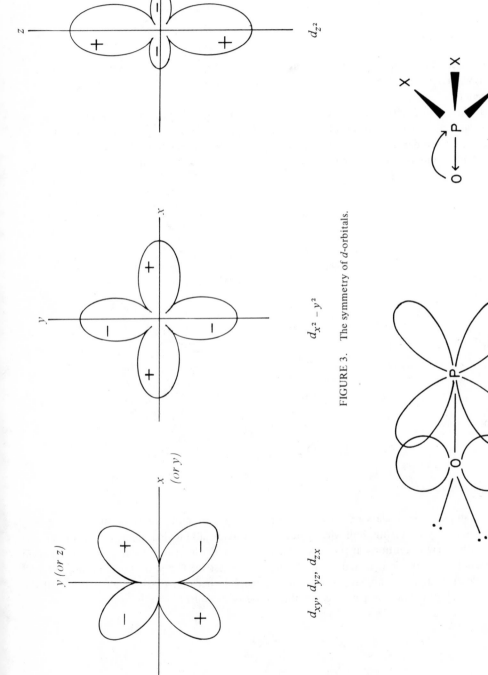

d_{xy}, d_{yz}, d_{zx}

$d_{x^2 - y^2}$

d_{z^2}

FIGURE 3.   The symmetry of d-orbitals.

$2p$ $3d$

FIGURE 4.   The $p\pi$-$d\pi$ bonding in phosphoryl group.

11

ii. Oxyacids

Esters of pentavalent phosphorus acids are the most important class of organophosphorus compounds. They are the derivatives of the following three parent acids: phosphinic acid (I) (often called phosphonous acid or hypophosphorous acid), phosphonic acid (II) (or phosphorous acid), and phosphoric acid (III).

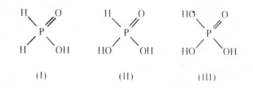

(I)   (II)   (III)

Naturally occurring organophosphorus compounds are all mono-or di-esters of phosphoric acid or its anhydride, with a few exceptions of phosphonic acid derivatives which were recently discovered in sheep rumen[39] and sea anemone.[40]

The majority of phosphorus pesticides are neutral esters or amides derived from phosphoric acid, its anhydride, or sulfur analogs. Certain fluorides have been used commercially, and several phosphonic acid derivatives have been recently developed as pesticides. Only one phosphinic acid derivative is known as an insecticide. Thus, these compounds may be classified into the following fundamental types (other combinations are, of course, possible):

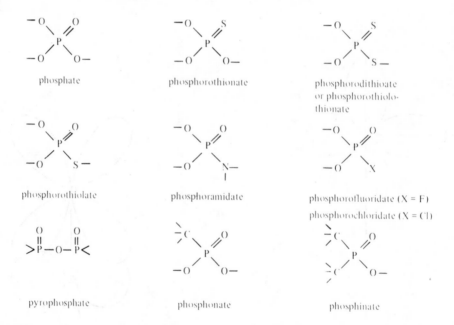

The representative examples of each class are shown in the preceding section on nomenclature.

Contrasting with naturally occurring, biologically active ester acids, all biocidal phosphorus compounds are neutral esters, with a few exceptions in the case of herbicides and fungicides. Ester acids derived from biocidal neutral esters by hydrolytic degradation are biologically inactive. The phosphorus atom in neutral esters is generally electron deficient and reactive as an electrophile because of the high polarity of the phosphoryl group. On the other hand, in partial esters this property diminishes greatly by dissociation.

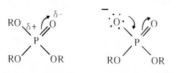

The dissociation constants of some oxyacids are shown in Table 4.[2,24,41,42] They are affected by the substituents on phosphorus, depending on the inductive effect of the substituents. The strength of the acid is in the order of the substituent: H> RO> HO> R. The substitution of oxygen with sulfur decreases acidity in water, but increases in alcohol solution.[2]

There is a thiono-thiolo tautomerism in thioic acids. Thiolo form is more stable than thiono form in *O, O*-dialkyl phosphorothioic acids, whereas the reverse is the case with phosphonothioic acids and phosphinothioic acids.[41]

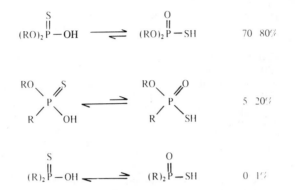

b. Compounds with Five σ-Bonds

The phosphorus atom in phosphorus pentachloride in the vapor phase is known to connect with each chlorine atom by σ-bonding. Similarly, phosphorus compounds with strongly electronegative elements, such as PF_5 and PF_3Cl_2, are also this case. These compounds exhibit the trigonal bipyramidal symmetry corresponding to sp^3d-hybridization (Figure 1c). This structure, having ten electrons in the valence shell of the phosphorus atom, is rather unstable, as expected from the octet theory. In fact, in the crystalline state, phosphorus pentachloride does not exist as trigonal bipyramidal, but as the mixture of two different ions, i. e., tetrahedral PCl_4^+ and octahedral PCl_6^- (Figure 1d).[24]

Recently, several stable organic compounds, such as (IV), in which phosphorus is quadruply connected, were synthesized.[43] They are considered to have the trigonal bipyramidal configuration (V).[44]

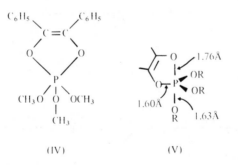

(IV) (V)

Two additional possible geometries (VI, VII) for quadruply connected phosphorus compounds have been proposed:

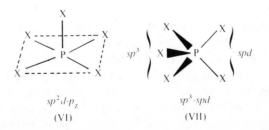

sp^2d-p_z sp^3-spd

(VI) (VII)

13

TABLE 4

pK Values of Some Pentavalent Phosphorus Acids[2,24,41,42]

Acid	pK_1	pK_2	pK_3
$H_2P(O)OH$	1.1	–	–
$HP(O)(OH)_2$	1.3	6.7	–
$CH_3P(O)(OH)_2$	2.3	–	–
$O=P(OH)_3$	2.0	7.1	12.3
$CH_3OP(O)(OH)_2$	1.54	–	–
$C_4H_9OP(O)(OH)_2$	1.88	6.84	–
$C_4H_9SP(O)(OH)_2$	~1	5.50	–
$(CH_3O)_2P(O)OH$	1.25*	–	–
$(C_2H_5O)_2P(O)OH$	1.37*	–	–
$(CH_3)_2P(O)OH$	3.13*	–	–
$(C_2H_5O)_2P(S)OH$	1.49	–	–
$(C_2H_5O)_2P(S)SH$	1.62*	–	–

* In 7% ethanol.

Only a few quadruply connected phosphorus compounds are known as stable compounds, but they are important as intermediates in the various reactions of phosphorus compounds.[45] Formation of such intermediates may lower the free energy of activation, and consequently facilitate the reaction. For example, in the course of the transformation from phosphonium (VIII) to phosphine oxide, the configuration is inverted by the action of hydroxide ion, giving X, whereas the configuration is retained through the Wittig reaction to yield XIII. The former reaction is analogous to the S_N2 type reactions on carbon atoms, and may proceed through a quadruply connected phosphorus intermediate with sp^3d-hybridization (IX). On the other hand, the Wittig reaction may proceed through a quadruply connected intermediate (XII) with tetragonal pyramidal geometry (VI) or sp^3-spd type hybridization (VII). This is analogous to the intermediate of the S_Ni type reaction in carbon compounds.

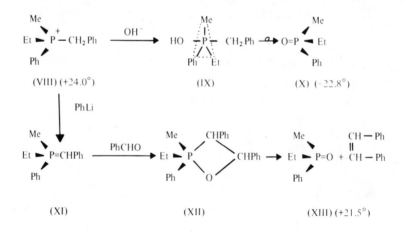

5. Optical Isomerism

Phosphonium salts and phosphoryl compounds have four σ-bondings through sp^3-hybrid orbitals as in tetrahedral carbon compounds. Therefore, the existence of optically active compounds in which the phosphorus atom serves as an asymmetric center is possible. The first resolution of an optically active phosphorus compound (ethylmethylphenylphosphine oxide) was reported by Meisenheimer in 1911.[46] There are many organophosphorus pesticides which have an asymmetric phosphorus atom, for example, crufomate, EPN and the related compounds, fonofos (Dyfonate®), methamidophos (Monitor®), and salithion. However, only several biologically active derivatives of phosphorus acids have been resolved into enantiomers, which differ in structure only in the left- and right-handedness (chirality) of their orientations, as shown by the figures of the enantiomers of sarin.

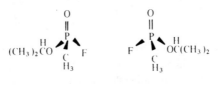

Enantiomers of sarin

The resolved compounds are the S-methyl isomer of parathion-methyl (XIV),[47] its methylphosphonate analogs (XV),[48] the ethylphosphonate analog of demeton-S (XVI),[49,50] the herbicide DMPA (Zytron®),[51] and sarin.[52] Although enantiomers have identical physical and chemical properties, except for the optical activity to rotate the plane of polarized light, the biological activities are, in general, greatly influenced by the chirality (see IV.A.4.c).

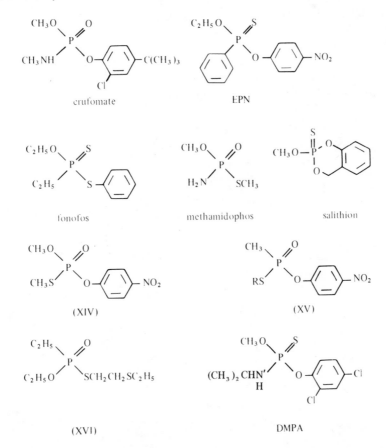

crufomate

EPN

fonofos

methamidophos

salithion

(XIV)

(XV)

(XVI)

DMPA

Moreover, asymmetric trivalent phosphines, in which one corner of the tetrahedron is occupied by an unshared pair of electrons, are far more optically stable than amines and can be resolved into enantiomers. Horner prepared several optically active phosphines by electrolytic reduction of the corresponding optically active benzylphosphonium salts.[53]

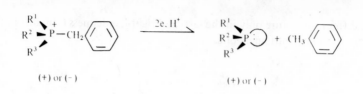

$$(+) \text{ or } (-) \qquad\qquad (+) \text{ or } (-)$$

Similarly, sulfoxides and sulfonium salts have a stable configuration in which an unshared pair of electrons occupies one corner of the tetrahedron.[54] Therefore, organophosphorus insecticides or metabolites having a sulfoxide group in the molecule are generally racemic mixtures or possibly one of the enantiomers in metabolites. The examples are the insecticides Aphidan, oxydisulfoton, oxydemeton-methyl, and the metabolite (XVII) of fenthion. Their resolution into enantiomers has not yet been reported. Moreover, Estox may be a mixture of diastereomers; it has two asymmetric centers. Malathion has an asymmetric carbon in the molecule, and was separated into dextro and levo isomers.[55] Recent development of organophosphorus stereochemistry has been reviewed by several authors.[56-58]

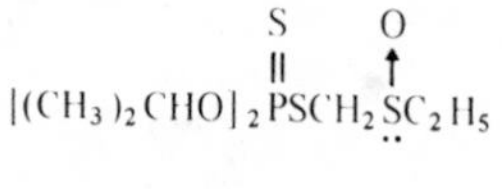

$$[(CH_3)_2CHO]_2 P\overset{S}{\overset{\|}{}}CH_2 S\overset{O}{\overset{\uparrow}{\underset{..}{}}}C_2H_5$$

Aphidan

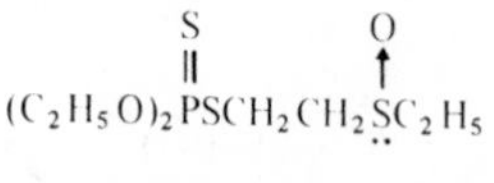

$$(C_2H_5O)_2 P\overset{S}{\overset{\|}{}}CH_2 CH_2 S\overset{O}{\overset{\uparrow}{\underset{..}{}}}C_2H_5$$

oxydisulfoton

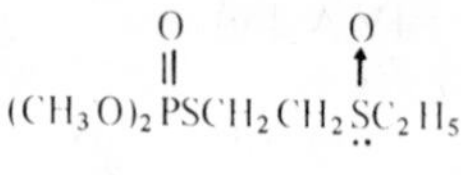

$$(CH_3O)_2 P\overset{O}{\overset{\|}{}}CH_2 CH_2 S\overset{O}{\overset{\uparrow}{\underset{..}{}}}C_2H_5$$

oxydemeton-methyl

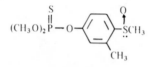

(XVII)

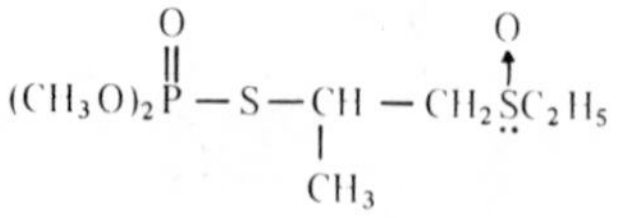

$$(CH_3O)_2 \overset{O}{\overset{\|}{P}} - S - \underset{CH_3}{\underset{|}{CH}} - CH_2 S\overset{O}{\overset{\uparrow}{\underset{..}{}}}C_2H_5$$

Estox

Chapter II

SYNTHESIS

More than one hundred organophosphorus compounds are, or were, actually used as pesticides in the world. Millions of candidate compounds have been and will be synthesized. The variety of organic moieties in organophosphorus pesticide molecules is too great to discuss all of their synthetic methods. This chapter deals only with the principal reactions, involving the phosphorus moiety, which are generally useful for the preparation of organophosphorus pesticides. For details, refer to more comprehensive books.[3,23,59]

A. PHOSPHATES

There were only a few practical insecticides of phosphate esters because of their relatively high toxicity and low stability, until the recent development of a unique class, phosphate esters of enols, the syntheses of which became easier by the finding of the Perkow reaction.[16] As phosphate esters are the active form of phosphorothionate insecticides, the preparation of corresponding phosphates is required for the in vitro study of the activity. On the other hand, great efforts have been directed towards the synthesis of many biologically interesting phosphate esters in living systems, and a number of synthetic methods have been developed.[8,35] As they are, however, all specifically designed to prepare mono- or di-esters of phosphoric acid, they are not suitable for the preparatory method of pesticidal neutral esters and will be not discussed in this book.

The general forms of pesticides in this class are dialkyl aryl phosphates (I) and dialkyl vinyl phosphates (II). Recently, another type of phosphates, the phosphorylated hydroxamic acids or oximes (III), was developed.

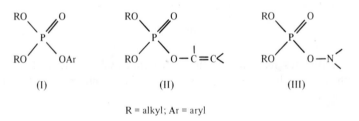

(I) (II) (III)

R = alkyl; Ar = aryl

They may be prepared by the following principal reactions:

1. the reaction of phosphorochloridates with hydroxy compounds,
2. the oxidative desulfuration of corresponding phosphorothionates,
3. the rearrangement of α-hydroxyalkylphosphonates, and
4. the reactions of phosphites.

1. Reaction of Phosphorochloridates with Hydroxy Compounds

$$(RO)_2\overset{O}{\overset{\|}{P}}Cl + R'OH \xrightarrow[-HCl]{} (RO)_2\overset{O}{\overset{\|}{P}}OR' \qquad (1)$$

This simple reaction is most commonly utilized for the synthesis of esters of not only phosphoric acid, but also of other phosphorus acids being applied to appropriate phosphorus chlorides. Certain bases, such as sodium carbonate, and tertiary amines are employed as dehydrochlorinating agents. Salts of hydroxy compounds are also used. Thus, paraoxon is prepared from diethyl phosphorochloridate and sodium p-nitrophenolate in acetonitrile (see Equation 2).[13]

$$(C_2H_5O)_2\overset{\overset{\displaystyle O}{\|}}{P}Cl + NaO-\!\!\!\left\langle\!\!\!\bigcirc\!\!\!\right\rangle\!\!\!-NO_2 \xrightarrow[75°\ 2hr]{} (C_2H_5O)_2\overset{\overset{\displaystyle O}{\|}}{P}-O-\!\!\!\left\langle\!\!\!\bigcirc\!\!\!\right\rangle\!\!\!-NO_2 + NaCl \qquad (2)$$

paraoxon   68%

The esters of enols are also synthesized similarly. For example, mevinphos is produced by the reaction of sodium enolate of methyl acetoacetate with dimethyl phosphorochloridate (see Equation 3).[13] By this procedure, the *trans*-crotonate isomer (β-form) is almost exclusively obtained (91–95% *trans*).[60]

$$(CH_3O)_2\overset{\overset{\displaystyle O}{\|}}{P}Cl + \underset{NaO}{\overset{CH_3}{\diagdown}}C=C\underset{CO_2CH_3}{\overset{H}{\diagup}} \longrightarrow \underset{(CH_3O)_2PO}{\overset{CH_3}{\diagdown}}C=C\underset{CO_2CH_3}{\overset{H}{\diagup}} + NaCl \qquad (3)$$

mevinphos (β-form)

In contrast, the *cis*-isomer (α-form) is preferably obtained (*cis*:*trans* = 2:1) by another synthetic method, employing the Perkow reaction of trimethyl phosphite with methyl α-chloroacetoacetate (Equation 4), which is utilized in the manufacture.[13] The *cis*-isomer is a more effective insecticide. (See Section A.4.a. for the Perkow reaction and Section E.1. for a possible reason why the *trans*-isomer is exclusively produced by reaction 3.)

$$(CH_3O)_3P + CH_3\overset{\overset{\displaystyle O}{\|}}{C}CHCl-CO_2CH_3 \xrightarrow[80°\ 12\ hr]{toluene} (CH_3O)_2PO-\underset{CH_3}{\overset{\displaystyle |}{C}}=CHCO_2CH_3 + CH_3Cl \qquad (4)$$

mevinphos 73% (*cis*:*trans*=2:1)

Dialkyl phosphorochloridates react also with oximes and hydroxamic acids in the presence of acid binding agents or with their alkali metal salts, giving a new class of phosphate esters (III). An anthelmintic drug, Maretin, is an example.[61]

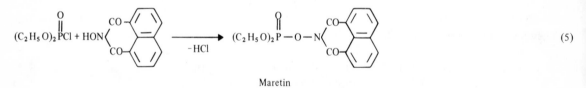

$$(C_2H_5O)_2\overset{\overset{\displaystyle O}{\|}}{P}Cl + HON\!\!\left\langle\begin{smallmatrix}CO-\\CO-\end{smallmatrix}\right. \xrightarrow{-HCl} (C_2H_5O)_2\overset{\overset{\displaystyle O}{\|}}{P}-O-N\!\!\left\langle\begin{smallmatrix}CO-\\CO-\end{smallmatrix}\right. \qquad (5)$$

Maretin

Several methods are known for the synthesis of the phosphorus intermediates, i.e., dialkyl phosphorochloridates and alkyl phosphorodichloridates.[3] They may be simply prepared by the reaction of alcohols and phosphorus oxychloride (Equation 6):

$$O=PCl_3 \xrightarrow[-HCl]{+ROH} (RO)_2\overset{\overset{\displaystyle O}{\|}}{P}Cl_2 \xrightarrow[-HCl]{+ROH} (RO)_2\overset{\overset{\displaystyle O}{\|}}{P}Cl \qquad (6)$$

The formation of either mono- or di-esters depends upon the relative proportions of the reagents used. Since secondary alcohols yield alkyl chlorides as by-products, the presence of equimolar amounts of tertiary amines is needed. Phenols react with phosphorus oxychloride by refluxing about 10 hr.

Dialkyl phosphorochloridates are also prepared by the reaction of trialkyl phosphites with chlorine (Equation 7):

$$(RO)_3P \xrightarrow{Cl_2} \underset{\underset{OR}{|}}{\overset{\overset{OR \curvearrowright Cl^-}{|}}{RO-P^+-Cl}} \longrightarrow (RO)_2PCl + RCl \tag{7}$$

Dialkyl phosphites are more useful for this purpose, because of ease of preparation (Equation 12) and smooth reaction with halogen (Equation 8). Thionyl chloride is also useful for a chlorinating agent.

$$(RO)_2POH + Cl_2 \longrightarrow (RO)_2\overset{\overset{O}{\|}}{P}Cl + HCl \tag{8}$$

Dialkyl phosphites also react with carbon tetrachloride to give corresponding dialkyl phosphorochloridates in high yields, provided that tertiary amines are present (Equation 9).[62] When tertiary amines are replaced with primary or secondary amines or ammonia, corresponding phosphoramidates are produced, as will be described later (Section II.E.2.).

$$(RO)_2POH + CCl_4 \xrightarrow{NR'_3} (RO)_2\overset{\overset{O}{\|}}{P}Cl + HCCl_3 \tag{9}$$

This reaction is expanded further to prepare phosphate triesters directly from dialkyl phosphites. Thus, the anthelmintic agent, Haloxon,[63] is prepared by an elegant method employing the reaction of bis-β-chloroethyl phosphite with 3-chloro-4-methyl-7-hydroxycoumarin and carbon tetrachloride in the presence of triethylamine (Equation 10).[2]

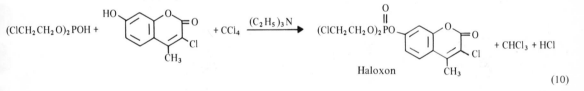

Haloxon

$$(10)$$

Alkyl phosphites are produced from phosphorus trichloride and alcohols. In the presence of a tertiary amine, the reaction finally proceeds to form trialkyl phosphites (Equation 11). In the absence of the base, however, the reaction stops to give dialkyl phosphites (Equation 12).

$$PCl_3 + 3ROH + 3NR'_3 \longrightarrow (RO)_3P + 3R'_3NH \cdot Cl \tag{11}$$

$$PCl_3 + 3ROH \longrightarrow (RO)_2POH + RCl + 2HCl \tag{12}$$

A part of the alcohol in the reaction 12 can be replaced by water (Equation 13).[64] The reaction is carried out in the presence of benzene or petroleum ether to treat phosphorus trichloride with an alcohol-water mixture (mole ratio = 2.6:0.4).

$$PCl_3 + 2ROH + H_2O \xrightarrow[15-20°]{} (RO)_2POH + 3HCl \tag{13}$$

Mixed dialkyl phosphites may be prepared by a transesterification reaction by the catalytic action of phosphoric acid (Equation 14).[65]

$$(RO)_2POH + R'OH \xrightarrow{H_3PO_4} \underset{R'O}{\overset{RO}{>}}POH + ROH \tag{14}$$

2. Oxidative Desulfuration of Phosphorothionates

Phosphorothionate esters are oxidatively transformed to phosphate esters. This reaction is specifically suitable to prepare in small quantities the radioactive reference compounds of active pesticide metabolites. The oxidative desulfuration reaction is also conveniently applied to the enzymatic analysis of organophosphorus pesticides (see Chapter IV).

For preparative purpose, nitric acid is often used as an oxidant. Thus, Coroxon is prepared in 87% yield by the action of nitric acid (d = 1.5) from its thiono analog, coumaphos (Equation 15).[13] Similarly, a phosphonothionate such as EPN is also converted to a corresponding phosphonate (Equation 16).[66]

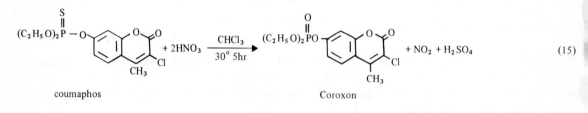

(15)

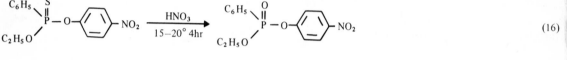

(16)

Dinitrogentetroxide gas is also utilized as an oxidant.[67] For example, parathion is converted to paraoxon in about 90% yield at room temperature (Equation 17). This procedure is also applicable to a phosphorodithioate, such as malathion, to convert it to the oxo-analog (Equation 18).

$$(C_2H_5O)_2\overset{\overset{S}{\|}}{P}-O-\langle\ \rangle-NO_2 \xrightarrow[\text{room temp, 16hr}]{N_2O_4/\text{ether}} (C_2H_5O)_2\overset{\overset{O}{\|}}{P}-O-\langle\ \rangle-NO_2 \tag{17}$$

parathion   paraoxon

$$(CH_3O)_2\overset{\overset{S}{\|}}{P}-S-\underset{\underset{CH_2CO_2C_2H_5}{|}}{CHCO_2C_2H_5} \xrightarrow{N_2O_4/CH_2Cl_2} (CH_3O)_2\overset{\overset{O}{\|}}{P}S\underset{\underset{CH_2CO_2C_2H_5}{|}}{CHCO_2C_2H_5} \tag{18}$$

(For the preparation of P = S compounds, see Sections B and C.)

3. Rearrangement of α-Hydroxyalkylphosphonates

Lorenz and his co-workers found that trichlorfon was converted to a vinyl phosphate, dichlorvos, by the action of alkali (Equation 19).[68] Trichlorfon is synthesized from dimethyl phosphite and chloral (see Section H.4.).

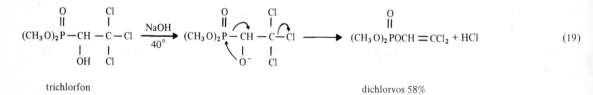

(19)

Similar rearrangement reactions take place also in α-hydroxyphosphonates having no halogen atom at the β-position as does α-hydroxy-α-phenylbenzylphosphonate,[69] which is produced by the reaction of dialkyl phosphite and benzophenone (Equation 20).

$$C_6H_5COC_6H_5 + HOP(OC_2H_5)_2 \longrightarrow \underset{\underset{C_6H_5}{|}}{C_6H_5-\overset{HO}{\underset{}{C}}-\overset{O}{\underset{}{P}}(OC_2H_5)_2} \overset{base}{\longrightarrow} \underset{\underset{C_6H_5}{|}}{C_6H_5\overset{H}{\underset{}{C}}-O-\overset{O}{\underset{}{P}}(OC_2H_5)_2} \qquad (20)$$

4. Reaction of Phosphites

a. Perkow Reaction

In 1952, Perkow[16] found that a trialkyl phosphite reacted with an α-halogenated carbonyl compound to give a vinyl phosphate (Equation 21) differently from the Michaelis-Arbuzov reaction, in which normally a phosphonate would be produced from a trialkyl phosphite and an alkyl halide (Equation 22) (see H.3.).

$$(RO)_3P + O=CR'-\underset{\underset{R'''}{|}}{\overset{\overset{R''}{|}}{C}}-X \begin{cases} \longrightarrow (RO)_2\overset{O}{\underset{}{P}}-O-CR'=CR''R''' + RX & (21) \\ \text{Perkow reaction} & \\ \longrightarrow (RO)_2\overset{O}{\underset{}{P}}-CR''R'''-CR'=O + RX & (22) \\ \text{Michaelis-Arbuzov reaction} & \end{cases}$$

The reactivity of α-halocarbonyl compounds decreases in the following order: α-haloaldehydes > α-haloketones > α-haloesters. With the latter two compounds, both the reactions occur competitively, whereas with α-haloaldehydes, the Perkow reaction takes place preferably. The kind of halogen also affects the competition of the reactions. The Perkow reaction is favored in the order Cl > Br > I, though chlorides are less reactive.

The Perkow reaction is very useful for the preparation of vinyl phosphates, a number of which are presently utilized as insecticides. They include chlorofenvinphos, crotoxyphos, dichlorvos, mevinphos, monocrotophos, phosphamidon, tetrachlorvinphos, and so on (see Chapter V). For example, dimethyl dichlorovinyl phosphate (dichlorvos) is produced from trimethyl phosphite and chloral (Equation 23).[13]

$$(CH_3O)_3P + O=CHCCl_3 \xrightarrow[60-70°]{benzene} (CH_3O)_2\overset{O}{\underset{}{P}}-OCH=CCl_2 + CH_3Cl \qquad (23)$$
$$\text{dichlorvos 70\%}$$

Four possible mechanisms have been proposed for the Perkow reaction according to the difference at the site where the phosphite attacks at the initial step:[26] 1) halogen atom, 2) α-carbon atom, 3) carbonyl carbon, or 4) carbonyl oxygen.

$$(RO)_3P: \quad \begin{matrix} 1 & \rightarrow X \\ 2 & \overset{}{C}-\overset{R'}{\underset{R''}{\diagdown}} \\ 3 & \overset{}{C}-R''' \\ & \overset{\|}{O} \\ & 4 \end{matrix}$$

1. When the phosphite attacks the halogen atom, a halophosphonium salt (IV) should be produced as an intermediate, which may be converted to a vinyl phosphate (Equation 24).

$$(RO)_3P: + X-C-C=O \quad \xrightarrow{\quad\quad} \quad (RO)_3P-X\cdot O-CR'''=CR'R'' \quad \longrightarrow \quad (RO)_3P-OCR'''=CR'R''\cdot X^-$$

with R' above C and R'', R''' below, labeled (IV)

$$\longrightarrow (RO)_2\overset{O}{\overset{\|}{P}}-OCR'''=CR'R'' + RX \tag{24}$$

In spite of the fact that halophosphonium salts produced from phosphines and bromoketones react with alcohols to give alkyl bromides and phosphine oxides, the Perkow reaction proceeds normally in alcohols.[70] Therefore, the formation of a halophosphonium salt intermediate appears unlikely to occur in the Perkow reaction.

2. If the reaction is initiated by the attack of the phosphite on α-carbon, a β-ketophosphonium salt (V) should be produced as an intermediate (Equation 25). However, no evidence has been found that such a β-ketophosphonium salt rearranges to an enol phosphonium salt.[70]

$$(RO)_3P: \quad + \quad \overset{R'\quad R''}{C-X} \quad \xrightarrow{\quad} \quad (RO)_3\overset{\oplus}{P}-\overset{R'}{C}-R''\cdot X^{\ominus} \quad \longrightarrow \quad (RO)_3\overset{\oplus}{P}-O-\overset{R'}{C}=\overset{}{C}-R''\cdot X^-$$

with $O=C-R'''$ groups, labeled (V)

$$\longrightarrow (RO)_2\overset{O}{\overset{\|}{P}}-CR'''=CR'R'' + RX \tag{25}$$

3. The initial nucleophilic attack of the phosphite on the carbonyl carbon (Equation 26), which is the most electrophilic center in the molecule, appears reasonable; it may be supported by the formation of an α-hydroxyalkylphosphonate under the condition of the Perkow reaction carried out in alcohols.[70] This mechanism involves a rearrangement process analogous to that in the base catalyzed transformation of trichlorfon into dichlorvos, as shown above in Equation 19.

$$(RO)_3P: \quad + \quad \overset{O}{\overset{\|}{C}}-R''' \quad \longrightarrow \quad (RO)_3\overset{\oplus}{P}-\overset{O^{\ominus}}{C}-R''' \quad \xrightarrow{X^{\ominus}\;R-O} \quad (RO)_2\overset{\oplus}{P}-O-CR'''=C-R'$$

with $R'-C-R''$ and X groups

$$\longrightarrow (RO)_2\overset{O}{\overset{\|}{P}}-O-CR'''=CR'R'' + RX \tag{26}$$

The Perkow reaction appears to be more or less stereospecific. In the reaction between a trialkyl phosphite and a derivative of 2-chloroacetoacetic acid, the formation of a *cis*-crotonic acid derivative is generally preponderant.[60,71] Kirby and Warren[26] have explained the mechanism by applying the principle of asymmetric induction. In accordance with Cram's rule,[72] the carbonyl bond is flanked by the two least bulky groups attached to the adjacent, asymmetric centers, and the phosphite may attack at the carbonyl carbon from the least hindered side of the configuration, as shown in the Newman's projections (Equation 27).

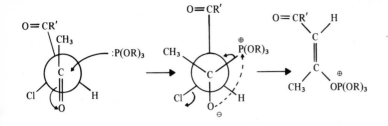

$$(27)$$

However, this mechanism cannot explain the *cis*-preponderance in phosphamidon,[73] which is synthesized from *N,N*-dimethyl α,α-dichloroacetoacetamide: H in the Newman's projections in Equation 27 is replaced by Cl, thus the phosphite may attack the carbonyl equally from both sides. This contradiction may be elucidated if the phosphite attacks directly on carbonyl oxygen as in the proposed mechanism 4.

4. As will be discussed in the following Section (A.4.b.), trialkyl phosphites attack the oxygen atoms of carbonyl groups of certain compounds such as quinones. Therefore, the carbonyl oxygen is one of the possible centers where the initial attack of the phosphite occurs. Through a simple mechanism analogous to the reaction of a phosphine with α-halocarbonyl compounds,[74] the enol phosphonium salts will be formed directly as shown in Equation 28. The possibility of nucleophilic attack by carbonyl oxygen on the phosphite has been suggested.[25]

$$(28)$$

b. Miscellaneous Reactions of Phosphites

Tertiary phosphorus compounds have a high affinity toward oxygen atoms. A *p*-quinone reacts with a trialkyl phosphite to give a corresponding *p*-alkoxyphenyl dialkyl phosphate (Equation 29).[75]

$$(29)$$

By the catalytic action of a base, dialkyl phosphites give *p*-hydroxyphenyl phosphates (Equation 30).[76]

$$(30)$$

By oxidation, phosphites are converted to phosphates. Trialkyl and triaryl phosphites are rapidly and quantitatively oxidized with ozone to their corresponding phosphates under mild conditions (Equation 31).[77]

$$2(RO)_3P + 2O_2 \longrightarrow 2(RO)_3P{=}O + O_2 \qquad (31)$$

Both trialkyl and dialkyl phosphites are smoothly oxidized at room temperature, giving dialkyl phosphates by the action of bromocyanoacetamide and benzyl alcohol. The reaction mechanism is postulated as follows:[78]

23

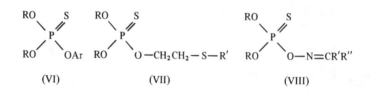

$$\text{(RO)}_2\text{PH} \rightleftharpoons \text{(RO)}_2\text{POH} \xrightarrow{\text{N≡CCHBr CONH}_2} \text{(RO)}_2\overset{\oplus}{\text{P}}\text{—OH} \quad \text{Br}^{\ominus}$$

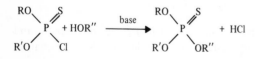

$$\xrightarrow[- \text{NCCH}_2\text{CONH}_2]{\text{C}_6\text{H}_5\text{CH}_2\text{OH}} \text{(RO)}_2\overset{\oplus}{\text{P}}\text{—OH} \quad \text{Br}^{\ominus} \xrightarrow[- \text{BrCH}_2\text{C}_6\text{H}_5]{} \text{(RO)}_2\text{POH} \tag{32}$$

B. PHOSPHOROTHIONATES

Phosphorothionate esters are generally less toxic and more stable than phosphate esters, and are one of the most important classes in organophosphorus pesticides. More than 30 different compounds of this class are used as insecticides throughout the world. Their general forms are dialkyl aryl (including the aromatic heterocyclic group) phosphorothionates (VI) and dialkyl β-alkylthioethyl phosphorothionates (VII). The esters of oximes (VIII) were recently developed as insecticides. The majority of phosphorothionate insecticides belong to type VI. A cyclic phosphorothionate, Salithion, is a modification of the type VI.

$$\begin{array}{ccc}
\underset{RO}{\overset{RO}{>}}\!\!P\!\!\underset{OAr}{\overset{S}{<}} & \underset{RO}{\overset{RO}{>}}\!\!P\!\!\underset{O-CH_2CH_2-S-R'}{\overset{S}{<}} & \underset{RO}{\overset{RO}{>}}\!\!P\!\!\underset{O-N=CR'R''}{\overset{S}{<}} \\
\text{(VI)} & \text{(VII)} & \text{(VIII)}
\end{array}$$

Phosphorothionates are easily converted to the S-alkyl phosphorothiolates by heat (~ 150°C) or by the action of certain reagents (Pistschimuka reaction). Therefore, the technical preparations of phosphorothionates are very often contaminated with the thiolo isomers, which are generally more toxic to mammals. Particularly, the compounds of type VII isomerize so readily that only the mixtures of both isomers could be prepared by the usual methods. Thus, for the compounds of this class, "phosphorothioates" are often used as nomenclature rather than "phosphorothionates."

The principal method to prepare phosphorothionates is based on the reaction between phosphorochloridothionates and hydroxy compounds. However, many methods are known for the preparation of the phosphorus intermediates, which are dialkyl phosphorochloridothionates or in certain cases alkyl phosphorodichloridothionates.

1. Preparation of Phosphorothionate Triesters

As phosphorochloridothionates are less reactive than corresponding phosphorochloridates, more drastic conditions are necessary, in general, for the reaction of hydroxy compounds with the former than for the analogous reaction with the latter, which was described in Section A.1.

$$\underset{R'O}{\overset{RO}{>}}\!\!P\!\!\underset{Cl}{\overset{S}{<}} + HOR'' \xrightarrow{\text{base}} \underset{R'O}{\overset{RO}{>}}\!\!P\!\!\underset{OR''}{\overset{S}{<}} + HCl \tag{33}$$

Organic or inorganic bases are necessary as dehydrochlorinating agents. The salts of the hydroxy compounds are also utilized. Copper powder is often used as a catalyst. As the O-methyl ester linkage is relatively unstable under certain conditions, special care must be taken in the preparation of phosphorothionates having O-methyl ester group(s) in the molecule. They more readily undergo the thermal isomerization into S-methyl phosphorothiolates than do higher alkyl homologs. Moreover, dimethyl phosphorochloridothionate is decomposed by reacting with tertiary amines used as dehydrochlorinating agents to form chloromethane and metaphosphorothioate (Equation 34).[79] Other alkyl homologs do not give rise to this reaction.

$$(CH_3O)_2\overset{\underset{\displaystyle \|}{S}}{P}Cl + NR_3 \longrightarrow CH_3Cl + CH_3OPOS \cdot NR_3 \qquad (34)$$

Thus, fenitrothion is prepared by the reaction of 3-methyl-4-nitrophenol with dimethyl phosphorochloridothionate, in the presence of potassium carbonate at 60–80° in methyl isobutyl ketone (Equation 35).[80]

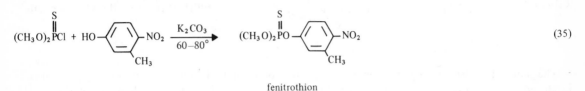

fenitrothion

Heterocyclic hydroxy compounds may react with phosphorochloridothionates under similar conditions. A new insecticide, isoxathion, developed in Japan is prepared from 3-hydroxy-5-phenylisoxazole and diethyl phosphorochloridothionate in the presence of anhydrous sodium carbonate (Equation 36).[81]

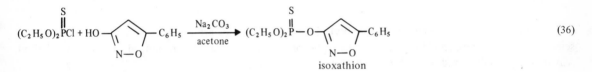

isoxathion

The phosphorothionates of oximes are also prepared similarly. The reaction is carried out with sodium salts of oximes or in the presence of potassium carbonate (Equation 37).[82] Phoxim is a new insecticide in this class.

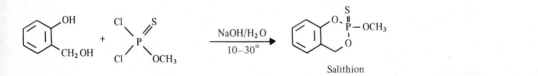

phoxim

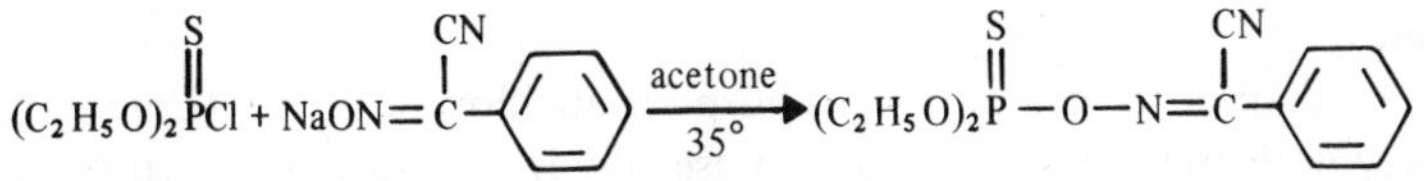

An aqueous caustic alkali solution is sometimes more efficient than other bases in organic solvents. Salithion was first synthesized by heating (80–90°) a mixture of saligenin and methyl phosphorodichloridothionate in toluene for 15 to 20 hr in the presence of potassium carbonate and copper powder.[83] It is now prepared more conveniently by applying the Schotten-Baumann acylating procedure: the mixture of the reactants in 20% aqueous sodium hydroxide solution is stirred for only 1 hr at room temperature (Equation 38).[84] Both the purity and yield are better than by the first procedure.

$$(38)$$

Salithion

Technical grade demeton is produced by the reaction of diethyl phosphorochloridothionate with 2-ethylthioethanol in the presence of a hydrochloric acid acceptor (Equation 39).[13] In the process, particularly at high temperature, the thionate (IX) produced is isomerized to the thiolate (X). Thus, the preparation obtained by this procedure is a mixture of both isomers of O,O-diethyl 2-ethylthioethyl phosphorothioate (70% thionate, 30% thiolate).

$$\underset{\underset{\text{(C}_2\text{H}_5\text{O})_2\overset{\parallel}{\text{P}}\text{Cl}}{\text{S}}}{} + \text{HOCH}_2\text{CH}_2\text{SC}_2\text{H}_5 \xrightarrow[\text{70--80}^\circ]{\text{K}_2\text{CO}_3} \underset{\text{IX: 70\%}}{\underset{}{(\text{C}_2\text{H}_5\text{O})_2\overset{\overset{\text{S}}{\parallel}}{\text{P}}-\text{OCH}_2\text{CH}_2\text{SC}_2\text{H}_5}} + \underset{\text{X: 30\%}}{(\text{C}_2\text{H}_5\text{O})_2\overset{\overset{\text{O}}{\parallel}}{\text{P}}-\text{SCH}_2\text{CH}_2\text{SC}_2\text{H}_5} \tag{39}$$

Fukuto and Metcalf[85] suggested that the isomerization might take place via the formation of a cyclic sulfonium intermediate (Equation 40).

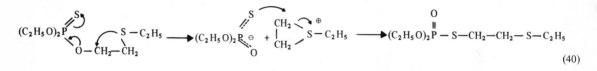

$$\tag{40}$$

The pure thiono isomer may be prepared by sulfuration of diethyl 2-ethylthioethyl phosphite, which is produced by the transesterification reaction of triethyl phosphite with 2-ethylthioethanol (Equation 41).[13] For the synthesis of the pure thiolate isomer, see Section D.2 of this Chapter.

$$(\text{C}_2\text{H}_5\text{O})_3\text{P} \xrightarrow[-\text{C}_2\text{H}_5\text{OH}]{\text{HOCH}_2\text{CH}_2\text{SC}_2\text{H}_5} (\text{C}_2\text{H}_5\text{O})_2\text{P}-\text{OCH}_2\text{CH}_2\text{SC}_2\text{H}_5 \xrightarrow{\text{S}} (\text{C}_2\text{H}_5\text{O})_2\overset{\overset{\text{S}}{\parallel}}{\text{P}}\text{OCH}_2\text{CH}_2\text{SC}_2\text{H}_5 \tag{41}$$

In a similar reaction procedure, aryl dialkyl phosphoroselenoates are prepared by addition of elemental selenium to the corresponding aryl dialkyl phosphite, which may be produced by heating dialkyl N,N-diethylphosphoramidites with phenols (Equation 42).[86]

$$(\text{C}_2\text{H}_5\text{O})_2\text{PN}(\text{C}_2\text{H}_5)_2 \xrightarrow[-(\text{C}_2\text{H}_5)\text{NH}]{\text{HO}-\langle\!\!\!\text{X}} (\text{C}_2\text{H}_5\text{O})_2\text{PO}-\langle\!\!\!\text{X} \xrightarrow{\text{Se}} (\text{C}_2\text{H}_5\text{O})_2\overset{\overset{\text{Se}}{\parallel}}{\text{PO}}-\langle\!\!\!\text{X} \tag{42}$$

2. Preparation of Phosphorus Intermediates

Phosphorus thiochloride reacts with an alcohol to give an alkyl phosphorodichloridothioate (Equation 43).[3] For the preparation of the secondary derivatives, it is necessary to use a hydrochloric acid acceptor or sodium (or magnesium) alcoholate. By these methods, mixed dialkyl phosphorochloridothionates may also be prepared (Equations 44 and 45),[34] and O-ethyl O-methyl O-p-nitrophenyl phosphorothionate (Thiophos-ME) was manufactured in the USSR until 1967.

$$\text{PSCl}_3 + \text{ROH} \longrightarrow \underset{}{\text{RO}\overset{\overset{\text{S}}{\parallel}}{\text{P}}\text{Cl}_2} + \text{HCl} \tag{43}$$

$$\underset{}{\text{RO}\overset{\overset{\text{S}}{\parallel}}{\text{P}}\text{Cl}_2} + \text{R}'\text{OH} \xrightarrow{\text{base}} \underset{\text{R}'\text{O}}{\overset{\text{RO}}{\diagdown}}\text{P}\underset{\text{Cl}}{\overset{\text{S}}{\diagup}} + \text{HCl} \tag{44}$$

$$\underset{}{\text{RO}\overset{\overset{\text{S}}{\parallel}}{\text{P}}\text{Cl}_2} + \text{NaOR}' \longrightarrow \underset{\text{R}'\text{O}}{\overset{\text{RO}}{\diagdown}}\text{P}\underset{\text{Cl}}{\overset{\text{S}}{\diagup}} + \text{NaCl} \tag{45}$$

Trivalent phosphorus compounds have an affinity for sulfur, and can be converted to corresponding pentavalent compounds having a thiophosphoryl group (P=S) by the reaction with sulfur or certain sulfides (see Equation 41). Thus halophosphites react with sulfur or phosphorus thiochloride to give the corresponding alkyl phosphorochloridothionates (Equations 46 and 47).[3,87]

$$ROPCl_2 + S \text{ (or } PSCl_3) \longrightarrow RO\overset{\overset{\displaystyle S}{\|}}{P}Cl_2 \tag{46}$$

$$(RO)_2PCl + S \text{ (or } PSCl_3) \longrightarrow (RO)_2\overset{\overset{\displaystyle S}{\|}}{P}Cl \tag{47}$$

Another method to prepare dialkyl phosphorochloridothionates, developed in the USA, is the chlorination of O,O-dialkyl phosphorodithioic acids (Equation 49), which may be produced by the reaction of appropriate alcohols with phosphorus pentasulfide (Equation 48)[88] and are very important intermediate compounds useful for the synthesis of phosphorodithioates too (see Section C).

$$P_2S_5 + 4ROH \longrightarrow 2(RO)_2\overset{\overset{\displaystyle S}{\|}}{P}SH + SH_2 \tag{48}$$

$$2(RO)_2\overset{\overset{\displaystyle S}{\|}}{P}SH + 3Cl_2 \longrightarrow 2(RO)_2\overset{\overset{\displaystyle S}{\|}}{P}Cl + 2HCl + S_2Cl_2 \tag{49}$$

Provided that the raw material, phosphorus pentasulfide, is of high quality, the dialkyl phosphorochloridothionates obtained are almost free from impurity, particularly of triesters,[2] which always may be formed as by-products in the reaction of alkyl phosphorodichloridothionates and alcohols, although Melnikov[34] has suggested that this method is less economical than the first method shown by Equations 43 to 45.

The chlorination of O,O-dialkyl phosphorothioites also gives dialkyl phosphorochloridothionates (Equation 50).

$$(RO)_2\overset{\overset{\displaystyle S}{\|}}{P}H \xrightarrow{Cl_2} (RO)_2\overset{\overset{\displaystyle S}{\|}}{P}Cl + HCl \tag{50}$$

C. PHOSPHOROTHIOLOTHIONATES

Since the discovery of the low toxic insecticide, malathion, in 1950, this class of phosphorus compounds has grown into one of the most important pesticide classes. Now, more than 40 compounds of the phosphorothiolothionate class are known as practical or promising experimental pesticides. The great majority of them are insecticides. Very recently, compounds with fungicidal or herbicidal activity were found among them.

In contrast with the classes of phosphorothionates and phosphates, aryl esters are not important as pesticides in this class. Many pesticides of this class have a hetero atom on the α- or β- carbon of an S-aliphatic group to construct a carboxy ester (XI), amide (XII), amine (XV), heterocycle (XVI), sulfide (XIII, XIV), or sulfoxide (XIII, XIV), as shown in the following general formulas:

$$(RO)_2\overset{\overset{\displaystyle S}{\|}}{P}-SCHR'\overset{\overset{\displaystyle O}{\|}}{C}OR'' \quad (XI) \qquad (RO)_2\overset{\overset{\displaystyle S}{\|}}{P}-SCH_2\overset{\overset{\displaystyle O}{\|}}{C}NR'R'' \quad (XII)$$

$$(RO)_2\overset{\overset{\displaystyle S}{\|}}{P}-SCH_2CH_2\overset{\overset{\displaystyle (O)}{\uparrow}}{S}R' \quad (XIII) \qquad (RO)_2\overset{\overset{\displaystyle S}{\|}}{P}-SCH_2\overset{\overset{\displaystyle (O)}{\uparrow}}{S}R' \quad (XIV)$$

$$(RO)_2\overset{\overset{\displaystyle S}{\|}}{P}-SCH_2CH_2NHR' \quad (XV) \qquad (RO)_2\overset{\overset{\displaystyle S}{\|}}{P}-SCH_2N\bigcirc \quad (XVI)$$

With a few exceptions of the dithiolate type, all the pesticidal esters of phosphorodithioic acid are the thiolothionate type. They are usually simply named as O,O-dialkyl S-substituted phosphorodithioates. These compounds are generally synthesized from O,O-dialkyl hydrogen phosphorodithioates by the formation of a PS-C bond by the following three methods: 1) alkylation with an alkyl halide; 2) addition to an aldehyde; or 3) addition to an olefin. They are also prepared by the formation of a P-S bond.

1. Alkylation of O,O-Dialkyl Phosphorodithioates

Many phosphorothiolothionate esters are prepared by this reaction:

$$\underset{\text{S}}{(RO)_2\overset{\text{S}}{\overset{\|}{P}}SM} + XR' \longrightarrow (RO)_2\overset{\text{S}}{\overset{\|}{P}}-S-R' + MX \tag{51}$$

where, M is metal, ammonium, or hydrogen, and X is halogen. For example, a systemic insecticide, thiometon, is synthesized by the reaction of 2-chlorodiethyl sulfide with sodium O,O-dimethyl phosphorodithioate (Equation 52).[13] As the sulfonate ion is a good leaving group as well as the halide ions in nucleophilic substitution reactions, the p-toluenesulfonate ester of 2-(ethylthio)ethanol is also employed in the preparation of thiometon. The sulfonate ester is synthesized by the reaction of the alcohol with p-toluenesulfonyl chloride in the presence of sodium hydroxide, and is used for the reaction with the sodium dithioate without isolation (Equation 53).[13]

$$(CH_3O)_2\overset{\text{S}}{\overset{\|}{P}}SNa + ClCH_2CH_2SC_2H_5 \longrightarrow (CH_3O)_2\overset{\text{S}}{\overset{\|}{P}}SCH_2CH_2SC_2H_5 + NaCl \tag{52}$$

thiometon 97%

$$HOCH_2CH_2SC_2H_5 \xrightarrow{+ CH_3 \langle\!\langle\rangle\!\rangle SO_2Cl} CH_3\langle\!\langle\rangle\!\rangle SO_3CH_2CH_2SC_2H_5 \xrightarrow[- CH_3\langle\!\langle\rangle\!\rangle SO_3Na]{+ (CH_3O)_2PSSNa} (CH_3O)_2\overset{\text{S}}{\overset{\|}{P}}SCH_2CH_2SC_2H_5 \tag{53}$$

An insecticide of type XI, acethion, is prepared by the reaction of sodium O,O-diethyl phosphorodithioate with ethyl chloroacetate in aqueous medium at 60–70° (Equation 54).[13] Dimethoate may be prepared similarly from dimethyl hydrogen phosphorodithioate and N-methylbromoacetamide in the presence of potassium carbonate. However, many methods to form the amide linkage after P-S-C bond formation have been developed.[2] The aminolysis of an ester or an anhydride is applied as shown in Equations 55 and 56.

$$(C_2H_5O)_2\overset{\text{S}}{\overset{\|}{P}}SNa + ClCH_2\overset{\text{O}}{\overset{\|}{C}}OC_2H_5 \longrightarrow (C_2H_5O)_2\overset{\text{S}}{\overset{\|}{P}}SCH_2\overset{\text{O}}{\overset{\|}{C}}OC_2H_5 + NaCl \tag{54}$$

acethion

$$(CH_3O)_2\overset{\text{S}}{\overset{\|}{P}}SNa \xrightarrow{ClCH_2COOC_6H_5} (CH_3O)_2\overset{\text{S}}{\overset{\|}{P}}SCH_2COOC_6H_5 \xrightarrow{CH_3NH_2} (CH_3O)_2\overset{\text{S}}{\overset{\|}{P}}SCH_2CONHCH_3 \tag{55}$$

dimethoate

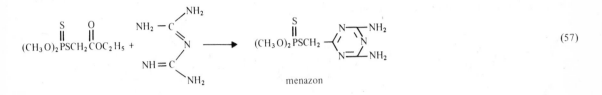

$$2(CH_3O)_2\overset{\overset{\displaystyle S}{\|}}{P}SNH_3CH_3 \xrightarrow[0-10°]{ClCH_2COCl} (CH_3O)_2\overset{\overset{\displaystyle S}{\|}}{P}SCH_2\overset{\overset{\displaystyle O}{\|}}{C}-S-\overset{\overset{\displaystyle S}{\|}}{P}(OCH_3)_2$$

$$\xrightarrow[0°]{2\ CH_3NH_2}(CH_3O)_2\overset{\overset{\displaystyle S}{\|}}{P}SCH_2CONHCH_3\ +\ (CH_3O)_2\overset{\overset{\displaystyle S}{\|}}{P}SNH_3CH_3 \tag{56}$$

This type of modification is further employed for the manufacture of menazon (Equation 57).[89] Carboxy esters yield s-triazines by condensation with biguanides.[90]

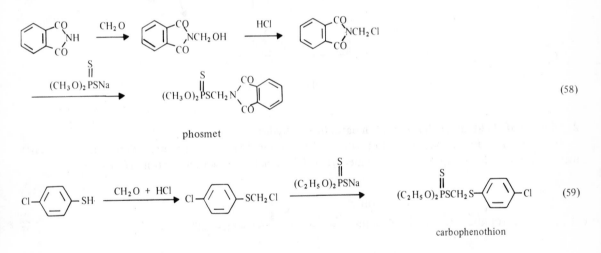

menazon

(57)

N- or S-halomethyl compounds are usually utilized as nonphosphorus intermediates to prepare the compounds of types XIV and XVI. They are synthesized by the reaction of formaldehyde with an appropriate compound containing a hetero atom. Thus, the addition product of phthalimide to formaldehyde is converted to the chloride, and the product is used for the preparation of the insecticide phosmet (Imidan®) (Equation 58). The insecticide carbophenothion is prepared from chloromethyl p-chlorophenyl sulfide, which is synthesized by chloromethylation of p-chlorothiophenol with formaldehyde and hydrochloric acid (Equation 59).[13]

phosmet

(58)

carbophenothion

(59)

Azinphosmethyl is also prepared in the same manner from the salt of O,O-dimethyl phosphorodithioic acid and N-chloromethylbenzazimide,[91] which is obtained by the addition of benzazimide to formaldehyde followed by chlorination with thionyl chloride (Equation 60). Recently, another procedure to prepare azinphosmethyl directly from benzazimide, formaldehyde, and the phosphorodithioic acid has been developed (Equation 61).[2]

(60)

azinphosmethyl

$$(CH_3O)_2\overset{\underset{\textstyle\|}{S}}{P}SH + CH_2O + \underset{N \underset{\textstyle N}{\diagdown}}{\overset{HN \diagup CO}{\underset{|}{\mid}}}\!\!\!\!\!\!\!\!\!\!\! \xrightarrow[-H_2O]{H^+} (CH_3O)_2\overset{\underset{\textstyle\|}{S}}{P}SCH_2N\underset{N \underset{\textstyle N}{\diagdown}}{\overset{CO}{\diagup}} \qquad (61)$$

The preparation of the insecticide methidathion (Superacide®) is also similar. It may be prepared from 2-methoxy-1,3,4-thiadiazol-5-(4*H*)-one (**XVII**) step by step via the methylol and chloromethyl derivatives (route *a* and *b*). Moreover, it is directly produced in higher yield by mixing the heterocycle **XVII** with paraformaldehyde and potassium *O,O*-dimethyl phosphorodithioate in sulfaric acid at 25–30° (route *c*) without isolation of any intermediate products.[92]

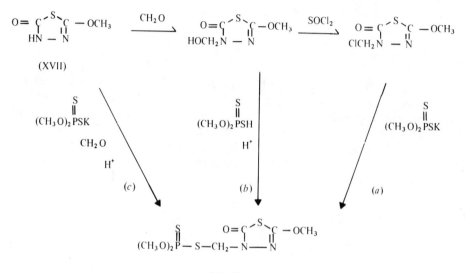

methidathion

2. Addition of *O,O*-Dialkyl Phosphorodithioates to Aldehydes

O,O-Dialkyl hydrogen phosphorodithioates add to carbonyl groups and the addition products (**XVIII**) may condense with mercaptans to form a methylene bridge between two sulfur atoms (Equation 62).[93]

$$(RO)_2\overset{\underset{\textstyle\|}{S}}{P}SH + O{=}CHR' \longrightarrow (RO)_2\overset{\underset{\textstyle\|}{S}}{P}{-}S{-}\overset{\underset{\textstyle|}{OH}}{C}HR' \xrightarrow[H_2O]{R''SH} (RO)_2\overset{\underset{\textstyle\|}{S}}{P}{-}S{-}\overset{\underset{\textstyle|}{R'}}{C}H{-}SR'' \qquad (62)$$

(**XVIII**)

In certain cases, *S*-(*S*-substituted thiomethyl) phosphorodithioates are prepared conveniently by the reactions of hydrogen phosphorodithioates, formaldehyde, and mercaptans without preliminary preparation of halomethyl derivatives, as described in the preceding section. For instance, phorate is prepared from diethyl phosphorodithioate, ethylmercaptan, and formaldehyde at room temperature without isolation of the hydroxymethyl phosphorodithioate intermediate (Equation 63).[93]

$$(C_2H_5O)_2\overset{\underset{\textstyle\|}{S}}{P}SH + CH_2O\ (aq.) + C_2H_5SH \longrightarrow (C_2H_5O)_2\overset{\underset{\textstyle\|}{S}}{P}SCH_2SC_2H_5 \qquad (63)$$

phorate 70% yield

3. Addition of *O,O*-Dialkyl Phosphorodithioates to Olefins

O,O-Dialkyl hydrogen phosphorodithioates also add to olefins to form a PS-C bond. Thus, malathion is produced by the addition of *O,O*-dimethyl hydrogen phosphorodithioate to diethyl maleate (Equation 64).[13] This reaction is catalyzed by small amounts of a base. In order to prevent the polymerization of maleate, a catalytic amount of hydroquinone is usually added to the reaction mixture.

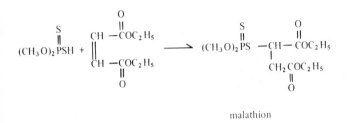

$$(64)$$

malathion

Similarly *O,O*-diethyl hydrogen phosphorodithioate adds to *N*-vinylphthalimide to give an insect fungicidal *S*-(α-phthalimidoethyl) phosphorodithioate according to Markownikoff's rule (normal addition) (Equation 65).[94]

$$(65)$$

Purified diethyl hydrogen phosphorodithioate adds, however, contrary to Markownikoff's rule, as do ordinary mercaptans (abnormal addition). This is due to the fact that ordinary samples of olefins contain enough amounts of peroxides to bring about the abnormal addition. On the other hand, crude *O,O*-dialkyl hydrogen phosphorodithioates prepared from phosphorus pentasulfide and alcohols contain a reducing agent, which is probably phosphorus sesquisulfide, and can destroy the peroxides in the olefin preparations so as to allow the normal addition to occur. Thus, under controlled conditions, *O,O*-dialkyl hydrogen phosphorodithioates may be made to add to unsymmetrical olefins to yield either the normal or the abnormal addition products. For example, pure *O,O*-diethyl hydrogen phosphorodithioate adds to styrene, forming *O,O*-diethyl *S*-(β-phenethyl) phosphorodithioate (Equation 66), while from the crude or pure dithioate with a small amount of phosphorus sesquisulfide, α-phenethyl ester is produced (Equation 67).[95]

$$(C_2H_5O)_2\overset{\text{S}}{\underset{\|}{P}}SH + CH_2{=}CHC_6H_5 \longrightarrow (C_2H_5O)_2\overset{\text{S}}{\underset{\|}{P}}SCH_2CH_2C_6H_5 \qquad (66)$$

pure

$$(C_2H_5O)_2\overset{\text{S}}{\underset{\|}{P}}SH + CH_2{=}CHC_6H_5 \xrightarrow{P_4S_3} (C_2H_5O)_2\overset{\text{S}}{\underset{\|}{P}}S{-}\underset{\underset{C_6H_5}{|}}{\overset{\overset{CH_3}{|}}{C}}H{-}C_6H_5 \qquad (67)$$

pure

In an analogous way, bis(diethoxyphosphinothioyl) disulfide adds to *p*-dioxene to give the insecticide dioxathion (Equation 68).[96] The reaction is carried out in the presence of catalytic quantities of iodine. It is also prepared by the reaction of 2,3-dichloro-*p*-dioxane with *O,O*-diethyl phosphorodithioic acid.[96] Both the preparations of dioxathion obtained by these different methods are the mixtures of *cis* and *trans* isomers, the ratio of which is approximately 2:3.

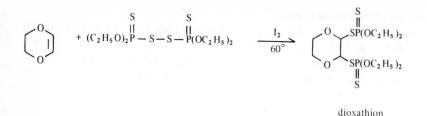

$$(68)$$

dioxathion

When phosphorodithioates are added to acetylenes, S-vinyl phosphorodithioates may be produced. Diethoxyphosphinothioyl ethyl disulfide adds to acetylene by irradiation with UV light, to give insecticidal O,O-diethyl S-(2-ethylthiovinyl) phosphorodithioate (Equation 69).[97]

$$\underset{\parallel}{\overset{S}{(C_2H_5O)_2PS}}-SC_2H_5 + HC\equiv CH \xrightarrow{h\nu} \underset{\parallel}{\overset{S}{(C_2H_5O)_2PS}}-CH=CH-SC_2H_5 \qquad (69)$$

4. Formation of P-S Bond

There are rather few examples for the preparation of phosphorodithioate pesticides by the formation of a P-S bond. This may be due to the easiness of the preparation of O,O-dialkyl hydrogen phosphorodithioates, the intermediates for the above-mentioned methods 1 to 3. Moreover, direct phosphorylation by phosphorus chlorides is not as suitable with thiol compounds as with hydroxy compounds, because the soft base mercaptide ion may preferably attack the soft carbon of the alkyl ester group rather than the hard phosphoryl center (Equation 70) (see II.D.2; III.B.1; III.C.2).

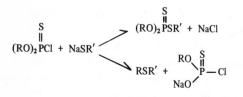

$$(70)$$

Thus, the reaction of sulfenyl chlorides with dialkyl phosphorothioites is applied to the P-S bond formation. This method is useful particularly for the synthesis of S-aryl esters (Equation 71).[98]

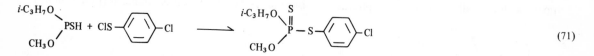

$$(71)$$

Alkyl thiosulfate monoesters and alkyl thiocyanates undergo similar reactions. As one of the methods to synthesize the insecticide dimethoate, the reaction of O,O-dimethyl phosphorothioite and methylcarbamoylmethyl thiosulfate was proposed, as shown by Equation 72, where M is a metallic or ammonium ion.[2] The insecticide thiodemeton is produced by the reaction of ethylthioethyl thiocyanate with diethyl phosphorothioite (Equation 73).[13]

$$(CH_3O)_2PSM + MOSO_2SCH_2CONHCH_3 \longrightarrow \underset{\parallel}{\overset{S}{(CH_3O)_2PSCH_2CONHCH_3}} + M_2SO_3 \qquad (72)$$

dimethoate

$$(C_2H_5O)_2PSH + NCSCH_2CH_2SC_2H_5 + C_2H_5ONa \longrightarrow \underset{\parallel}{\overset{S}{(C_2H_5O)_2PSCH_2CH_2SC_2H_5}} + NaCN + C_2H_5OH \qquad (73)$$

thiodemeton

The phosphorus intermediates, dialkyl phosphorothioites, may be produced by the sulfuration of corresponding phosphites with phosphorus pentasulfide (Equation 74).

$$(RO)_2POH \xrightarrow{P_2S_5} (RO)_2PSH \tag{74}$$

5. Formation of S-X Bond

Recently, S-chloro phosphorodithioates have been utilized for the preparation of S-X bonds. Thus, O,O-dialkyl S-aziridinyl phosphorodithioates are prepared according to the following equation:[99]

$$(RO)_2 \overset{\overset{\displaystyle S}{\|}}{P}S-Cl + HN\!\!\triangleleft \xrightarrow[-HCl]{R_3N} (RO)_2\overset{\overset{\displaystyle S}{\|}}{P}-S-N\!\!\triangleleft \tag{75}$$

D. PHOSPHOROTHIOLATES

Phosphorothiolate esters are considered as the active form of the corresponding phosphorodithioate insecticides and of some readily isomerizable phosphorothionate insecticides such as demeton. More than ten compounds of these types (XIX, XX) have been developed as insecticides. A variety of interesting biological activities were recently found in other types of phosphorothiolate esters. Low toxic fungicides were found in S-benzyl phosphorothiolates (XXI) and in bis(S-aryl) phosphorodithiolates (XXII; n = 2). Nematocides and defoliants are also known in the type XXII compounds. The general formulas of this class are as follows:

$$(RO)_2\overset{\overset{\displaystyle O}{\|}}{P}-S-CH_2CY \quad (XIX) \qquad (RO)_2\overset{\overset{\displaystyle O}{\|}}{P}-S-CH_2CH_2YR' \quad (XX)$$

$$(RO)_2\overset{\overset{\displaystyle O}{\|}}{P}-S-CH_2-A \quad (XXI) \qquad (RO)_{3-n}\overset{\overset{\displaystyle O}{\|}}{P}(SR')_n \quad (XXII)$$

where Y is an alkoxy or alkylamino group, A is an aromatic or aromatic heterocycle group, R' is an alkyl or aromatic group, and n is a number from 1 to 3.

Various methods for the preparation of phosphorothiolates have been developed. They may be classified into the following five principal methods: 1) reaction between dialkyl phosphorochloridates and mercaptans; 2) the alkylation of O,O-dialkyl phosphorothioates with alkyl halides; 3) the reaction of dialkyl or trialkyl phosphites with sulfur compounds; 4) the thiono-thiolo rearrangement of phosphorothionates; and 5) the oxidation of phosphorothiolothionates and phosphorothioites.

1. Reaction of Dialkyl Phosphorochloridates with Mercaptans

$$(RO)_2\overset{\overset{\displaystyle O}{\|}}{P}Cl + HSR' \longrightarrow (RO)_2\overset{\overset{\displaystyle O}{\|}}{P}SR' + HCl \tag{76}$$

Though the analogous reaction between phosphorochloridothionates and hydroxy compounds is very important for the preparation of phosphorothionate pesticides, this reaction is rarely applied to the preparation of phosphorothiolates because an unfavorable alkylation reaction may also take place as mentioned above (see Equation 70).

Demeton-S is produced from diethyl phosphorochloridate and the sodium salt of 2-ethylthioethylthiol (Equation 77).

$$(C_2H_5O)_2\overset{\overset{\displaystyle O}{\|}}{P}Cl + NaSCH_2CH_2SC_2H_5 \longrightarrow (C_2H_5O)_2\overset{\overset{\displaystyle O}{\|}}{P}SCH_2CH_2SC_2H_5 + NaCl \tag{77}$$

demeton-S

The fungicide edifenphos (Hinosan[®]) is synthesized from ethyl phosphorodichloridate and thiophenol by the action of sodium ethoxide in benzene at $10-15°$ (Equation 78).[100]

$$C_2H_5O\overset{\overset{\displaystyle O}{\|}}{P}Cl_2 + 2HSC_6H_5 + 2NaOC_2H_5 \longrightarrow C_2H_5O\overset{\overset{\displaystyle O}{\|}}{P}(SC_6H_5)_2 + 2NaCl_2 + 2C_2H_5OH \tag{78}$$

edifenphos

2. Alkylation of *O,O*-Dialkyl Phosphorothioates

$$(RO)_2\overset{\overset{\displaystyle O}{\|}}{P}SM + XR' \longrightarrow (RO)_2\overset{\overset{\displaystyle O}{\|}}{P}SR' + MX \tag{79}$$

where M is metal, ammonium, or hydrogen, and X is halogen. This reaction, analogous to Equation 51, is a common reaction to form an S-C bond and is employed for the preparation of many phosphorothiolate pesticides. For instance, a fungicide, Kitazin P[®], is prepared from ammonium *O,O*-diisopropyl phosphorothioate and benzyl chloride (Equation 80).[101] The reaction occurs selectively on the sulfur atom, and no thiono isomer is produced.

Kitazin P[®]

$$\tag{80}$$

O,O-Dialkyl hydrogen phosphorothioates are the tautomeric mixtures of thiono and thiolo isomers in the preponderance of the latter, as discussed in Chapter I.

$$(RO)_2\overset{\overset{\displaystyle S}{\|}}{P}-OH \rightleftharpoons (RO)_2\overset{\overset{\displaystyle O}{\|}}{P}-SH$$

In the reaction with diazomethane, *O,O*-diethyl phosphorothioate gives both methyl thiolate and thionate isomers in the ratio of 4 to 1. However, its sodium salt produces only the methyl thiolate in 89% yield by the reaction with methyl iodide.[102] The salts of *O,O*-dialkyl phosphorothioates appear to exist as

$$(RO)_2P\overset{\nearrow O}{\underset{\searrow S}{\ominus}} \quad M^{\oplus}$$

and the more nucleophilic sulfur atom reacts with alkyl halides to form preferentially the thiolate esters.

According to Pearson's concept of hard and soft acids and bases, hard acids prefer to bind to hard or nonpolarizable bases, and soft acids prefer to bind to soft or polarizable bases.[103] Typical hard acids are positively charged centers such as:

$$\overset{O}{\underset{\|}{R\overset{\|}{C}}}-\ ,\ \ R-\overset{O}{\underset{\underset{\|}{O}}{\overset{\|}{S}}}-\ ,\ \ R_2\overset{O}{\overset{\|}{P}}-\ ,\ \ H^+,\ \text{and}\ R_3C^+.$$

The electronically saturated center is regarded as a typical soft acid, as follows:

$$RCH_2{}^-,\ R_2P^-,\ RS^-,\ Br^-,\ \text{and}\ H^-.$$

The softness of a base (nucleophile) increases with increasing radius and with increasing electron affinity. Thus, it is in the following order:[25]

$$R_3C^- > R_2N^- > RO^- > F^-;\ I^- \gtrsim Br^- > Cl^- > F^-;\ R_3Si^- > R_2P^- > RS^- > Cl^-.$$

Therefore, in the reaction between an *O,O*-dialkyl phosphorothioate ion and an alkyl halide which has a soft acid center, the softer nucleophile S^-, being more reactive with the soft acid than the harder O^-, may attack the alkyl group to result in the formation of a thiolate ester (Equation 81).

$$(RO)_2P\overset{O\ \text{(hard base)}}{\underset{S\ \text{(soft base)}\ \text{(soft acid)}}{\Big\langle}} \overset{\frown}{R'\!-\!X} \longrightarrow (RO)_2\overset{O}{\overset{\|}{P}}-S-R' + X^- \qquad (81)$$

Miller calculated that the "carbon basicity" of a thiol anion is more than 10^4 times as great as the corresponding oxide anion.[104] This means that the equilibrium constant for Reaction 82 is at least 10^4 times as large as that for Reaction 83.

$$(RO)_2\overset{O}{\overset{\|}{P}}-S^- + CH_3X \rightleftharpoons (RO)_2\overset{O}{\overset{\|}{P}}-S-CH_3 + X^- \qquad (82)$$

$$(RO)_2\overset{O}{\overset{\|}{P}}-O^- + CH_3X \rightleftharpoons (RO)_2\overset{O}{\overset{\|}{P}}-O-CH_3 + X^- \qquad (83)$$

Reaction 79 is, therefore, useful for the preparation of pure thiolate isomers in demeton-type compounds which are readily isomerizable (see B.1.). Thus, demeton-S-methyl is prepared in good yield from potassium *O,O*-dimethyl phosphorothioate and 2-chlorodiethyl sulfide (Equation 84).[13]

$$(CH_3O)_2\overset{O}{\overset{\|}{P}}SK + ClCH_2CH_2SC_2H_5 \longrightarrow (CH_3O)_2\overset{O}{\overset{\|}{P}}SCH_2CH_2SC_2H_5 + KCl \qquad (84)$$

demeton-S-methyl

The sulfoxide analog of demeton-S-methyl, oxydemeton-methyl, is prepared in the same manner by replacing the sulfide with 2-bromodiethyl sulfoxide. The alternative method is the oxidation of demeton-S-methyl. The oxidation is performed with bromine-water or hydrogen peroxide (Equation 85).[13]

$$(CH_3O)_2\overset{O}{\overset{\|}{P}}-S-CH_2CH_2-S-C_2H_5 + Br_2 + H_2O \longrightarrow (CH_3O)_2\overset{O}{\overset{\|}{P}}-S-CH_2CH_2-\overset{O}{\overset{\uparrow}{S}}-C_2H_5 + 2HBr \qquad (85)$$

oxydemeton-methyl

Under such conditions of reaction, phosphorothiolate-sulfur is insensitive to oxidation. Thiono-sulfur is less susceptible than sulfide, but is oxidizable with an excess of bromine. Thus, demeton-O gives 2-ethylsulfinylethyl phosphate by oxidation with bromine (Equation 86).[13]

$$(C_2H_5O)_2\overset{\overset{S}{\|}}{P}-OCH_2CH_2SC_2H_5 \ + \ 5Br_2 \ + \ 6H_2O \ \longrightarrow \ (C_2H_5O)_2\overset{\overset{O}{\|}}{P}OCH_2CH_2\overset{\overset{O}{\uparrow}}{S}C_2H_5 \ + \ H_2SO_4 \ + \ 10HBr \tag{86}$$

demeton-O

When ethyleneimine is applied as an alkylating agent, O,O-dialkyl phosphorothioic acids yield S-(β-aminoethyl) phosphorothioates, which may be acetylated to give the oxo-analogs of the insecticide amiphos (Equation 87).[105]

$$(RO)_2\overset{\overset{O}{\|}}{P}-SH \ + \ HN\triangleleft \ \longrightarrow \ (RO)_2\overset{\overset{O}{\|}}{P}S-CH_2CH_2NH_2 \ \xrightarrow{(CH_3CO)_2O} \ (RO)_2\overset{\overset{O}{\|}}{P}SCH_2CH_2NHCOCH_3 \tag{87}$$

Phosphorus intermediates, O,O-dialkyl hydrogen phosphorothioates or their salts, are synthesized as follows:

1. Hydrolysis of trialkyl phosphorothionates (Equation 88) or more conveniently dialkyl phosphoro-chloridothionates (Equation 89) with alkali;

$$(RO)_3P{=}S \ + \ KOH \ \longrightarrow \ (RO)_2\overset{\overset{O}{\|}}{P}SK \ + \ ROH \tag{88}$$

$$(RO)_2\overset{\overset{S}{\|}}{P}Cl \ + \ KOH \ \longrightarrow \ (RO)_2\overset{\overset{O}{\|}}{P}SK \ + \ KCl \tag{89}$$

2. Reaction of dialkyl phosphorochloridates and sulfides (Equation 90);

$$(RO)_2\overset{\overset{O}{\|}}{P}Cl \ + \ K_2S \ \longrightarrow \ (RO)_2\overset{\overset{O}{\|}}{P}SK \ + \ KCl \tag{90}$$

3. Addition of elemental sulfur to dialkyl phosphites;

$$(RO)_2POH \ + \ S \ \longrightarrow \ (RO)_2\overset{\overset{O}{\|}}{P}SH \tag{91}$$

The reaction is performed in dioxane at 100° in 70% yield of the addition product.[38] In the presence of a base, the reaction proceeds almost quantitatively at room temperature (Equation 92).

$$(RO)_2POH \ + \ S \ + \ NH_3 \ \longrightarrow \ (RO)_2\overset{\overset{O}{\|}}{P}SNH_4 \tag{92}$$

For the preparation of dialkyl phosphites see Section A.1. of this Chapter.

3. Reaction of Phosphites with Sulfur Compounds

Trialkyl phosphites are active nucleophilic agents and perform substitution at a bivalent sulfur connected to a good leaving group. Dialkyl phosphites are much less active, as expected from their character as pentavalent phosphonates (see I.C.2.). However, their salts behave as trivalent phosphorus compounds and are active enough to react with certain sulfur compounds.

Trialkyl phosphites react with alkyl- and aryl-sulfenyl chlorides to give corresponding phosphoro-thiolates in good yields,[106] probably through a mechanism analogous to that of the Michaelis-Arbusov reaction (Equation 93). (For the Michaelis-Arbuzov reaction refer to Section H.3.)

$$(RO)_3P: \quad + \quad \overset{Cl}{\underset{S-R'}{\vert}} \quad \longrightarrow \quad RO-\overset{R-O}{\underset{RO}{\overset{\vert}{P^+}}}-S-R' \quad \longrightarrow \quad (RO)_2\overset{O}{\overset{\Vert}{P}}-SR' + RCl \qquad (93)$$

The phosphite attacks the sulfur atom to displace halide. The intermediate quasi phosphonium salt may be subsequently converted to a phosphorothiolate ester by a nucleophilic displacement of the chloride ion, accompanied by elimination of alkyl chloride.

Alkyl- and aryl-sulfonyl chlorides also react with triethyl phosphite to give the corresponding phosphorothiolate, instead of the expected sulfonylphosphonates (Equation 94). The sulfonyl chloride appears to be deoxygenated, at first, by the phosphite yielding the corresponding sulfenyl chloride.[107] Phosphites have a high tendency to transform into phosphates by abstracting an oxygen atom from other molecules.

$$RSO_2Cl \xrightarrow[-2(C_2H_5O)_3P=O]{2(C_2H_5O)_3P} RSCl \xrightarrow{(C_2H_5O)_3P} (C_2H_5O)_2\overset{O}{\overset{\Vert}{P}}SR + C_2H_5Cl \qquad (94)$$

Phosphorothioites react similarly with sulfenyl chlorides to give phosphorodithiolates. This is proposed as an alternative method for the preparation of the fungicide edifenphos (Equation 95).[2]

$$(C_2H_5O)_2P-S- \!\!\left\langle \rule{0.5em}{0em} \right\rangle \quad + \quad ClS- \!\!\left\langle \rule{0.5em}{0em} \right\rangle \quad \longrightarrow \quad C_2H_5O\overset{O}{\overset{\Vert}{P}}(S-\!\!\left\langle \rule{0.5em}{0em} \right\rangle)_2 + C_2H_5Cl \qquad (95)$$

A similar reaction occurs between sulfenyl compounds and dialkyl phosphites (Equation 96). For instance, a fungicidal phosphorothiolate (Cerezin) was prepared as follows:[108]

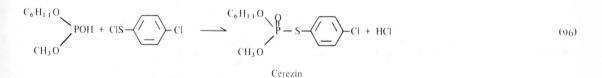

$$\underset{CH_3O}{\overset{C_6H_{11}O}{\diagdown}}\!\!POH + ClS-\!\!\left\langle \rule{0.5em}{0em} \right\rangle\!\!-Cl \longrightarrow \underset{CH_3O}{\overset{C_6H_{11}O}{\diagdown}}\!\!\overset{O}{P}-S-\!\!\left\langle \rule{0.5em}{0em} \right\rangle\!\!-Cl + HCl \qquad (96)$$

Cerezin

Dialkyl phosphorochloridites also react with alkylsulfenyl chlorides to give O,S-dialkyl phosphoro-chloridothiolates, which may be applied for the preparation of thiolate triesters. For example, a thiol isomer of parathion can be prepared by the following sequence of reactions.[109]

$$(RO)_2PCl \xrightarrow[-RCl]{R'SCl} \underset{R'S}{\overset{RO}{\diagdown}}\!\!\overset{O}{P}\!\!\overset{\diagup}{\diagdown}\!\!Cl \xrightarrow[Na_2CO_3]{HO-\!\!\left\langle \rule{0.5em}{0em} \right\rangle\!\!-NO_2} \underset{R'S}{\overset{RO}{\diagdown}}\!\!\overset{O}{P}\!\!\overset{\diagup}{\diagdown}\!\!O-\!\!\left\langle \rule{0.5em}{0em} \right\rangle\!\!-NO_2 \qquad (97)$$

Thiocyanates react with trialkyl phosphites, giving phosphorothiolates and nitriles.[110] By this reaction, demeton-S may be produced (Equation 98).[13]

$$(C_2H_5O)_3P + NCSCH_2CH_2SC_2H_5 \longrightarrow (C_2H_5O)_2\overset{\displaystyle O}{\overset{\|}{P}}SCH_2CH_2SC_2H_5 + C_2H_5CN \tag{98}$$

demeton-S

In the presence of bromotrichloromethane, trialkyl phosphites react with mercaptans or thiophenols, producing the corresponding O,O-dialkyl S-alkyl or aryl phosphorothiolates in good yields.[111] A radical-chain transfer mechanism has been postulated for this reaction (Equation 99).

$$(RO)_3P \xrightarrow{R'S\cdot} (RO)_3\overset{\cdot}{P}SR' \xrightarrow[-\overset{\cdot}{C}Cl_3]{BrCCl_3} (RO)_2\overset{O-R\,\,Br^-}{\underset{+}{\overset{|}{P}}} -SR' \longrightarrow (RO)_2\overset{\displaystyle O}{\overset{\|}{P}}SR' \tag{99}$$

S-Alkyl, but not aryl, esters of aliphatic and aromatic sulfonothioic acids behave like sulfenyl halides: they react with trialkyl phosphites or the salts of dialkyl phosphites to give corresponding O,O,S-trialkyl phosphorothiolates (Equations 100 and 101).[112]

$$(RO)_3P: + \overset{R'}{\underset{}{\overset{|}{S}}}-SO_2R'' \longrightarrow \overset{RO}{\underset{RO}{>}}\overset{O-R\,\,^-SO_2R''}{\underset{SR'}{P+}} \longrightarrow (RO)_2\overset{\displaystyle O}{\overset{\|}{P}}SR' + RSO_2R'' \tag{100}$$

$$(RO)_2PONa + R'SSO_2R'' \longrightarrow (RO)_2\overset{\displaystyle O}{\overset{\|}{P}}SR' + R''SO_2Na \tag{101}$$

Disulfides also react with phosphites, resulting in the cleavage of the S-S bond and the formation of a P-S bond. Trialkyl phosphites react with aliphatic disulfides at elevated temperature (Equation 102).[113] The salts of dialkyl phosphites react similarly with disulfides at low temperature (0–40°) and in solvents such as benzene to yield thiolate triesters, but at elevated temperature, as in boiling benzene, a dealkylation reaction proceeds further to result in the production of diesters and dialkyl sulfides (Equation 103).[114]

$$(RO)_3P + R'SSR' \longrightarrow (RO)_3\overset{+}{P}SR' \cdot\,^-SR' \longrightarrow (RO)_2\overset{\displaystyle O}{\overset{\|}{P}}SR' + RSR' \tag{102}$$

$$(RO)_2PONa + R'SSR' \xrightarrow[40°]{benzene} (RO)_2\overset{\displaystyle O}{\overset{\|}{P}}SR' + R'SNa \xrightarrow{80°} \overset{RO}{\underset{R'S}{>}}\overset{O}{\underset{ONa}{P\!\!\nearrow}} + R'SR \tag{103}$$

Sodium dialkyl phosphorothioites undergo the same reaction to form O,O,S-trialkyl phosphorodithioates at low temperature. However, the further process requires a more polar solvent such as ethanol.

An interesting reaction of disulfides with cyclic phosphoramidites is applied for the preparation of phosphoramidothiolates. By reaction with diaryl or alkyl aryl disulfides, the cyclic structure opens, forming O-(ω-substituted thioalkyl) phosphoramidothiolates (Equation 104).[115] An alkyl aryl disulfide gives exclusively an S-alkyl O-arylthioalkyl phosphoramidothiolate (Equation 105).

$$R_2N-P\overset{O-CH_2}{\underset{O-CH_2}{<}} + ArSSAr \longrightarrow \overset{Ar-S\overset{\delta-}{\cdots\cdots}S-Ar}{\underset{R_2N}{\overset{|\delta+}{P}+}\overset{}{\underset{O-CH_2}{\diagdown\,}}\overset{CH_2}{\underset{O}{\diagup}}} \longrightarrow \overset{Ar-S}{\underset{R_2N}{>}}\overset{O}{\underset{O-CH_2CH_2SAr}{P\!\!\nearrow}} \tag{104}$$

$$R_2N-P\underset{O}{\overset{O}{\big<}} + RS-SAr \longrightarrow \quad \begin{matrix} \delta^- \\ RS----SAr \\ \delta^+ | \\ R_2N \end{matrix} \quad \longrightarrow \quad R-S\underset{R_2N}{\overset{O}{\big\backslash}}P\underset{O-CH_2CH_2CH_2SAr}{\overset{O}{\big<}} \qquad (105)$$

An addition reaction of dialkyl phosphites to sulfinylamines is known. The reaction occurs in the presence of alkali metal or alkali alkoxide catalyst, and is believed to proceed according to Equation 106. The addition compounds are patented for pesticides.[116]

$$(CH_3)_3CNH_2 \xrightarrow{SO_2Cl_2} (CH_3)_3C-N=SO \xrightarrow{(C_2H_5O)_2POH} (C_2H_5O)_2\overset{O\ O}{\overset{\|\ \|}{P}}-SNHC(CH_3)_3 \qquad (106)$$

4. Thiono-thiolo Rearrangement of Phosphorothionates

Phosphorothionates are isomerized to corresponding phosphorothiolates by heat at about 150° as already described (II.B). This often causes the contamination of phosphorothionate preparations with impurity. Though this reaction is not generally applicable for the preparation of phosphorothiolates, certain easily isomerizable phosphorothionates, which have a hetero atom on β-carbon, are exceptionally suitable for preparation of their thiolate isomers by the thermal isomerization method.

For example, Amiton is prepared by warming its thiono isomer, which was produced from diethyl phosphorochloridothionate and sodium 2-(diethylamino)ethoxide (Equation 107).[13]

$$(C_2H_5O)_2\overset{S}{\overset{\|}{P}}Cl + NaOCH_2CH_2N(C_2H_5)_2 \xrightarrow[-NaCl]{} (C_2H_5O)_2\overset{S}{\overset{\|}{P}}OCH_2CH_2N(C_2H_5)_2$$

$$\xrightarrow{70-80°} (C_2H_5O)_2\overset{O}{\overset{\|}{P}}-SCH_2CH_2N(C_2H_5)_2 \qquad (107)$$

Amiton

The rearrangement may proceed through the formation of a cyclic intermediate accompanying the O-C bond cleavage and the subsequent alkylation of the sulfur atom (softer nucleophile) by the aziridinium ion (XXIII) (Equation 108.)[117]

$$(C_2H_5O)_2\overset{S}{\overset{\|}{P}}-O-CH_2-CH_2\overset{\frown N(C_2H_5)_2}{} \longrightarrow (C_2H_5O)_2P\overset{S}{\underset{O}{\big<}}^{\ominus} \begin{matrix} CH_2 O^{\oplus} \\ | \\ N(C_2H_5)_2 \\ | \\ CH_2 \end{matrix} \longrightarrow (C_2H_5O)_2\overset{O}{\overset{\|}{P}}-SCH_2CH_2N(C_2H_5)_2 \quad (108)$$

(XXIII)

The thiono-thiolo rearrangement reaction takes place more generally by the action of alkyl halides on phosphorothionates (Pistschimuka reaction) (Equation 109).

$$\underset{B}{\overset{OR}{\overset{|}{A-P=S}}} \ R'-X \longrightarrow \underset{B}{\overset{O-R \leftarrow X^\ominus}{\overset{|}{A-P^\oplus-S-R'}}} \longrightarrow \underset{B}{\overset{O}{\overset{\|}{A-P-SR'}}} + RX \qquad (109)$$

The rearrangement requires rather vigorous conditions, such as heating to more than 100°, and its preparative value is often limited by poor yields. However, these disadvantages are overcome by the use of strongly polar solvents.[118] Thus, salithion is smoothly converted to its thiolate isomer by the action of methyl iodide in the presence of dimethylformamide (DMF) and potassium carbonate (Equation 110).[119]

$$\text{(structure)} - OCH_3 \ + \ CH_3I \ \xrightarrow[50°]{DMF, K_2CO_3} \ \text{(structure)} - SCH_3 \tag{110}$$

When other alkyl iodides are used, the corresponding alkyl phosphorothiolates are produced.[120] The sulfur atom of trialkyl phosphorothionates may be alkylated not only by alkyl iodides, but also by alkyl chlorides. For instance, trimethyl phosphorothionate reacts with β-chlorodiethyl sulfide, by heating above 100°, to yield demeton-S-methyl (Equation 111).[34]

$$(CH_3O)_3P{=}S \ + \ ClCH_2CH_2SC_2H_5 \ \longrightarrow \ (CH_3O)_2\overset{\displaystyle O}{\overset{\|}{P}}SCH_2CH_2SC_2H_5 \ + \ CH_3Cl \tag{111}$$

The Pistschimuka reaction can be further extended to the preparation of phosphoramidothiolates (Equation 112).[121] O,S-Dimethyl phosphoramidothiolate (methamidophos, Monitor®) is a recently developed insecticide, which is prepared by this reaction.

$$(CH_3O)_2\overset{\displaystyle S}{\overset{\|}{P}}NH_2 \ + \ CH_3I \ \longrightarrow \ \underset{\text{methamidophos}}{\overset{\displaystyle \underset{CH_3O}{\overset{CH_3S}{\diagdown}}\underset{NH_2}{\overset{O}{P}\diagup}}{}} \ + \ CH_3I \tag{112}$$

An N-methylated product is also formed as a by-product. When the anions of phosphoramidothioates are employed, alkylation occurs preferentially on the nitrogen rather than on the sulfur atom.[122]

5. Oxidation of Phosphorodithioates and Phosphorothioites

The oxidative desulfuration of phosphorodithioates has been discussed in Section A.2. of this Chapter. It is utilized for the preparation of phosphorothiolates in small quantities only. On the other hand, trivalent phosphorus compounds have a high affinity for oxygen to form a phosphoryl bond (P=O). Thus, phosphorothioites are oxidized to give phosphorothiolates. For example, S,S,S-tributyl phosphorotrithiolate, a cotton defoliant (DEF), is prepared by the oxidation of tributyl phosphorotrithioite (Equation 113).

$$\underset{\text{DEF}}{(C_4H_9S)_3P} \ \xrightarrow{[O]} \ (C_4H_9S)_3P{=}O \tag{113}$$

6. Addition of Phosphorylsulfenyl Chlorides to Unsaturated Hydrocarbons

Phosphorylsulfenyl chlorides add to olefins or acetylenes to yield β-chloroalkyl or β-chlorovinyl esters, respectively. The addition reaction is nonselective so that both n- and iso-propyl phosphorothiolates (1:1) are formed by the addition of dimethoxyphosphinylsulfenyl chloride to propene (Equation 114).[126]

$$(CH_3O)_2\overset{\displaystyle O}{\overset{\|}{P}}SCl \ + \ CH_2{=}CH{-}CH_3 \ \longrightarrow \ (CH_3O)_2\overset{\displaystyle O}{\overset{\|}{P}}SCH_2CHClCH_3 \ + \ (CH_3O)_2\overset{\displaystyle O}{\overset{\|}{P}}S\overset{CH_3}{\overset{|}{C}}HCH_2Cl \tag{114}$$

Examples applied to get insecticidal compounds are presented in Equations 115 and 116.[123-125]

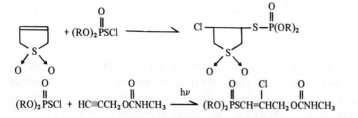

$$(RO)_2\overset{\displaystyle O}{\overset{\|}{P}}SCl \ + \ HC{\equiv}CCH_2OCNHCH_3 \ \xrightarrow{h\nu} \ (RO)_2\overset{\displaystyle O}{\overset{\|}{P}}SCH{=}CHCH_2OCNHCH_3 \tag{116}$$

Bis(alkoxy)phosphinylsulfenyl chlorides are synthesized by the reaction of trialkyl phosphorothionates with chlorine, or more preferably with sulfuryl chloride (Michalski reaction) (equation 117).[126,127]

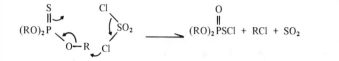

$$(RO)_2PSCl + RCl + SO_2 \qquad (117)$$

E. DERIVATIVES OF PHOSPHORAMIDIC ACID

This section deals with phosphoric acid derivatives which have P-N bond(s) in the molecule, but neither the P-F nor P-O-P bond (for which, refer to Sections F and G). The principal formulas of pesticides in this class are as follows:

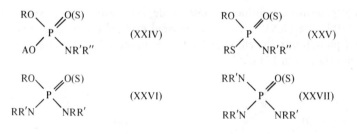

where, A is an aryl or alkyl, R is an alkyl, and R′ and R″ are hydrogen or alkyl group.

Very recently, a high insecticidal activity was found in simple phosphoramidothiolates, such as Monitor. Besides insecticides, a variety of pesticides are found in this class: acaricides, anthelmintics, nematocides, fungicides, herbicides, rodenticides, sterilants, and antitumor agents.

In this section, the formation of a P-N linkage and the modification of the amide group are discussed. The P-N linkage is generally formed by 1) the reactions of phosphorus chlorides with amines, and 2) the reactions of trivalent phosphorus compounds and amines. (See preceding sections on the formation of ester linkages.)

1. Reaction of Phosphorus Chlorides with Amines

The amides of phosphorus acids are generally prepared by the reaction of phosphorus chlorides with amines (or ammonia), according to Michaelis.[4] Hydrochlorides of amines are rather often used (Equation 118). For the preparation of ethylenimides, tertiary amines are employed to purge hydrochloric acid liberated in the reaction of ethylenimine with the chlorides.[27]

$$PXCl_3 \xrightarrow[-2HCl]{+RNH_2 \cdot HCl} RNHPCl_2 \xrightarrow[-2HCl]{+RHN_2 \cdot HCl} (RNH)_2 PCl \qquad (118)$$

From the phosphoramid(othio)ic chlorides, any desired amido esters may be produced according to the above mentioned reactions for ester formation. For example, the helminthicide crufomate (Ruelene®) may be prepared by the following scheme (Equation 119).

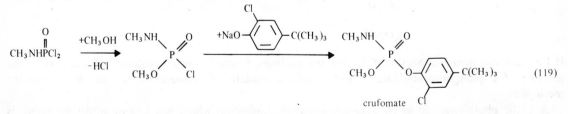

$$(119)$$

However, it is more usual to introduce a P-N bond at the last stage of the reaction sequence. Thus, the herbicide DMPA (Zytron®) is prepared by the following sequence of reactions (Equation 120):

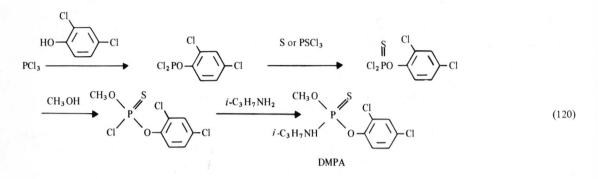

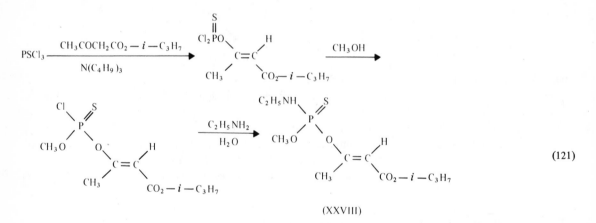

(120)

DMPA

An interesting vinyl phosphoramidothionate insecticide, O-(1-carboisopropoxy-1-propen-2-yl) O-methyl N-ethylphosphoramidothionate (XXVIII), is also prepared by finally introducing an amide linkage to the product derived from phosphorus thiochloride by the sequence of reactions with isopropyl acetoacetate followed with methanol (Equation 121).[128]

(121)

(XXVIII)

Bases catalyze the first step of the reactions. When the enolic metal salt of acetoacetate ester is used, the *trans*-crotonate isome is almost exclusively obtained, whereas the *cis*-isomer is exclusively obtained from the enolic tetraalkylammonium salts. The *cis*-crotonate isomer is preferred because the final *cis*-amido ester XXVIII is less toxic to mammals than the *trans*-isomer. The metallic salts of acetoacetate have a chelating ring structure (XXIX) and form the *trans* product from the reaction with phosphorus thiochloride. In contrast, in the tetraalkylammonium salts, the negatively charged oxygen atom and the carbonyl group of the ester are on opposite sides as a result of repulsion to give the *cis*-crotonate (XXX).

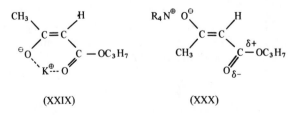

(XXIX)   (XXX)

It has been mentioned in Section A.1. that the exclusive formation of *cis*-crotonate also occurs in the reaction of a phosphorochloridate with the sodium enolate of acetoacetate for the preparation of mevinphos.

A cyclic phosphoramide ester, cyclophosphamide (Endoxan), which is a known antitumor agent, is prepared from *N,N*-bis(β-chloroethyl) phosphoramidic dichloride and γ-aminopropanol in the presence of triethylamine as the acceptor of hydrochloric acid (Equation 122).[129]

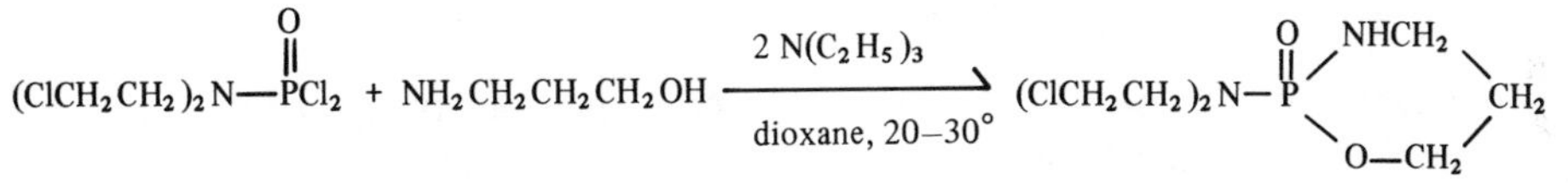

$$(ClCH_2CH_2)_2N-\overset{\overset{O}{\|}}{P}Cl_2 + NH_2CH_2CH_2CH_2OH \xrightarrow[\text{dioxane, 20--30}^\circ]{2\ N(C_2H_5)_3} (ClCH_2CH_2)_2N-\overset{\overset{O}{\|}}{P}\overset{NHCH_2}{\underset{O-CH_2}{\diagup}}CH_2 \qquad (122)$$

cyclophosphamide

The phosphoryl derivatives of weak bases such as phthalimide are synthesized by the use of the alkali metal salts. Thus, the fungicide Dowco 199 is prepared by the reaction of diethyl phosphoro-chloridothionate with the potassium salt of phthalimide (Equation 123).[130,131]

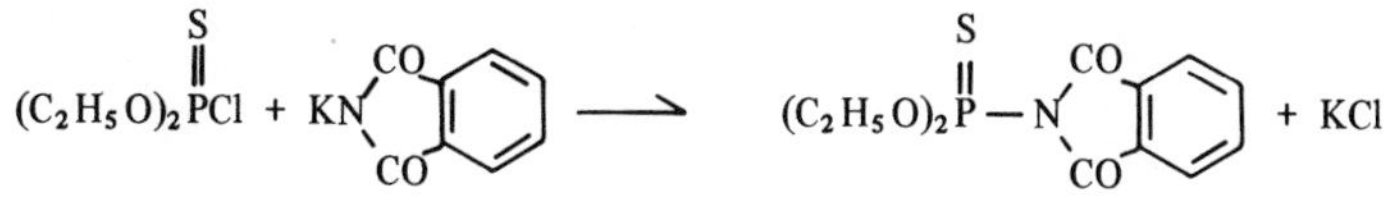

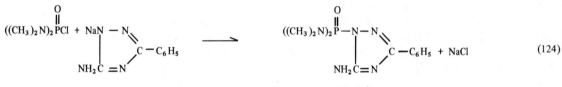

$$(123)$$

Dowco 199

Similarly, the sodium salt of aminophenyltriazole reacts with bis(dimethylamino)phosphoric chloride to yield triamiphos (Wepsin), the first organophosphorus fungicide (Equation 124).[132]

$$(124)$$

triamiphos

Certain phosphoramidothiolate esters are prepared conveniently by the thiono-thiolo rearrangement reaction from the corresponding phosphoramidothionates, as mentioned above (Equation 112).

2. Reactions of Trivalent Phosphorus Compounds with Nitrogen Compounds

Dialkyl phosphorochloridites react with amines to give corresponding amidites. A nerve gas, tabun, was synthesized by Saunders as follows:[11]

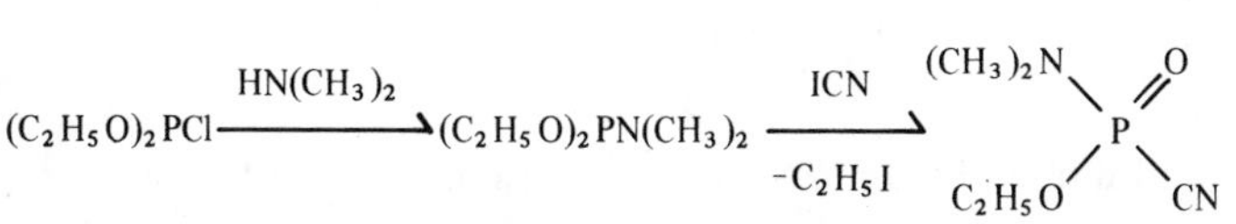

$$(C_2H_5O)_2PCl \xrightarrow{HN(CH_3)_2} (C_2H_5O)_2PN(CH_3)_2 \xrightarrow[-C_2H_5I]{ICN} \underset{C_2H_5O}{\overset{(CH_3)_2N}{\diagdown}}P\overset{\diagup O}{\underset{CN}{\diagdown}} \qquad (125)$$

tabun

Dialkyl phosphites react smoothly with primary or secondary amines or ammonia in carbon tetrachloride to give dialkyl phosphoramidates in excellent yields (Equation 126).[133] This reaction was first considered to proceed by the intermediate formation of trichloromethylphosphonate, but later the formation of a phosphorochloridate intermediate was demonstrated. Phosphorochloridates are obtained by use of tertiary amines (see Equation 9).

$$(RO)_2POH + CCl_4 \xrightarrow[-CHCl_3]{} (RO)_2\overset{\overset{O}{\|}}{P}Cl \xrightarrow[-HCl]{+R'NH_2} (RO)_2\overset{\overset{O}{\|}}{P}NHR' \qquad (126)$$

Trialkyl phosphites react with chlorourea to give dialkoxylphosphinylureas, which have some pesticidal activities including a herbicidal effect (Equation 127).

$$(RO)_3P + ClNHCONH_2 \longrightarrow (RO)_2\overset{\oplus}{P}-NHCONH_2 \longrightarrow (RO)_2\overset{\overset{O}{\parallel}}{P}NHCONH_2 + RCl \qquad (127)$$

Trialkyl phosphites undergo an interesting reaction with azides. The reaction with hydrazoic acid appears to proceed through a trialkyl phosphorimidate intermediate, which rearranges subsequently to an N-alkylphosphoramidate (Equation 128).[134,135]

$$(CH_3O)_3P: + H\overset{\oplus}{N}=N=\overset{\ominus}{N} \xrightarrow[-N_2]{benzene} (CH_3O)_2P=NH \longrightarrow (CH_3O)_2\overset{\overset{O}{\parallel}}{P}NHCH_3 \qquad (128)$$

When this reaction is applied to azidoalkyl carbonates, a new class of insecticides, urethanephosphates, is produced.[136,137] The insecticide demuphos is an example.

$$(CH_3O)_3P + N_3\overset{\overset{O}{\parallel}}{C}O - i - C_3H_7 \longrightarrow (CH_3O)_2\overset{\overset{O}{\parallel}}{P} - \underset{\underset{CH_3}{|}}{N} - \overset{\overset{O}{\parallel}}{C}O - i - C_3H_7 \qquad (129)$$

demuphos

When an azide of dialkyl phosphoric acid reacts with a trialkyl phosphite, an O,O,O-trialkyl N-dialkoxyphosphinylphosphorimidate (**XXXI**) may be produced. The phosphorimidate reacts with dry hydrogen to form a diphosphoramide (**XXXII**) and an alkyl chloride (Equation 130).[138] The diphosphoramide is a tautomeric mixture of the amido form and the imidoic acid form (**XXXIII**) (Equation 131). Therefore, diazomethane reacts to give both N-methylated and O-methylated products (Equation 131).[139]

$$(RO)_3P: + N_3\overset{\overset{O}{\parallel}}{P}(OR)_2 \xrightarrow[-N_2]{} (RO)_3P=N-\overset{\overset{O}{\parallel}}{P}(OR)_2 \xrightarrow[15-20°]{dry\ HCl} (RO)_2\overset{\overset{O}{\parallel}}{P}-NH-\overset{\overset{O}{\parallel}}{P}(OR)_2 + RCl \qquad (130)$$

(XXXI)   (XXXII)

$$(RO)_2\overset{\overset{O}{\parallel}}{P}-NH-\overset{\overset{O}{\parallel}}{P}(OR)_2 \rightleftharpoons (RO)_2\overset{\overset{HO}{|}}{P}=N-\overset{\overset{O}{\parallel}}{P}(OR)_2 \xrightarrow[-N_2]{CH_2N_2} (RO)_2\overset{\overset{O}{\parallel}}{P}-\underset{\underset{CH_3}{|}}{N}-\overset{\overset{O}{\parallel}}{P}(OR)_2 + (RO)_2\overset{\overset{CH_3O}{|}}{P}=N-\overset{\overset{O}{\parallel}}{P}(OR)_2 \qquad (131)$$

(XXXII)   (XXXIII)

These N-phosphorylated phosphorimidates and diphosphoramides have been evaluated as insecticides.[140]

3. Modification of Phosphoramides

The amino group of phosphoramidates can be acylated by acyl chlorides or anhydrides in the presence of a strong base. Thus, a new fungicide, N-(diethoxyphosphinothioyl)phthalimide (Dowco 199), is synthesized from diethyl phosphoramidothionate and phthalyl chloride or, more preferably, phthalic anhydride (Equation 132).[131]

$$(C_2H_5O)_2\overset{\overset{S}{\parallel}}{P}NH_2 + 2\ O\overset{CO}{\underset{CO}{\diagup}}\diagdown \text{(ring)} + 2NaH \xrightarrow{DMF} (C_2H_5O)_2\overset{\overset{S}{\parallel}}{P}-N\overset{CO}{\underset{CO}{\diagup}}\diagdown \text{(ring)} + \text{(ring)}\overset{COONa}{\underset{COONa}{\diagdown}} + 2H_2 \qquad (132)$$

Dowco 199

The lithium salts of phosphoramidates react similarly with chloroformate to form urethane phosphates, new candidates for selective insecticides (Equation 133).[141]

$$(RO)_2 \overset{\overset{\displaystyle S(O)}{\|}}{P} - NHR' \xrightarrow[-C_6H_6]{C_6H_5Li} (RO)_2 \overset{\overset{\displaystyle S(O)}{\|}}{\underset{\underset{\displaystyle R'}{|}}{P}} -NLi \xrightarrow{\quad ClCO_2 \text{-Ar-}R''\quad} (RO)_2 \overset{\overset{\displaystyle S(O)}{\|}}{\underset{\underset{\displaystyle R'}{|}}{P}} -N - \overset{\overset{\displaystyle O}{\|}}{C} O-\text{Ar-}R'' \tag{133}$$

The carbonylation does not occur on S (or O), but preferentially on N. The direct phosphorylation of carbamates with phosphoryl chlorides or anhydrides does not produce the desired products. The oxime carbamate derivatives of urethane phosphates are prepared by the transesterification of a nitrophenyl carbamate ester with an oxime as follows:

$$(CH_3O)_2 \overset{\overset{\displaystyle S}{\|}}{\underset{\underset{\displaystyle CH_3}{|}}{P}} -N - \overset{\overset{\displaystyle O}{\|}}{C}O - \text{Ar-}NO_2 + NaON{=}CHC(CH_3)_2 SCH_3 \longrightarrow$$

$$(CH_3O)_2 \overset{\overset{\displaystyle S}{\|}}{\underset{\underset{\displaystyle CH_3}{|}}{P}} -N - \overset{\overset{\displaystyle O}{\|}}{C}O -N{=}CHC(CH_3)_2 SCH_3 + NaO - \text{Ar-}NO_2 \tag{134}$$

The oxime preferably attacks the carbonyl carbon rather than the phosphorus atom.

Another method for the preparation of urethane phosphates is the addition of alcohols to isocyanatophosphate esters.[142] A new insecticide, avenin, is prepared as follows:[34]

$$Cl_2 \overset{\overset{\displaystyle O}{\|}}{P}NCO \xrightarrow[Et_3N]{2MeOH} (MeO)_2 \overset{\overset{\displaystyle O}{\|}}{P}NCO \xrightarrow[Et_3N]{i\text{-PrOH}} (MeO)_2 \overset{\overset{\displaystyle O}{\|}}{P}NH - \overset{\overset{\displaystyle O}{\|}}{C}O - i - Pr \tag{135}$$

avenin

Aromatic amines also add to isocyanatophosphoric acid derivatives to give N-arylureides. Thus, a new insect sterilant, 1-(bis(1-aziridinyl)phosphinyl)-3-(3,4-dichlorophenyl)urea,[143] is prepared as follows:[144]

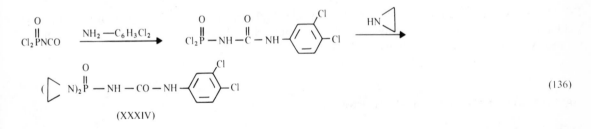

(XXXIV)

Dialkyl isothiocyanatophosphates react similarly with sodium thiolates, yielding insecticidal N-phosphoryl dithiocarbamates (XXXV):[145]

$$2(RO)_2 \overset{\overset{\displaystyle O}{\|}}{P}NCS + NaSCH_2CH_2SNa \longrightarrow (RO)_2 \overset{\overset{\displaystyle O}{\|}}{P}NH\overset{\overset{\displaystyle S}{\|}}{C}SCH_2CH_2S\overset{\overset{\displaystyle S}{\|}}{C}NH\overset{\overset{\displaystyle O}{\|}}{P}(OR)_2 \tag{137}$$

(XXXV)

Isocyanate and isothiocyanate derivatives of phosphoric acid may be prepared by the treatment of the chlorides with an alkali metal cyanate of thiocyanate, as shown by the following example:[146]

$$Cl{-}\overset{\overset{\displaystyle S}{\|}}{P}{\langle}\text{...}{\rangle}\overset{\overset{\displaystyle S}{\|}}{P}{-}Cl + 2NaSCN \longrightarrow SCN{-}\overset{\overset{\displaystyle S}{\|}}{P}{\langle}\text{...}{\rangle}\overset{\overset{\displaystyle S}{\|}}{P}{-}NCS + 2NaCl \tag{138}$$

Isocyanatophosphoric dichloride is synthesized by the reaction of phosphorus pentachloride with ethyl carbamate:[144]

$$PCl_5 + H_2NCO_2C_2H_5 \xrightarrow[75°]{} \overset{\overset{\displaystyle O}{\|}}{Cl_2P}-NCO + C_2H_5Cl + 2HCl \qquad (139)$$

Isocyanatophosphate derivatives are also prepared by the following reactions:[146a]

$$>PCl + (RO)_2C=NCl \longrightarrow >\overset{\overset{\displaystyle O}{\|}}{P}NCO + 2RCl \qquad (140)$$

$$>\overset{\overset{\displaystyle O(S)}{\|}}{P}NH_2 + COCl_2 \longrightarrow >\overset{\overset{\displaystyle O(S)}{\|}}{P}NCO + 2HCl \qquad (141)$$

F. DERIVATIVES OF PHOSPHOROFLUORIDIC ACID

This section is concerned with a class of phosphorus compounds containing a P-F bond. It involves phosphorodiamidic fluorides (XXXVI), phosphorofluoridate esters (XXVII), and phosphonofluoridate esters (XXXVIII). Only the compounds of type XXXVI were employed for agricultural uses in the early stage of development of organophosphorus systemic insecticides. Others are not pesticides, but nerve gases. Detailed discussions for those appear in Saunder's monograph.[11]

$$\underset{(XXXVI)}{(RR'N)_2\overset{\overset{\displaystyle O}{\|}}{P}-F} \qquad \underset{(XXXVII)}{(RO)_2\overset{\overset{\displaystyle O}{\|}}{P}-F} \qquad \underset{(XXXVIII)}{\underset{RO}{\overset{R}{>}}P\underset{F}{\overset{O}{<}}}$$

Phosphoryl fluorides are generally prepared by the displacement of the chlorine atom in phosphoryl chlorides with fluorine. Tetramethyl phosphorodiamidic chloride reacts with sodium fluoride to give dimefox, a systemic insecticide (Equation 142).[13]

$$((CH_3)_2N)_2\overset{\overset{\displaystyle O}{\|}}{P}Cl + NaF \xrightarrow[75°]{} ((CH_3)_2N)_2\overset{\overset{\displaystyle O}{\|}}{P}F + NaCl \qquad (142)$$
dimefox

Diisopropyl phosphorofluoridate (DFP) is similarly synthesized from its chloridate analog (Equation 143).[11] Phosphonofluoridates are also prepared in a similar way from appropriate phosphonic chlorides (Equation 144).[11]

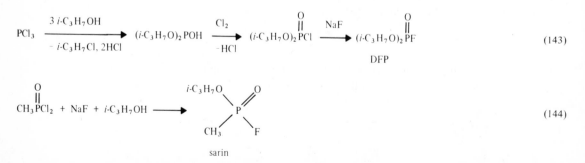

$$PCl_3 \xrightarrow[- i\text{-}C_3H_7Cl, 2HCl]{3\ i\text{-}C_3H_7OH} (i\text{-}C_3H_7O)_2POH \xrightarrow[-HCl]{Cl_2} (i\text{-}C_3H_7O)_2\overset{\overset{\displaystyle O}{\|}}{P}Cl \xrightarrow{NaF} (i\text{-}C_3H_7O)_2\overset{\overset{\displaystyle O}{\|}}{P}F \qquad (143)$$
DFP

$$CH_3\overset{\overset{\displaystyle O}{\|}}{P}Cl_2 + NaF + i\text{-}C_3H_7OH \longrightarrow \underset{CH_3}{\overset{i\text{-}C_3H_7O}{>}}P\underset{F}{\overset{O}{<}} \qquad (144)$$
sarin

G. DERIVATIVES OF PYROPHOSPHORIC ACID

Since tetraethyl pyrophosphate was found as the first organophosphorus insecticide in 1944, active studies on the derivatives of pyrophosphoric acid have been carried out. However, only a rather small number of the compounds in this class are practically used now.

There are a variety of methods for the preparation of the esters and amides of pyrophosphoric acid and its thio analogs. The best method for practical preparation of symmetric esters may be the reaction of a dialkyl phosphorochloridate with water in the presence of a base.[147] A part of phosphorochloridates is at first hydrolyzed and the product reacts with another part of ynhydrolyzed phosphorochloridates (Equation 145). This reaction is also suitable for the synthesis of dithionopyrophosphates such as sulfotep.[13]

$$2(RO)_2 \overset{\overset{X}{\parallel}}{P}Cl \xrightarrow[-HCl]{H_2O/base} (RO)_2 \overset{\overset{X}{\parallel}}{P}OH + Cl\overset{\overset{X}{\parallel}}{P}(OR)_2 \xrightarrow[-HCl]{base} (RO)_2 \overset{\overset{X}{\parallel}}{P}-O-\overset{\overset{X}{\parallel}}{P}(OR)_2 \qquad (145)$$

This reaction is also applicable to the preparation of phosphoramide derivatives; the reaction product of dimethylamine with phosphorus oxychloride reacts without isolation with water in the presence of tertiary amine to give schradan (octamethylpyrophosphoramide) (Equation 146).[13]

$$2POCl_3 \xrightarrow[-4(CH_3)_2NH\cdot HCl]{8(CH_3)_2NH} 2((CH_3)_2N)_2\overset{\overset{O}{\parallel}}{P}Cl \xrightarrow[-2NR_3\cdot HCl]{H_2O, 2NR_3} ((CH_3)_2N)_2\overset{\overset{O}{\parallel}}{P}O\overset{\overset{O}{\parallel}}{P}(N(CH_3)_2)_2 \qquad (146)$$
$$\text{schradan}$$

When a dialkyl phosphorochloridate reacts with a dialkyl phosphorothioate ion, an ambident nucleophile having two possible attacking atoms of sulfur and oxygen, only the oxygen of the ion is always phosphorylated to give a monothionopyrophosphate (Equation 147).[2,148,149]

$$(RO)_2P\overset{S}{\underset{O}{\diagdown}}{}^{\ominus} + Cl-\overset{\overset{O}{\parallel}}{P}(OR)_2 \longrightarrow (RO)_2\overset{\overset{S}{\parallel}}{P}-O-\overset{\overset{O}{\parallel}}{P}(OR)_2 \qquad (147)$$

It is interesting to note that the alkylation reaction with alkyl halides always occurs on the sulfur atom of the phosphorothioate ion (Equation 81). As mentioned in D.2. of this Chapter, phosphoryl-P is a typical hard acid center, and thus tends to react more preferably with the hard base (O^-) than with the soft base (S^-).

$$(RO)_2\overset{\overset{O}{\parallel}}{P}-SR' \xleftarrow[\text{soft acid}]{X-R'} (RO)_2P\overset{S}{\underset{O}{\diagdown}}{}^{\ominus} \longrightarrow (RO)_2\overset{\overset{O}{\parallel}}{P}-O-\overset{\overset{O}{\parallel}}{P}(OR)_2 \qquad (148)$$
$$\text{soft base} \qquad Cl-\overset{\overset{O}{\parallel}}{P}(OR')_2$$
$$\text{hard base} \quad \text{hard acid}$$

Therefore, phosphorylation on the sulfur atom may be performed only in such a molecule as *O,O*-dialkyl phosphorodithioates which have no reactive charged oxygen.[2]

$$(RO)_2\overset{\overset{S}{\parallel}}{P}Cl + KS\overset{\overset{S}{\parallel}}{P}(OR)_2 \longrightarrow (RO)_2\overset{\overset{S}{\parallel}}{P}-S-\overset{\overset{S}{\parallel}}{P}(OR)_2 \qquad (149)$$

Phosphites are also employed for the synthesis of pyrophosphates. Dialkyl phosphites react with

O,O-dialkyl S-morpholinodithiophosphate to give dithiopyrophosphate esters (Equation 150).[150]

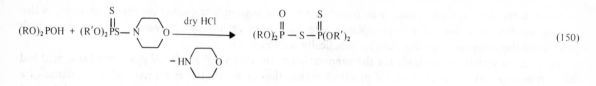

$$(RO)_2\overset{\overset{\displaystyle S}{\|}}{P}OH + (R'O)_2\overset{\overset{\displaystyle S}{\|}}{P}S-N\underset{}{\bigcirc}O \xrightarrow[-HN\bigcirc O]{dry\ HCl} (RO)_2\overset{\overset{\displaystyle O}{\|}}{P}-S-\overset{\overset{\displaystyle S}{\|}}{P}(OR')_2 \quad (150)$$

TEPP is also synthesized by the reaction of diethyl phosphite ion with diethyl phosphorochloridate followed by oxidation (Equation 151).[13]

$$(C_2H_5O)_2\overset{\overset{\displaystyle O}{\|}}{P}ONa + ClP(OC_2H_5)_2 \xrightarrow[-NaCl]{} (C_2H_5O)_2P-O-\overset{\overset{\displaystyle O}{\|}}{P}(OC_2H_5)_2 \xrightarrow{O_2} (C_2H_5O)_2\overset{\overset{\displaystyle O}{\|}}{P}-O-\overset{\overset{\displaystyle O}{\|}}{P}(OC_2H_5)_2 \quad (151)$$

TEPP

As discussed in D.3., trialkyl phosphites react with disulfides (Equation 102). Similar displacement with trialkyl phosphites also occurs at the sulfur atom of bis-(dialkoxyphosphinothioyl) disulfides (Equation 152).[34]

$$(RO)_3P + (R'O)_2\overset{\overset{\displaystyle S}{\|}}{P}S-S\overset{\overset{\displaystyle S}{\|}}{P}(OR')_2 \longrightarrow (RO)_2\overset{\overset{\displaystyle O}{\|}}{P}-S-\overset{\overset{\displaystyle S}{\|}}{P}(OR')_2 + RS\overset{\overset{\displaystyle S}{\|}}{P}(OR')_2 \quad (152)$$

The disulfides are prepared by oxidation of O,O-dialkyl phosphorodithioates (Equation 153), which are produced by the reaction of phosphorus pentasulfide and alcohols (Equation 48). A practical insecticide of this type, Phostex®, is prepared by using a mixture of ethanol and isopropanol.[13]

$$P_2S_5 \xrightarrow[-SH_2]{4ROH} 2(RO)_2\overset{\overset{\displaystyle S}{\|}}{P}SH \xrightarrow[-2H_2O]{H_2O_2} (RO)_2\overset{\overset{\displaystyle S}{\|}}{P}-S-S-\overset{\overset{\displaystyle S}{\|}}{P}(OR)_2 \quad (153)$$

H. PHOSPHONATES

Since the first insecticide derived from phosphonic acid, EPN, and a different type of phosphonate insecticide, trichlorfon, were found in 1949 and 1952, respectively, efforts have been directed towards discovering new pesticides from phosphonate derivatives, in which an organic part connects directly with the phosphorus atom by a P-C bond.[151] At present, about ten different compounds derived from phosphonic acid have been mainly developed as insecticides. Some of these are fungicides, and one is utilized as a plant growth regulator.

Typical structures of this class are as follows:

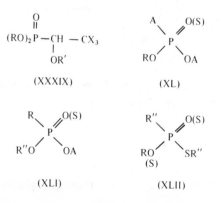

(XXXIX) (XL)

(XLI) (XLII)

where R is an alkyl group, R' is hydrogen or an acyl group, R'' is an alkyl or an aryl group, A is an aryl group, and X is a halogen atom.

Besides these structures, amides, fluoridates, and free acids are also involved in this class. This section deals only with the formation of a P-C bond. Ester and amide linkages may be produced in a way similar to the derivatives of phosphoric acid. A number of reactions for the preparation of a P-C bond are known.[23,59,152] The following reactions are relatively common examples: 1) the catalytic action of Lewis acids on the reaction of phosphorus trichloride with aromatic hydrocarbons or with alkyl halides; 2) the reaction of phosphorus trichloride with organometallic compounds; 3) the reaction of phosphites with alkyl halides; and 4) the addition of dialkyl phosphites to unsaturated compounds.

1. Catalysis of Lewis Acids on the Reaction of Phosphorus Trichloride

When a mixture of an aromatic hydrocarbon and an excess amount of phosphorus trichloride is heated for several hours in the presence of aluminum chloride, the Friedel-Crafts reaction takes place, and the complex of an arylphosphonous dichloride with aluminum chloride is obtained. The complex is decomposed by phosphorus oxychloride or pyridine to give the arylphosphonous dichloride.[153] The phosphonous dichloride reacts with phosphorus thiochloride or elemental sulfur to be converted to an arylphosphonothionic dichloride,[87] which may be employed for the further synthesis of ester derivatives such as EPN (Equation 154).[13,154]

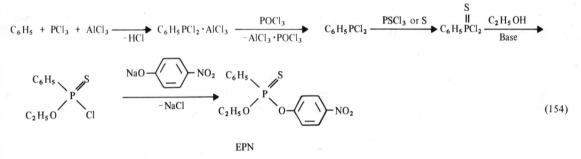

(154)

EPN

Phosphorus trichloride also reacts with alkyl halides by the catalytic action of aluminum chloride to yield a complex (XLIII), which is then hydrolyzed with water to give an alkylphosphonic dichloride (Equation 155).[155] The alkylphosphonic dichloride may be converted to its thiono analog by the action of phosphorus pentasulfide.[156] When the complex XLIII is decomposed with sulfur in the presence of freshly roasted potassium chloride, the corresponding alkylphosphonothionic dichloride is directly obtained (Equation 156).[157] However, an alkylphosphorous dichloride is produced by the reductive decomposition of the complex XLIII with aluminum in the presence of potassium chloride (Equation 157).[158]

$$\text{PCl}_3 \ + \ \text{RCl} \ + \ \text{AlCl}_3 \longrightarrow \underset{\text{(XLIII)}}{\text{RPCl}_4 \cdot \text{AlCl}_3} \xrightarrow[0^\circ]{\text{H}_2\text{O}} \overset{\overset{\text{O}}{\|}}{\text{RPCl}_2} \qquad (155)$$

$$\text{RPCl}_4 \cdot \text{AlCl}_3 \ + \ \text{S} \ + \ \text{KCl} \xrightarrow{\text{heat}} \overset{\overset{\text{S}}{\|}}{\text{RPCl}_2} \qquad (156)$$

$$\text{RPCl}_4 \cdot \text{AlCl}_3 \ + \ \text{Al} \ + \ \text{KCl} \longrightarrow \text{RPCl}_2 \qquad (157)$$

In the reactions 156 and 157, the addition of potassium chloride is required for binding the resulting aluminum chloride, whereby a higher yield of the final products is obtained. Besides aluminum chloride, zinc chloride or ferric chloride is also utilized as the catalyst.

Phosphonic dichlorides are generally more reactive than corresponding phosphorodichloridates, and react with dihydroxy compounds such as saligenin in the presence of tertiary amines to give pesticidal cyclic phosphonates (Equation 158).[159]

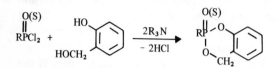

$$(158)$$

2. Reaction of Phosphorus Trichloride with Organometallic Compounds

Certain organometallic compounds react with phosphorus trichloride to form a P-C bond. Using trialkylaluminums, Szabo and Menn developed an elegant method to prepare alkylphosphonothioic dichlorides.[160] A trialkylaluminum reacts with 3 moles of phosphorus trichloride and sulfur, forming an alkylphosphothioic dichloride-aluminum chloride complex, which is then decomposed by ice water:

$$PCl_3 \xrightarrow[\substack{hexane \\ 30-50°}]{1/3\ AlR_3} RPCl_2 \cdot 1/3 AlCl_3 \xrightarrow[40°]{S} \overset{S}{\overset{\|}{R}}PCl_2 \cdot 1/3 AlCl_3 \xrightarrow[0-10°]{H_2O} \overset{S}{\overset{\|}{R}}PCl_2 \tag{159}$$

From the alkylphosphonothioic dichlorides, several insecticides have been derived. For example, the new phosphonate insecticide fonofos (Dyfonate) is prepared from ethylphosphonothioic dichloride, as shown in Equation 160.[161]

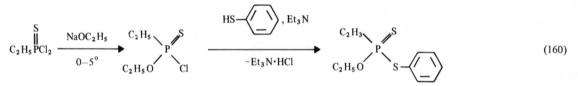

$$(160)$$

Similarly, insecticidal alkylphosphonate analogs of phosmet were prepared:[160]

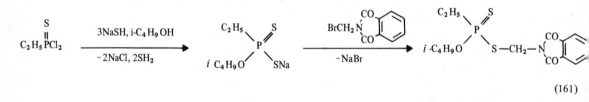

$$(161)$$

Phosphonodithioates also may be prepared in a different way through dithioanhydrides (XLIV), as shown by the following examples. Alkyl halides add to form S-alkyl phosphonohalogenodithioates (Equation 162).[162] Alcohols also add to dithioanhydrides (XLIV) to yield O-alkyl hydrogen phosphonodithioates (Equation 163).[163] They are utilized for the preparation of final phosphonodithioate esters.

$$2\ RPCl_2 + 2SH_2 \xrightarrow{-4HCl} \underset{(XLIV)}{RP\diamondsuit PR} \xrightarrow{2R'X} 2\ \overset{S}{\overset{\|}{R}}P-X \xrightarrow{2R''ONa} 2\ \overset{S}{\overset{\|}{R}}P-OR'' \tag{162}$$
$$\qquad\qquad SR' \qquad\qquad SR'$$

$$XLIV + 2\ R'OH \longrightarrow 2\ \underset{R'O}{R-\overset{S}{\overset{\|}{P}}-SH} \longrightarrow 2\ \underset{R'O}{R-\overset{S}{\overset{\|}{P}}-SR''} \tag{163}$$

For the preparation of phosphonothiolates, *O,O*-dialkyl phosphonothionates, produced from phosphonothioic dichlorides and alcohols, are partially hydrolyzed with alkali, and the resulting phosphonothioic salts are alkylated with alkyl halides; alkylation occurs preferably on the sulfur atom (see D.2.).

$$
\underset{\substack{\| \\ \text{RPCl}_2}}{\text{S}} + 2\text{NaOR}' \longrightarrow \underset{\substack{\| \\ \text{RP(OR}')_2}}{\text{S}} \xrightarrow{\text{KOH}} \underset{\substack{\| \\ \text{RP}-\text{SK} \\ | \\ \text{R}'\text{O}}}{\text{O}} \xrightarrow{\text{ClR}''} \underset{\substack{\| \\ \text{RP}-\text{SR}'' \\ | \\ \text{R}'\text{O}}}{\text{O}}
\tag{164}
$$

Organic compounds of cadmium, lead, lithium, mercury, and tin are also utilized for the formation of the P-C bond by reaction with phosphorus trihalides or halophosphites. Selected examples of their reactions are shown as follows:[164]

$$\text{PCl}_3 + (n\text{-C}_4\text{H}_9)_2\text{Cd} \longrightarrow n\text{-C}_4\text{H}_9\text{PCl}_2 \tag{165}$$

$$\text{PCl}_3 + (\text{C}_2\text{H}_5)_4\text{Pb} \longrightarrow \text{C}_2\text{H}_5\text{PCl}_2 \tag{166}$$

$$\text{PCl}_3 + \text{R}_2\text{Hg} \longrightarrow \text{RPCl}_2 \tag{167}$$

$$(n\text{-C}_4\text{H}_9\text{O})_2\text{PCl} + \text{C}_4\text{H}_9\text{Li} \longrightarrow (n\text{-C}_4\text{H}_9\text{O})_2\text{PC}_4\text{H}_9 \tag{168}$$

$$\text{PCl}_3 + 2(\text{C}_4\text{H}_9)_3\text{SnCH}_2\text{CO}_2\text{CH}_3 \longrightarrow \text{ClP(CH}_2\text{CO}_2\text{CH}_3)_2 \tag{169}$$

On the other hand, in general, Grignard reagents give tertiary phosphines by the reaction with phosphorus trihalides.[3]

$$\text{PX}_3 + \text{RMgX} \longrightarrow \text{R}_3\text{P} \tag{170}$$

Aluminum chloride catalyzes the exchange of halogen atoms with alkyl groups of organometallic compounds. Thus, phosphorothioic chlorides, which are less reactive than phosphorus trichloride, may be alkylated with tetraalkyl lead in the presence of aluminum chloride, and this reaction is usable for the direct preparation of alkylphosphonothioic and phosphinothioic chlorides and their derivatives (Equations 171 to 173).[165]

$$\text{PSCl}_3 + \text{R}_4\text{Pb} \xrightarrow{\text{AlCl}_3} \underset{\substack{\| \\ \text{RPCl}_2}}{\text{S}} + \underset{\substack{\| \\ \text{R}_2\text{PCl}}}{\text{S}} \tag{171}$$

$$\underset{\substack{\| \\ ((\text{CH}_3)_2\text{N})_2\text{PCl}}}{\text{S}} + (\text{CH}_3)_4\text{Pb} \xrightarrow{\text{AlCl}_3} \underset{\substack{\| \\ ((\text{CH}_3)_2\text{N})_2\text{PCH}_3}}{\text{S}} \tag{172}$$

$$\underset{\substack{\| \\ (\text{C}_2\text{H}_5\text{O})_2\text{PCl}}}{\text{S}} + (\text{C}_2\text{H}_5)_4\text{Pb} \xrightarrow{\text{AlCl}_3} \underset{\substack{\| \\ (\text{C}_2\text{H}_5\text{O})_2\text{PC}_2\text{H}_5}}{\text{S}} \tag{173}$$

3. Reaction of Phosphites with Alkyl Halides

Trialkyl phosphites have a high nucleophilic reactivity, and react with alkyl halides to produce phosphonate esters.[166] This reaction is one of the most versatile methods for the formation of P-C bonds and is well known as the Michaelis-Arbuzov reaction, or simply the Arbuzov reaction.[167] It is generally accepted that the reaction proceeds by the initial nucleophilic attack of a phosphite on the carbon of an alkyl halide, and the subsequent nucleophilic substitution of the halide anion displaced on an alkyl group of the alkoxyphosphonium intermediate, as shown in Equation 174. The driving force of the reaction is attributed to the formation of a phosphoryl bond with high bonding energy (see Chapter I).

$$\begin{array}{c} RO \\ RO - P: \\ RO \end{array} + \overset{\frown}{R'-X} \longrightarrow \overset{X^{\ominus} \curvearrowleft R - O \downarrow}{\underset{RO}{\overset{RO}{\diagdown}} P^{\oplus} - R'} \longrightarrow \overset{O}{\underset{RO'}{\overset{\parallel}{\underset{RO}{\diagup}}} P} + RX \qquad (174)$$

If the alkyl groups of the two reagents in Equation 174 are identical, the reaction is an isomerization of the phosphite into a phosphonate and may be called an Arbusov rearrangement. As the phosphoryl form is more stable, the reaction proceeds in favor of the phosphoryl derivatives.

The Michaelis-Arbusov reaction generally occurs in compounds of the formula $\overset{A}{\underset{B}{\diagup}}P-O-R$. Phosphites are less reactive than phosphinites and phosphonites, usually requiring many hours at reflux temperature. The reactivity order is $R_2POR > RP(OR)_2 > P(OR)_3$ for phosphorus compounds, and $CH_3X > RCH_2X > RR'CHX$; $X = I > Br > Cl$ for alkyl halides.

A dialkyl phosphite is less nucleophilic than the above mentioned compounds, but its anion is active enough. Thus, the salts of dialkyl phosphites react similarly with alkyl halides to give phosphonate esters (Michaelis-Becker reaction) (Equation 175).[168]

$$(RO)_2POM + R'X \longrightarrow (RO)_2\overset{O}{\overset{\parallel}{P}}-R' + MX \qquad (175)$$

The dialkyl alkylphosphonates may be converted to chlorides by the reaction with phosphorus chlorides (Equation 176), oxalyl chloride,[169] or phosgene (Equation 177).[170]

$$R'-\overset{O}{\overset{\parallel}{P}}(OR)_2 \xrightarrow{PCl_5} \overset{R'}{\underset{RO}{\diagdown}}\overset{O}{\overset{\parallel}{P}}-Cl \xrightarrow{PCl_5} R'PCl_2 \qquad (176)$$

$$R'-\overset{O}{\overset{\parallel}{P}}(OR)_2 + COCl_2 \longrightarrow \overset{R'}{\underset{RO}{\diagdown}}\overset{O}{\underset{Cl}{\diagup}} \qquad (177)$$

The insecticide butonate is manufactured by the reaction of dimethyl phosphite with α-chloro-β-trichloroethyl n-butyrate, which is prepared from chloral and butyryl chloride in the presence of zinc chloride.[171]

$$Cl_3CCH{=}O + C_3H_7COCl \xrightarrow{ZnCl_2} Cl_3CCH\overset{Cl}{\underset{}{|}}-OCOC_3H_7 \xrightarrow{(CH_3O)_2POH} (CH_3O)_2\overset{O}{\overset{\parallel}{P}}-\underset{\underset{O{=}C-C_3H_7}{\overset{|}{O}}}{CH}-CCl_3 \qquad (178)$$

butonate

Acyl halides also react with dialkyl phosphites to form a P-C bond. For example, the reaction of trichloroacetyl chloride with diethyl phosphite gives diethyl trichloroacetylphosphonate (Equation 179), which is an oxidized analog of the insecticide trichlorfon, and shows strong insecticidal properties.[172] The acetylphosphonates react with amines, such as morpholine, giving α-amino derivatives of trichlorfon (Equation 180), which also have insecticidal activity.

$$(C_2H_5O)_2POH + ClCCCl_3 \xrightarrow[60-90°]{} (C_2H_5O)_2\overset{O}{\overset{\parallel}{P}}-\overset{O}{\overset{\parallel}{C}}-CCl_3 + HCl \qquad (179)$$

$$(RO)_2\overset{\overset{O}{\|}}{P}-\overset{\overset{O}{\|}}{C}-CCl_3 \ + \ HN\diagdown O \quad \xrightarrow{\text{ether}} \quad (RO)_2\overset{\overset{O}{\|}}{P}-\overset{\overset{OH}{|}}{\underset{\underset{CCl_3}{|}}{C}}-N\diagdown O \tag{180}$$

Similarly, carbamoyl chlorides react with alkyl phosphorodiamidites, yielding carbamoylphosphonic diamides, which are active as acaricides and herbicides (Equation 181).[173]

$$((C_2H_5)_2N)_2POC_2H_5 \ + \ Cl\overset{\overset{O}{\|}}{C}N(C_2H_5)_2 \xrightarrow[170°]{} ((C_2H_5)_2N)_2\overset{\overset{O}{\|}}{P}-\overset{\overset{O}{\|}}{C}N(C_2H_5)_2 \ + \ C_2H_5Cl \tag{181}$$

4. Addition of Dialkyl Phosphites to Carbonyl Compounds

Dialkyl phosphites (also phosphines) having a P-H bond react with aldehydes or ketones to form hydroxyalkylated products:[166]

$$(RO)_2\overset{\overset{O}{\|}}{P}H \ + \ O{=}CHR' \longrightarrow (RO)_2\overset{\overset{O}{\|}}{P}-\overset{\overset{OH}{|}}{C}H-R' \tag{182}$$

The insecticide trichlorfon is prepared by this reaction from dimethyl phosphite and chloral (Equation 183).[174] It is also produced by the simultaneous reaction of phosphorus trichloride, chloral, and methanol (Equation 184).[175] In this reaction, dichlorvos (dimethyl dichlorovinyl phosphate) is not produced, suggesting that trimethyl phosphite does not form as an intermediate and, consequently, no Perkow reaction occurs (see Equations 12 and 23).

$$(CH_3O)_2\overset{\overset{O}{\|}}{P}H \ + \ O{=}CH{-}CCl_3 \longrightarrow (CH_3O)_2\overset{\overset{O}{\|}}{P}-\underset{\underset{OH}{|}}{CH}-CCl_3 \tag{183}$$

trichlorfon

$$PCl_3 \ + \ O{=}CH{-}CCl_3 \ + \ 3CH_3OH \xrightarrow[5-10°]{} (CH_3O)_2\overset{\overset{O}{\|}}{P}-\underset{\underset{OH}{|}}{CHCl_3} \ + \ CH_3Cl \ + \ 2HCl \tag{184}$$

When the reaction is carried out by mixing phosphorus trichloride with an aldehyde at $-70°$, then with 3 moles of a mercaptan, the corresponding S,S-dialkyl 1-alkylthioalkylphosphonodithiolate is produced (Equation 185).[176]

$$PCl_3 \ + \ O{=}CHR' \ + \ 3RSH \longrightarrow (RS)_2\overset{\overset{O}{\|}}{P}-CHR'-SR \ + \ 3HCl \tag{185}$$

These two reactions (184 and 185) appear to proceed through a hypothetical common intermediate (XLV). Alcohols prefer to attack the phosphorus atom of the intermediate, resulting in the formation of α-hydroxyphosphonates (Equation 186), whereas mercaptans prefer to attack the carbon atom to give sulfides (Equation 187).

$$PCl_3 \ + \ O{\Rightarrow}CHR' \longrightarrow Cl_3\overset{+}{P}-\underset{\underset{CHR'}{|}}{\overset{\overset{O^-}{|}}{}} \longrightarrow \begin{bmatrix} \underset{Cl}{\overset{Cl}{\diagdown}}P\diagdown_{\diagup}^{O}\diagdown CHR' \end{bmatrix}$$

(XLV)

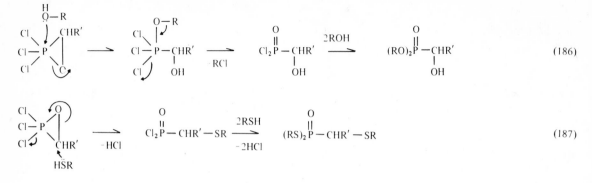

$$Cl_2P-CHR'-SR \xrightarrow{2RSH} (RS)_2P-CHR'-SR \tag{187}$$

The hydroxy group of trichlorfon can be acylated with an acyl anhydride or halide. The insecticide butonate was prepared originally by the butyrylation of trichlorfon (Equation 188.)[177] (For another synthetic method, see Equation 178.)

$$(CH_3O)_2P-CH-CCl_3 + (C_3H_7CO)_2O \longrightarrow (CH_3O)_2P-CH-CCl_3 + C_3H_7CO_2H \tag{188}$$

butonate

The esters of α-keto acids also react with dialkyl phosphites to give corresponding α-hydroxy-phosphonates, which have insecticidal activity (Equation 189).[178]

$$(RO)_2PH + CH_3C-CO_2R' \xrightarrow{100°} (RO)_2P-C-CO_2R' \tag{189}$$

Dialkyl phosphites also add to isocyanates to form carbamoylphosphonates. Thus, a fungicidal arylsulfonylcarbamoylphosphonate is prepared by the reaction of a sulfonyl isocyanate with a dialkyl phosphite (Equation 190).[179]

$$(CH_3O)_2POH + CH_3-\langle\!\rangle-SO_2NCO \longrightarrow (CH_3O)_2P-CNHSO_2\langle\!\rangle CH_3 \tag{190}$$

Furthermore, dialkyl phosphites add to a Schiff base to give α-aminophosphonates. For example, N-cyclohexylideneaniline reacts with diethyl phosphite to yield an herbicidal product (Equation 191).[180]

$$(C_2H_5O)_2POH + \langle\!\rangle{=}N-C_6H_5 \longrightarrow (C_2H_5O)_2P-\langle\!\rangle \tag{191}$$

5. Addition of Phosphorus Pentachloride to Olefins

Phosphorus pentachloride adds readily to olefins, in the sense of $Cl^-PCl_4{}^+$, to form β-chloroalkyl derivatives of the type $RPCl_4$, which are converted to phosphonic acids by treatment with water.[181] Under the conditions of hydrolysis, β-chloroalkylphosphonic acids or further dehydrochlorinated products, vinylphosphonic acids, are produced (Equation 192). β-Chloroethylphosphonic acid (Ethrel: XLVI; R = H) is utilized as a plant growth regulator.

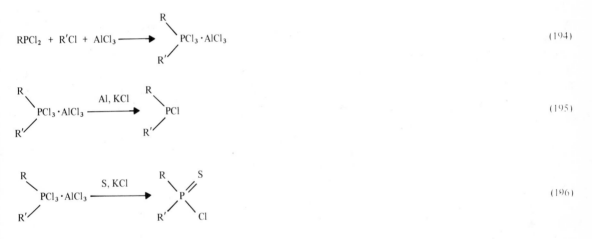

$$RCH{=}CH_2 + PCl_5 \longrightarrow \underset{(XLVI)}{RCH-CH_2PCl_4} \xrightarrow{H_2O} R-CH-CH_2-P(OH)_2 \xrightarrow{-HCl} RCH{=}CH-P(OH)_2 \qquad (192)$$

I. PHOSPHINATES

Some efforts have been made to develop pesticides in the series of phosphinate esters, which have two P-C bonds.[151] Only one practical insecticide, Agvitor, is known in this class.

Phosphinates are produced in essentially the same manner as phosphonates. Phosphinous halides may be synthesized by the reaction of phosphonous halides with tetraalkyllead:[182]

$$RPX_2 + R'_4Pb \longrightarrow RR'PX \qquad (193)$$

An alkylphosphonous dichloride reacts with an alkyl chloride in the presence of aluminum chloride giving a reactive crystalline complex (Equation 194).[183] The complex is decomposed by heating with aluminum powder in the presence of potassium chloride, to give a dialkylphosphinous chloride in high yield (Equation 195).[158] In the presence of sulfur, the complex gives a dialkylphosphinothioic chloride (Equation 196).[183]

$$RPCl_2 + R'Cl + AlCl_3 \longrightarrow \underset{R'}{\overset{R}{\diagdown}} PCl_3 \cdot AlCl_3 \qquad (194)$$

$$\underset{R'}{\overset{R}{\diagdown}} PCl_3 \cdot AlCl_3 \xrightarrow{Al,\ KCl} \underset{R'}{\overset{R}{\diagdown}} PCl \qquad (195)$$

$$\underset{R'}{\overset{R}{\diagdown}} PCl_3 \cdot AlCl_3 \xrightarrow{S,\ KCl} \underset{R'}{\overset{R}{\diagdown}} \underset{Cl}{\overset{S}{\diagup}} P \qquad (196)$$

Another interesting method to prepare phosphinothioic chloride is the reaction of tetraalkyldiphosphine disulfides with sulfuryl chloride (Equation 197).[184] When thionyl chloride is used, the oxo analogs may be obtained (Equation 198).[185] The diphosphine disulfides are prepared by the reaction of Grignard reagents with phosphorus thiochloride (Equation 199).[186]

$$\underset{R'}{\overset{S}{R-P}} - \underset{R'}{\overset{S}{P-R}} + SO_2Cl_2 \xrightarrow{benzene} 2R-\underset{R'}{\overset{S}{P}}-Cl + SO_2 \qquad (197)$$

$$77{-}88\%$$

$$\underset{R'}{\overset{S}{R-P}} - \underset{R'}{\overset{S}{P-R}} + 2SOCl_2 \xrightarrow{CCl_4} R-\underset{R'}{\overset{O}{P}}-Cl + 2S + S_2Cl_2 \qquad (198)$$

$$2PSCl_3 + 4RMgX \xrightarrow[0-5°]{ether} \overset{\overset{S}{\|}}{R_2P} - \overset{\overset{S}{\|}}{PR_2} \qquad (199)$$

Insecticide Agvitor is prepared from diethylphosphinothioic chloride and sodium 2,4,5-trichlorophenoxide (Equation 200).

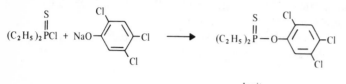

$$(200)$$

Agvitor

Recently, an improved synthesis for phosphinate esters from Grignard reagents and alkyl or aryl phosphorodichloridates and their thiono esters was patented (Equation 201).[187] The phosphinate esters have pesticidal properties including herbicidal activity.

$$C_6H_5OP\overset{O(S)}{\overset{\|}{Cl_2}} + 2C_4H_9MgBr \xrightarrow[\text{room temp}]{ether} C_6H_5OP\overset{O(S)}{\overset{\|}{(C_4H_9)_2}} + 2MgBrCl \qquad (201)$$

Like dialkyl phosphites, monoalkyl phosphonites add to carbonyl to form a P-C bond. Thus, the phosphinate analogs of trichlorfon are produced by the reaction of alkyl phosphonites with chloral (Equation 202).[188] They are further converted to vinyl phosphonates by the action of alkali, as in the conversion of trichlorfon to dichlorvos (Equation 203) (see II.A.3.).

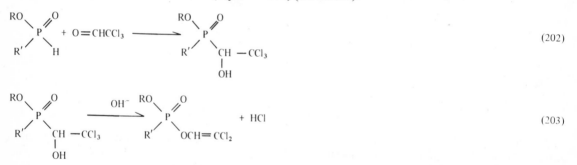

$$(202)$$

$$(203)$$

Monoalkyl phosphonites also add to unsaturated hydrocarbons. Thus, ethyl methylphosphonite reacts with propenamide giving an insecticidal phosphinate:[189]

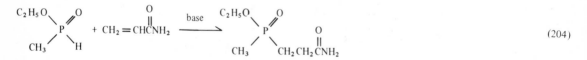

$$(204)$$

The Michaelis-Arbuzov reaction is also useful to prepare phosphinate esters. Insecticidal S-propyl benzylmethylphosphinothioate is prepared by heating O-methyl S-propyl methylphosphonite with benzyl chloride:[190]

$$(205)$$

Chapter III

CHEMICAL REACTIONS

With only a very few exceptions, all organophosphorus pesticides are neutral phosphoryl or thiophosphoryl compounds. A great majority of them are ester derivatives. The polarized phosphoryl group creates a positive charge on the phosphorus atom, which consequently becomes highly electrophilic and reactive with nucleophiles. This is the basic principle of the various reactions of organophosphoryl compounds.

Trivalent phosphorus compounds having an unshared electron pair on the phosphorus atom are quite different in chemical properties from phosphoryl compounds; they are highly nucleophilic agents except for the halides, which have some electrophilic character. Trivalent phosphorus compounds are rarely utilized as pesticides but are very useful as starting materials for the preparation of organophosphorus pesticides. In Chapter II, many reactions of trivalent phosphorus compounds are described in connection with the preparation of pesticides. In this chapter, the chemical reactions of phosphoryl compounds will be discussed.

A. HYDROLYSIS

Fully esterified phosphorus acids, like carboxylic esters, are susceptible to hydrolysis. The hydrolysis rates of organophosphorus pesticides and their metabolites are of great importance, because the hydrolysis results in the detoxication of the pesticides and, moreover, their susceptibility to alkaline hydrolysis relates to their biological activity.

The reaction between a phosphorus ester (A) and water, base, or acid (B) obeys second order kinetics.

$$A + B \longrightarrow C + D$$
$$\text{(products)}$$

The rate equation may be represented as:

$$dx/dt = k_2 (a-x) (b-x)$$

where a and b are the initial concentration of reactants A and B, and x is the decrease in concentration after time t. In conditions where one reactant B is in large excess, or where its concentration is held constant, the reaction may be regarded as a pseudo first order reaction and the equation reduces to:

$$dx/dt = k_1 (a-x)$$

Besides hydrolysis rate constants k_1 or k_2, the rate is very often shown by the half-life(t_{50}) of the ester. That is the time in which 50% of the ester originally present has hydrolyzed. The half-life value is given by the equation:

$$t_{50} = \frac{1}{k_1} \ln 2 = 0.693/k_1$$

The half-lives of common organophosphorus pesticides including some metabolites at pH 6 and 70° are listed in Table 5.[191] The hydrolysis rate is dependent upon the chemical structure and reaction conditions such as pH, temperature, the kind of solvent used, and the existence of catalytic reagents. In aqueous solutions, many organophosphorus pesticides are most stable between pH 1 and 5.[192] In this range, the variation in pH of the solution has practically no effect on the hydrolysis rate. The organophosphates are much more unstable under alkaline conditions. The hydrolysis rate increases steeply at pH higher than 7 to 8. Since the hydrolysis is primarily catalyzed by the hydroxide ion under alkaline conditions, the hydrolysis rate increases, roughly speaking, almost tenfold with each additional pH unit.[32]

The effect of temperature is also important, and the average temperature quotient ($k_{30°}/k_{20°}$) of the hydrolysis rates of 21 phosphorus esters is about 3.75.[192] This means that the hydrolysis rate increases about 4 times by the temperature rise of each 10°. Thus, the half-life values at 20° will be some hundred times larger than the values in Table 5.

TABLE 5

Hydrolysis Rates of Some Organophosphorus Pesticides at 70° in Ethanol – pH 6.0 Buffer Solution (1:4)[191]

Name	Structure	Half-life (h)
Paraoxon	$(C_2H_5O)_2P(O)O$—⟨benzene⟩—NO_2	28.0
Parathion	$(C_2H_5O)_2P(S)O$—⟨benzene⟩—NO_2	43.0
Parathion-methyl	$(CH_3O)_2P(S)O$—⟨benzene⟩—NO_2	8.4
Fenitrothion	$(CH_3O)_2P(S)O$—⟨benzene, CH_3⟩—NO_2	11.2
Fenthion	$(CH_3O)_2P(S)O$—⟨benzene, CH_3⟩—SCH_3	22.4
Fenchlorphos	$(CH_3O)_2P(S)O$—⟨benzene, Cl, Cl, Cl⟩	10.4
Demeton-S-methyl	$(CH_3O)_2P(O)SC_2H_4SO_2C_2H_5$	7.6
Oxydemeton-methyl	$(CH_3O)_2P(O)SC_2H_4SOC_2H_5$	12.4
Dioxydemeton-S-methyl	$(CH_3O)_2P(O)SC_2H_4SC_2H_5$	5.1
Demeton-S	$(C_2H_5O)_2P(O)SC_2H_4SC_2H_5$	18.0
Thiometon	$(CH_3O)_2P(S)SC_2H_4SC_2H_5$	17.0
Disulfoton	$(C_2H_5O)_2P(S)SC_2H_4SC_2H_5$	32.0
Phorate	$(C_2H_5O)_2P(S)SCH_2$	1.75
Phorate-O-analog	$(C_2H_5O)_2P(O)SCH_2SC_2H_5$	0.5

TABLE 5 (continued)

Hydrolysis Rates of Some Organophosphorus Pesticides at 70° in Ethanol — pH 6.0 Buffer Solution   (1:4)[191]

Name	Structure	Half-life (h)
Carbophenothion	$(C_2H_5O)_2P(S)SCH_2S$—⟨benzene ring⟩—Cl	110.0
Phenkapton	$(C_2H_5O)_2P(S)SCH_2S$—⟨benzene ring with Cl, Cl⟩	92.0
Ethion	$(C_2H_5O)_2P(S)SCH_2SP(S)(OC_2H_5)_2$	37.5
Azinphosmethyl	$(CH_3O)_2P(S)SCH_2$—⟨benzotriazinone ring with N, N=N, O⟩	10.4
Dimethoate	$(CH_3O)_2P(S)SCH_2CONHCH_3$	12.0
Morphothion	$(CH_3O)_2P(S)SCH_2CON$⟨morpholine ring, O⟩	18.4
Mecarbam	$(C_2H_5O)_2P(S)SCH_2CON(CH_3)CO_2C_2H_5$	5.9
Menazon	$(CH_3O)_2P(S)SCH_2$—⟨triazine ring with NH$_2$, NH$_2$⟩	27.6
Malathion	$(CH_3O)_2P(O)SCHCO_2C_2H_5$ $\vert$ $CH_2CO_2C_2H_5$	7.8
Malaoxon	$(CH_3O)_2P(S)SCHCO_2C_2H_5$ $\vert$ $CH_2CO_2C_2H_5$	7.0
Dichlorvos	$(CH_3O)_2P(O)OCH{=}CCl_2$	1.35
Trichlorphon	$(CH_3O)_2P(O)CH(OH)-CCl_3$	3.2
Chlorfenvinphos	$(C_2H_5O)_2P(O)OC{=}CHCl$ ⟨dichlorophenyl ring with Cl, Cl⟩	93.0

TABLE 5 (continued)

Hydrolysis Rates of Some Organophosphorus Pesticides at 70° in Ethanol — pH 6.0 Buffer Solution (1:4)[191]

Name	Structure	Half-life (h)
Mevinphos α	CH_3 $\|$ $(CH_3O)_2P(O)O-C=CCO_2CH_3$ $\|$ H	3.7
Mevinphos β	CH_3 H $\|$ $\|$ $(CH_3O)_2P(O)OC=CCO_2CH_3$	4.5
Phosphamidon α*	CH_3 Cl $\|$ $\|$ $(CH_3O)_2P(O)OC=C-CON(C_2H_5)_2$	10.5
Phosphamidon β*	CH_3 $\|$ $(CH_3O)_2P(O)OC=CCON(C_2H_5)_2$ $\|$ Cl	14.0
Diazinon	$(C_2H_5O)_2P(S)O$ with pyrimidine ring, CH_3, N, $i-C_3H_7$	37.0
Thionazin	$(C_2H_5O)_2P(S)O$ — pyrazine ring with two N	29.2
Thionazin-O-analog	$(C_2H_5O)_2P(O)O$ — pyrazine ring with two N	8.2
Dimefox	$((CH_3)_2N)_2P(O)F$	212
Schradan	$((CH_3)_2N)_2P=O)_2O$	Negligible hydrolysis in 96 hr

*For structure, see Reference 193.

1. Hydrolysis of Simple Phosphate Esters

The triesters of phosphoric acid are susceptible to alkaline hydrolysis. In the hydrolysis of the phosphate ester linkage P-O-C, there are two possibilities: a nucleophile attacks either the phosphorus atom accompanied with P-O bond fission or the carbon atom with O-C bond fission. As phosphoryl-P is a hard acid center and tetrahedral-C is a soft acid, the hydroxide ion, which is a hard base, may preferably react with phosphorus atom (see II.D.2). With phosphorus, the hydroxide ion is a better nucleophile than water by a factor of ca 10^8 but with saturated carbon, by a much smaller factor of ca 10^4.[194] Thus, water attacks the carbon atom and the hydroxide ion selectively attacks the phosphorus atom, accompanied by the fission of the P-O bond.

The most acidic group (AO) in the phosphate molecule (I) is most susceptible to alkaline hydrolysis. Thus, the P-O-A bond is cleaved and the acidic group displaced with the hydroxide ion to form a diester and an anion AO⁻, the most stable of three possible anions (AO⁻, RO⁻, R'O⁻) which could be produced by hydrolysis of the triester. For example, the hydrolysis of a dialkyl aryl phosphate proceeds by the fission of the aryl ester bond. The reaction is analogous to an ordinary S_N2 displacement at carbon. The nucleophilic displacement reactions at phosphoryl centers occur with high stereospecificity, and take place with inversion of configuration at phosphorus:[57]

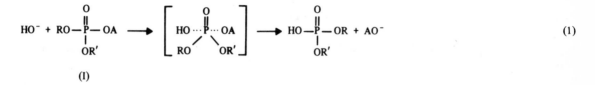

(1)

(I)

It is probable that the reaction proceeds via a metastable pentacoordinate intermediate (II), although the evidence for its formation is not as compelling (see Sections I.C.4.b and III.A.2.).[194a]

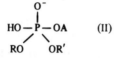

(II)

In acid solutions, the hydrolysis of trimethyl phosphate is slow and acid catalysis is rarely, if at all, observed.[195] The reactive species appears to be not a protonated conjugate acid, but a neutral molecule. Isotope studies using water-^{18}O indicate that the reaction proceeds entirely with C-O bond fission in acid and neutral solutions (Equation 2). In mixed esters and amides, the most basic group is removed, with some exceptions, contrary to the alkali-catalyzed hydrolysis. Thus, phosphoramidic compounds yield amines, and alkyl aryl phosphate esters alcohols, by acid hydrolysis.

$$H_2O: \quad + CH_3-O-\overset{\overset{\displaystyle O}{\|}}{P}(OR)_2 \longrightarrow H_2\overset{+}{O}CH_3 + {}^-O\overset{\overset{\displaystyle O}{\|}}{P}(OR)_2$$

(2)

If the hydrolyzed products are diesters, they are stable and greatly resist alkaline hydrolysis. Phosphoric diesters dissociate in media of pH higher than 1.5. The anion greatly decreases the electrophilic property of phosphorus and protects it from the attack of anionic nucleophiles by electrostatic repulsion. For instance, diethyl phosphate has a half-life of 5 billion days at 100° and pH 10. Under acid conditions, the hydrolysis of dimethyl phosphate is catalyzed, but even in $5M$ perchloric acid at 100° the half-life is 320 min. The hydrolysis of the diesters proceeds largely with C-O bond fission of both the neutral species (Equation 3) and the conjugate acids (Equation 4).[196]

$$H_2O: \quad R-O-\overset{\overset{\displaystyle O}{\|}}{\underset{\underset{\displaystyle OH}{|}}{P}}-OR$$

(3)

$$H_2O: \quad R-\overset{+}{O}-\overset{\overset{\displaystyle O}{\|}}{\underset{\underset{\displaystyle OH}{|}}{P}}-OR$$
$$\qquad\quad\; \underset{H}{|}$$

(4)

On the other hand, the main reacting species of monoesters is a monoanion. The monoanion of methyl phosphate undergoes hydrolysis about 8,000 times faster than the monoanion of dimethyl phosphate.[196] It is assumed that the reaction proceeds through the formation of an unstable monomeric metaphosphate ion intermediate (III) accompanied by P-O bond fission. The intermediate III reacts rapidly with water to give inorganic phosphate (Equation 5).[194]

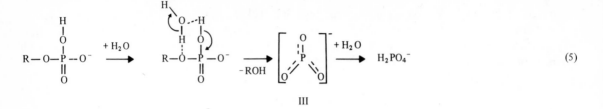

$$(5)$$

III

Under strongly acid conditions, the neutral molecule and conjugate acid may also contribute to the hydrolysis mainly by C-O bond fission:

$$(6)$$

2. Chemical Structure and Hydrolizability

The alkaline hydrolysis is initiated by the nucleophilic attack of the hydroxide ion at phosphorus. The reaction depends upon the electron deficiency of the phosphorus atom, which may be affected by the electronic properties of substituents on phosphorus. Thus, the hydrolyzability of the esters is increased by the presence of electron-withdrawing groups and is decreased by the presence of electron-releasing groups. The electron-withdrawing property of acidic substituents in ordinary phosphorus pesticides increases in the order $OR < OPh < SPh < OP(O)(OR)_2 < F$. This is the same order of the acid strengths of the acids displaced except for hydrofluoric acid. The dissociation of HF is rather lower than that of diethyl phosphoric acid because it exists as the associated form $(HF)_2$ in water. There is a positive correlation between the rates of alkaline hydrolysis (log k_{OH}) of phosphate esters $(C_2H_5O)_2P(O)X$ and the acid strengths (pK_a) of the acids, HX (Table 6).[33]

For a series of diethyl substituted phenyl phosphates, paraoxon analogs, a detailed investigation was conducted by Fukuto and Metcalf.[197] They found a linear relationship between hydrolyzability as logarithms of hydrolysis constants and Hammet's σ constants of substituents on the phenyl ring. This means that the more electron attractive the substituent is, the more reactive the phenyl phosphate is. Electron-releasing substituents make the ester less hydrolyzable. Thus, the p-nitro derivative, whose σ constant is +1.27, is hydrolyzed 29 times faster than the nonsubstituted phenyl phosphate ($\sigma=0$), and 142 times faster than the m-dimethylamino derivative ($\sigma = -0.211$). A similar relationship was also observed with diethyl S-(p-substituted)phenyl phosphorothiolates.[198]

In general, organophosphorus pesticides have bonds connecting phosphorus with hetero atoms such as oxygen, nitrogen, sulfur, and halogens, which all possess the lone pair of electrons. Such a lone pair of electrons can be donated into the vacant $3d$ orbitals of the phosphorus atom (see Section I.C.4.a). By virtue of this $p\pi$-$d\pi$ contribution, the bondings are fortified, the electron density of phosphorus increases, and consequently, the phosphorus compounds become less susceptible to the attack of nucleophiles. Electron-withdrawing groups make the bond deficient in π-electrons and the compounds more reactive.

TABLE 6

Relative Hydrolysis Rates of $(C_2H_5O)_2P(O)X$ and Acid pK_a Values of HX

X	pK_a	Relative hydrolysis rate
OC_2H_5	19	1
OC_6H_5	9.95	1×10^2
SC_6H_5	8.3	1.3×10^4
$OP(O)(OC_2H_5)_2$	1.37	4.0×10^5
F	4.73	5.6×10^5

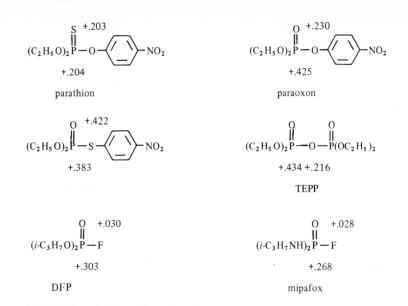

FIGURE 5.   Calculated formal positive charges on the hydrolyzable bonds of several organophosphorus compounds.[199]

Using Hückel's molecular orbital calculations, Pullman found in highly reactive organophosphorus compounds a labile bond on which both the phosphorus atom and the bonded atom are charged positively, as shown in Figure 5.[199]

Certain classes of organophosphorus compounds are much more reactive than expected from the pK values or electron attractivity of the acidic groups. These include phosphorothiolates, phosphoramidates, enol phosphates, and cyclic phosphates, and will be discussed later. Steric factors, activation, and other factors should also be considered.

The hydrolyzability of phosphorus esters is also subject to the influence of nonacidic groups. Alkyl groups have an inductive effect to release electrons. The effect overlaps with the $p\pi$-$d\pi$ contribution of lone pair electrons on the oxygen of the ester group and increases in the order: methyl < ethyl < propyl. Thus, methyl esters are generally more unstable than corresponding ethyl and higher alkyl esters; demeton-S methyl and parathion-methyl are hydrolyzed 2 and 5 times faster, respectively, than the corresponding ethyl homologs (Table 5).

Moreover, the methyl ester group is susceptible to dealkylation reactions (see Section C of this Chapter). In the course of hydrolysis, the methyl esters often yield dealkylated products, besides the normal hydrolysis product by the cleavage of a bond connected to the most acidic group. For example, dimethoate (dimethyl S-(N-methylcarbamoylmethyl) phosphorothiolothionate) decomposes in alkaline medium (pH 11) to give des-methyl dimethoate (49% of total hydrolyses products) and dimethyl phosphorothioic acid (38%).[200] The dealkylation reaction takes place much more readily in a methyl ester bond than in an ethyl ester bond. When the reaction was carried out with equimolar KOH (0.025M) in 95% ethanol for 20 hr at room temperature, parathion yielded 11% of the dealkylated product, whereas a much higher percentage (46%) of degradation at the alkyl phosphate bond was observed with parathion-methyl.[201] When the concentration of KOH is increased or the concentration of ethanol is decreased, normal hydrolysis is more prevalent than dealkylation. This suggests that the hydroxide ion attacks phosphorus to cause hydrolysis, and alcohol attacks the methyl group preferably to cause dealkylation.

a. Derivatives of Phosphorothioic Acids

The majority of organophosphorus pesticides are phosphorothionate esters, which are generally 2 to 20 times more stable than the corresponding phosphate esters. Sulfur is less electronegative or less electron attractive than oxygen, as indicated in Table 7. Therefore, the phosphorus atom of phosphorothionate esters is less electrophilic and consequently less reactive with the hydroxide ion.

In contrast, phosphorothiolate esters are much more reactive than the corresponding phosphate esters.

TABLE 7

H 2.1			
C 2.5	N 3.0	O 3.5	F 4.0
Si 1.8	P 2.1	S 2.5	Cl 3.0
			Br 2.8
			I 2.5

The *S*-aryl thiolate analog of paraoxon is hydrolyzed 22 times faster than paraoxon in 0.01N sodium hydroxide solution.[198] The hydrolysis rate of diethyl *S*-phenyl phosphorothiolate is about 100 times as fast as that of its phosphate analog, and 4 times as fast as that of the *p*-nitro substituted phosphate, paraoxon. Diisopropyl methylphosphonodithiolate reacts 25,000 times faster in alkali than the corresponding oxygen ester, with a corresponding decrease in activation energy (from 14.9 kcal mole^{-1} for diisopropyl methylphosphonate to 11.4 kcal for the dithiolate).[202] This high reactivity of phosphorothiolate esters is more than expected from the acidity of thiols (approximate pKa values are alkylthiols, 12 to 13; alcohols, 16 to 18; phenols, 10; thiophenols, 8; and *p*-nitrophenol, 7). The fission usually occurs at the P-S bond,[202] provided the thiol is the most acidic substituent.

The thiol phosphate ester linkage is not always disrupted; if a more acidic substituent is involved in the molecule, the acidic ester linkage is hydrolyzed. In this case, the thiolate esters are more than 100 times more susceptible to alkaline hydrolysis than corresponding phosphate esters. For example, *O,S*-diethyl *O*-*p*-nitrophenyl phosphorothiolate, which often exists as an unwanted impurity in technical preparations of parathion, is hydrolyzed at the *p*-nitrophenyl ester linkage 470 times faster than the corresponding phosphate ester, paraoxon.[33]

Despite the lower electronegativity of the sulfur atom as compared with oxygen, phosphorothiolate esters are generally much more reactive than corresponding oxygen analogs, as mentioned above. This is attributed to (1) the polarizability of the sulfur atom,[33] (2) the reduced $p\pi$-$d\pi$ contribution in the P-S bond,[26] because $p\pi$-$d\pi$ overlap is less efficient with $3p$ than with $2p$ orbitals, and (3) the lower bond strength of the P-S bond.[202,203]

In certain types of phosphorus esters having a hetero atom at the β-position of an alkyl ester group, the alkaline hydrolysis occurs more rapidly in *O*-alkyl esters than in *S*-alkyl analogs. For example, the phosphate analog of demeton (diethyl 2-ethylthioethyl phosphorothioate) is hydrolyzed twice as fast as the thiolate analog, demeton-S.[204] Amiton (diethyl *S*-(2-diethylaminoethyl) phosphorothiolate) is more slowly hydrolyzed in alkaline medium than even its thiono isomer.[117] This is due to the difference in the mechanism of hydrolysis. β-Alkylthio-(or amino)-ethyl phosphate esters are readily decomposed by forming cyclic onium ion intermediates, accompanied with C-O bond fission (Equation 7), in the same way as in the thiono-thiolo rearrangement of their thiono analogs (reactions 40 and 108 in Chapter II).

$$(RO)_2\overset{X}{\underset{}{P}}-O-CH_2-CH_2 \longrightarrow (RO)_2\overset{X}{\underset{}{P}}O^- + \overset{\overset{+}{Y}-R'}{CH_2-CH_2} \xrightarrow{OH^-} HOCH_2CH_2YR' \qquad (7)$$

X = O or S; Y = S or NR

On the other hand, *S*-alkyl esters are usually hydrolyzed by P-S bond fission: the C-S bond is more stable than the P-S bond. *S*-Alkyl groups of phosphorothiolates are so poor in alkylating or electrophilic property that they are some hundred times less active towards nucleophiles than *O*-alkyl groups of corresponding phosphate esters.[104] Actually, amiton is one thousand times less able to alkylate thiosulfate than its thiono isomer.[117] Kinetic studies show that the alkaline hydrolysis of amiton is a bimolecular reaction, in contrast with that of its thiono isomer which obeys first order kinetics indicating that the unimolecular formation of the intermediate is the rate determining step. Therefore, a normal nucleophilic displacement at the phosphorus atom (Equation 8) may be more important than the intramolecular alkylation as shown in

Equation 7, for the hydrolysis of S-(β-alkylthio-(or amino)-ethyl) phosphorothiolates. The P-S bond splits in the hydrolysis of phosphorothiolothionates too (Equation 9).

$$(RO)_2 \overset{\overset{\text{O}}{\|}}{P}-S-CH_2 CH_2-Y-R' + NaOH \longrightarrow (RO)_2 \overset{\overset{\text{O}}{\|}}{P}ONa + HSCH_2 CH_2 YR' \tag{8}$$

$$(RO)_2 \overset{\overset{\text{S}}{\|}}{P}-S-CH_2 CH_2 SC_2 H_5 \xrightarrow{H_2 O} (RO)_2 \overset{\overset{\text{O}}{\|}}{P}SOH + HSCH_2 CH_2 SC_2 H_5 \tag{9}$$

These facts do not necessarily imply that the neighboring hetero atom never participates in the hydrolysis of β-alkylthioethyl phosphorothiolates. The oxidation of the thioether group increases the stability towards hydrolysis. The half-life of demeton-S-methyl in an aqueous solution between pH 1 and 5 at 20°C is 88 days, while that of its sulfoxide (oxydemeton-methyl) and sulfone (dioxydemeton-S-methyl) derivatives is 236 days. The increase in stability by oxidation is more pronounced by branching at the α-carbon, i.e., with S-(2-ethylthio)isopropyl phosphorothiolates. This branching causes the ester to be unstable to hydrolysis to a half-life of 1.8 days. This effect disappears in the sulfoxide (Estox) and sulfone derivatives, which have a half-life of 240 days.[192] The increased stability towards hydrolysis of the oxidation products should be due to the impossibility of the neighboring participation.

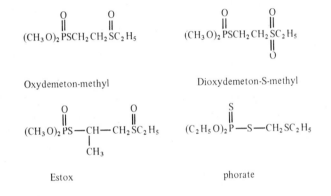

Oxydemeton-methyl

Dioxydemeton-S-methyl

Estox

phorate

On the contrary, the sulfoxide and sulfone derivatives of phorate and its oxygen analog are 40 to 130 times more susceptible to hydrolysis in 0.1 M sodium carbonate solution than the sulfides.[205] This is apparently because the sulfur atom at the α-carbon cannot participate in the hydrolysis as can the sulfur atom at the β-carbon and the electron-withdrawal effect of sulfoxide and sulfone groups accelerates the hydrolysis.

An unusual reaction accompanying C-S fission takes place in certain phosphorus esters having an electron attractive group such as carboxyl, cyano, or sulfonyl group in the β-position. As the β-hydrogen is activated by the electron attractive group, β-elimination occurs by the action of alkali.* Thus, dioxydemeton-S-methyl (dimethyl S-2-ethylsulfonylethyl phosphorothiolate) is rapidly decomposed with alkali, giving dimethyl phosphorothioic acid and ethyl vinyl sulfone (Equation 10),[192] though it is rather stable under acidic conditions.

$$(CH_3 O)_2 \overset{\overset{\text{O}}{\|}}{P}-S-CH_2-CH-SO_2 C_2 H_5 \longrightarrow (CH_3 O)_2 \overset{\overset{\text{O}}{\|}}{P}S^- + C_2 H_5 SO_2 CH=CH_2 + H_2 O \tag{10}$$

*The β-elimination in fully esterified phosphorus acids by the action of base occurs not only in thioates but also in oxygen esters. This is, however, rather less common, because the highly electrophilic phosphorus reacts readily with the base.

This reaction is applied to the detection of demeton-S-methyl with 2,6-dibromo-*N*-chloro-*p*-quinoneimine after previous oxidation to sulfone on paper chromatography; this reagent gives only pale yellow color with demeton, but intense red color with the phosphorothioic acid.[206]

By the action of alkali in organic solvents, malathion is also decomposed through the β-elimination process into fumarate and diethyl phosphorodithioic acid (Equation 11).[207] This reaction is applied for the colorimetric estimation of malathion by converting the dithioic acid into a copper complex.[207]

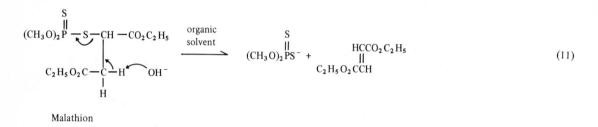

$$ (11) $$

Malathion

However, in aqueous conditions above pH 9, it appears to decompose by P-S bond fission, i.e., by substitution reaction at phosphorus:[208]

$$ (CH_3O)_2\overset{\text{S}}{\underset{HO^-}{P}}-S-\underset{CH_2CO_2C_2H_5}{\underset{|}{CH}}-CO_2C_2H_5 \xrightarrow{\text{water}} (CH_3O)_2PSOH + \underset{CH_2CO_2C_2H_5}{\underset{|}{HS-CH}}-CO_2C_2H_5 \qquad (12) $$

Since the dispersion of charges is hindered by solvation, ionizing media generally suppress elimination (E2) more than substitution (S_N2). Charge dispersion in the former is larger than in the latter at the transition state, as shown by Equations 13 and 14.

$$ X-\overset{|}{\underset{|}{C}}-\overset{|}{\underset{|}{C}}-H + Y^- \longrightarrow \overset{\delta-}{X}\cdots\overset{|}{\underset{|}{C}}-\overset{|}{\underset{|}{C}}\cdots H\cdots\overset{\delta-}{Y} \qquad E2 \qquad (13) $$

$$ X-\overset{O}{\overset{||}{P}}- + Y^- \longrightarrow \overset{\delta-}{X}\cdots\overset{O}{\overset{||}{P}}\cdots\overset{\delta-}{Y} \qquad S_N2 \qquad (14) $$

In acid solutions (>pH 2), malathion is relatively stable but, by refluxing in 1*N* hydrochloric acid, it undergoes hydrolysis to give mercaptosuccinate with P-S bond fission:[13]

$$ (CH_3O)_2\overset{\text{S}}{\underset{H_2O:}{P}}-S\overset{H^+}{-}\underset{CH_2CO_2C_2H_5}{\underset{|}{CH}}-CO_2C_2H_5 \longrightarrow (CH_3O)_2PSOH + \underset{CH_2CO_2C_2H_5}{\underset{|}{HS-CH}}CO_2C_2H_5 \qquad (15) $$

Phosphorothiolate esters with a methylene bridge between sulfur and a hetero atom, as shown by a general formula, $>P(Y)S-\overset{|}{\underset{|}{C}}-X$, where X is N or S and Y is S or O, behave differently on alkaline hydrolysis. For example, azinphosmethyl is hydrolyzed in alkaline media accompanied with C-S bond fission.[13] By acid hydrolysis with concentrated hydrochloric acid, formaldehyde is liberated (Equation 16).[209] Anthranilic acid is also produced by further alkaline hydrolysis.

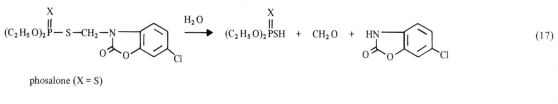

The hydrolysis products, phosphorodithioic acid, formaldehyde, and anthranilic acid, are used for the colorimetric determination of aginphosmethyl.

Phosalone and its oxo analog are similarly hydrolyzed by C-S bond fission, giving formaldehyde (Equation 17).[210] Phorate and carbophenothion, having a partial structure P-S-C-S-R, also liberate formaldehyde by both alkaline and acid hydrolysis.[209]

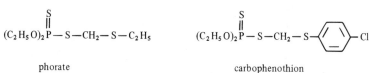

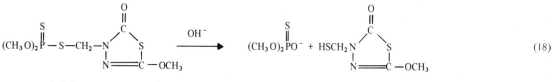

However, alkaline hydrolysis of methidathion (Supracid) proceeds in a different way, accompanied with P-S bond fission, but does not liberate formaldehyde:[92]

$$(CH_3O)_2 P—S—CH_2—N \quad \overset{OH^-}{\longrightarrow} \quad (CH_3O)_2 PO^- + HSCH_2 N \qquad (18)$$

methidathion

b. Derivatives of Phosphoramidic Acids

Phosphoramides of secondary amines are extremely resistant to hydrolysis (Table 5), owing to the electromeric donation of electrons on nitrogen to form $p\pi$-$d\pi$ bonding between P and N. This reduces the positive charge of phosphorus, tending to make the phosphoramides less reactive toward the hydroxide ion. As electron-donating ability should decrease in the order $R_2N > RHN \gg RO$, the phosphoramides of primary amines are expected to be somewhat more susceptible than N,N-dialkylphosphoramides, but much more stable than phosphates. However, they are unexpectedly very susceptible to alkaline hydrolysis. For example, mipafox (N,N'-diisopropylphosphorodiamidic fluoride) is hydrolyzed as fast as DFP (Diisopropyl phosphorofluoridate). Their bimolecular hydrolysis rates k_{OH} ($25°$) (l. mole^{-1} min^{-1}) are 49 to 50 respectively, whereas N,N,N',N'-tetramethylphosphorodiamidic fluoride is 14,000 times less hydrolyzable.[33] The difference in alkaline hydrolysis rates of the corresponding chloride analogs is much greater: N,N'-diisopropylphosphorodiamidic chloride is hydrolyzed at least 4 million times faster than N,N,N',N'-tetramethylphosphorodiamidic chloride in alkaline solutions, though in neutral or slightly acidic conditions their hydrolysis rates are comparable.[211] In other nucleophilic substitution reactions with fluoride, azide, or pyridine, no such difference is observed between the reaction rates of these diamidic chlorides. This indicates that the alkaline hydrolysis of phosphorodiamidic halides of monoalkylamines must proceed by a special mechanism.

N-alkylphosphoramidates behave as weak acids and make salts with strong bases, as has been suggested in reaction 133 or 132 in Chapter II.

$$\begin{array}{ccc} \underset{R_2P-NR'}{\overset{O\ \ H}{\|\ \ |}} & \overset{\text{base}}{\longrightarrow} & \underset{R_2P-NR'}{\overset{O}{\|}}{\overset{\ominus}{}} \longleftrightarrow \underset{R_2P=NR'}{\overset{O^{\ominus}}{|}} \end{array} \tag{19}$$

This type of reaction may take place in aqueous alkaline solutions too, provided that the pKa of the amido proton is less than 14. Traylor and Westheimer proposed that halogen was eliminated from the anion formed by removal of the amido proton, to form a very reactive monomeric metaphosphate type intermediate (IV), to which water rapidly adds to give a diamidic acid:[211]

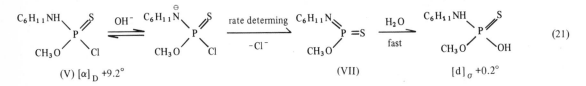

(IV)

Any attempt to trap the metaphosphate intermediate has been unsuccessful. However, stereochemical investigations provide more direct evidence of such intermediate formation, since monomeric metaphosphates are expected to be planar and to yield racemic addition products. Optically active methyl *N*-cyclohexylphosphoramidochloridothionate (V) is slowly hydrolyzed in a neutral solution at 1/13 the rate of dimethyl phosphorochloridothionate (VI), and the reaction proceeds stereospecifically, indicating that normal S_N2 reaction occurs on phosphorus. On the contrary, under alkaline conditions the mono-alkylamidochloridate (V) is hydrolyzed 45,000 times as fast as the dialkyl ester chloride (VI), and yields a racemic hydrolysis product, which must result from a symmetric metaphosphate intermediate (VII):[212]

$$\begin{array}{ccccccc} \underset{CH_3O}{\overset{C_6H_{11}NH}{\diagdown}}\underset{Cl}{\overset{S}{\diagup}}P & \underset{\longleftarrow}{\overset{OH^-}{\rightleftharpoons}} & \underset{CH_3O}{\overset{C_6H_{11}\overset{\ominus}{N}}{\diagdown}}\underset{Cl}{\overset{S}{\diagup}}P & \underset{-Cl^-}{\overset{\text{rate determing}}{\longrightarrow}} & \underset{CH_3O}{\overset{C_6H_{11}N}{\diagdown}}P=S & \underset{\text{fast}}{\overset{H_2O}{\longrightarrow}} & \underset{CH_3O}{\overset{C_6H_{11}NH}{\diagdown}}\underset{OH}{\overset{S}{\diagup}}P \\ & & & & & & \end{array} \tag{21}$$

$$\text{(V) } [\alpha]_D +9.2° \qquad\qquad\qquad\qquad\qquad \text{(VII)} \qquad\qquad\quad [d]_\sigma +0.2°$$

A similar intervention of monomeric metaphosphate type intermediates is also probable in the reactions of phosphoramidate diesters. Methyl *p*-nitrophenyl *N*-cyclohexylphosphoramidothionate, an amido analog of parathion-methyl, is more susceptible to alkaline hydrolysis than parathion-methyl.[213] The high rate of reaction of the deprotonated amido anion with water suggests that the aryl ester bond is abnormally weak, and there is a distinct tendency for unimolecular elimination to occur.

O,S-Dialkyl *N*-alkylphosphoramidothiolates are hydrolyzed under alkaline conditions faster than anticipated. *O,S*-Dimethyl *N*-methylphosphoramidothiolate and the *N*-unsubstituted homolog (insecticide methamidiphos, Monitor) gave pseudo first order hydrolysis rate constants (k_o) of 1.8×10^{-2} and 3.2×10^{-2} min^{-1}, respectively, in a phosphate buffer at pH 11.5 and 30°.[121] These values are comparable with that of methyl paraoxon ($k_o = 1.2 \times 10^{-2}$ min^{-1}). Under this condition, the P-SCH$_3$ bond cleavage occurs predominantly, whereas in an aqueous potassium hydroxide solution, the cleavage of P-OCH$_3$ bond predominates.[214] The *N,N*-dimethyl homolog is completely resistant to alkaline hydrolysis. The high reactivity of the *N*-monoalkyl and unsubstituted phosphoramidothiolates, compared to the *N,N*-dialkyl amidate, is difficult to explain on the basis of polar or steric grounds, and the following metaphosphate mechanism was proposed (Equation 22).[121] However, *S*-phenyl phosphoramidothiolates appear to be hydrolyzed by a direct displacement reaction on phosphorus.[215]

$$CH_3S-\overset{\displaystyle O}{\underset{\displaystyle CH_3O}{\overset{\|}{P}}}-N\overset{H}{\underset{CH_3}{}} + OH^- \longrightarrow \left[\overset{\displaystyle O}{\underset{\displaystyle CH_3O}{\overset{\|}{P}}}=NCH_3\right] \xrightarrow{H_2O} HO-\overset{\displaystyle O}{\underset{\displaystyle CH_3O}{\overset{\|}{P}}}-NHCH_3 + CH_3SH \qquad (22)$$

On the other hand, the phosphoramides of secondary amines are rather susceptible to acid hydrolysis. The 50% hydrolysis times of schradan in $1N$ HCl, $1N$ NaOH, and neutral aqueous solutions are 3.3 hr, 70 days, and 100 years, respectively.[216] Dialkylamides have a basic property and are protonated to form adjacent positive charges (VIII) which cause the disruption of the P-N bond. The resulting amido acid is more susceptible to the electrophilic attack of a proton and decomposes further. The hydroxide ion, however, hardly attacks the phosphorus atom of phosphoramides which is less electrophilic, as mentioned above.

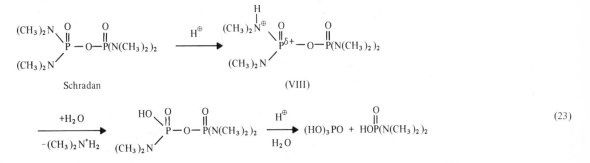

Schradan (VIII)

(23)

Recently, some O-alkyl O-(substituted phenyl) N-alkylphosphoramidates and phosphoramidothionates have been developed as insecticides and herbicides. They are also susceptible to acid hydrolysis. The acid catalyzed hydrolysis of the oxo-analog of the herbicide DMPA (Zytron) and related compounds takes place with P-N bond fission, and the liberation of phenols and alcohols is only negligible.[217]

$$i\text{-}C_3H_7NH\diagdown\overset{\displaystyle O}{\underset{\displaystyle CH_3O\diagup}{\overset{\|}{P}}}-O-\!\!\!\bigcirc\!\!\!\underset{Cl}{\overset{}{}}-Cl \xrightarrow{H^\oplus/H_2O} HO\diagdown\overset{\displaystyle O}{\underset{\displaystyle CH_3O\diagup}{\overset{\|}{P}}}-O-\!\!\!\bigcirc\!\!\!\underset{Cl}{\overset{}{}}-Cl + i\text{-}C_3H_7\overset{+}{N}H_3 \qquad (24)$$

DMPA-oxo analog

The initial protonation of nitrogen or phosphoryl oxygen is believed to be the first step of the reaction:

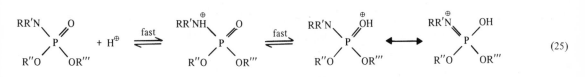

(25)

A kinetic study shows that the following mechanisms are most probable.[217]
1. A simple S_N2 type substitution of water:

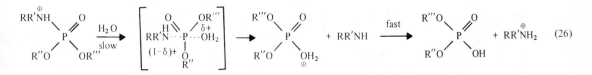

(26)

2. A two-step, addition-elimination sequence involving the formation of an intermediate having five σ-bondings and the slow subsequent unimolecular cleavage of the P-N bond:

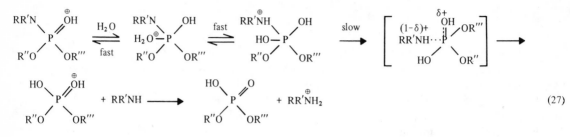

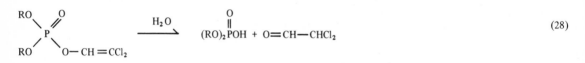

$$(RO)_2POH + RR'NH \longrightarrow \qquad + RR'NH_2 \qquad (27)$$

c. Enol and Imidoyl Phosphates

Phosphate esters of enols undergo hydrolysis, proceeding with the cleavage of the enol ester linkage, to yield the corresponding carbonyl compounds by the catalytic action of both bases and acids:

$$\underset{RO}{\overset{RO}{\diagdown}}\underset{O-CH=CCl_2}{\overset{O}{\diagup}}P \xrightarrow{\;H_2O\;} (RO)_2POH + O=CH-CHCl_2 \qquad (28)$$

The mesomeric effect of the double bond system and the inductive effect of substituents overlap to accelerate the hydrolysis. In comparison with the hydrolysis rate of triethyl phosphate, those of the enol phosphates are much greater, as shown in Table 8.[218] Diethyl vinyl phosphate is hydrolyzed 12 times as fast as triethyl phosphate in an acid solution. An enol ester derived from ethyl bromomalonate (IX) is 3,000 times more reactive than the triethyl ester. Substitution on the α-position of the vinyl group accelerates the hydrolysis in the order $H < CO_2C_2H_5 < C_6H_5 < CH_3$.

The rate of hydrolysis is affected by a steric factor too. A geometrical isomer (α) of mevinphos (cis-mevinphos), which has a bulky group in trans position to the phosphate ester grouping, is hydrolyzed faster than another isomer (β) (trans-mevinphos), in which these groupings are in cis position* (Table 9).[219] Trans-mevinphos yields only dimethyl phosphate, while the cis isomer yields dimethyl phosphate and "phosdrin acid," which is the hydrolysis product of the carboxymethyl ester group. This difference may be due to the geometrical configuration; in the trans isomer, the phosdrin acid formed may be further decomposed by the intramolecular catalytic action of the carboxyl group, which is in a sterically suitable position to attack the phosphorus.[221]

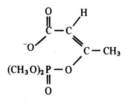

trans phosdrin acid

In acid catalyzed hydrolysis of such a reactive enol phosphate as IX, protonation may lead to electron withdrawal from the double bond and activation of the ester:[222]

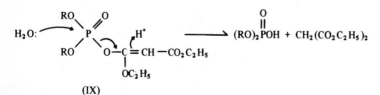

$$\longrightarrow (RO)_2POH + CH_2(CO_2C_2H_5)_2 \qquad (29)$$

(IX)

*The geometry of mevinphos was first assigned reversely.[71,220]

TABLE 8

Acid Hydrolysis of Diethyl Vinyl Phosphates, $(C_2H_5O)_2P(O)OX$ (0.1N HCl-40% Ethanol 85°)[218]

X	$k \times 10^3$ (min^{-1})	t_{50} (hr)
$-\underset{\underset{OC_2H_5}{\vert}}{C}=CHCO_2C_2H_5$ (IX)	130.0	0.1
$-\underset{\underset{CH_3}{\vert}}{C}=CH_2$	14.88	0.78
$-\underset{\underset{C_6H_5}{\vert}}{C}=CH_2$	6.46	1.79
$-\underset{\underset{CO_2C_2H_5}{\vert}}{C}=CH_2$	2.38	4.86
$-CH=CCl_2$	1.34	8.59
$-CH=CHCO_2C_2H_5$	0.99	11.71
$-CH=CH_2$	0.46	25.21
$-CH_2CH_3$	0.037	308.7

TABLE 9

Hydrolysis of Dialkyl Vinyl Phosphates (pH 11, 28°)[219]

	Half-life (hr)	
Ester	β-Isomer	α-Isomer
$(CH_3O)_2\overset{\overset{O}{\Vert}}{P}O-CH=CCl_2$ (dichlorvos)		0.2
$(CH_3O)_2\overset{\overset{O}{\Vert}}{P}O-\underset{\underset{}{}}{\overset{\overset{CH_3}{\vert}}{C}}=CHCO_2CH_3$ (mevinphos)	3.0	1.8
$(C_2H_5O)_2\overset{\overset{O}{\Vert}}{P}O-\overset{\overset{CH_3}{\vert}}{C}=CHCO_2C_2H_5$	8.9	3.4
$(C_2H_5O)_2\overset{\overset{O}{\Vert}}{P}O-CH=CHCl$	11.0	6.0

Similar acid catalyzed hydrolysis is observed in enol phosphates of acid amides (imidoyl phosphates). Some aromatic heterocycle derivatives having a partial structure P-O-C=N show this property. For example, diazinon (diethyl 2-isopropyl-6-methyl-4-pyrimidinyl phosphorothionate) is very stable in neutral aqueous solutions (t_{50} at pH 7.4 and 20°C is 4,436 hr) but is hydrolyzed 12 times as rapidly as parathion in acid medium (t_{50} at pH 3.1 is 12 hr), though in alkaline medium the hydrolysis of both the insecticides goes on at almost the same rate (t_{50} at pH 10.4 is 145 hr).[13,223] The acid catalysis of hydrolysis is probably due to the protonation on the nitrogen atom in the heterocyclic ring (X) (Equation 30). Imidoyl phosphates are known as powerful phosphorylating agents in the presence of proton donors.[222]

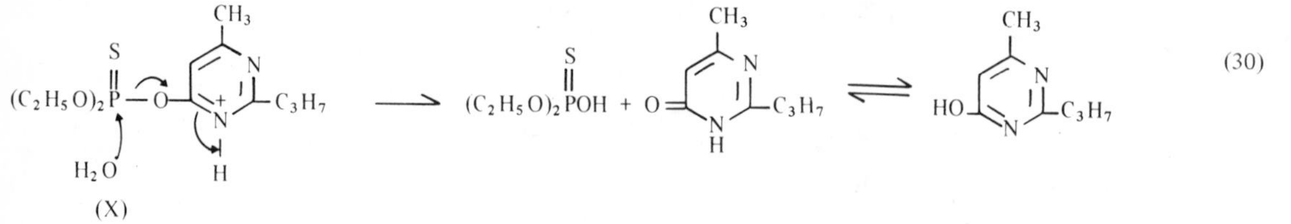

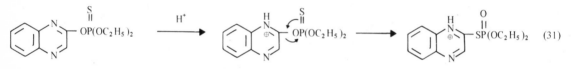

(30)

(X)

The new insecticide diethquinalphion (Bayrusil®: diethyl quinoxalyl-2 phosphorothionate) is another example. This compound is very susceptible to acid hydrolysis, while stable under basic conditions; it has a half-life 10 times that of parathion at pH 11. The susceptibility to acids is attributable to the protonation of a nitrogen atom. Since bis(2-quinoxalyl)sulfide was obtained as a degradation product, it was proposed that the protonated diethquinalphion rearranges to a thiolate isomer, which may be labile and hydrolyzed rapidly:[224]

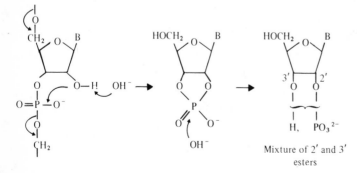

(31)

diethquinalphion

As the technical preparations usually contain acidifying by-products, the acid hydrolysis runs autocatalytically. Some tertiary amines are effective in preventing the acid catalyzed degradation of its emulsifiable concentrations.

d. Cyclic Phosphate Esters

β-Hydroxyalkyl phosphates are highly susceptible to alkaline hydrolysis. This is attributed to an intramolecular catalysis by the β-hydroxyl group through rapid cyclization and subsequent displacement. A typical example is the alkaline hydrolysis of ribonucleic acids (Equation 32). The intermediates, five-membered cyclic phosphate esters, are very unstable, though phosphate diesters generally resist alkaline hydrolysis.

(32)

Mixture of 2′ and 3′ esters

In comparison with open-chain analogs, the hydrolysis of five-membered cyclic phosphates is greatly accelerated under both alkaline and acid conditions. The relative hydrolysis rates of some cyclic phosphate esters are listed in Table 10.[26],[225-228] The half-life of ethylene phosphate is 50 min in 0.5N sodium hydroxide at 25°, and is 10^7 to 10^8 times as fast as that of dimethyl phosphate.

This extraordinary enhancement of the reactivity is due to the strain in the five-membered ring. Molecular orbital calculations[229] and NMR studies[230] indicate that the strain results in lowered $p\pi$-$d\pi$ contribution in P-O bonds, and consequently makes the phosphorus atom more positively charged (Figure 6).

As shown in Table 10, the great enhancement of the reactivity of five-membered cyclic esters is observed not only in the ring opening, but also in the hydrolysis of the exocyclic ester linkage except for the phosphonate. From these results and other observations, Westheimer proposed the transition state in the hydrolysis of cyclic esters involving "pseudorotation" of trigonal bipyramidal intermediates:[231]

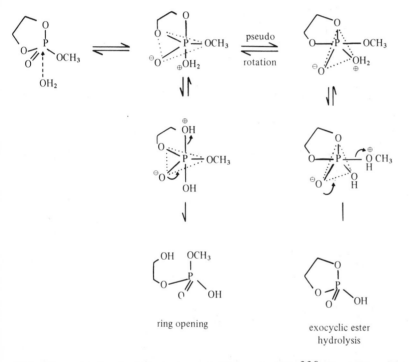

ring opening

exocyclic ester hydrolysis

(33)

Molecular orbital calculations support this mechanism.[229] Thus, the nucleophile approaches the phosphate ester on the back side of one of the endocyclic P-O bonds. This and the newly formed bonds consist of apical (or axial) bonds, and other P-OC bonds and P-O⁻ (formerly phosphoryl) become basal (radial or equatorial) bonds of a trigonal bipyramidal intermediate. The intermediate undergoes pseudorotation easily, provided the phosphoryl oxygen serves as "pivot" which remains in a basal position. The apical oxygens prefer to be protonated, and then move away from phosphorus. The lack of acceleration in the exocyclic ester hydrolysis of cyclic phosphonates is observed (Table 10). A carbon atom (less electronegative than the oxygen atom) is too energetically unfavorable to occupy an apical position, and consequently the pseudorotation is restricted. This results in the exocyclic ester group not being able to occupy the apical position from which leaving groups always depart.

Unfortunately, five-membered cyclic phosphate neutral esters are too unstable to be utilized as pesticides.[232] Though six-membered cyclic esters derived from alkanediols are not so very reactive, those derived from o-hydroxybenzyl alcohol (saligenin) are considerably reactive, and methyl saligenin cyclic phosphorothionate (2-methoxy-4H-1,3,2-benzodioxaphosphorin-2-sulfide) has been actually utilized as an insecticide, named Salithion, in Japan since 1968.[233] The hydrolysis rate of Salithion-O-analog is much faster than that of paraoxon, as shown in Table 11.[234] The acidity of saligenin (pK$_a$ = 9.92) is almost the same as that of phenol, and much weaker than that of p-nitrophenol (pK$_a$ = 7.14). Diethyl phenyl

TABLE 10

Relative Rates of Hydrolysis of Cyclic Esters

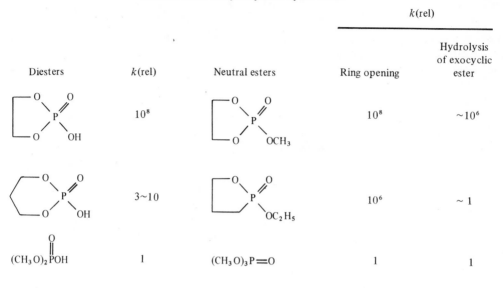

Diesters	k(rel)	Neutral esters	Ring opening	Hydrolysis of exocyclic ester
(cyclic diester)	10^8	(cyclic neutral ester, OCH₃)	10^8	$\sim 10^6$
(6-membered cyclic diester)	$3 \sim 10$	(5-membered, OC₂H₅)	10^6	~ 1
$(CH_3O)_2\overset{O}{\overset{\|}{P}}OH$	1	$(CH_3O)_3P{=}O$	1	1

Data from References 225—228.

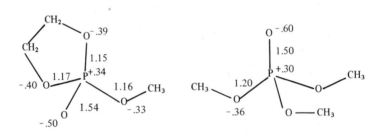

FIGURE 6. Electronic structures of cyclic and acyclic phosphates.[229] Signed numbers: net atomic charges; unsigned numbers: overlap populations.

phosphate is only 1/300 as reactive as the *p*-nitro substituted analog, paraoxon. Thus, the high reactivity of the saligenin cyclic phosphate is much greater than expected. Strain in the ring appears to be not so much; the endocyclic O-P-O angle of Salithion (104°) is in the range of the angle of acyclic phosphate esters (102° to 108°).[235] Five-membered cyclic phosphates have a strained O-P-O angle of 98° to 99°.[236] The reactivity of saligenin cyclic phosphate is also affected by the exocyclic substituent on phosphorus by virture of their electronic characteristics. The relative reaction rate decreases in the following order: RS ≃ Ar > ArO > R > RO > NHAr > NHR > NR₂.

The hydrolysis of the cyclic esters proceeds with the initial fission of the endocyclic aryl ester bond (Equation 34). The cleavage of the exocyclic ester bond does not take place by alkaline hydrolysis, though it does in dealkylation (see III.C.2). This is the case even in the derivatives of a stronger acid such as *p*-cyanophenol (pK$_a$ = 7.95). On the contrary, in the ethylene cyclic phosphate of *p*-nitrophenol, the hydrolysis of the exocyclic ester occurs preferably prior to ring opening (Equation 35).[232]

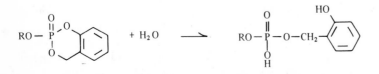

$$\text{(34)}$$

TABLE 11

Hydrolysis Rates of Some Cyclic and Acyclic Phosphate Esters of Phenols

Esters	pK_a of phenols	$k \times 10^5$ (min^{-1})	Conditions	Reference
CH$_3$O—P (cyclic catechol phosphate, methyl)	9.92	140	pH 7.7, 25°	234
C$_2$H$_5$O—P (cyclic catechol phosphate, ethyl)	9.92	50	pH 7.7, 25°	234
(C$_2$H$_5$O)$_2$PO—⟨ ⟩—NO$_2$	7.14	27	pH 9.5, 37°	197
(C$_2$H$_5$O)$_2$P—O—⟨ ⟩	9.94	0.92	pH 9.5, 37°	197

$$\left[\begin{array}{c}O\\O\end{array}\right\rangle P - O - \langle\ \rangle - NO_2 + H_2O \longrightarrow \left[\begin{array}{c}O\\O\end{array}\right\rangle P - OH + HO - \langle\ \rangle - NO_2 \tag{35}$$

Cyclophosphamide is gradually hydrolyzed in water at both the endo- and exo-cyclic phosphoramide linkages without the cleavage of the ester bond:[129]

$$(ClCH_2CH_2)_2N-P \overset{O}{\underset{O-CH_2}{\overset{NH-CH_2}{\diagup}}}CH_2 \xrightarrow[37°,\ 32\ days]{H_2O} (ClCH_2CH_2)_2N-\overset{O}{\underset{O^-}{\overset{||}{P}}}-OCH_2CH_2CH_2\overset{+}{N}H_3$$

$$\longrightarrow (ClCH_2CH_2)_2NH + \overset{^-O}{\underset{HO}{\diagdown}}\overset{O}{\overset{||}{P}}-OCH_2CH_2CH_2\overset{+}{N}H_3 \tag{36}$$

e. Phosphonates

In alkylphosphonates, there is no $p\pi$-$d\pi$ contribution to the P-C bond because the carbon has no lone pair of electrons. Therefore, despite the fact that carbon atom has lower electronegativity than oxygen atom, the phosphorus atom of phosphonates is more electrophilic than that of corresponding phosphate esters, in which lone pair electrons possessed by oxygen can be donated to phosphorus by $p\pi$-$d\pi$ overlap. Thus, phosphonate esters are usually more susceptible to alkaline hydrolysis than phosphate esters, and phosphinate esters are more susceptible than phosphonates. For example, ethyl p-nitrophenyl ethylphosphonate is hydrolyzed 39 times faster in pH 8.3 buffer than is the corresponding phosphate, paraoxon.[237] EPN (ethyl p-nitrophenyl phenylphosphonothionate) is 5 times more susceptible to alkaline hydrolysis than parathion.[33] The hydrolysis rate of 2-methyl-4H-1,3,2-benzodioxaphosphorin-2-oxide (XI) is 16 times faster than that of the corresponding 2-methoxy analog (Salithion-O-analog).[159] Under acidic conditions, the relationship of the hydrolyzability is reversed; phosphinates are more stable than phosphonates, which are more stable than phosphates.

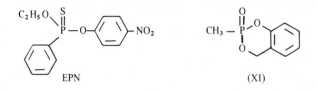

EPN   (XI)

In all of these reactions, the P-C bond remains intact, and the ester linkage of the most acidic group is always disrupted. However, in certain phosphonates the cleavage of the P-C bond takes place. As mentioned in Chapter II (Section A.3), α-hydroxyalkylphosphonates undergo rearrangement into phosphate by the action of alkali, accompanied with P-C bond fission. Trichlorfon is converted into dichlorvos by this reaction. The produced dichlorvos is, of course, susceptible to alkali, and decomposition proceeds further (Equation 37). After a half hour warming at 70° in an aqueous solution of pH 8, trichlorfon disappears completely to form dichlorvos (54%) and hydrolysis product (46%).[13] Trichlorfon is rather stable in acid region, but is slowly hydrolyzed by the action of acid, being accompanied with the cleavage of the P-O-CH$_3$ bond (Equation 38).[13]

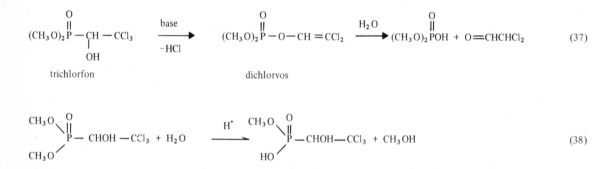

(37)

(38)

It is very interesting to note that a methylphosphonate ester (XII) having a carbonyl group on the β-position of an alkyl ester group is decomposed in a different way from its corresponding phosphate ester analog (XIII) in alkaline hydrolysis: the phosphonate releases methanol (Equation 39), but the phosphate yields a ketoalcohol (Equation 40).

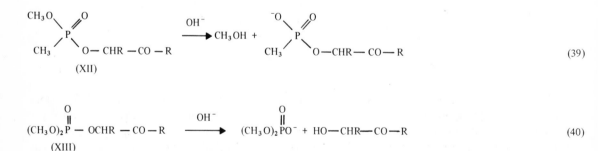

(39)

(40)

This selectivity in hydrolysis is rationally explained by the following mechanism (Equation 41) involving cyclization and pseudorotation.[238] Hydrated β-carbonyl participates to form a cyclic intermediate which has trigonal bipyramidal structure. According to a general rule for pseudorotation, that less electronegative substituents (carbon atom in this case) prefer to occupy basal positions and more electronegative substituents (oxygen atom) to occupy apical positions, the hydrolysis of the phosphonate proceeds selectively through path a without pseudorotation of the intermediate ($a:b = 20:1$). On the other hand, the hydrolysis of the phosphate proceeds through another way, b, after pseudorotation ($a:b = 1:30$).

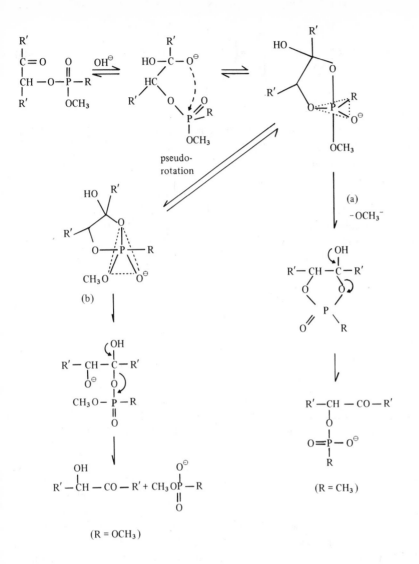

$$(R = OCH_3)$$

$$(41)$$

β-Haloalkylphosphonic acids decompose rapidly in aqueous solutions at pH higher than 5, yielding olefins quantitatively (Equation 42).[181,239] 2-Chloroethylphosphonic acid is utilized as a plant growth regulator (ethephon). It releases ethylene gas, which has a fruit ripening hormone activity.[240]

$$Cl-CH_2CH_2-\overset{\overset{\displaystyle O}{\|}}{P}(OH)_2 \xrightarrow{OH^-} Cl^- + CH_2=CH_2 + O=P(OH)_3 \qquad (42)$$

ethephon

Two possible mechanisms have been proposed for this elimination reaction.[181] The first mechanism (Equation 43) is an internal fragmentation leading to monomeric metaphosphate, which then reacts with water. The second one (Equation 44) is a fragmentation initiated by the attack of a water molecule on the phosphonate ion. These β-haloalkylphosphonic acids may be used as phosphorylating agents.[181]

$$Cl-\overset{\overset{\displaystyle R}{|}}{CH}-CH_2-\overset{\overset{\displaystyle O}{\|}}{P}\overset{\displaystyle C^{O^-}}{\underset{\displaystyle OH}{}} \longrightarrow Cl^- + R-CH=CH_2 + \left[\begin{array}{c} O \\ \diagdown \\ O \end{array} P-OH \right] \xrightarrow{H_2O} O=P(OH)_3 \qquad (43)$$

$$Cl-\overset{R}{\underset{}{CH}}-CH_2-\overset{O}{\underset{OH}{P}}\overset{O^-}{\underset{}{}}:OH_2 \longrightarrow Cl^- + RCH=CH_2 + \left[O=\overset{O^-}{\underset{OH}{P}}-\overset{+}{O}H_2 \right] \rightarrow O=P(OH)_3 \qquad (44)$$

3. Catalysis

a. Metallic Ions

Some metallic ions catalyze the hydrolysis of organophosphorus compounds. The activity as a catalyst for the hydrolysis of tabun increases in the order: $Zn^{2+} < Co^{2+} < Ni^{2+} < Ag^+ < Au^{3+} < Pd^{2+} < Cu^{2+}$.[241] Cupric ion is the most active catalyst among metallic ions. The catalytic effect of the cupric ion increases in the presence of certain chelating agents.[242] The chelates of copper with α,α'-dipyridyl and L-histidine increase the rate of hydrolysis of DFP about 600 and 300 times, respectively, at pH 7.6 and 38°. The dipyridyl-copper (1:1) chelate also accelerates the hydrolysis of TEPP. The half-life of TEPP (10 μmole) is reduced from approximately 3 hr to 4 min in the presence of 50 μmole of the chelate. However, fully coordinated copper, like the 1:2 chelation of dipyridyl and the chelate with ethylenediaminetetraacetic acid, is much less effective as a catalyst. The mechanism of the action of the copper-chelate catalysts is suggested as follows:

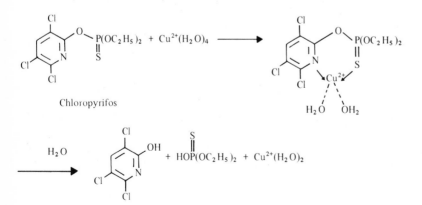

The electrophilic Cu^{++} probably increases the polarization of the P=O and the P-F bonds, and thereby facilitates the approach of a hydroxide ion to the phosphorus atom, followed by the expulsion of the F anion.

The catalysis of copper is much more effective on phosphorothionate esters than on their oxygen analogs; the hydrolysis of EPN and parathion is accelerated by the presence of the cupric ion 47.6 and 20 times, respectively, whereas that of paraoxon is only 1.7 times.[243] The formation of a complex between Cu^{++} and the sulfur aotm (P=S → Cu^{++}) may greatly increase the positive charge of phosphorus. An extreme enhancement of hydrolysis rate by cupric salts is observed in an imidoyl phosphorothionate, chloropyrifos (Dursban®; diethyl 3,5,6-trichloro-2-pyridyl phosphorothionate). It is stable in a water-methanol mixture (1:1) at 20°, but is rapidly hydrolyzed by the addition of cupric salt (half-life 0.9 hr). A mechanism forming a bidentate chelation through nitrogen in the ring structure and the sulfur on the thiophosphoryl group has been suggested (Equation 45).[244]

Chloropyrifos

$$(45)$$

b. Bases with α-Effect

Some basic oxy-anions having lone pairs of electrons on the adjacent atoms are more reactive as nucleophiles than expected from their basicity. Anions of hydroxamic acids, oximes, hypochlorite, and peroxides are typical examples. They catalyze the hydrolysis of phosphorus compounds. (Many of them are not true catalysts, because they decompose in the course of reactions.) Table 12 shows their pK_a values and second-order rate constants for the reaction with TEPP and paraoxon.[245]

The high nucleophilic reactivity of these bases is due to an electronic shift of the lone pair electrons on the adjacent atom (α-effect):

$$\ddot{B}-O^\ominus + \quad\overset{O}{\underset{}{\underset{}{>}}}P-X \longrightarrow B\overset{\frown}{-}O\cdots\overset{O}{\underset{}{P}}\cdots \overset{\delta-}{X} \longrightarrow BO-\overset{O}{\underset{}{P}}< \ + \ X^\ominus \tag{46}$$

$$B= \ R_2C=\ddot{N}-, \quad :\ddot{C}l-, \quad R-\ddot{O}- \quad etc.$$

The sequence of reactions with an oxime anion and sarin, for instance, involves a rate determining O-phosphorylation of the oxime anion, followed by rapid hydrolysis of the oxime phosphate:

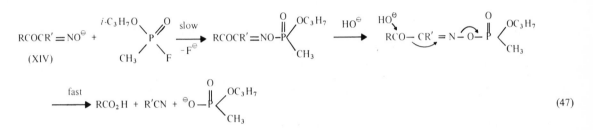

$$\xrightarrow{\text{fast}} \quad RCO_2H + R'CN + \ ^\ominus O-\overset{O}{\underset{CH_3}{P}}\overset{OC_3H_7}{} \tag{47}$$

Reactions of other hydroxylamine derivatives, hypochlorite, and peroxides are essentially the same: a nucleophilic attack on phosphorus, yielding unstable O-phosphorylated intermediates, which are hydrolyzed rapidly.[33] Certain oximes are useful for the antidotes of phosphate poisons; hypochlorites are used to remove contamination from toxic phosphorus compounds, and the reaction of peroxide is applied for the detection of phosphorus compounds.[33] As mononitrosoacetone (XIV: R = CH_3; R'= H) yields, quantitatively, hydrogen cyanide via reaction 47, it may be utilized for the determination of phosphate esters.

TABLE 12

Rate Constants for Hydrolysis of TEPP and Paraoxon by Several Bases at 25°

Base	pK_a	k (1 mole^{-1} min^{-1})	
		TEPP	Paraoxon
H_2O	0	0.0017	10^{-6}
NH_2OH	6	26	—
ClO^-	7.4	267	—
$CH_3COCH=NO^-$	8.3	59	0.2
$C_6H_5CONHO^-$	8.8	160	1.0
$CH_3COC(CH_3)=NO^-$	9.3	16	0.3
HOO^-	11.8	2180	—
HO^-	14	21	0.52

Data from Reference 245.

B. PHOSPHORYLATING PROPERTIES

1. Selectivity in Nucleophilic Substitution

The substitution reactions of pentavalent phosphorus esters with nucleophiles may be classified into two groups, namely, phosphorylation and alkylation. In the former reaction, a nucleophile attacks the electrophilic phosphorus atom, as shown in Equation 48. In some cases, however, nucleophiles react on the α-carbon atom of the ester to be alkylated in accordance with Equation 49.

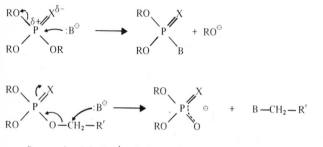

$$B = \text{nucleophile}; R,R' = \text{alkyl or aryl}; X = O \text{ or } S$$

For example, alkoxide ions are preferentially phosphorylated, whereas mercaptide ions are alkylated by phosphate or phosphorothioate esters. This difference in behavior of the nucleophilic agents suggests that there are two different nucleophilicity scales applicable to phosphorylation and alkylation, respectively.[247] The decreasing order in the reactivity of nucleophiles towards tetrahedral phosphorus is $F^- > OH^- > PhO^- > PhS^- > Cl^-, Br^-, I^-, S_2O_3{}^{2-}$; while towards tetrahedral carbon it is $S_2O_3{}^{2-} > SH^- > I^- \cong (RO)_2POS^- > OH^- > Br^- > PhO^- > Cl^- > F^-$. Hard bases apparently prefer to attack the phosphorus, and soft bases prefer the carbon atom.

Further discussions on the alkylation reactions will be presented in the following Section C. Phosphorylation is an important reaction for the prepration and biological actions of organophosphorus compounds. The reactions for the purpose of pesticide preparation have been discussed in the preceding Chapter II. The biological activities of organophosphorus insecticides, such as esterase inhibition and animal toxicity, are attributed to the phosphorylation reaction (Equation 53). Moreover, the reactions of many biologically important, natural phosphorus compounds such as ATP are also basically phosphorylation. By phosphorylation, alcohols, amines, and phosphates are converted to phosphate esters, phosphoramides, and pyrophosphates, respectively (Equations 50 to 52).

$$R'OH + (RO)_2\overset{O}{\overset{\|}{P}}OA \longrightarrow R'O-\overset{O}{\overset{\|}{P}}(OR)_2 + AO^\ominus + H^\oplus \tag{50}$$

$$R'NH_2 + (RO)_2\overset{O}{\overset{\|}{P}}OA \longrightarrow R'NH-\overset{O}{\overset{\|}{P}}(OR)_2 + AO^\ominus + H^\oplus \tag{51}$$

$$(R'O)_2\overset{O}{\overset{\|}{P}}O^\ominus + (RO)_2\overset{O}{\overset{\|}{P}}OA \longrightarrow (R'O)_2\overset{O}{\overset{\|}{P}}-O-\overset{O}{\overset{\|}{P}}(OR)_2 + AO^\ominus \tag{52}$$

$$\text{Enzyme-OH} + (RO)_2\overset{O}{\overset{\|}{P}}OA \longrightarrow \text{Enzyme}-O-\overset{O}{\overset{\|}{P}}(OR)_2 + AO^\ominus + H^\oplus \tag{53}$$

As described in the preceding Section A, the alkaline hydrolysis of organophosphorus esters proceeds by the nucleophilic attack of hydroxide ion on the phosphorus atom. Therefore, many parts of the discussion on the hydrolysis are applicable to the phosphorylation reaction. Organophosphorus compounds, being susceptible to hydrolysis, generally have a high phosphorylating activity, for example, towards cholinesterases.

2. Structure and Reactivity

Schrader proposed a general formula (XV) in 1950 for biologically active phosphorus compounds, compounds possessing an acid-anhydride linkage.[13]

(XV)

The "acyl" does not mean $R-\overset{\overset{\textstyle O}{\|}}{C}-$ in this case, but acid radicals of any proton acids including HF, HCN, phosphoric acid, enols, mercaptans, and so on. As the phosphorus atom of an "anhydride" containing a strong acid radical becomes more positive in charge, the "acyl" group becomes more readily displaced as an anion by the attack of a nucleophile on phosphorus. Therefore, the stronger the acid is, the better a phosphorylating agent the phosphorus compound becomes.

$$\underset{R^2}{\overset{R^1}{\diagdown}}\overset{\overset{\textstyle O}{\|}}{P}-Acyl + B^{\ominus} \longrightarrow \underset{R^2}{\overset{R^1}{\diagdown}}\overset{\overset{\textstyle O}{\|}}{P}-B + Acyl^{\ominus} \tag{54}$$

Recent investigations indicate that certain classes of organophosphorus compounds are much more active than expected from the strength of the acid. Such examples are given in the preceding Section A for hydrolysis. A modification of the "acyl rule" was presented in 1964 by Clark and his co-workers for phosphorylating agents.[222] According to them, numerous phosphorylating agents may be classified into some P-XYZ systems, where X, Y, and Z are atoms or groups consisting of any element, but usually H, C, N, O, S, and halogens. Z must be capable of accommodating the electrons of the P-X bond. In good phosphorylating agents, the P-X bond must be weak, and Z must be strongly electron attractive or must become so by the action of any other agents. The bonds involving phosphorus and hetero atoms are generally rather strong, since the lone pair of electrons on the hetero atoms can be donated into the vacant $3d$ orbitals of the phosphorus atom by virtue of a $p\pi$-$d\pi$ overlapping. The $p\pi$-$p\pi$ contribution in the P-X bond can be reduced by introducing pure $p\pi$-bonding with an sp^2-hybridized Y atom.

$$\overset{\diagup}{\underset{\diagup}{}}\overset{\|}{P}-\ddot{X}-Y\equiv Z \quad \longleftrightarrow \quad \overset{\diagup}{\underset{\diagup}{}}\overset{\|}{P}-\overset{\oplus}{X}=Y-\ddot{Z}^{\ominus}$$

If X has no lone pair of electrons, there is no $p\pi$-$d\pi$ contribution.

Thus, phosphorylating agents are classified into class A compounds, having no $p\pi$-$d\pi$ contribution to the P-X bond, and class B compounds, in which $p\pi$-$d\pi$ contribution is attenuated by $p\pi$-bonding. Examples of the class A compounds are β-chloroalkylphosphonates (see reactions 43 and 44). Class B is subdivided into four types of compounds: the attenuation of the $p\pi$-$d\pi$ contribution is performed 1) in the ground state; 2) by electrophilic attack on Z; 3) by oxidative attack on Z; and 4) by the loss of Z. Hence, class B1 corresponds to the classic "acyl" phosphates, such as TEPP and parathion. The latter three classes, B2, B3, and B4, involve potential "acyl" phosphates which become "acyl" phosphates by the actions of certain reagents.

Parathion (or paraoxon) is a typical example of the B1 group:

$$(RO)_2\overset{\overset{\textstyle O}{\|}}{P}OR' + HO-\underset{}{\underset{}{}}-NO_2 \tag{55}$$

(X)   (Y)(Z)

The transesterification is catalyzed by strong bases. This type of reaction may be applied to preparative purposes. Thus, demeton-methyl is elegantly prepared from parathion-methyl by reaction with β-hydroxy-ethyl ethyl sulfide in the presence of 50% aqueous sodium hydroxide at room temperature.[248]

$$(CH_3O)_2\overset{\underset{\|}{S}}{P}-O-\!\!\left\langle\!\!\bigcirc\!\!\right\rangle\!\!-NO_2 + HOCH_2CH_2SC_2H_5 \xrightarrow{\text{NaOH}} (CH_3O)_2\overset{\underset{\|}{S}}{P}OCH_2CH_2SC_2H_5 + NaO-\!\!\left\langle\!\!\bigcirc\!\!\right\rangle\!\!-NO_2 \qquad (56)$$

The compounds belonging to B2, B3, and B4 are required to be activated before phosphorylation. The examples of B2 group compounds are enol phosphates. They are activated by protonation:

$$\text{B: } + \underset{(X)\,(Y)\,(Z)}{\overset{RO}{\underset{RO}{>}}\!\!P\!\!\overset{O}{\underset{O-CR'=CR''R'''}{}}}\!\!H^{\oplus} \longrightarrow (RO)_2\overset{\underset{\|}{O}}{P}B + O=CR'-CHR''R''' \qquad (57)$$

Phosphate esters of hydroquinone resist alkaline hydrolysis, but act as phosphorylating agents under oxidizing conditions (group B3) (Equation 58). This reaction is regarded as a model of oxidative phosphorylation in biological systems.[7] However, no actual organophosphorus pesticide belonging to B3 group is known.

$$(RO)_2\overset{\underset{\|}{O}}{P}-O-\!\!\underset{(X)}{\left\langle\!\!\bigcirc\!\!\right\rangle}\!\!-\underset{(Y)\,(Z)}{O-H} \xrightarrow{-2e} (RO)_2\overset{\underset{\|}{O}}{P}-O-\!\!\left\langle\!\!\bigcirc\!\!\right\rangle\!\!-O^{\oplus} \xrightarrow{+B^{\ominus}} (RO)_2\overset{\underset{\|}{O}}{P}B + O=\!\!\left\langle\!\!\bigcirc\!\!\right\rangle\!\!=O \qquad (58)$$

Trichlorfon belongs to group B4. Its conversion to dichlorvos is attributed to intramolecular phosphorylation, initiated by the loss of chlorine ion (Z):

$$(CH_3O)_2\overset{\underset{\|}{O}}{P}\!\!\overset{OH}{\underset{}{-CH}}\!-\!C\!\!\overset{Cl}{\underset{Cl}{\overset{|}{-}Cl}} \xrightarrow{OH^{\ominus}} (CH_3O)_2\overset{\underset{\|}{O}}{P}OCH=CCl_2 + HCl \qquad (59)$$

Schrader's "acyl rule" and the P-XYZ classification of phosphorylating agents are very useful for the design of new pesticides and also for understanding the mode of action. However, they cannot be applied to some classes of organophosphorus insecticides or phosphorylating agents. For example, the high reactivity of cyclic phosphates cannot be explained by these concepts: ribonucleoside-2',3' cyclic phosphates,[35] catechol cyclic phosphates,[249] and saligenin cyclic phosphorothiolates[119] have been applied as phosphorylating agents. In five-membered cyclic phosphates, the $p\pi$-$d\pi$ contribution to P-O bonds is reduced by steric reason (see p. 73,). In conclusion, increased positive charge on the phosphorus atom and decreased $p\pi$-$d\pi$ contribution to the P-X bond are requisites for phosphorylating activity.

C. ALKYLATING PROPERTIES

Because phosphorus acids are such strong acids that their anions can serve as good leaving groups, phosphorus esters have alkylating properties. The important role of the alkylation reaction by phosphate esters may be recognized in the biogenesis of isoprenoids.[250] With organophosphorus pesticides, the alkylation reaction is significant in their chemical and biochemical degradations.[251] Certain biological activities of phosphorus esters appear to be attributable to the alkylating properties.[251] The reaction is also applied to preparative purposes.

1. Structure and Alkylating Properties

Since allyl and benzyl esters give relatively stable carbonium ions, they react readily with nucleophilic agents to alkylate them by an $S_N 1$ mechanism. The C-C bond formation in terpene biogenesis is due to the alkylation of an olefin with an allyl phosphate:[250]

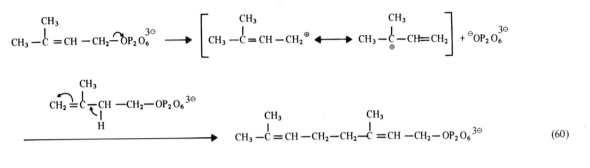

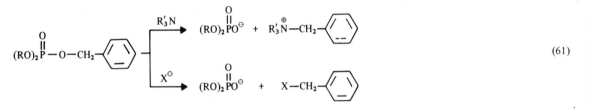

The readily removable benzyl group is used as a convenient protective group of phosphorylating agents for the preparation of biological phosphate esters. Tertiary amines such as 4-methylmorpholine[252] and salts such as lithium chloride[253] are utilized for debenzylation.

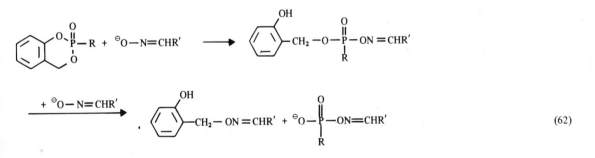

Cyclic phosphorus esters of saligenin are regarded as special benzyl derivatives. After the ring opening at P-O-C (aryl) by an initial nucleophilic attack, the resultant o-hydroxybenzyl esters are readily debenzylated by the action of nucleophiles. Thus, certain oximes such as monoisonitrosoacetone (MINA), pyridine-aldoxime methiodide (PAM), and salicylaldoxime react with saligenin cyclic phosphates to decompose them, as shown in the following scheme:[254]

Furthermore, in aqueous conditions the cyclic esters react with thiol compounds such as cysteine to give corresponding o-hydroxybenzyl thioethers (Equation 63).[255] The cyclic esters inhibit SH-enzymes like yeast alcohol dehydrogenase, probably due to a similar reaction. The reaction rate of saligenin derivatives decreases in the order: R = SR > Ar > OAr > OR >> OH. The thiono analogs are almost inactive. This order agrees with that of the susceptibility to hydrolysis. The time-course of the alkylation of cysteine shows that the reaction occurs after a considerable time lag, while the hydrolysis of the cyclic esters takes place immediately. The partially hydrolyzed products, the o-hydroxybenzyl esters, react immediately with cysteine. Therefore, the reaction mechanism is proposed as shown in Equation 63.[256]

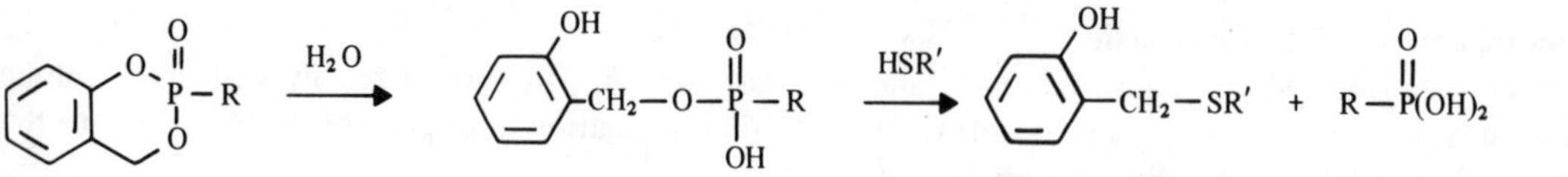

$$\text{(structure)} \xrightarrow{H_2O} \text{(structure)} \xrightarrow{HSR'} \text{(structure)} + R-\overset{O}{\underset{\|}{P}}(OH)_2 \qquad (63)$$

The presence of an electron-releasing hydroxyl group at the ortho position makes the benzyl ester much more active as an alkylating agent than unsubstituted benzyl esters, possibly due to a resonance effect stabilizing the carbonium ion.

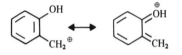

In connection with ordinary organophosphorus pesticides, the alkylation reaction is most significant in the methyl ester group, which is much more susceptible towards nucleophilic agents than ethyl and higher *n*-alkyl ester groups. It was mentioned in the preceding section that some methyl esters are dealkylated readily by solvolysis in ethanol (p. 63). In the course of the manufacturing of organophosphorus insecticides, particularly those having the methyl ester group, unwanted modifications of molecules often take place by the transfer of an alkyl group to a sulfur or nitrogen atom. When demeton-S-methyl is stored, one molecule methylates the sulfide group of another one to give a much more toxic sulfonium compound (almost 1,000 times as toxic as the original compound):[257]

$$2(CH_3O)_2\overset{O}{\underset{\|}{P}}SCH_2CH_2SC_2H_5 \longrightarrow (CH_3O)_2\overset{O}{\underset{\|}{P}}SCH_2CH_2\overset{CH_3}{\underset{+}{S}}C_2H_5 + \overset{-}{O}-\overset{O}{\underset{\underset{CH_3O}{|}}{P}}SCH_2CH_2SC_2H_5 \qquad (64)$$

demeton-S-methyl

In the presence of pyridine, which is often used as a catalyst for the preparation of phosphorus esters, parathion-methyl decomposes to form the *N*-methylpyridinium salt of *p*-nitrophenyl phosphorothioic acid:[79]

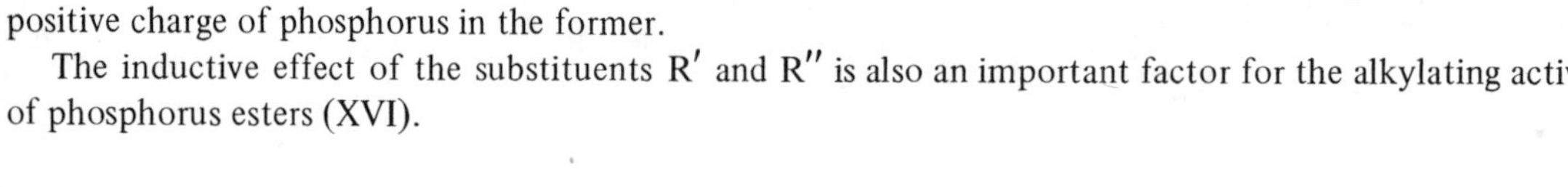

$$NO_2-\text{(aryl)}-O-\overset{S}{\underset{\|}{P}}(OMe)_2 + 2N\text{(py)} \longrightarrow \left[NO_2-\text{(aryl)}-OPSO_2\right]^{2\ominus}\left[Me-\overset{\oplus}{N}\text{(py)}\right]_2 \qquad (65)$$

Phosphate esters are generally more reactive as alkylating agents than the corresponding phosphorothionate esters, for the C-O bond of the former is weaker than that of the latter, being due to the more positive charge of phosphorus in the former.

The inductive effect of the substituents R' and R'' is also an important factor for the alkylating activity of phosphorus esters (XVI).

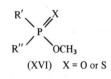

(XVI)   X = O or S

The greater the electron attractivity of the substituents is, the higher the alkylating activity of the ester. Thus, the methylating power of XVI decreases in the following order:[247]

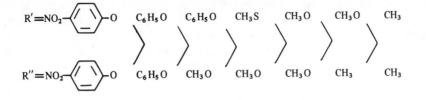

$R'=NO_2-$⟨⟩$-O$ C_6H_5O C_6H_5O CH_3S CH_3O CH_3O CH_3

$R''=NO_2-$⟨⟩$-O$ C_6H_5O CH_3O CH_3O CH_3O CH_3 CH_3

Nucleophilic displacement occurs neither on the carbon atom of an alkyl thiolate ester group nor that of an aryl ester group.

The insecticide dichlorvos has two different kinds of alkylating groups: methyl ester group and dichlorovinyl group. Iodide ion reacts readily with the methyl ester group, yielding desmethyl dichlorvos, whereas mercaptides replace the chlorine atoms (Equation 66). By reaction with potassium p-nitrophenoxide, one chlorine atom is displaced to give dimethyl 2-chloro-2-p-nitrophenoxyvinyl phosphate.[258]

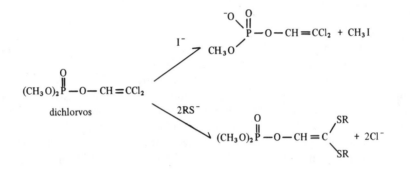

$$(CH_3O)_2\overset{O}{\overset{\|}{P}}-O-CH=CCl_2 \tag{66}$$

dichlorvos

2-Chloroethyl sulfides and 2-chloroethylamines are well known alkylating agents. Analogously, phosphorus esters of 2-alkylthio- or 2-dialkylamino-ethanols show alkylating behavior by forming cyclic ethylenesulfonium or immonium ions by intramolecular alkylation of the hetero atom (Equation 67). The cyclic onium ions are highly active alkylating agents.

$$(RO)_2\overset{X}{\overset{\|}{P}}-O-CH_2-CH_2 \longrightarrow (RO)_2\overset{X}{\overset{\|}{P}}=O + \overset{\oplus}{CH_2-CH_2}\overset{Y-R}{} \tag{67}$$

X = O or S; Y = S or NR

When the onium ions react with the sulfur atom of phosphorothioate anions (X = S), thiono-thiolo isomerization results. The cyclic ions also react with the hetero atom of the original and isomerized thiolate esters. Thus, demeton-O-methyl in water may yield a small quantity of S-(ethylthioethyl) demeton-S-methyl, which is about 10,000 times more toxic than demeton-O-methyl (Equation 68).[259]

$$(CH_3O)_2\overset{O}{\overset{\|}{P}}SCH_2CH_2\overset{C_2H_5}{\overset{|}{S}}{:} + \overset{\oplus}{S}{-}C_2H_5 \quad CH_2-CH_2 \longrightarrow (CH_3O)_2\overset{O}{\overset{\|}{P}}SCH_2CH_2\overset{C_2H_5}{\overset{|}{\underset{\oplus}{S}}}{-}CH_2CH_2SC_2H_5 \tag{68}$$

In the course of the preparation of amiton from its thiono isomer (see Chapter II, Equation 108), N-(diethylaminoethyl) amiton is similarly formed as a by-product by the reaction of amiton with the aziridinium intermediate (Equation 69).[260]

$$(C_2H_5O)_2\overset{O}{\overset{\|}{P}}SCH_2CH_2\overset{C_2H_5}{\overset{|}{N}}{:} + \overset{\oplus}{N(C_2H_5)_2}\quad CH_2-CH_2 \longrightarrow (C_2H_5O)_2\overset{O}{\overset{\|}{P}}SCH_2CH_2\overset{C_2H_5}{\overset{|\oplus}{N}}CH_2CH_2N(C_2H_5)_2 \tag{69}$$

Many chemosterilants are aziridine derivatives of phosphoric acid. The ethyleneimides are known as alkylating agents, and are more reactive than uncharged N-alkylaziridine because electron withdrawal by the phosphoryl group makes the methylene group more susceptible to the attack of nucleophiles.[261] For example, thiosulfate ion, which is one of the strongest nucleophilic agents, reacts readily with metepa in an aqueous solution:

$$O = P \left[\underset{\overset{|}{H^{\oplus}}}{N} \overset{\overset{CH_3}{|}}{\underset{CH_2}{\overset{CH}{\diagdown}}} \right]_3 + 3 S_2 O_3{}^{2\ominus} \xrightarrow{3 H_2 O} O = P \left[NHCH_2 \overset{\overset{CH_3}{|}}{CH} - S_2 O_3{}^{\ominus} \right]_3 + 3 OH^{\ominus} \qquad (70)$$

metepa

This reaction is applied for the determination of aziridine compounds. Amines, mercaptans, carboxylic acids, halogens, and alkyl halides react similarly to cause ring-opening.

2. Nucleophilic Agents

A variety of nucleophilic agents attack the alkyl group of phosphorus esters. Soft bases prefer to react with the carbon atom rather than the phosphorus atom. Amines and sodium iodide are often used for the preparation of desmethyl derivatives of organophosphorus pesticides. Various sulfur compounds are also utilized for dealkylation. O,O-Dialkyl phosphorodithioates, dithiocarbamates, mercaptans, sulfides, thiocyanates, and thiourea are useful for this purpose. For example, desmethyl fenitrothion is prepared by the reaction of fenitrothion with sodium iodide in acetone (Equation 71).[262] Phosphamidon reacts with mercaptans in an alkaline solution, resulting in the cleavage of a methyl ester linkage (Equation 72). This reaction can be applied for quantitative determination of the insecticide.[193] Dimethoate is demethylated readily by the reaction with O,O-dimethyl phosphorodithioate (Equation 73).[263] The methyl group of Salithion is removed without disruption of the cyclic ester structure by the reaction with cyclohexylamine or, more preferably, with dithiocarbamate (Equation 74).

$$(CH_3 O)_2 \overset{\overset{S}{\|}}{P} - O - \underset{CH_3}{\diagdown} - NO_2 + I^- \longrightarrow O \overset{\overset{S}{\|}}{\underset{CH_3 O}{\diagup}} P - O - \underset{CH_3}{\diagdown} - NO_2 + CH_3 I \qquad (71)$$

fenitrothion

$$(CH_3 O)_2 \overset{\overset{O}{\|}}{P} - O - \overset{\overset{CH_3}{|}}{\underset{Cl}{C}} = C - \overset{\overset{O}{\|}}{C} N(C_2 H_5)_2 + RS^- \longrightarrow {}^- O - \overset{\overset{O}{\|}}{\underset{CH_3 O}{P}} - O - \overset{\overset{CH_3}{|}}{\underset{Cl}{C}} = C - \overset{\overset{O}{\|}}{C} N(C_2 H_5)_2 + RSCH_3 \qquad (72)$$

phosphamidon

$$(CH_3 O)_2 \overset{\overset{S}{\|}}{P} - SCH_2 CONHCH_3 + (CH_3 O)_2 \overset{\overset{S}{\|}}{P} S^- \longrightarrow {}^- O \overset{\overset{S}{\|}}{\underset{CH_3 O}{P}} SCH_2 CONHCH_3 + (CH_3 O)_2 \overset{\overset{S}{\|}}{P} SCH_3 \qquad (73)$$

dimethoate

$$\underset{}{\diagdown} \overset{O}{\diagdown} \overset{\overset{S}{\|}}{P} - OCH_3 + R_2 N\overset{\overset{S}{\|}}{C} S^- \longrightarrow \underset{}{\diagdown} \overset{O}{\diagdown} \overset{\overset{S}{\|}}{P} = O + R_2 N\overset{\overset{S}{\|}}{C} SCH_3 \qquad (74)$$

Salithion

Alkoxide ions attack both phosphorus and carbon atoms in a different way than amines and sulfur compounds:[264]

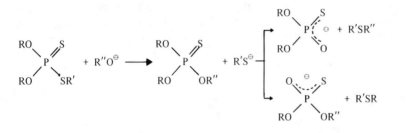

$$(75)$$

Thus, methoxide ion reacts with O,O,S-trimethyl phosphorothiolate to give an ether and O,S-dimethyl phosphorothioic acid by direct displacement at carbon atom (path a) and, furthermore, to give dimethyl sulfide and dimethyl phosphate, which are apparently formed through the reaction of trimethyl phosphate and mercaptide ion formed by initial transesterification (path b). Similarly, O,O,S-trialkyl phosphorodithioates react with sodium alkoxides to form sulfides probably through the following steps of reaction:[95]

$$(76)$$

Silver nitrate is also utilized for dealkylation of phosphorothionate esters.[265] In this reaction, a soft acid Ag^+ coordinates with the sulfur atom of the thiophosphoryl group to make the charge of phosphorus more positive and the α-carbon more susceptible to nucleophilic attack:

$$\longrightarrow (RO)_2POSAg + RONO_2 \qquad (77)$$

The aryl ester linkage of triaryl phosphorothionate is also disrupted by the action of silver nitrate. This is explained by an attack on the phosphorus atom with nitrate ion (Equation 78).[127] In the reaction of alkyl aryl phosphorothionates, however, only the alkyl ester linkage is cleaved.

$$(78)$$

By the reaction of mercuric chloride with the insecticide dioxathion, a thioacetal derivative, chlorine atoms substitute at carbon atoms to form dioxane dichloride, which is readily hydrolyzed to dioxal (Equation 79).[266] This reaction is utilized for the colorimetric determination of dioxathione.

dioxathion

$$(79)$$

3. Isomerization of Phosphorothioates: Thiono-thiolo Rearrangement

Since the phosphorothioate anion formed by the dealkylation of a phosphorothionate ester is a strong nucleophilic agent, it can react again with an alkylated product. As mentioned in Section II·D·2, dialkyl phosphorothioic acids are preferentially alkylated at sulfur atom. Thus, the realkylated product is a phosphorothiolate (step b in Equation 80). Therefore, in the course of dealkylation, phosphorothionate esters are often isomerized to phosphorothiolates, as shown in Equation 80.[247]

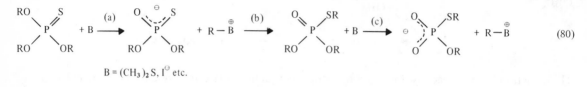

$$B = (CH_3)_2 S, I^{\ominus} \text{ etc.}$$

(80)

The phosphorothiolate can be further dealkylated by C-O bond fission, but not by C-S bond fission (step c), because carbon and sulfur have almost the same electronegativity (see Table 7) and consequently the heterolysis of the C-S bond is much more difficult than that of the C-O bond.

This type of realkylation occurs readily when dialkyl sulfides or iodide ion is used as a nucleophilic agent. Tertiary phosphines act similarly. Thus, parathion-methyl reacts with diethylphenylphosphine, by keeping for two weeks at room temperature, to form the phosphonium salt of S-methyl O-p-nitrophenyl phosphorothioic acid (Equation 81).[267] Dimethoate reacts similarly with triaminophosphines (Equation 82).[268] The thiono-thiolo isomerization is complete only if the phosphorus ester has at least two methoxy groups.

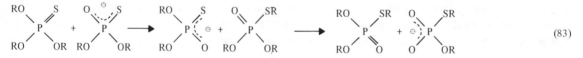

(81)

(82)

When amines such as triethylamine are used, no such realkylation occurs, for the alkyl group binds firmly with such amines. Even in such a case, the isomerization occurs by another mechanism shown in Equation 83: unreacted triesters alkylate the diester anions.[247]

(83)

This type of isomerization is observed in a stored emulsifiable preparation of parathion-methyl containing dimethylformamide.[262] The amide carbonyl is able to accept a methyl group from parathion-methyl, to form desmethyl parathion-methyl and dimethylformimidium acid methyl ester. The latter cannot realkylate the former, but is decomposed by water to dimethylamine and methyl formate. However, the desmethyl ester is methylated at sulfur atom with another molecule of parathion-methyl to form the S-methyl thiolate isomer:

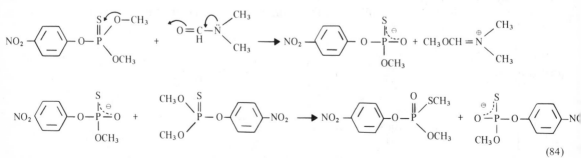

(84)

In the absence of nucleophilic agents, the sulfur atom of the thiophosphoryl group can be alkylated by the ester itself (self-alkylation). Thus, the thiono-thiolo rearrangement takes place at high temperature (120 to 180°). The reaction appears to proceed intermolecularly in accordance with Equation 85.[247]

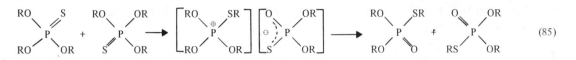

$$\tag{85}$$

One molecule of alkyl phosphorothionate ester may alkylate the sulfur atom of another molecule. The primary product, alkoxy alkylthiophosphonium ion, is disrupted at the O-alkyl bond and alkylates the sulfur atom of the phosphorothioate ion. The O-methyl ester group is transferred much more readily than ethyl and higher alkyl groups. For example, parathion-methyl is isomerized 91% into S-methyl phosphorothiolate by heating the pure crystalline compound at 150° for 6.5 hr (Equation 86).[269] Parathion is isomerized 90% by heating at 150° for 24 hr.

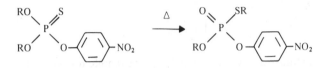

$$\tag{86}$$

Aryl ester groups do not transfer. The inductive effects of substituents influence the reactivity in the same way as other alkylation reactions. Thus, the rate of the thermal isomerization of an ester $(CH_3O)_2 \overset{\underset{\|}{S}}{P}R$ decreases in the following order:[247]

$$R = NO_2 \text{—} \langle \rangle \text{—} O > C_6H_5O > CH_3O > CH_3$$

The isomerization may occur in the course of the preparation of phosphorothionate insecticides through some processes with heat. Therefore, technical parathion may contain 5 to 20% S-ethyl isomer. This reaction also occurs even at room temperature upon long storage of methyl phosphorothionates in polar solvents. The knowledge of the thermal isomerization is very important in order to avoid being misled in the evaluation of phosphorothionate insecticides. The phosphorothiolate isomers have quite different chemical and biological properties from thiono isomers: the former are more susceptible to hydrolysis and have much higher anticholinesterase activity than the latter.

Alkyl iodides that are stronger alkylating agents than phosphorothionate esters accelerate more efficiently the thiono-thiolo conversion (Pistschimuka reaction):

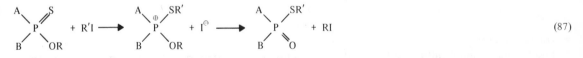

$$\tag{87}$$

When R in Equation 87 is the same as R', the reaction results in thiono-thiolo isomerization. The substituents A and B favor the reaction in the order $Cl < RS < RO < R < R_2N$, by increasing the nucleophilic activity of the sulfur atom.[127] The Pistschimuka reaction proceeds smoothly in strongly polar solvents such as dimethylformamide.[118,119] It is applicable for preparative use as described in Chapter II (D.4).

Some Lewis acids, such as $FeCl_3$, $AlCl_3$, and $SnCl_4$, also accelerate the thiono-thiolo isomerization at low temperature (Equation 88).[247]

$$2 (RO)_3PS + SnCl_4 \longrightarrow \left[\begin{array}{c} RO \\ \diagdown \\ RO \end{array} P \begin{array}{c} O \\ \diagup \\ SR \end{array} \right]_2 SnCl_4 \tag{88}$$

As discussed in Chapter II, some phosphorothionates such as demeton-O or the thiono isomer of amiton, which have a β-alkylthio- or dialkylamino-ethyl ester group, isomerize much more readily than unsubstituted alkyl or alkyl aryl phosphorothionates. In this case, the β-substituted ethyl group migrates onto the sulfur atom through the formation of a cyclic onium intermediate due to a neighboring group participation (Equation 89). The cyclic intermediate realkylates the phosphorothioate ion to form the phosphorothiolate.

$$(RO)_2 \overset{S}{\overset{\|}{P}}-O-CH_2-CH_2 \cdots :X-R \longrightarrow \left[(RO)_2 P \overset{S}{\underset{O}{\ominus}} \right]\left[\overset{\oplus}{\overset{X-R}{CH_2-CH_2}} \right] \longrightarrow (RO)_2 \overset{O}{\overset{\|}{P}}SCH_2 CH_2 XR \qquad (89)$$

X = S or NR

In the presence of a stronger nucleophile than the nitrogen or sulfur atom in the ester molecule, the self-alkylation to form the cyclic intermediate is suppressed and the smallest alkyl group is split off:[247]

$$\begin{matrix} CH_3O \\ \\ CH_3O \end{matrix} \overset{S}{\underset{OCH_2CH_2SC_2H_5}{P}} + N(CH_3)_3 \longrightarrow \left[\begin{matrix} O \cdots \cdots S \\ P \\ CH_3O \quad OCH_2CH_2SC_2H_5 \end{matrix} \right]^- \left[N(CH_3)_4 \right]^+ \qquad (90)$$

Amiton itself is relatively stable, but by prolonged storage at high temperatures (for example for 6 months at 60° or for 2 days at 100°) decomposes into tetraethylpiperazinium salts (Equation 91),[270] which may be formed from the cycloimmonium ion. The thermal decomposition of amiton at higher temperature than 165° takes place in a different way, as will be discussed in the following Section.

$$2(C_2H_5O)_2 \overset{O}{\overset{\|}{P}}SCH_2 CH_2 N(C_2H_5)_2 \overset{\Delta}{\longrightarrow} \left[(C_2H_5O)_2 \overset{O}{\overset{\|}{P}} \overset{\oplus}{=} S \right]_2, \ (C_2H_5)_2 \overset{\oplus}{N} \overset{\oplus}{N}(C_2H_5)_2 \overset{\Delta}{\longrightarrow}$$

$$\left[\begin{matrix} C_2H_5O \\ \\ C_2H_5S \end{matrix} \overset{O}{\underset{O}{P}} \ominus \right]_2 \ (C_2H_5)_2 \overset{\oplus}{N} \overset{\oplus}{N}(C_2H_5)_2 \qquad (91)$$

The sulfone derivatives of demeton-O and demeton-O-methyl are very stable and do not isomerize because the formation of the onium intermediates is impossible.

$$(C_2H_5O)_2 \overset{S}{\overset{\|}{P}}-OCH_2CH_2-\overset{O}{\underset{O}{\overset{\uparrow}{S}}}-C_2H_5 \qquad \text{demeton-O sulfone}$$

4. Transalkylation

Besides the thiono sulfur, sulfide and amino groups in phosphorus ester molecules also serve as nucleophilic centers to accept alkyl groups. Thus, demeton-S and demeton-S-methyl change into very toxic salt-like products on standing at room temperature. The principal reaction is a transalkylation, forming sulfonium salts as shown in Equation 64. The reaction is more rapid for the methyl ester than the ethyl homolog. The formation of triethylamine from amiton by heating at 165° is also explained by transalkylation of O-ethyl onto the nitrogen atom:[271]

$$2\,(C_2H_5O)_2 \overset{O}{\overset{\|}{P}}SCH_2 CH_2 N(C_2H_5)_2 \overset{\Delta}{\longrightarrow} \left[(C_2H_5O)_2 \overset{O}{\overset{\|}{P}}SCH_2 CH_2 \overset{+}{N}(C_2H_5)_3 \right]\left[\begin{matrix} ^-O \\ \\ C_2H_5O \end{matrix} \overset{O}{\underset{SCH_2CH_2N(C_2H_5)_2}{P}} \right] \longrightarrow$$

$$N(C_2H_5)_3 + 1/n\,(SCH_2CH_2)_n + ? \qquad (92)$$

Another interesting similar example of transalkylation is observed in the thermal decomposition of dimethyl isoxazolyl phosphorothionate. In this case, the methyl group does not migrate to thiophosphoryl sulfur atom, but to the nitrogen atom of isoxazole ring:

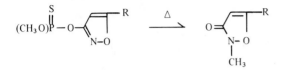

$$(93)$$

The sulfur atom of the thiolate ester group can also accept an alkyl group in transalkylation reactions.[247] Thus, dialkyl sulfide and alkyl aryl sulfide are liberated in high yields from *O*-aryl *O,S*-dialkyl phosphorothiolates (Equation 94) and *O,O*-dialkyl *S*-aryl phosphorothiolates (Equation 95) by heating, respectively.

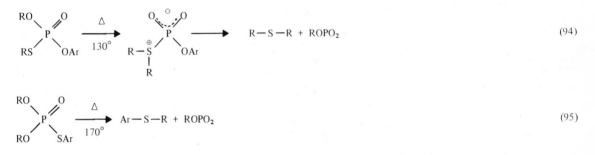

$$(94)$$

$$(95)$$

This type of rearrangement reaction also appears to occur in nature: ethyl propyl sulfide and its oxidized products were discovered as the degradation products of the nematocide-insecticide MOCAP (*O*-ethyl *S,S*-dipropyl phosphorodithiolate) in corn plants grown on soil treated with the pesticide.[272]

$$\overset{O}{\underset{\parallel}{C_2H_5OP(SC_3H_7)_2}}$$

MOCAP

The thermal intramolecular transalkylation to a thiolate sulfur atom occurs readily in *O*-β-arylthioethyl *S*-aryl phosphoramidothiolates. The participation of the neighboring sulfur atom makes the C-O bond weak, and the alkyl group transfers to the thiolate sulfur, probably through the quasi four-membered ring transition state:[273]

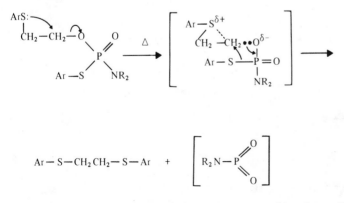

$$(96)$$

The homologous γ-thiopropyl derivatives are stable above 200°, possibly owing to the fact that the formation of a four-membered sulfonium ion is energetically much less favorable.

The *o*-hydroxylbenzyl group transfers similarly onto the thiolate sulfur atom. The thiolo isomer of

Salithion reacts with an alcohol at 60° in the presence of a tertiary amine to form hydroxybenzyl methyl sulfide and dialkyl phosphate.[119] This reaction is explained rationally by the following mechanism:

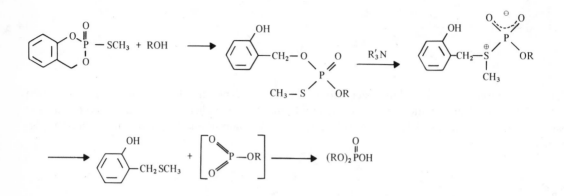

$$ (97) $$

The cyclic structure may be opened at first at the P-O (aryl) bond by the nucleophilic attack of the alcohol on phosphorus to give O-alkyl O-hydroxybenzyl S-methyl phosphorothiolate. The electron-releasing hydroxyl group at the ortho position of the benzyl ester may promote the formation of a carbonium ion. This benzyl cation may rearrange to the thiolate sulfur atom, giving the sulfide and consequently a metaphosphate, which is active enough to phosphorylate another molecule of alcohol. When a primary or secondary amine is used as a catalyst, the hydroxybenzyl group transfers to the amine:

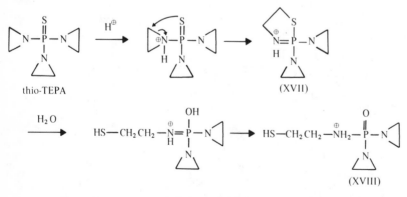

$$ (98) $$

Another type of intramolecular transalkylation reaction is proposed for the mechanism of formation of an SH-compound (XVIII) by the acid catalyzed hydrolysis of *tris*-(1-aziridinyl) phosphine sulfide (thio-TEPA).[274] The protonated aziridine ring may react with thiophosphoryl sulfur atom, forming a thiazaphospholidine intermediate (XVII) followed by hydrolytic cleavage of the P-S bond, as shown in Equation 99.

$$ (99) $$

D. OXIDATION AND REDUCTION

The oxidation of neutral organophosphorus esters usually involves the sulfur atoms of thiophosphoryl and thioether groups in the molecule. Oxidation generally makes organophosphorus pesticides more reactive or stronger inhibitors of esterases. The thiophosphoryl group is oxidatively desulfurized to phosphoryl group with various oxidants.

Nitric acid and dinitrogen tetraoxide are utilized for the preparative purpose of corresponding oxo-analogs from phosphorothionate esters (see Chapter II.A). Bromine water is very often used for the conversion of inactive phosphorothionate insecticides into potent anticholinesterases, mainly for analytical purposes. Peroxy acids are also used for the same purpose. The irradiation with ultraviolet light accelerates the oxidation and will be discussed in the following Section.

The reaction products of phosphorothionate esters with these oxidants are not always the corresponding oxo-analogs: the oxidation initiates further degradation of the esters and any susceptible groups in the side chains suffer the effects of the oxidants. For example, dimethoate does not transform to its oxo-analog, dimethoxon, by the action of bromine, but to as many as five unknown esterase inhibitors that are less polar than dimethoxon.[275] The oxidation of thioether group in side chains will be discussed later.

An attempt to isolate an oxidation intermediate which decomposes to an oxo-analog and a cleavage product was undertaken by McBain et al.[276] By a careful treatment with m-chloroperbenzoic acid under anhydrous conditions, fonofos (Dyfonate®; O-ethyl S-phenyl ethylphosphonothiolothionate) was converted to the oxo-analog (22%) and an oxygenated product (30%). The latter was supposed to be the hypothetical oxidation intermediate (XIX). However, it was later characterized as the phosphinyl disulfide (XX).[277] Herriott proposed the structure XXI as a peroxy acid oxidation intermediate of thiophosphoryl compounds.[278]

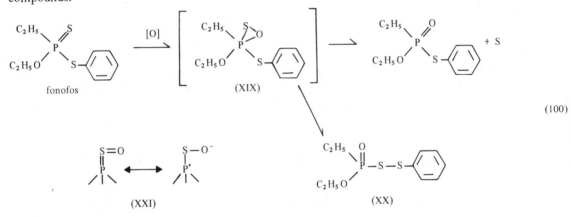

(100)

The thioether group in a side chain is oxidized to sulfoxide and then sulfone:

$$P\cdot\cdot\,C-S-C \xrightarrow{[O]} P\cdot\cdot\,C-\overset{O}{S}-C \xrightarrow{[O]} P\cdot\cdot\,C-\underset{O}{\overset{O}{S}}-C$$

(101)

A thioether group is more susceptible to oxidation than a thiophosphoryl group. Thus, fenthion is oxidized to the sulfoxide with hydrogen peroxide and to the sulfone with potassium permanganate without alteration of the thiophosphoryl group.[13] The insecticide fensulfothion is actually manufactured by the oxidation of the corresponding sulfide with hydrogen peroxide in the presence of sulfuric acid.

$$(C_2H_5O)_2\overset{S}{P}-O-\!\!\left\langle\!\!\bigcirc\!\!\right\rangle\!\!-SCH_3 \xrightarrow[H_2SO_4]{H_2O_2} (C_2H_5O)_2\overset{S}{P}-O-\!\!\left\langle\!\!\bigcirc\!\!\right\rangle\!\!-SOCH_3$$

fensulfothion

(102)

93

Demeton-O is similarly oxidized to its sulfoxide with a restricted amount of bromine in water or with 30% hydrogen peroxide at room temperature:[204]

$$(C_2H_5O)_2\overset{\overset{\displaystyle S}{\|}}{P}OCH_2CH_2SC_2H_5 + Br_2 + H_2O \longrightarrow (C_2H_5O)_2\overset{\overset{\displaystyle S}{\|}}{P}OCH_2CH_2\overset{\overset{\displaystyle O}{\uparrow}}{S}C_2H_5 + 2HBr \qquad (103)$$

demeton-O

Both sulfur atoms are oxidized with excess amounts of bromine (Chapter II, Equation 85) or with nitric acid. Phosphorothiolate sulfur atom greatly resists oxidation. Thus, demeton-S is oxidized to its sulfone with potassium permanganate, keeping the thiolate sulfur atom intact:[204]

$$(C_2H_5O)_2\overset{\overset{\displaystyle O}{\|}}{P}SCH_2CH_2SC_2H_5 \xrightarrow{K_2MnO_4} (C_2H_5O)_2\overset{\overset{\displaystyle O}{\|}}{P}SCH_2CH_2\overset{\overset{\displaystyle O}{\uparrow}}{\underset{\underset{\displaystyle O}{\downarrow}}{S}}C_2H_5 \qquad (104)$$

Phorate is oxidized to its sulfoxide and phorate O-analog sulfoxide with hydrogen peroxide, and to its sulfone and O-analog sulfone with peroxy acids or potassium permanganate.[13] Carbonphenothion is similarly oxidized with peracetic acid, forming the sulfoxide and sulfone of the O-analog and an additional three anticholinesterases.[279]

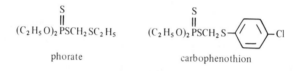

phorate   carbophenothion

In certain cases, thiolate sulfur can be oxidized. By the reaction with bromine in acetic acid, phenkapton is decomposed and the sulfur atom linked to the aromatic ring is converted to aryl sulfonic acid, while both the thiono and thiolo sulfur atoms are oxidized to sulfuric acid.[280]

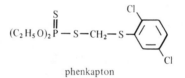

phenkapton

Under the same reaction condition, O,O-diethyl phosphorodithioic acid gives two moles of sulfuric acid, whereas O,O,S-triethyl phosphorodithioate gives only one mole of the acid. Therefore, the thiolo sulfur atom of phencapton is probably oxidized after the hydrolytic cleavage of the PS-CS bond (see Section A for the hydrolysis of this bond). The thiolate sulfur atom of S-trichloroethyl phosphorodichloridothiolate is oxidized without bond-cleavage by ozone to the sulfinylphosphoric dichloride, which was patented for preparation of pesticides:[281]

$$Cl_3CCH_2 - S - \overset{\overset{\displaystyle O}{\|}}{P}Cl_2 \xrightarrow{O_3} Cl_3CCH_2 - \overset{\overset{\displaystyle O}{\uparrow}}{S} - \overset{\overset{\displaystyle O}{\|}}{P}Cl_2 \qquad (105)$$

The S-alkyl thiolate sulfur atom in partial esters of phosphorothioic acid is very susceptible to oxidation, in contrast with that in neutral esters, probably due to high electron density of the ions. Thus, S-alkyl and O,S-dialkyl phosphorothiolates are readily decomposed oxidatively by the action of iodine.[119,282,283] In the presence of water, the phosphorothiolate esters are hydrolyzed, whereas they phosphorylate alcohols under anhydrous conditions (Equation 106). This latter reaction can be applied for the preparation of biologically important phosphate esters.

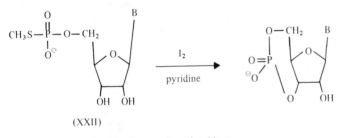

$$\text{RS-}\overset{\overset{O}{\|}}{\underset{\underset{O^{\ominus}}{|}}{P}}\text{-OR'} \xrightarrow{I_2} \left[R-\overset{\oplus}{\underset{\underset{I}{|}}{S}}-\overset{\overset{O}{\|}}{\underset{\underset{O^{\ominus}}{|}}{P}}\text{-OR'} \right] I^{\ominus} \xrightarrow{R''OH} RSI + R''O-\overset{\overset{O}{\|}}{\underset{\underset{O^{\ominus}}{|}}{P}}\text{-OR'} + HI$$

$$RSI + RS-\overset{\overset{O}{\|}}{\underset{\underset{O^{\ominus}}{|}}{P}}\text{-OR'} \longrightarrow RS^{\ominus} + R-\overset{\oplus}{\underset{\underset{I}{|}}{S}}-\overset{\overset{O}{\|}}{\underset{\underset{O^{\ominus}}{|}}{P}}\text{-OR'} \xrightarrow{R''OH} RS-SR + R''O\overset{\overset{O}{\|}}{\underset{\underset{O^{\ominus}}{|}}{P}}\text{OR'} + HI \qquad (106)$$

R', R'' = H or alkyl group

The most interesting example may be the application of the insecticide Salithion for the preparation of cyclic 3',5'-adenosine monophosphate (cyclic AMP) and its related compounds.[284] Salithion is isomerized to the more reactive S-methyl thiolate by the Pistschimuka reaction (see reaction 110 in Chapter II). The thiolate isomer reacts with adenosine (protected with borate), removing the o-hydroxybenzyl group with cyclohexylamine to form adenosine 5'-S-methyl phosphorothiolate (XXII) (see reaction 98, where R is adenosine), which is then oxidatively desulfurized with iodine under anhydrous conditions, resulting in the formation of cyclic AMP:

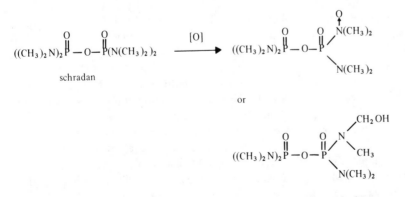

(XXII)

B = adensine or other related base

(107)

The phosphoramide group also can be oxidized. Schradan (octamethyl pyrophosphoramide) is converted by the action of neutral permanganate or other various oxidizing agents to a potent inhibitor of cholinesterase, which liberates formaldehyde upon decomposition. The oxidized anticholinesterase product was suggested as the N-oxide by Casida and his co-workers.[285] However, this is not fully characterized yet, and a methylol structure has also been proposed:[31]

$$((CH_3)_2N)_2\overset{\overset{O}{\|}}{P}-O-\overset{\overset{O}{\|}}{P}(N(CH_3)_2)_2 \xrightarrow{[O]} ((CH_3)_2N)_2\overset{\overset{O}{\|}}{P}-O-\overset{\overset{\overset{O}{\uparrow}}{\underset{N(CH_3)_2}{}}}{\underset{N(CH_3)_2}{\overset{\|}{P}}}$$

schradan

or

(108)

$$((CH_3)_2N)_2\overset{\overset{O}{\|}}{P}-O-\overset{\overset{O}{\|}}{P}\overset{CH_2OH}{\underset{N(CH_3)_2}{\overset{\diagup}{N}\diagdown CH_3}}$$

Chlorine reacts with schradan to form chlorinated anticholinesterase products.[286] The chlorinated products decompose hydrolytically to yield formaldehyde (Equation 109). The stability decreases drastically by increasing the number of chlorine atoms substituted.

$$((CH_3)_2N)_2\overset{\overset{\displaystyle O}{\|}}{P}-O-\overset{\overset{\displaystyle O}{\|}}{P}(N(CH_3)_2)_2 \quad \xrightarrow[-HCl]{Cl_2} \quad ((CH_3)_2N)_2\overset{\overset{\displaystyle O}{\|}}{P}-O-\overset{\overset{\displaystyle O}{\|}}{P}\overset{\displaystyle N \nearrow \searrow CH_3}{\underset{\displaystyle N(CH_3)_2}{}}\overset{\displaystyle CH_2Cl}{}$$

$$\xrightarrow{5H_2O} \quad 3(CH_3)_2NH + 2H_3PO_4 + CH_3NH_2 + CH_2O + HCl \tag{109}$$

It is known that some phosphate esters are rather resistant to alkaline hydrolysis, but are decomposed rapidly by the addition of an oxidizing agent. Quinol phosphates are typical examples: they are immediately decomposed, liberating quinone by the action of an oxidizing agent such as bromine.[7] The reaction may proceed as shown in the following scheme:

$$(RO)_2\overset{\overset{\displaystyle O}{\|}}{P}-O-\underset{\underline{}}{\bigcirc}-\overset{.}{O}H \; + \; Br-Br \quad \longrightarrow \quad O=\bigcirc=O + 2Br^{\ominus} + H^{\oplus} + \left[(RO)_2P=O\right]^{\oplus}$$

$$\xrightarrow{H_2O} \quad (RO)_2\overset{\overset{\displaystyle O}{\|}}{P}OH \tag{110}$$

In an aqueous sodium hypobromite solution (pH 9.5, 20°), mevinphos is rapidly hydrolyzed to dimethyl hydrogen phosphate.[287] In the absence of bromine at pH 9.5 and 20°, the hydrolysis of mevinphos is negligible. The oxidative breakdown is presumably a consequence of a process shown in XXIII.

$$(CH_3O)_2\overset{\overset{\displaystyle O}{\|}}{P}-O-\overset{\overset{\displaystyle CH_3}{|}}{\underset{\overset{\oplus}{Br}}{C}}-CHCOOCH_3 \qquad (XXIII)$$

Bromine adds to dichlorvos to give a dibromide which is utilized as an insecticide under the name of naled (Dibrom®).

$$(CH_3O)_2\overset{\overset{\displaystyle O}{\|}}{P}-O-CH=CCl_2 \; + \; Br_2 \quad \longrightarrow \quad (CH_3O)_2\overset{\overset{\displaystyle O}{\|}}{P}-O-CHBr-CBrCl_2 \tag{111}$$

$$\text{dichlorvos} \qquad\qquad\qquad\qquad\qquad \text{naled}$$

Reduction generally causes organophosphorus pesticides to lose reactivity. Benzyl esters are debenzylated by catalytic hydrogenation. Thiophosphoryl sulfur is reduced to hydrogen sulfide upon boiling the compound with 48% hydrobromic acid. This reaction is applied for the residue analysis of diazinon, dimethoate, and azinphos: the H_2S is converted into methylene blue by reaction with dimethyl-p-phenylenediamine and ferric chloride.[288] Phosphorodithioates give two moles of H_2S.

The aromatic nitro group in organophosphorus pesticides such as parathion, fenitrothion, and EPN is readily reduced to an amino group. This renders the pesticides inactive. The reduction occurs chemically, electrochemically, and biologically. After the reduction of parathion into the amine with zinc and hydrochloric acid, diazotization or diazotization followed by azo-coupling is applied for the determination of the pesticide.[288] Recently, thiourea dioxide was proposed as a reducing agent suitable for the automated parathion analysis.[289] It rapidly reduces nitro group at mild alkaline pH to preclude parathion hydrolysis. It has been suggested that thiourea dioxide is converted by the action of alkali to a hypothetical sulfoxylate which is the actual reducing agent:

$$(NH_2)_2C=SO_2 \quad \xrightarrow{2OH^-} \quad (NH_2)C=O + SO_2{}^{2-} + H_2O \tag{112}$$

The reduction of the nitro group also occurs at the dropping mercury electrode and gives two polarographic waves in acid solutions.[290] The first wave corresponds to the reduction of the nitro group to hydroxylamine and the second wave is the reduction of the hydroxylamine to the amine.

$$R-NO_2 + 4e + 4H^+ \longrightarrow R-NHOH + H_2O \qquad (113)$$

$$R-NHOH + 2e + 2H^+ \longrightarrow R-NH_2 + H_2O \qquad (114)$$

The sulfoxide group is readily reduced to sulfide with some reducing agents in acid solutions. Zinc in acetic acid, hydroiodic acid, and hydrochloric acid in ethanol are often utilized for the reduction. Titanous chloride also reduces sulfoxides and is applied for the determination of oxydemeton-methyl;[288] the residual amount of titanous chloride is titrated with ferric chloride after the reaction is completed. Ammonium thiocyanate acts as a catalyst for the reduction. Under certain environmental conditions, the reduction of sulfoxides can take place in plants and soils (see IV.B.1.c and 2.c).

E. PHOTOCHEMICAL REACTIONS

Among the physical forces that are responsible for the chemical change of pesticides in the environment, solar radiation is the most powerful. Although the atmosphere effectively absorbs ultraviolet light of short wavelengths (shorter than 290 nm), sufficient energy exists within the range of ultraviolet (UV) sunlight wavelengths (290 to 450 nm) to bring about many chemical transformations of organophosphorus pesticides:[291] oxidation of thiophosphoryl and thioether groups, cleavage of ester and other linkages, thiono-thiolo and *cis-trans* isomerizations, and polymerization. The products of photochemical reactions and their rate of formation are greatly affected by many factors such as light intensity, the wavelength of light, irradiation time, the state of the chemicals (for example as a thin film or in solution), the kind of supporting medium or solvent, pH of solution, and the presence of water, air, and photosensitizers.

The irradiation of UV light causes many organophosphorothionate pesticides to be converted to more potent enzyme inhibitors.[292,293] This is due to photooxidation and photoisomerization. The photochemical conversion of the thiophosphoryl group to the phosphoryl group has been demonstrated with parathion-methyl, dimethoate, and EPN.[293,294] For dimethoate there is a discrepancy in literatures, probably due to a difference in experimental conditions. Mendoza and his co-workers reported that the esterase-inhibition photoproduct of dimethoate was chromatographically different from dimethoxon.[275] Dauterman described the conversion of dimethoate to dimethoxon by exposing the thin film to air even in the absence of UV light.[295]

The thioether group in the side chain is photocatalytically oxidized to sulfoxide and sulfone. The photooxidation is observed with dialkyl thioethers (phorate, disulfoton, thiometon), alkyl aryl thioethers (fenthion, carbophenothion), and diaryl thioethers (Abate®).[296,297] This appears to occur more rapidly than the oxidative desulfuration of the thiophosphoryl group. Thus, with fenthion on plants, photooxidation takes place at the thioether group while oxidative desulfuration of the thiophosphoryl group is caused by enzymatic oxidation.[298]

Recently, a photochemical oxidation of an alkyl side chain was reported. By irradiation with ultraviolet light, diazinon in a film is converted into hydroxydiazinon (diethyl 2-(2'-hydroxy-2'-propyl)-4-methyl-6-pryrimidinyl phosphorothionate).[299] Hydroxydiazinon is as biologically active as the parent compound.

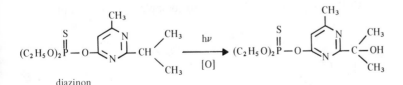

diazinon

$$(115)$$

Although some of the UV-irradiation products are identical with chemical oxidation products, some other irradiation products do not correspond to known oxidation products.[277,296]

Thiono-thiolo isomerization is also induced by UV-irradiation, at least with methyl esters such as parathion-methyl,[269,295,300] though it is not so extensive as that due to heat exposure.

$$(CH_3O)_2 \overset{\overset{S}{\|}}{P}-O-\langle\ \rangle-NO_2 \xrightarrow{h\nu} \overset{CH_3S}{\underset{CH_3O}{>}}\overset{O}{\underset{}{\overset{\|}{P}}}-O-\langle\ \rangle-NO_2 \qquad (116)$$

parathion-methyl

Moreover, the production of *S-p*-nitrophenyl isomer has also been suggested.[300a] A different kind of photocatalytic thiono-thiolo rearrangement is observed with phoxim, in which the oxyimino form changes into the thioimino form:[301]

$$(C_2H_5O)_2 \overset{\overset{S}{\|}}{P}-O-N=\overset{\overset{CN}{|}}{C}-\langle\ \rangle \xrightarrow{h\nu} (C_2H_5O)_2 \overset{\overset{O}{\|}}{P}-S-N=\overset{\overset{CN}{|}}{C}-\langle\ \rangle \qquad (117)$$

phoxim

An unusual rearrangement from thiol esters to thiono esters is also reported with *S*-benzyl phosphorothiolates.[302] Organophosphorus fungicides Kitazin P (*S*-benzyl diisopropyl phosphorothiolate) and Inezin (*S*-benzyl ethyl phenylphosphonothiolate) are converted into corresponding thiono esters by the irradiation of UV light of 254 nm (Equations 118, 119). However, the isomerization scarcely takes place when the fungicides are irradiated at 360 nm. This type of isomerization does not take place with S-phenyl phosphorothiolates such as the fungicide edifenphos (*S,S*-diphenyl ethyl phosphorodithiolate).

$$(i\text{-}C_3H_7O)_2 \overset{\overset{O}{\|}}{P}-S-CH_2-\langle\ \rangle \xrightarrow[254\ nm]{h\nu} (i\text{-}C_3H_7O)_2 \overset{\overset{S}{\|}}{P}-O-CH_2-\langle\ \rangle \qquad (118)$$

Kitazin P

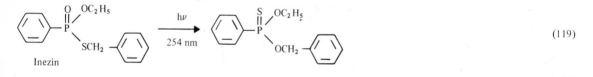

$$(119)$$

Inezin

It is well known that *cis-trans* isomerization is induced by UV-irradiation. Such an example in organophosphate esters is the photocatalyzed isomerization of mevinphos. Starting with either the *cis* or *trans* crotonate isomer of mevinphos, ultraviolet irradiation yields a mixture of approximately 30% *cis* and 70% *trans* isomers.[60]

$$(C_2H_5O)_2 \overset{\overset{O}{\|}}{P}-O-\overset{\overset{CH_3}{|}}{C}=CHCOOC_2H_5 \xrightarrow{h\nu} (C_2H_5O)_2 \overset{O}{\underset{}{\overset{\|}{P}}}O-\overset{\overset{CH_3}{|}}{C}=\overset{\overset{COOC_2H_5}{|}}{CH} + (C_2H_5O)_2 \overset{O}{\underset{}{\overset{\|}{P}}}O-\overset{\overset{CH_3}{|}}{C}=CH$$
$$\qquad\overset{|}{COOC_2H_5}$$

mevinphos   *cis* 30%   *trans* 70% (120)

Under natural conditions, the *cis-trans* conversion of chlorofenvinphos (XXIV; R = C_2H_5, Cl_n = 2,4-dichloro) and tetrachlorvinphos (XXIV; R = CH_3, Cl_n = 2,4,5-trichloro) took place on foliage, probably due to sunlight.[303]

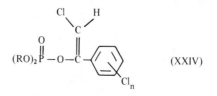

(XXIV)

In the presence of water or moisture, UV-irradiation causes hydrolysis of phosphorus esters. The hydrolysis usually occurs in the most acidic ester linkage. Parathion and EPN yield *p*-nitrophenol. When chloropyrifos (Dursban®; diethyl 3,5,6-trichloro-2-pyridyl phosphorothionate) is exposed to ultraviolet light or sunlight, it undergoes hydrolysis in the presence of water, with the liberation of 3,5,6-trichloro-2-pyridinol.[304] The pyridinol is further photolytically dehalogenated with the formation of a series of diols, triols, and tetraol, which are readily oxidized, resulting in the cleavage of the pyridinol ring with the liberation of carbon dioxide (Equation 121). In the dry state, the photolytic decomposition of chloropyrifos is less than 2% after 1,200 hr irradiation.

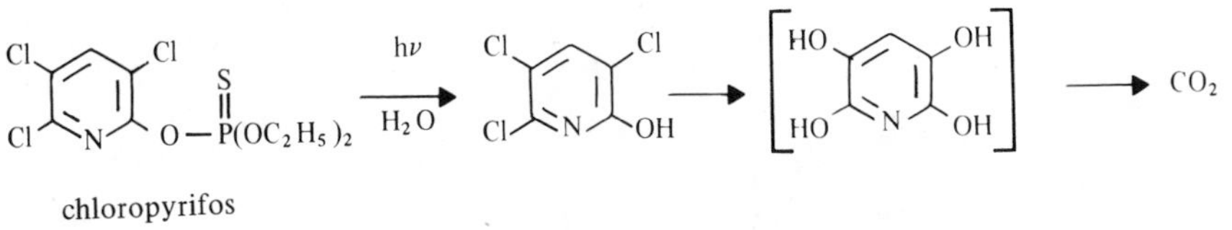

$$\tag{121}$$

chloropyrifos

Irradiation with UV light around 360 nm causes edifenphos, an *S*-phenyl phosphorothiolate, to degrade mainly by cleavage of the P-S linkage under a variety of conditions that include aqueous solution, hexane solution, and dry film (Equation 122).[302] *S*-Benzyl phosphorothiolates are, however, disrupted at the S-C bond as well as at the P-S bond.[302] Thus, Inezin yields both *O*-ethyl phenylphosphonothioate and phosphonate as photodecomposition products (Equation 123). Similarly, Kitazin P gives *O,O*-iisopropyl phosphorothioate and phosphate (Equation 124).

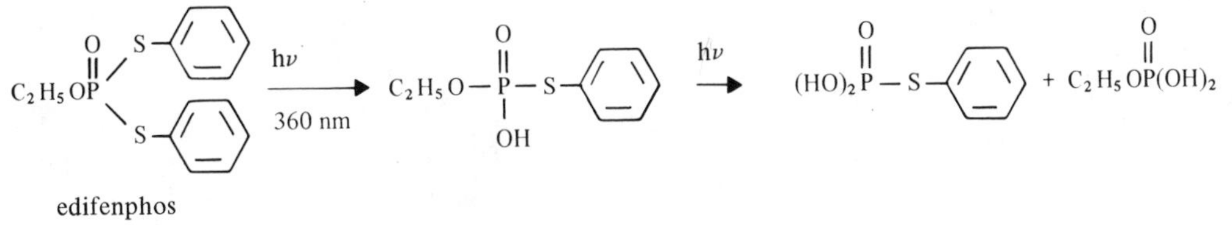

$$\tag{122}$$

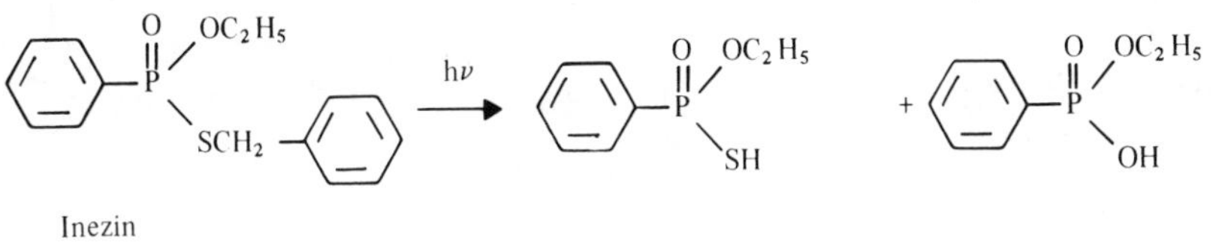

edifenphos

$$\tag{123}$$

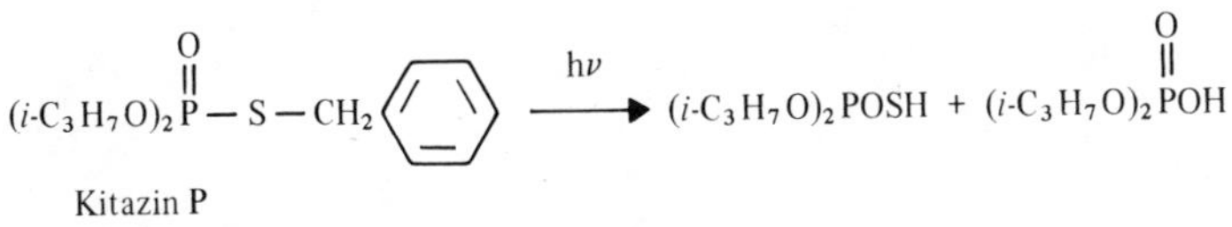

Inezin

$$\tag{124}$$

Kitazin P

Solvent interaction during photolysis is also important because photodegradation products are sometimes different, depending on the solvent used. For example, by UV-irradiation in a hexane solution, azinphosmethyl decomposes reductively to benzazimid and trimethyl phosphorothioate, together with small amounts of other products, by abstracting hydrogen from the solvent (Equation 125). However, irradiation in methanol produces a large proportion of the compound XXV.[291]

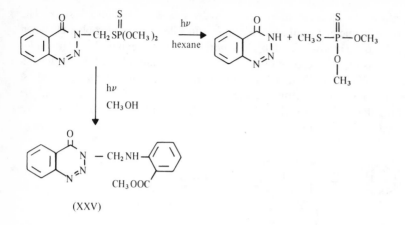

(125)

(XXV)

F. THERMAL DECOMPOSITION

The thiono-thiolo rearrangement is the most important reaction induced by heat, and was described in the preceding Section (III.C.3). Another type of reaction taking place by heating phosphorus esters is thermal decomposition. The reaction is principally *cis*-elimination to form olefins, and may proceed through cyclic transition states as do carboxylic esters and xanthates (Chugaev reaction) (Equation 126). Thus, the *trans* form of dioxathion undergoes pyrolysis more readily at 135 to 140° than the *cis* isomer, which decomposes at 160 to 165°.[96]

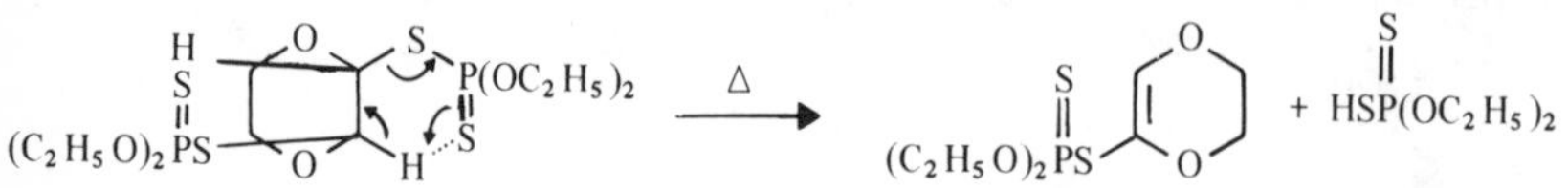

$$(C_2H_5O)_2PS \text{—} \quad \xrightarrow{\Delta} \quad (C_2H_5O)_2PS \quad + \quad HSP(OC_2H_5)_2$$ (126)

trans dioxathion

Heating alkyl diaryl phosphates similarly results in the formation of an olefin and diaryl phosphate.[305]

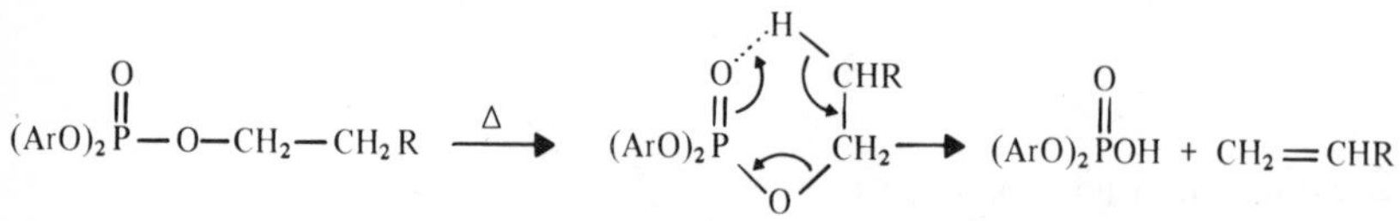

$$(ArO)_2\overset{O}{\underset{\|}{P}}\text{—}O\text{—}CH_2\text{—}CH_2R \xrightarrow{\Delta} (ArO)_2P \quad \longrightarrow \quad (ArO)_2\overset{O}{\underset{\|}{P}}OH + CH_2\text{=}CHR$$ (127)

Pyrophosphate esters undergo a different type of thermal rearrangement, as shown in Equation 128:[295]

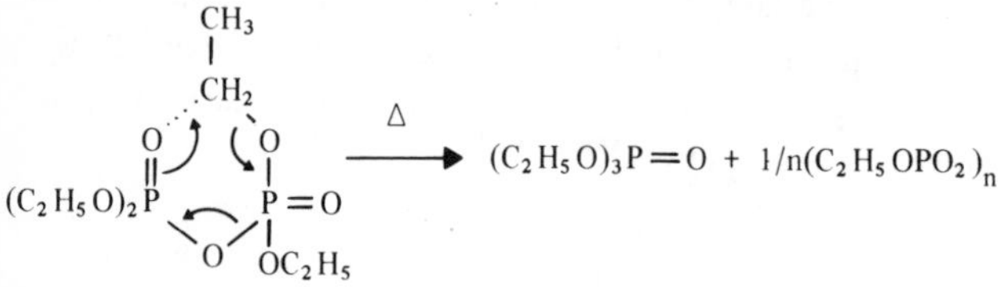

$$(C_2H_5O)_2P \quad \xrightarrow{\Delta} \quad (C_2H_5O)_3P\text{=}O + 1/n(C_2H_5OPO_2)_n$$ (128)

Beckman-type rearrangements take place readily by heat with the phosphate esters of acetophenoxime. The rearranged products, *N*-phenylacetimidoyl phosphates, are potent inhibitors of acetylcholinesterase (see IV.A.3.b).[306]

$$(RO)_2 \overset{\displaystyle C \equiv O}{\underset{O}{\underset{|}{\overset{|}{P}}}} \longrightarrow (RO)_2 PO_2^- \quad \left[\cdots \right] \longrightarrow (RO)_2 \overset{O}{\overset{\|}{P}} - O - \overset{CH_3}{\underset{|}{C}} = N - \text{—} \quad (129)$$

G. REACTIONS USEFUL FOR ANALYSIS

The analytical methods of pesticides are reviewed comprehensively in a series of the books *Analytical Methods for Pesticides, Plant Growth Regulators, and Food Additives*, edited by Zweig. This book deals only with the principles of the chemical reactions useful for analysis.

1. Color Reactions and Some Related Reactions
a. Total Phosphorus and Active Esters

Phosphomolybdate — Organophosphorus compounds may be degraded into orthophosphoric acid, which can be colorimetrically determined. The digestion may be performed by one of the following methods: oxygen flask combustion[307] or heating with a mixture of acids or oxidizing agents such as perchloric acid,[308] sulfuric acid,[309] nitric acid,[310] and hydrogen peroxide.[311,312]

Phosphoric acid reacts with ammonium molybdate to form yellow ammonium phosphomolybdate:

$$PO_4^{-3} + 12MoO_4^{-2} + 3NH_4^+ + 24H^+ \longrightarrow (NH_4)_3PO_4 \cdot 12MoO_3 + 12H_2O \quad (130)$$

The phosphomolybdate complex yields an intense blue color (molybdenum blue) by reduction in an acid solution. The molybdenum blue is a reduced molybdenum oxide complex, for example, $Mo_2O_5 \cdot 3MoO_3$, but the constitution and color change depends on the conditions ($\lambda_{max} = 650$ to 820 nm). A variety of reducing agents are utilized: amidol (2,4-diaminophenol),[312] 1-amino-2-naphthol-4-sulfonic acid,[308] benzidine,[314] hydrazine,[315] hydroquinone,[316] and stannous chloride.[317] This reaction is also applied for visualization of phosphorus esters on paper and thin-layer chromatograms. The reduction of phosphomolybdenum on the chromatogram is often performed by hydrogen sulfide[318] or irradiation with ultraviolet light,[319] besides the above mentioned reducing agents. The decomposition of phosphorothionates which resist hydrolysis is accelerated by preliminary treatment of chromatograms with N-bromosuccinimide.

In the presence of safranine, phosphomolybdate forms a complex with safranine which precipitates. The precipitate is soluble in acetone and colorimetrically determined at 535 nm.[320] A mixture of ammonium molybdate and ammonium metavanadate reacts with phosphoric acid to form yellow phosphomolybdovanadate ($\lambda_{max} = 400$ to 470 nm), which is more stable than molybdenum blue and is useful for colorimetric determination of organophosphorus pesticides.[321,322]

Amine-peroxide reaction — Various organophosphorus compounds accelerate the rate of oxidation of amines such as benzidine and o-dianisidine to form a colored substance in an alkaline peroxide solution (Schönemann reaction). In practice, sodium perborate is used as peroxide reagent. The reaction probably proceeds through the formation of a perphosphoric acid which is responsible for the oxidation of the amine.[323]

$$2 \underset{R'}{\overset{R}{>}}\!\!\overset{O}{\overset{\|}{P}}\!\!-X + 2 OOH^{\ominus} \xrightarrow{-X^{\ominus}} 2 \underset{R'}{\overset{R}{>}}\!\!\overset{O}{\overset{\|}{P}}\!\!-OOH \xrightarrow{2RNH_2} R-N=N-R + 2 \underset{R'}{\overset{R}{>}}\!\!\overset{O}{\overset{\|}{P}}OH + 2H_2O \quad (131)$$

The amine-peroxide reaction may be applicable to many organophosphorus pesticides having a partial positive charge on the phosphorus atom allowing the nucleophilic displacement of labile perhydroxyl ion.

The oxidation of indole also takes place in an alkaline peroxide solution in the presence of organophosphorus compounds. It is converted into a fluorescent product, indoxyl, which is further oxidized to indigo.[324] The first step of the oxidation of indole requires rigorous oxidation conditions, although the following steps do not.

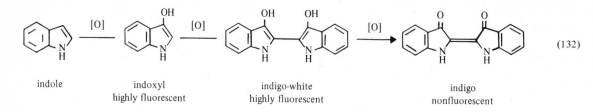

indole   indoxyl   indigo-white   indigo
    highly fluorescent highly fluorescent nonfluorescent

Luminol is similarly utilized for the determination of organophosphorus pesticides by its chemilumine-scence reaction.[325]

4-(*p*-Nitrobenzyl)pyridine – 4-(*p*-Nitrobenzyl)pyridine reacts with phosphorus esters to form a blue dye (λ_{max} 625 nm) in the presence of a base.[326] This is also useful as a general chromogenic reagent for detection of μg amounts of thio- and nonthio organophosphate pesticides on flat-bed chromatograms[327] and for colorimetry.[328] The mechanism of this reaction was proposed by Kramer and Gamson as follows:[326]

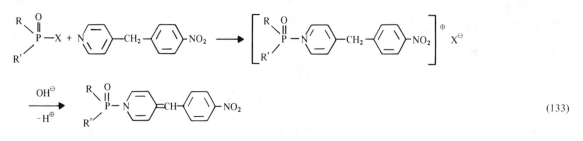

(133)

As nitrobenzylpyridine is a highly sensitive reagent for alkylating agents,[329] and phosphorus esters have alkylating activity as mentioned in Section C, it is possible that the blue dye formed is an *N*-alkylated product rather than a phosphorylated one.

b. Phosphorothioates

The sulfur atom of phosphorothionates is liberated as hydrogen sulfide by heating with hydrobromic acid. The hydrogen sulfide reacts with dimethyl-*p*-phenylenediamine in the presence of an oxidizing agent, ferric chloride, to form methylene blue with a maximum absorption at 670 nm.[288]

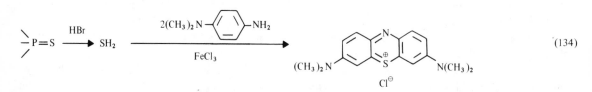

(134)

Sulfur containing compounds involving phosphorothioate esters form yellow complexes with palladium chloride. This reaction is utilized for visualization on thin-layer chromatograms[330] and also for colorimetric determination of phosphorothioates.[331] 2,6-Dibromo-*N*-chloro-*p*-quinoneimine (DCQ) reacts with thiophosphoryl compounds to give a reddish brown color.[332] Phosphorothiolates give no color or a yellow color. The thio ether group gives a yellowish color. Thiophenols give an orange color, while phenols give a blue color. This color reaction is useful for flat-bed chromatography and also for colorimetric determination.

Thiophosphoryl sulfur atom reacts readily with bromine or *N*-bromosuccinimide. Therefore, when the chromatogram is sprayed with a fluorescein solution after treating it with the bromine reagent, fluorescein is brominated to form eosine as the background, whereas the spot of phosphorothionate is visualized under ultraviolet light due to unchanged fluroescein.[333]

Thiophosphoryl compounds give blue or magenta spots on the chromatogram by spraying with an acetone solution of silver nitrate and bromocresol green followed with a 0.01% citric acid solution.[334]

Although sulfur compounds will react with the reagent to give brown spots, no compounds other than organophosphorothionates are known to give a blue or magenta spot. For chromogenic spray reagents, a comprehensive review has appeared.[333]

O,O-Dialkyl phosphorodithioic acids react with copper ion to form a yellow colored chelate which is extractable into carbon tetrachloride. The color is measured at 420 nm. As malathion, ethion, and azinphos form corresponding phosphorodithioic acids by alkali hydrolysis (see Section III.A.2a), they can be determined by this color reaction.[207]

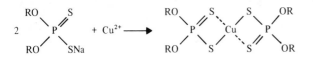

(135)

Certain phosphorodithioates, such as dimethoate, are hydrolyzed with alkali to form thiols, for which a variety of color reactions are known. For example, they react with nitrite to form *S*-nitroso compounds, which diazotize sulfanilamide in the presence of mercury acetate and form an azo dye.[335]

c. Organic Moiety

Functional groups on the organic moiety give specific color reactions which may be useful for identification and determination of organophosphorus pesticides.

Amines and nitro compounds — Aliphatic secondary amines react with carbon disulfide and cupric ion in an alkali solution to form brown cupric dithiocarbamates, which are soluble in organic solvents. Thus, schradan is hydrolyzed to dimethylamine and is determined by this color reaction.[336]

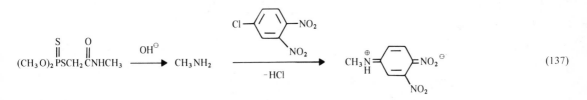

(136)

Dimethoate is hydrolyzed to liberate methylamine. The amine reacts with 2,4-dinitrochlorobenzene to give a yellow product with a maximum absorption at 530 nm.[288]

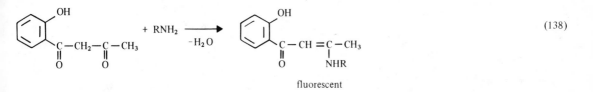

(137)

Primary aliphatic amines react with *o*-acetoacetylphenol to form a strong fluorescent product.[337] At residue levels, no reaction was observed with secondary amines, primary aromatic amines, ammonia, or amino acids. The reaction does not take place in the presence of water and, therefore, it is necessary to add a water scavenger, 2,2-dimethoxypropane. This fluorogenic reaction may be applicable for the determination of pesticides containing a primary amine as a carboxy amide or a phosphoramide such as dimethoate, crufomate (Ruelene®), and DMPA (Zytron®).

$$
\begin{array}{c}
\text{OH} \\
\text{C—CH}_2\text{—C—CH}_3 \\
\text{O} \qquad \text{O}
\end{array}
+ \text{RNH}_2 \xrightarrow[-\text{H}_2\text{O}]{}
\begin{array}{c}
\text{OH} \\
\text{C—CH} = \text{C—CH}_3 \\
\text{O} \qquad\quad \text{NHR}
\end{array}
\qquad (138)
$$

fluorescent

Aromatic nitro compounds are readily reduced to amines, which give an intense color by coupling with phenols or aromatic amines after diazotization. Parathion and similar nitrophenyl phosphorus esters can be converted into azo dyes without hydrolysis by coupling with N-naphthylethylenediamine after reduction with zinc and acid followed by diazotization (Averell-Norris' method).[338] The maximum absorption wavelength of the dye derived from parathion (555 nm) is different from that derived from p-nitrophenol (585 nm). The yellow color of nitrophenolate ion (λ_{max} = 400 nm) can also be utilized for the colorimetric determination of parathion and related pesticides after alkaline hydrolysis.

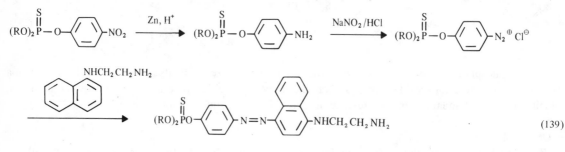

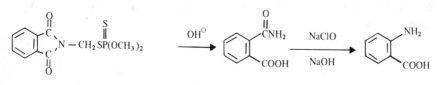

(139)

The azo coupling reaction is applicable to the colorimetric determination of azinphosmethyl, too. As mentioned in Section A.2.a, azinphosmethyl is hydrolyzed in dilute alkali at room temperature to form anthranilic acid, an aromatic amine, which couples with N-(1-naphthyl)ethylenediamine after diazotization.[339] Anthranilic acid itself is highly fluorescent and can also be determined fluorimetrically.

Another type of coupling reaction of anthranilic acid is proposed for the colorimetric determination of phosmet (Imidan®).[315] Phosmet is hydrolyzed to phthalamic acid, which is converted to anthranilic acid by the action of hypochlorite (Hofmann rearrangement). Anthranilic acid is coupled with 3-methyl-2-benzothiazolone-hydrazone in the presence of ferric chloride, an oxidizing agent, to produce a magenta colored product with an absorption maximum at 570 nm.

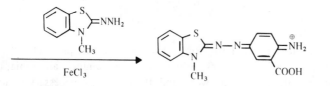

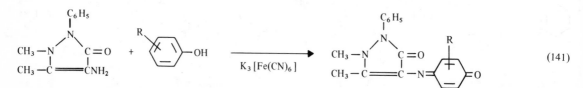

(140)

Phenols — Phenols liberated from phosphorus esters by hydrolysis serve various color reactions. One of the most useful reagents for this purpose is 4-aminoantipyrine, which yields a red dye with phenols in the presence of an oxidizing agent, potassium ferricyanide.[340]

(141)

Substituents on the para position to the hydroxyl group prevent the reaction, except for halogen, carboxyl, sulfonic acid, hydroxyl, and methoxyl, which are probably expelled in the course of reaction. Many phenolic phosphorus insecticides, including abate, bromophos, ronnel, and Salithion, are determined by this reaction after hydrolysis.

For visualization of phenols on flat-bed chromatograms the azo coupling reaction is convenient (Equation 142). Diazotized sulfanilic acid or p-nitraniline is often used. The color deepens by alkali treatment. Dibromo-N-chloroquinoneimine (DCQ) is also utilized: it condenses with phenols to give blue indophenols (Equation 143).[314]

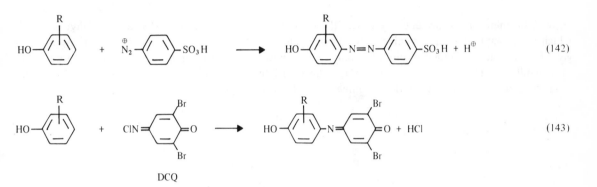

Thiols — Mercaptides that are typical soft bases combine readily with palladium, a soft acid, to make stable compounds. Thus, butyl mercaptan produced by the hydrolysis of the defoliant DEF (S,S,S-tributyl phosphorotrithiolate) reacts with a palladium chelate of 8-hydroxy-5-quinoline sulfonic acid (nonfluorescent) to tie up part of the palladium, freeing a corresponding amount of the complexing agent. On addition of magnesium chloride, a fluorescent magnesium chelate is formed.[341]

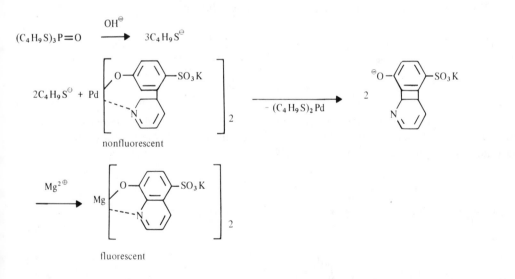

Thiophenols give a variety of color reactions. DCQ gives an orange color with thiophenols and is applied to the determination of carbophenothion and phenkapton: p-chlorothiophenol and 2,5-dichlorothiophenol, respectively, are produced by alkaline hydrolysis. Thiophenols may be derivatized into a green dye, thioindigo, by the reaction with chloroacetic acid followed by condensation in sulfuric acid.[342] The reaction process from phenkapton is shown by the following scheme:

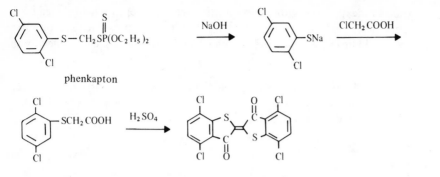

With bromine in the presence of water, thiophenols are oxidatively converted to a sulfonyl bromide, which reacts with potassium cyanide to form cyanogen bromide. Cyanogen bromide reacts with pyridine to give the unstable *N*-cyanopyridinium bromide, which, with an aromatic amine, gives the deep colored anile of pentadienal. This reaction is utilized for the colorimetric determination of phenkapton.[342] As the aromatic amine, benzidine, phenylenediamine, or aniline may be used.

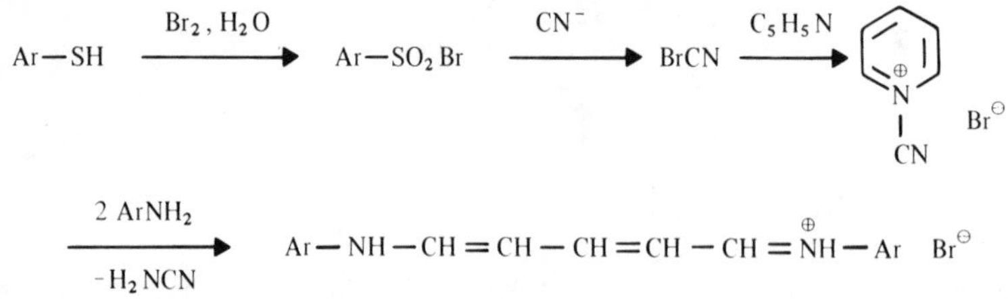

$$Ar-SH \xrightarrow{Br_2, H_2O} Ar-SO_2Br \xrightarrow{CN^-} BrCN \xrightarrow{C_5H_5N}$$

(146)

$$\xrightarrow[-H_2NCN]{2\ ArNH_2} Ar-NH-CH=CH-CH=CH-CH=\overset{\oplus}{N}H-Ar \quad Br^{\ominus}$$

Carbonyl compounds — The color reaction of cyanogen bromide is also applied to the determination of mevinphos.[343] Mevinphos is hydrolyzed with alkali, giving methyl acetoacetate, which is brominated on the active methylene. After removing excess bromine with phenol, the brominated compound reacts with sodium cyanide to give cyanogen bromide. Cyanogen bromide gives a dye, as mentioned above, with Aldridge's reagent consisting of pyridine and benzidine or phenylenediamine.

$$(CH_3O)_2\overset{O}{\underset{}{P}}-O-\overset{CH_3}{\underset{}{C}}=CH-\overset{O}{\underset{}{C}}OCH_3 \xrightarrow{NaOH} CH_3\overset{O}{\underset{}{C}}-CH_2-\overset{O}{\underset{}{C}}OCH_3 \xrightarrow{Br_2} CH_3\overset{O}{\underset{}{C}}-\underset{Br}{\overset{}{C}}H-\overset{O}{\underset{}{C}}OCH_3 \xrightarrow{CN^\ominus} BrCN$$

(147)

mevinphos

Another vinyl phosphate insecticide, phosphamidon, 2-chloro-2-diethylcarbamoyl-1-methylvinyl dimethyl phosphate, is hydrolyzed with alkali to form a readily oxidizable α-hydroxyketone, which reduces tetrazolium chloride to give a blue formanzane.[288]

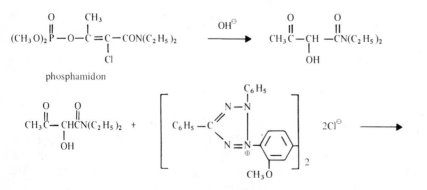

tetrazolium chloride
(colorless, water soluble)

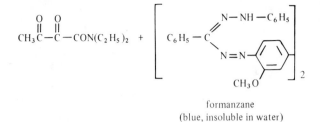

$$(148)$$

formanzane
(blue, insoluble in water)

Carbonyl compounds show many characteristic reactions which may be useful for identification and determination. Hydroxylamine hydrochloride reacts with a carbonyl compound to form an oxime, liberating an equivalent of hydrochloric acid, which may be determined by titration. This can be applied for the formulation analysis of mevinphos, which is decomposed by heating with 30% sulfuric acid to split off one mole of acetone.[288]

$$(CH_3O)_2\overset{O}{\overset{\|}{P}}O-\overset{CH_3}{\overset{|}{C}}=CHCO_2CH_3 \xrightarrow{H^\oplus} CH_3COCH_3 \xrightarrow{NH_2OH \cdot HCl} (CH_3)_2C=NOH + HCl \qquad (149)$$

2,4-Dinitrophenylhydrazine reacts with a carbonyl compound to form a colored phenylhydrazone, which may be determined colorimetrically. As mentioned in Section III.C.2, dioxathion is decomposed by the action of mercuric chloride, giving dioxal, which is converted to 2,4-dinitrophenylhydrazone (λ_{max} = 614 nm).[266]

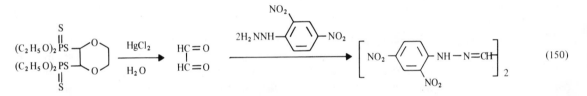

$$(150)$$

Several phosphorus esters produce formaldehyde by hydrolysis (p. 67). Formaldehyde reacts with chromotropic acid by warming in a strong sulfuric acid solution to form a violet-pink color (λ_{max} = 570 nm). The reaction probably proceeds through the initial condensation of phenolic chromotropic acid with formaldehyde, followed by oxidation to a quinoidal compound:[314]

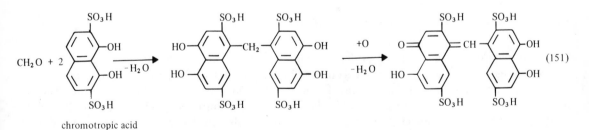

$$(151)$$

chromotropic acid

The chromotropic acid reaction is applied to the colorimetric determination of phorate. The residue of phorate, including oxidized metabolites (O-analog, sulfoxide and sulfone), is first oxidized to O-analog sulfone with perbenzoic acid before alkaline hydrolysis to form formaldehyde in order to estimate the total residue.[288]

Miscellaneous — The pyrolysis of trichlorfon at 550° produces chloroform, which reacts with pyridine in the presence of aqueous caustic alkali to give a bright red color (Fujiwara reaction).[344] The reaction probably proceeds through the cleavage of the pyridine ring after addition of chloroform to the nitrogen atom (compare with the reaction of cyanogen bromide with pyridine in the presence of amine) (Equation 146). This color reaction was proposed for the analysis of trichlorfon.[345]

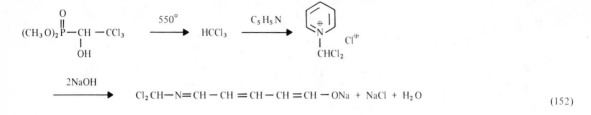

$$(CH_3O)_2\overset{\overset{\displaystyle O}{\|}}{P}-\underset{\underset{\displaystyle OH}{|}}{CH}-CCl_3 \quad \xrightarrow{550°} \quad HCCl_3 \quad \xrightarrow{C_5H_5N}$$

$$\xrightarrow{2NaOH} \quad Cl_2CH-N=CH-CH=CH-CH=CH-ONa + NaCl + H_2O \qquad (152)$$

Coumaphos is hydrolyzed in alkali to produce a highly fluorescent compound, which may be either the coumaric or coumarilic acid.[341] The fluorimetry is 1,250 times more sensitive than colorimetry, which is based on hydrolysis and the reaction of the liberated hydroxycoumarin with 4-aminoantipyrine.

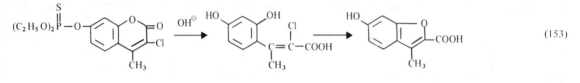

(153)

fluorescent

d. Esterase Inhibitors

Organophosphorus pesticides are essentially the inhibitors of esterases including cholinesterases (ChE). They can thus be determined by enzymatic methods.[346,347] For this purpose, an excess of suitable esterase is first allowed to react with the inhibitor: phosphorylation occurs on the active site of the esterase, resulting in the inactivation of the enzyme. After the incubation period, the residual activity of uninhibited esterase is measured.

With the exception of the active metabolites of organophosphorus pesticides the organophosphorus pesticides themselves are generally poor inhibitors of esterases, unless they are previously converted into the active form. The activation is performed by oxidation, for which bromine, vapor or aqueous solution, and peracids are usually used. Bromine water is superior to the vapor for the activation. Recently, silver oxide was utilized for the quantitative conversion of parathion to paraoxon in aqueous media.[348] Irradiation with ultraviolet light is also utilized for the activation on paper and thin-layer chromatograms. The activated products from an organophosphorus pesticide by different methods are not always the same.

Cholinesterases hydrolyze acetylcholine (ACh) into acetic acid and choline (other substrates also may be used):

$$CH_3COOCH_2CH_2\overset{\oplus}{N}(CH_3)_3 + H_2O \xrightarrow{ChE} CH_3COOH + HOCH_2CH_2\overset{\oplus}{N}(CH_3)_3 \qquad (154)$$

The activity of ChE may be measured by determining either the remaining substrate or one of the products. The pH change due to the production of acetic acid is measured potentiometrically. pH Indicators, such as bromthymol blue, are applied for the visualization of inhibitors on flat-bed chromatograms. The acid produced is also determined titrimetrically. Furthermore, when a buffer containing carbonate is used, the production of the acid results in the evolution of carbon dioxide, which can be measured with a Warburg manometer.

The unchanged ACh can be determined colorimetrically by reacting with hydroxylamine to form a hydroxamic acid, which gives a purple color with ferric ion with a maximum absorption at 540 nm (Hestrin method).

$$CH_3\overset{\overset{\displaystyle O}{\|}}{C}OCH_2CH_2\overset{\oplus}{N}(CH_3)_3 + NH_2OH \longrightarrow CH_3\overset{\overset{\displaystyle O}{\|}}{C}NHOH + HOCH_2CH_2\overset{\oplus}{N}(CH_3)_3$$

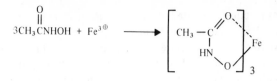

$$3CH_3\overset{O}{\overset{\|}{C}}NHOH + Fe^{3\oplus} \longrightarrow \left[\begin{array}{c} CH_3-C \overset{O}{\diagdown} \\ \overset{|}{HN} \diagup O \end{array} Fe \right]_3 \qquad (155)$$

Acetylthiocholine is a more convenient substrate for the assay of ChE activity because thiocholine formed by the hydrolysis can be readily determined colorimetrically; it reacts rapidly and irreversibly with 5-dithiobis-2-nitrobenzoic acid, releasing a colored product, 2-nitro-5-thiobenzoic acid, with a maximum absorption at 412 nm (Ellman method).[349]

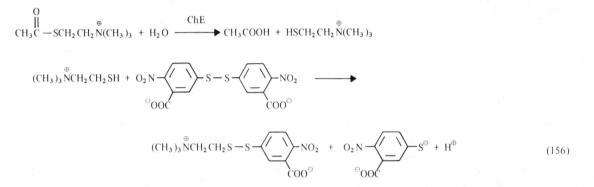

$$CH_3\overset{O}{\overset{\|}{C}}-SCH_2CH_2\overset{\oplus}{N}(CH_3)_3 + H_2O \xrightarrow{\text{ChE}} CH_3COOH + HSCH_2CH_2\overset{\oplus}{N}(CH_3)_3$$

Because this method is based on the measurement of a reaction product instead of the remaining intact substrate, it is more sensitive than the Hestrin method. Similarly, phenyl thioacetate, recently proposed as a substrate for cholinesterases,[350] liberates thiophenol by enzymatic hydrolysis.

For esterases, several substrates which are visualizable after hydrolysis are known. Alpha- and β-naphthyl acetates are utilized for the colorimetric measurement of esterase activity[351] and also for enzymatic detection of esterase inhibitors on paper and thin-layer chromatograms.[293] Naphthyl acetates are enzymatically hydrolyzed to liberate naphthols, which are readily converted to azo dyes with, for example, diazo blue B (tetrazotized o-dianisidine).

Indophenyl acetate is hydrolyzed by the action of esterases to form indophenol, which is blue-purple with an absorption maximum at 625 nm at pH 8.0. The acetate has an absorption maximum at 500 nm in all pH ranges and does not interfere with the colorimetric measurement of indophenol ion.[352]

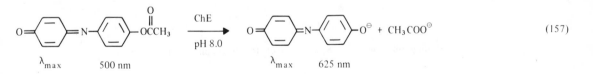

$$(157)$$

$$\lambda_{max} \quad 500 \text{ nm} \qquad \qquad \lambda_{max} \quad 625 \text{ nm}$$

Indophenyl acetate can be used as the substrate of acetylcholinesterase. However, its behavior as the substrate is quite different from the true substrate, acetylcholine. Acetylcholinesterase is inhibited irreversibly with an alkylating agent MCP (2-chloro-N-(chloroethyl)-N-methyl-2-phenylethylamine) when the substrate is acetylcholine, but is activated for indophenyl acetate.[353] A rational explanation for this phenomenon has appeared (see IV.A.1.c).[354]

$$\begin{array}{c} \overset{Cl}{\overset{|}{CH}}-CH_2-\overset{CH_3}{\overset{|}{N}}-CH_2CH_2Cl \end{array}$$

MCP

The acetate of indoxyl or substituted indoxyl was recently evaluated as the fluorogenic or chromogenic substrate of esterases. These substrates are hydrolyzed by esterases to form highly fluorescent indoxyl, which can be determined fluorimetrically.[355] N-Methylindoxyl acetate is a more suitable fluorogenic substrate for the trace analysis of cholinesterase inhibitors.[356] The fluorescent hydrolysis product, N-methylindoxyl, has a greater fluorescence stability than indoxyl, and subnanogram amounts of paraoxon are detectable on the thin-layer chromatogram.[357] Indoxyls are further converted into indigo blue by autooxidation. For the enzymatic detection of organophosphorus pesticides on thin-layer chromatograms, Mendoza asserted that 5-bromoindoxyl acetate was superior to naphthyl acetate.[358]

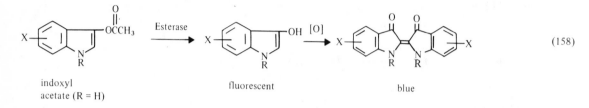

indoxyl acetate (R = H)   fluorescent   blue   (158)

2. Derivatization for Gas Chromatography

Recent advances in the detectors of gas chromatographs have enabled analysts to determine nanogram and picogram ranges of pesticide residues. Some excellent reviews for the application of gas chromatography to the analysis of pesticides have appeared.[359-361]

A variety of detection systems may be applied for the analysis of organophosphorus pesticides. Their sensitivity and selectivity are listed in Table 13. The last four detectors in the table are highly sensitive to phosphorus compounds. They are generally based on the extensive degradation of the compounds. In the microcoulometric detector, the column effluent is reduced with molecular hydrogen at 950°; phosphorus, sulfur and chlorine are converted to phosphine, hydrogen sulfide, and hydrogen chloride, respectively. These reduced products precipitate silver ion and can be measured by the silver/silver acetate electrode in a microcoulometric titration cell.[362]

In the emission spectrometric detector, the effluents are swept into an intense microwave powered argon discharge region, where the compound is completely fragmented and excited, resulting in emission at frequencies characteristic of the elemental composition of the compound. A phosphorus line at 2,536 Å and a sulfur line at 2,576 Å may be used for specific detection.[363]

The alkali-flame or thermionic detector is the most widely used detector for determining organophosphorus compounds. A modification of a flame-ionization detector, installing an alkali metal salt like Na_2SO_4, KCl, KBr, RbCl, Rb_2SO_4, and CsBr, enhances the sensitivity to phosphorus or halogen containing compounds.[364] Its mechanism has not been fully clarified. Vaporized alkali metal may be excited and ionized by reacting with the combustion products of such compounds in the flame.

TABLE 13

Gas Chromatographic Detection Systems Usable for Organophosphorus Pesticides

Detection system	Responding element	Approx. min. detectability	
		Ideal	Practical
Flame-ionization	C	ng	µg
Electron-capture	Halogen, NO_2	pg	ng
Electrolytic conductivity	Halogen, S, N	ng	ng
Microcoulometric	Halogen, S, P	ng	µg
Emission spectrometric	P, S, halogen	ng	ng
Thermionic	P, halogen	pg	ng
Flame-photometric	P, S	pg	ng

The flame-photometric detector is also a widely used detector for determining phosphorus compounds. Light emitted by effluents burning in a hydrogen rich flame is monitored through a 526 nm and 394 nm filter for phosphorus and sulfur compounds, respectively. Both compounds can be simultaneously detected by a dual-flame-photometric detector.[365]

As gas chromatography is only effective to vaporizable substances, the direct use of gas chromatography is limited to volatile, nonpolar, and thermally stable compounds. The majority of organophosphorus pesticides can be directly subjected to gas chromatography without modification. However, some of the organophosphates and many of their naturally produced oxidation and hydrolysis products are difficult to detect gas chromatographically, owing to their high polarity and low volatility. When the gas chromatographic separation or measurement of such compounds is desired, it is usual to prepare volatile and thermally stable derivatives. Derivatives are also prepared in order to change the retention time of the compound, to separate it from interferences, to make it responsive to a particular detection system, and to convert a complicated mixture into a simpler one.

The hydrolytic metabolites of organophosphorus insecticides and their oxygen analogs are successfully chromatographed after methylation with diazomethane.[366,367] For example, one such metabolite of parathion-methyl, O,O-dimethyl phosphorothioic acid, is converted with diazomethane into O,O,S-trimethyl phosphorothiolate (90%), accompanied by the formation of small amounts of by-products, which consist mainly of the thiono isomer (Equation 159).[367] The corresponding oxidized metabolite, dimethyl phosphoric acid, gives a single product, trimethyl phosphate. The esterification of phosphorus acids occurs also in the column during the gas chromatographic process, provided that the acids are dissolved in 10% methanolic hydrogen chloride.[368] Using methanol-hydrogen chloride, the methylation occurs only on the sulfur atom of dialkyl phosphorothioic acid due to the poorer alkylating activity of the methyloxonium ion (Equation 160). The much higher reactivity of the thiol anion toward alkylating agents, in comparison with that of corresponding oxide anion, has already been discussed in section II.D.2. As diazomethane is a much stronger alkylating agent than methyloxonium ion, even the oxygen atom of the phosphorothioic acid can be alkylated.

$$(RO)_2PSOH + CH_2N_2 \longrightarrow (RO)_2\overset{\overset{O}{\|}}{P}SCH_3 + (RO)_2\overset{\overset{S}{\|}}{P}OCH_3 \qquad (159)$$
$$ \underset{90\%}{} \quad \underset{10\% >}{}$$

$$(RO)_2PSOH + CH_3\overset{\oplus}{\underset{|}{\underset{H}{O}}}H \longrightarrow (RO)_2\overset{\overset{O}{\|}}{P}SCH_3 \qquad (160)$$

Another type of important ionic metabolites, desalkyl products, also may be chromatographed after methylation with diazomethane or methanol-hydrogen chloride.[367] Diazomethane again methylates mainly the sulfur atom: desmethyl parathion-methyl gives S-methyl parathion-methyl isomer 90% and parathion-methyl 10% (Equation 161). Again, methanol-hydrogen chloride gives only the S-methyl product (Equation 162). The conversion of desmethyl paraoxon-methyl into paraoxon-methyl with methanol-hydrogen chloride is, however, so insufficient that it is not detectable in an amount less than 100 µg. When diazomethane is used as an alkylating agent, it can be detected at the 5 µg level.

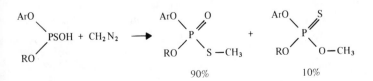

$$(161)$$

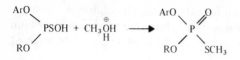

$$\text{(162)}$$

Thus, methylation with diazomethane is generally applied for the gas chromatographic analysis of partial esters of phosphorus acids. However, for the analysis of malathion mono- and di-carboxylic acids, methanol with boron trifluoride is more useful than diazomethane.[369] By reaction with the former, the carboxylic acids are quantitatively converted into their methyl esters, which can be readily gas chromatographed.

$$\underset{\underset{CH_2COOH}{|}}{(CH_3O)_2}\overset{\overset{S}{\|}}{P} - S - CHCOOH + 2CH_3OH \xrightarrow{\;BF_3\;} \underset{\underset{CH_2COOCH_3}{|}}{(CH_3O)_2}\overset{\overset{S}{\|}}{P} - S - CHCOOCH_3 \qquad \text{(163)}$$

There are too many organophosphorus pesticides to determine each residue by a simple gas chromatographic procedure. For a rapid screening purpose, it was recently proposed to determine them after converting into simple trialkyl esters by the hydrolysis of the pesticides followed by alkylation with diazoalkane.[370-372] By heating at 70° for 15 min in $5N$ sodium hydroxide methanolic solution, almost all common organophosphorus pesticides decompose, accompanied by the cleavage of the bond linking the phosphorus atom with the oxygen or sulfur atom adjoining the most acidic group.[371] Thus, two kinds of products that are dialkyl phosphates and O,O-dialkyl phosphorothioates are produced from the four types of ordinary organophosphorus pesticides (Equations 164 and 165). Hydrolysis conditions to get a similar result, using aqueous sodium hydroxide solutions for a wide variety of organophosphorus pesticides, were also reported. The hydrolysis products can be readily chromatographed after alkylation with diazomethane or diazoethane.

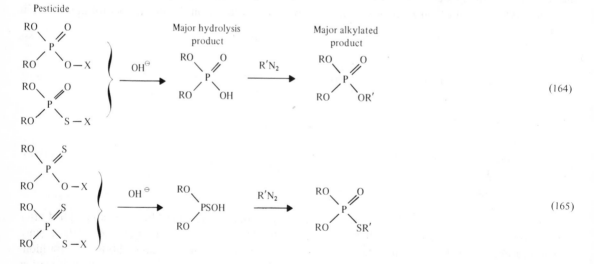

R = methyl, ethyl, propyl etc. R' = methyl or ethyl

Polar phosphoramidates are very difficult to pass through the gas chromatograph column without serious loss by adsorption, even though the corresponding thiono analogs are readily gas chromatographed. Thus, the primary oxidation metabolite of a new cotton insecticide, isocarbophos (o-isopropoxycarbonylphenyl methyl phosphoramidothionate), cannot be detected by gas chromatography, unless it was previously modified to a less polar derivative. The oxygen analog is hydrolyzed with $2N$ hydrochloric acid, accompanied with the cleavage of the P-N bond to form a diester (cf. Section III.A.2.b), which is easily converted into a gas chromatographically detectable triester by methylation with diazomethane.[373]

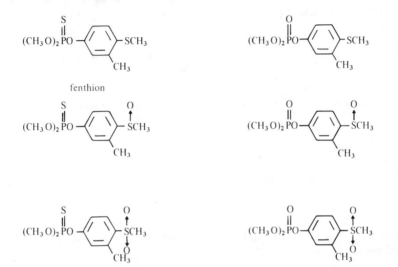

$$\text{(166)}$$

Phosphorothionate pesticides having a thioether group, such as fenthion, phorate, and disulfoton, each form five toxic metabolites: the thiophosphoryl group (P=S) may be oxidized to a phosphoryl (P=O) and the thioether group may be oxidized to a sulfoxide (S→O), and further to a sulfone (SO$_2$). Structures of fenthion and its toxic metabolites are shown as an example.

fenthion

Though this complicated mixture can be analyzed separately by the combination of liquid and gas chromatographies,[374] it takes a rather long time and, in certain cases, the determination of the total residues is more convenient for monitoring purposes. Thus, by oxidizing with m-chloroperbenzoic acid, they are converted to a single compound, the oxygen analog sulfone, making the analysis simpler and much more rapid.[374]

3. Fragmentation in Mass Spectrometry

In a mass spectrometer, a substance under investigation is bombarded with an electron beam, and a spectrum of the resultant positive ion fragments is recorded. A molecular ion (M$^+$) is formed by interaction with the beam electrons at an energy of about 9 to 15 eV.

$$M + e^- \longrightarrow M^+ + 2e^- \tag{167}$$

In practice, an electron beam energy of 70 eV is usually used for mass spectrometry, and many fragment ions are formed. The positive ions produced are separated on the basis of mass/charge (the majority of ions are singly charged). From the molecular ion, the molecular weight of the substance is accurately obtained, and the empirical formula is also obtained from an isotopic abundance ratio or, more precisely, by high resolution mass spectrometry. The fragmentation pattern is specific to the compound and serves to characterize the compound. The mass detection system is highly sensitive, permitting detection at levels as low as 10 ng of compound. Mass spectrometry in tandem with gas chromatography has been used to advantage for the separation and identification of pesticide residues.[375]

The probability of cleavage of a particular bond by bombardment with electrons is related to the bond strength, to the possibility of low energy transitions, and to the stability of both the charged and uncharged fragments formed in the fragmentation process. The decomposition proceeds exclusively by unimolecular reaction, because the mass spectrometry is operated under extremely low vapor pressure (10^{-5} mmHg), being accompanied, consequently, by very few collisions of fragments. In addition to simple cleavages, intramolecular rearrangements take place very often. The mass spectra of a number of organophosphorus pesticides have been reported by Damico,[376] and Jorg et al.[377] A comprehensive review on the mass spectrometry of phosphorus compounds appeared recently.[378]

a. Phosphate and Phosphorothionate Esters

A significant rearrangement of two hydrogen atoms to phosphorus connected oxygens with the elimination of an alkyl group occurs at the first step of the fragmentation of trialkyl phosphates. For example, triethyl phosphate decomposes with the following consecutive elimination of two ethylene molecules, giving finally the intense peak of a fragment ion $P(OH)_4{}^+$ (m/e 99).[379,380]

$$(C_2H_5O)_3P{=}O\Big]^{+\cdot} \xrightarrow[m^*\,132]{-C_2H_3} (C_2H_5O)_2\overset{+}{P}(OH)_2 \xrightarrow[m^*\,104]{-C_2H_4} C_2H_5\overset{+}{O}P(OH)_3 \xrightarrow[m^*\,77]{-C_2H_4} \overset{+}{P}(OH)_4 \qquad (168)$$
$$\text{m/e 182} \qquad\qquad \text{m/e 155} \qquad\qquad \text{m/e 127} \qquad\qquad \text{m/e 99}$$

The elimination of C_2H_3 is believed to take place by a double hydrogen rearrangement:

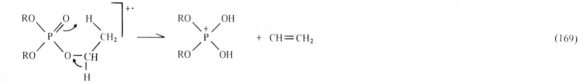

$$(169)$$

The fragment ion m/e 127 also decomposes in another way, accompanied by the elimination of water, as shown by the following scheme:

$$C_2H_5\overset{+}{O}P(OH)_3 \xrightarrow{-H_2O} C_2H_5\overset{O}{\underset{+}{\overset{\|}{O}P}OH} \xrightarrow[m^*\,60]{-C_2H_4} O{=}\overset{+}{P}(OH)_2 \qquad (170)$$
$$\text{m/e 127} \qquad\qquad \text{m/e 109} \qquad\qquad \text{m/e 81}$$

In these fragmentation schemes, m* means a metastable peak. The presence of a metastable peak is generally regarded, with few exceptions, as the direct evidence of the transition m_1 to m_2 by ejection of neutral atoms of mass $m_1 - m_2$ as a single entity in a one step process. The position of the metastable peak is given by the following equation.

$$m^* = m_2{}^2/m_1$$

Thus, the metastable peak observed at 132 in the spectrum of triethylphosphate indicates the transformation from mass 182 to mass 155, ejecting C_2H_3.

$$m^* = 155^2/182 = 132.0$$

As the Z moiety of common organophosphorus pesticides, shown by the general structure

$$\begin{array}{c} RO \\ \diagdown \\ \diagup \\ RO \end{array}\!\!\overset{S(O)}{\underset{(S)}{\overset{\|}{P}}}{-}O{-}Z$$

is often complex, the successive rearrangement-fragmentation does not always give the final ion $P(OH)_4{}^+$ (m/e 99) or $(HO)_3PSH^+$ (m/e 115).

The double rearrangement of hydrogen atoms also takes place, as with triethyl phosphate, in the fragmentation of tetraethyl pyrophosphate (TEPP).[380]

$$(C_2H_5O)_2\overset{\text{O}}{\overset{\|}{P}}-O-\overset{\text{O}}{\overset{\|}{P}}(OC_2H_5)_2\Big]^{+\cdot} \xrightarrow{-C_2H_3} \underset{C_2H_5O}{\overset{(HO)_2}{>}}\overset{+}{P}-O-\overset{\text{O}}{\overset{\|}{P}}(OC_2H_5)_2 \xrightarrow{-C_2H_4}$$

TEPP m/e 290   m* 239   m/e 263   m* 210

$$C_4H_{13}O_7P_2^+ \xrightarrow{-C_2H_4} C_2H_9O_7P_2^+ \xrightarrow{-C_2H_4} (HO)_2\overset{\text{O}}{\overset{\|}{P}}\overset{+}{O}P(OH)_3 \xrightarrow{-H_2O} (HO)_2\overset{\text{O O}}{\overset{\|\ \|}{P}}\overset{+}{O}POH \quad (171)$$

m/e 235   m* 182   m/e 207   m* 155   m/e 179   m/e 161

The preparations of TEPP are generally contaminated with triethyl phosphate. They are readily analyzed separately with mass spectrometry carried out at room temperature for triethyl phosphate and at 150° for tetraethyl pyrophosphate.[380]

The thiono analog of TEPP, sulfotep, decomposes differently from TEPP. The double rearrangement does not take place because the sulfur atom does not accept hydrogen. Thus, M^+-27 does not appear, but a peak at M^+-28 does. The latter is characteristic of organophosphorus pesticides having a bisethoxy group, and is formed by the β-cleavage of the ethoxy group with hydrogen rearrangement.

$$\underset{C_2H_5O}{\overset{C_2H_5O}{>}}\overset{O(S)}{\underset{}{\overset{\|}{P}}}-O-Z\Big]^{+\cdot} \longrightarrow \underset{C_2H_5O}{\overset{HO}{>}}\overset{O(S)}{\underset{}{\overset{\|}{P}}}-O-Z\Big]^{+\cdot} \quad (172)$$

$$M^+-28$$

This peak appears commonly in diethyl phosphate and phosphorothionate esters, while it does not appear in diethyl phosphorodithioates such as disulfoton, carbophenothion, ethion, and dioxathion. A significant intensity of a rearrangement fragment ion, $(HO)_2PS^+$ (m/e 97), is observed in all O,O-diethyl phosphorothionate and phosphorodithioate esters, but not in all corresponding methyl esters, indicating that a hydrogen atom rearranges only from β-carbon of an ethyl ester group.

$$(C_2H_5O)_2\overset{S}{\overset{\|}{P}}-\underset{(S)}{\overset{}{O}}-Z\Big]^{+\cdot} \longrightarrow (HO)_2\overset{+}{PS} \quad (173)$$

$$\text{m/e 97}$$

The base peak ion (m/e 127) for mevinphos and phosphamidon is postulated to be formed by the double hydrogen rearrangement from eliminated Z moiety to the phosphorus connected oxygen atoms.

$$(CH_3O)_2\overset{O}{\overset{\|}{P}}-O-\underset{CH_3}{\overset{}{C}}=CHCO_2CH_3\ \Big]^{+\cdot} \longrightarrow (CH_3O)_2\overset{+}{P}(OH)_2 \quad (174)$$

mevinphos   m/e 127

The mass spectrum of paraoxon is similar to parathion, and their base peak ion (m/e 109) is probably formed by the reaction shown in the following scheme:

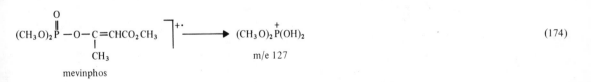

$$(C_2H_5O)_2\overset{O(S)}{\overset{\|}{P}}-O-\!\!\langle\!\!\langle\ \rangle\!\!\rangle\!-NO_2\Big]^{+\cdot} \longrightarrow C_2H_5O-\overset{O}{\overset{\|}{P}}=\overset{+}{O}H \longleftrightarrow C_2H_5\overset{+}{O}=\overset{O}{\overset{\|}{P}}-OH \quad (175)$$

$$\text{m/e 109}$$

O,O-dimethyl phosphorothionate esters such as fenitrothion and parathion-methyl also give a fragment ion of m/e 109, which should be formed by a different process. Moreover, these methyl esters, almost without exception, give fragment ions of m/e 125, 93, 79, 63, and 47. Thus, they appear to be formed by the following processes:[378]

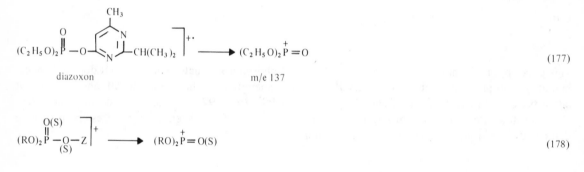

Paraoxon also gives relatively intense peaks of rearrangement fragment ions $(HO)_4P^+$ (m/e 99) and $O=P(OH)_2^+$ (m/e 81). Another fragment ion, m/e 139 ($HOC_6H_4NO_2^+$), may be formed due to α-cleavage of the molecular ion with hydrogen rearrangement.

Diazoxon forms a base peak ion (m/e 137) which is probably due to α-cleavage with the charge on the phosphorus moiety (Equation 177). This type of fragmentation is most common with almost all phosphorus esters, including O,O-dialkyl phosphates, phosphorothionates, phosphorothiolates, phosphorodithioates, and also O-alkyl alkylphosphonodithioates such as fonofos (Equation 178).[381] The only exception in the organophosphorus pesticides studied was paraoxon.

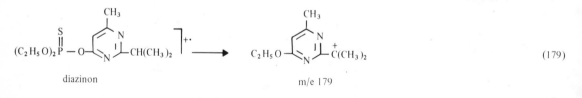

The mass spectrum of diazinon is quite different from that of diazoxon. The base peak of the former (m/e 179), which is absent in its oxo-analog, may be formed by migration of an ethyl group from the ethoxy group to the oxygen atom connecting to the Z moiety, accompanied with removal of a hydrogen from the isopropyl group.

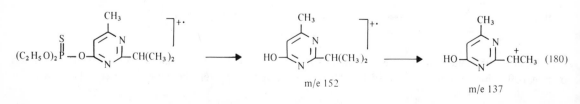

Another intense peak of diazinon (m/e 137) was suggested to be produced by the loss of the propylene group of the ion m/e 179. Pardue indicated, however, from metastable data that it is formed by the loss of a methyl group from another significant fragment ion (m/e 152) which may be formed by the α-cleavage of the molecular ion with hydrogen rearrangement.[299]

The cleavage at branched carbon atoms is a common pattern of fragmentation. Thus, $M^+ - 15$ (CH_3) ion peaks are characteristic for diazinon and diazoxon.

Some characteristic rearrangement reactions induced by electron impact in phosphorothionates were clarified by Cooks and Gerrard.[382] The thiono-thiolo rearrangement may be induced before any fission occurs:

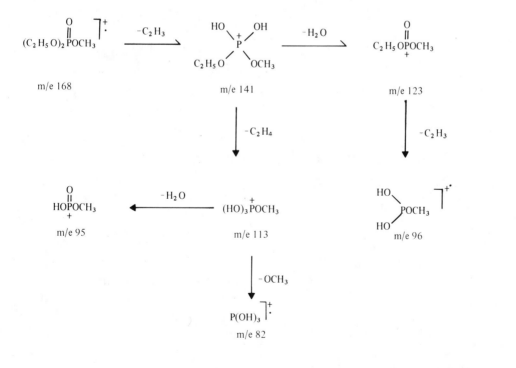

(181)

However, substantial differences in fragmentation behavior are observed between thiono and thiolo isomers, indicating that equilibrium between them is not complete in the ion source. For example, diethyl methyl phosphorothionate undergoes a fragmentation process leading to the formation of ions $C_2H_5O\overset{+}{P}OCH_3$ (m/e 107) and $CH_3O\overset{+}{P}OH$ (m/e 79), which are not detected with the corresponding thiolate isomer.[375] This is of interest from the viewpoint of residue analysis for the hydrolytic products of organophosphorus pesticides, because these isomers are generally produced by derivatization with diazomethane of dialkyl phosphorothioates (see III.G.2). The fragmentation schemes for the three possible derivatization products, namely, diethyl methyl phosphate, diethyl methyl phosphorothionate, and diethyl S-methyl phosphorothiolate are given (Equations 182 to 184):[383]

Rearrangement reactions may also result in the combination of two substituents. Thus, diphenyl methyl phosphorothionate, for example, gives ions of the composition $C_6H_5SC_6H_5$, $C_6H_5SCH_3$, and C_7H_7. Furthermore, the loss of SH from the molecule ion gives a base peak in the mass spectrum of dimethyl N-cyclohexylphosphoramidothionate (Equation 185). The structure of the fragment ion was posturated as XXVI.[382] The SH elimination is favored wherever stereochemically suitable hydrogen is present.

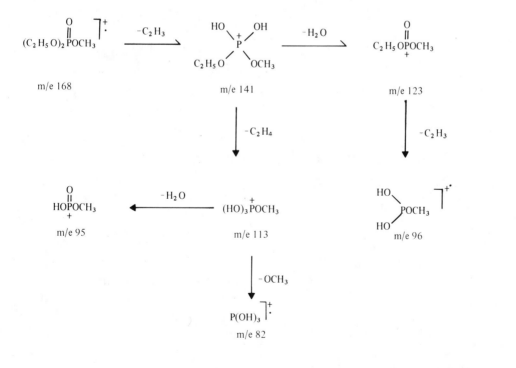

(182)

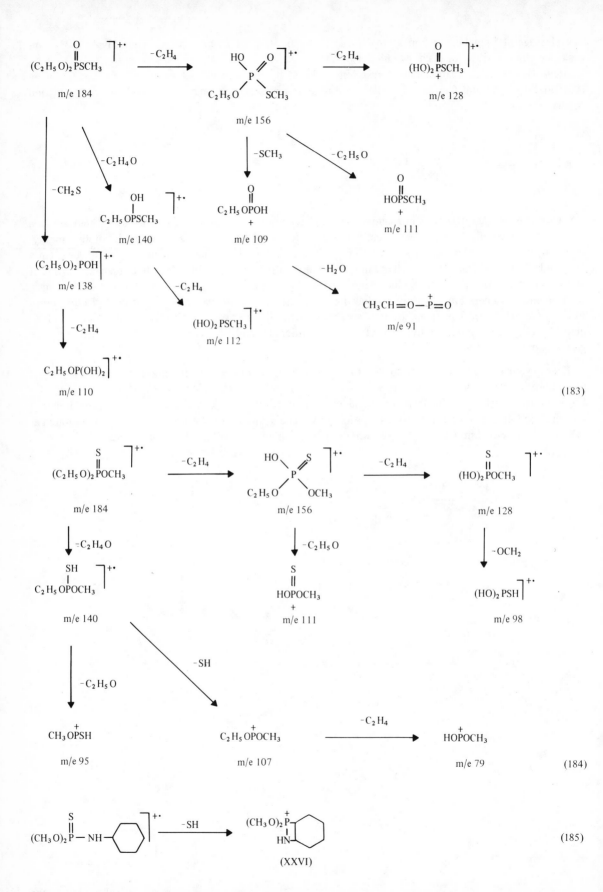

(183)

(184)

(185)

(XXVI)

b. Phosphorothionothiolate and Phosphorothiolate Esters

In phosphorus esters having a P-S-alkyl(Z) bond, β-cleavage at the S-alkyl bond takes place very often to form a relatively intense peak with the charge on the Z moiety, in contrast with phosphate and phosphorothionate esters, with which such fragmentation is not usually observed. The base peaks of malathion (m/e 173), phosmet (Imidan[®]) (m/e 160), azinphosmethyl (m/e 160), carbophenothion (m/e 157), and Kitazin[®] (m/e 91) are all due to the β-cleavage.

$$(RO)_2 \overset{S(O)}{\overset{\|}{P}} - S - Z \Big]^{+\cdot} \longrightarrow Z^+ \tag{186}$$

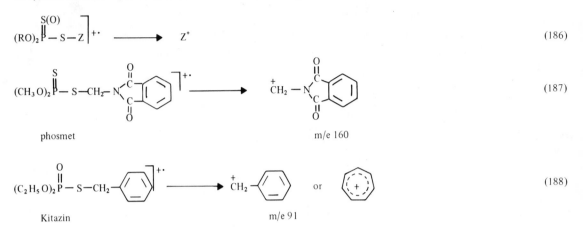

phosmet   m/e 160   (187)

Kitazin   m/e 91   (188)

However, this is not the case with S-aryl compounds. Fonofos (Dyfonate[®]: ethyl S-phenyl ethylphosphonothiolothionate) gives a base peak due to α-cleavage with hydrogen rearrangement.[381]

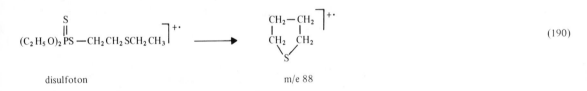

fonofos   m/e 110   (189)

In some compounds, the β-cleavage occurs accompanied with rearrangement. The base peak of disulfoton (m/e 88) may be due to such a rearrangement forming tetrahydrothiofuran ion.

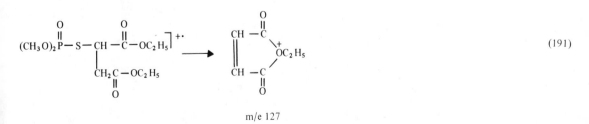

disulfoton   m/e 88   (190)

The base peak of malaoxon is postulated as the ethyl oxonium ion of maleic anhydride formed by rearrangement.

$$(CH_3O)_2 \overset{O}{\overset{\|}{P}} - S - \underset{\underset{\underset{O}{\|}}{\underset{CH_2C-OC_2H_5}{|}}}{CH} - \overset{O}{\overset{\|}{C}} - OC_2H_5 \Big]^{+\cdot} \longrightarrow \tag{191}$$

m/e 127

The fragmentation of dimethoate proceeds in a very different way. Its base peak (m/e 87) appears to be due to the migration of O-methyl group to the Z moiety, forming N-methylpropionamide. However, the

119

elemental composition of the fragment ion determined by high resolution mass spectrometry is not consistent with methylpropionamide, but with C_3H_5NS, methyl thioacetonitrile.

$$\left[(CH_3O)_2 \overset{\overset{S}{\|}}{P} -SCH_2 - \overset{\overset{O}{\|}}{C}NHCH_3 \right]^{+\cdot} \longrightarrow C_3H_5N_5S^+ \qquad (192)$$

dimethoate   m/e 87

On the other hand, dimethoxon gives a characteristic rearrangement ion peak (m/e 156) due to the migration of a hydrogen from nitrogen to phosphoryl oxygen, accompanied with loss of N-methyl isocyanate as the neutral species.

$$\left[\begin{array}{c} H-NCH_3 \\ \end{array} \right]^{+\cdot} \longrightarrow (CH_3O)_2 \overset{OH}{\underset{+}{P}} -SCH_2 \; + \; O=C=NCH_3 \qquad (193)$$

dimethoxon   m/e 156

This type of rearrangement does not occur with dimethoate, probably owing to the lower affinity of thiophosphoryl for hydrogen. Another intense peak of dimethoxon (m/e 110) may be due to α-cleavage accompanied by a rearrangement of hydrogen.

$$\left[(CH_3O)_2 \overset{\overset{O}{\|}}{P} -SCH_2CONHCH_3 \right]^{+\cdot} \longrightarrow (CH_3O)_2 \overset{+}{P}OH \qquad (194)$$

m/e 110

Saligenin cyclic phosphorus esters show interesting patterns of fragmentation.[235] The methyl phosphorothionate (Salithion) and methyl phosphate give an intense peak of M^+-15, indicating that β-cleavage takes place at the exocyclic ester group, whereas the S-methyl phosphorothiolate gives a characteristic base peak at m/e 169, suggesting that α-cleavage of the exocyclic ester linkage occurs more easily than in the corresponding phosphate and phosphorothionate (Equation 195). Another characteristic fragmentation process of Salithion is the direct loss of SH followed by the elimination of formaldehyde (Equation 196).

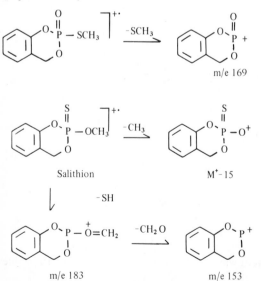

c. Chemical Ionization Mass Spectrometry

Chemicals may also be ionized upon collision with reagent ions, which are generated by electron bombardment of a suitable gas such as methane in the mass spectrometer at pressure of about 1 mm Hg. The most abundant ions formed from methane are CH_5^+ and $C_2H_5^+$. These ions function to donate sample molecules a proton to form a "quasi" molecular ion (MH^+) at a mass one atomic mass unit higher than the molecular weight.

$$M + CH_5^+ \longrightarrow MH^+ + CH_4 \tag{197}$$

This chemical ionization (CI) technique has been recently utilized for mass spectrometry.[383a] Since the energy involved in the proton transfer reaction is relatively low (less than 20 eV) and the quasi molecular ion, which contains an even number of electrons, is more stable than the odd electron molecular ion, chemical ionization mass spectra generally give the abundant quasi molecular ions and a smaller number of fragment ions in comparison with conventional electron-impact (EI) mass spectra.

The CI mass spectra of 15 organophosphorus insecticides and their major metabolites were recently reported.[383b] The fragmentation by CI is sometimes different from that by EI, and simpler spectra are usually obtained by CI technique. The initial fragmentation of organophosphorus esters occurs by α, β, or γ cleavage, as shown by Equation 198.

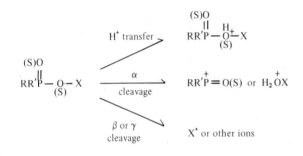

$$\tag{198}$$

Chapter IV

BIOCHEMISTRY

As discussed in the preceding chapter, pentavalent phosphorus esters have phosphorylating and alkylating properties. The insecticidal activity and mammalian toxicity are generally accepted as due to the phosphorylation of acetylcholinesterase. Besides acetylcholinesterase, they also inhibit cholinesterase, aliesterases, and so-called "serine proteinases" by phosphorylating the serine hydroxyl group in the active zone of the enzyme molecule. A variety of physiological effects induced by organophosphorus esters may be attributed to the inhibition of such enzymes.

In certain cases, the alkylating property also appears to be biologically important.[251] Some phosphorus esters inhibit "SH enzymes," probably owing to alkylation of the SH group; the inhibition of protein biosynthesis by the alkylation of DNA is believed to be the primary effect of organophosphorus chemosterilants. The fungicidal activity of certain organophosphorus esters is due to the inhibition of chitin biosynthesis, although the reaction mechanism is not yet known. These will be discussed in Chapter V. The alkylating property may also contribute to certain types of detoxication mechanisms. In addition to these reactions, biological oxidation is very important in the metabolism of organophosphorus pesticides with respect to their activation and detoxication.

A. INHIBITION OF ESTERASES

1. Biological Significance of Cholinesterases

a. Nerve Function

The target of organophosphorus esters, including insecticides and nerve gases, is acetylcholinesterase, which is one of the hydrolytic enzymes for acetylcholine. The inhibition of the enzyme disturbs the normal nervous function, finally resulting in the death of animals. The nervous system of a vertebrate is shown schematically in Figure 7. The nerve cell called the neuron, the component of the nervous system, consists of an elongated axon and short branched dendrites. The axon ending connects with another neuron through a synapse, or with a muscular fiber through a special synapse named the neuromuscular junction. The axon is covered with the nerve membrane, which has a selective permeability to ions. In general, the concentration of the potassium ion is higher inside the axon than outside, and that of the sodium ion is the reverse; for example, the internal and external ion concentration of the giant axon of squid for potassium is 410 and 22 mM, respectively, and that for sodium is 49 and 440 mM, respectively. Owing to the concentration gradients of these ions, the resting nerve membrane is polarized at the equilibrium potential according to Nernst's equation,

$$E = \frac{RT}{F} \ln \frac{P_{K^+}[K^+]_o + P_{Na^+}[Na^+]_o + P_{Cl^-}[Cl^-]_i}{P_{K^+}[K^+]_i + P_{Na^+}[Na^+]_i + P_{Cl^-}[Cl^-]_o} \tag{1}$$

where R, T, and F are the gas constant, the absolute temperature, and the Faraday constant, respectively, $[\;]_o$ and $[\;]_i$ are the external and internal indicated ion concentrations, and P is the permeability coefficient of the indicated ion. The resting membrane is selectively permeable to potassium ion, and the permeability coefficients of Na$^+$ and Cl$^-$ are much smaller in comparison to that of K$^+$;

$$P_{K^+} : P_{Na^+} : P_{Cl^-} = 1 : 0.04 : 0.45$$

Thus, the resting membrane potential approaches approximately the equilibrium potential for potassium,

$$E_K = \frac{RT}{F} \ln \frac{[K^+]_o}{[K^+]_i} \tag{2}$$

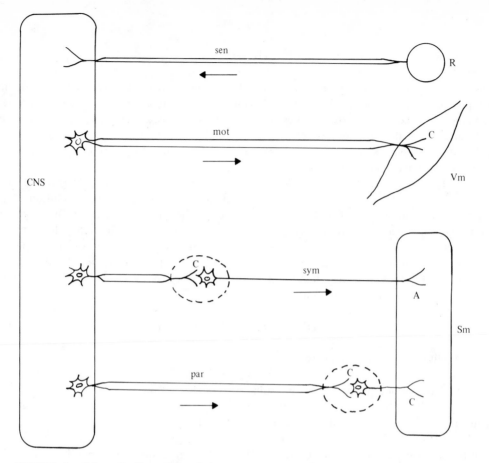

FIGURE 7. Schematic illustration of the nervous system of vertebrates. CNS: central nervous system, sen: sensory nerve, mot: motor nerve, sym: sympathetic nerve, par: parasympathetic nerve, R: receptor, Vm: voluntary muscle, Sm: smooth muscle, A: adrenergic synapse, C: cholinergic synapse.

The potential usually has a value of 60 to 70 mv, the inside of the membrane being negative with respect to the outside. When the nerve is excited, the membrane depolarizes and the membrane potential changes rapidly toward the reverse direction from the resting potential. This is due to the rapid and great increase in the conductivity of the membrane to sodium ion: the permeability coefficients change to:

$$P_{K^+} : P_{Na^+} : P_{Cl^-} = 1 : 20 : 0.45$$

This allows the membrane potential to approach the sodium equilibrium potential,

$$E_{Na} = \frac{RT}{F} \ln \frac{[Na^+]_o}{[Na^+]_i} \tag{3}$$

The increased sodium conductivity soon decreases, and the potassium conductivity now starts increasing, resulting in the recovery of the resting membrane potential, as shown in Figure 8. This rapid change in the membrane potential is called an action potential. The action potential accompanies a nerve impulse, which propagates along the axon towards the nerve ending.

The electric nerve impulse conducted along the axon cannot cross the synaptic cleft (200 to 300Å in interneuron synapses and 500 to 600 Å in neuromuscular junctions) in the junction between cells (Figure 9), but causes the release of a chemical substance called transmitter or neurohormone from the axon ending. The transmitter migrates to the receptor on the postsynaptic membrane of another neuron or

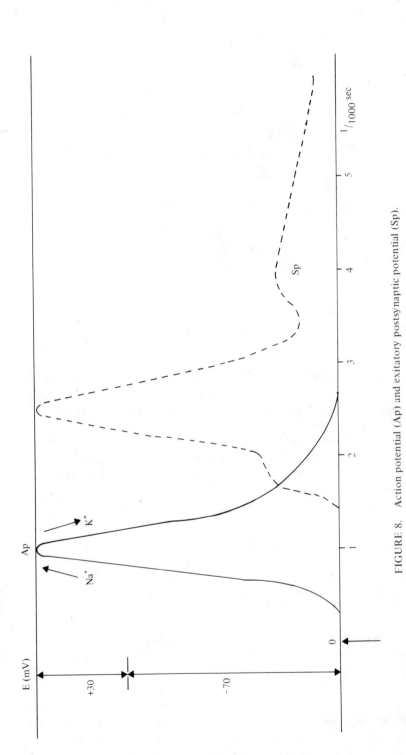

FIGURE 8. Action potential (Ap) and exitatory postsynaptic potential (Sp).

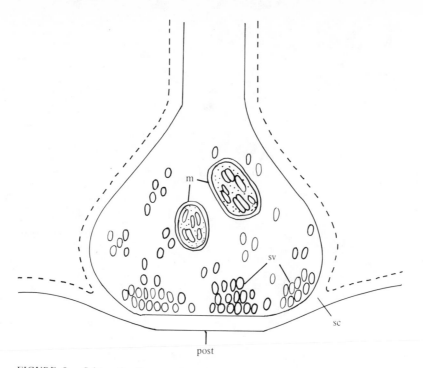

FIGURE 9. Schematic diagram of synapse; m: mitochondria, post: postsynaptic membrane, sc. synaptic cleft, sv: synaptic vesicles.

muscle fiber. The interaction of the transmitter with the receptor causes a change in the cation conductance of the postsynaptic membrane, followed by the depolarization of the membrane (Figure 10). It produces the excitatory postsynaptic potential or end-plate potential. When the potential achieves a threshold, an action potential rises rapidly to excite the neuron or muscle fiber.

The best known transmitters are acetylcholine and noradrenaline (norepinephrine). In addition, other biogenic amines, such as dopamine and 5-hydroxytryptamine (5-HT; serotonin), and amino acids such as γ-aminobutyric acid, glutamic acid, and glycine have also been recognized to act as possible transmitters in certain synapses. γ-Aminobutyric acid (GABA) was recently confirmed as a transmitter in inhibitory neuromuscular junctions present in arthropods.[384]

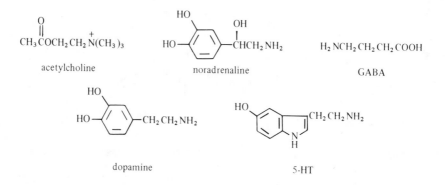

In the nervous system of vertebrates, noradrenaline serves as the transmitter at most of the postganglionic sympathetic nerve terminals (adrenergic synapse). Acetylcholine is a transmitter operating in the cholinergic synapses, which include the synapses of the central nervous system, the neuromuscular junctions of motor nerves, sensory nerve endings, ganglionic synapses of both sympathetic and parasympathetic nerves, all postganglionic parasympathetic nerve terminals, and sympathetic nerve terminals on the sweat glands, blood vessels, and the adrenal medulla (Figure 7).

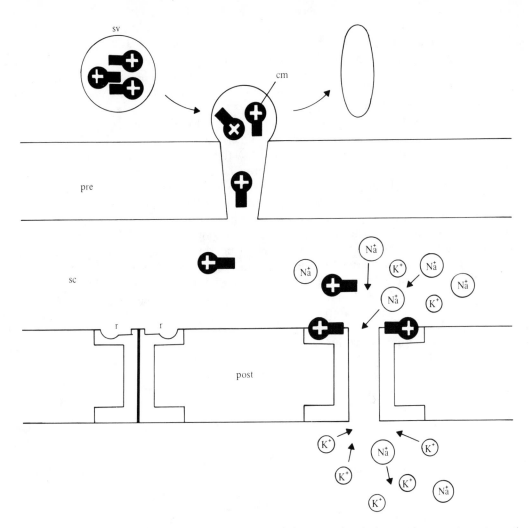

FIGURE 10. Schematic illustration of synaptic transmission; cm: chemical transmitter, post: postsynaptic membrane, pre: presynaptic membrane, sc: synaptic cleft, sv: synaptic vesicles, r: receptor.

In insects, acetylcholine is accepted as the transmitter in the synapses of the central nervous system, which is a chain of ventral ganglia.[385] Catecholamines are also found in insects, and the possibility of the existence of an adrenergic synaptic mechanism in the insect central nervous system, particularly in the sixth abdominal ganglion of the cockroach, was suggested.[386] It is believed that the neuromuscular junctions of insects are not cholinergic, although the presence of cholinesterase has been demonstrated histochemically in motor end-plates of cricket muscle (but not of any other insect muscles).[350] The transmitter in insect neuromuscular junctions is not known,[387] though there are hypotheses that an indolealkylamine, a nicotinamide derivative, and glutamic acid participate in the synaptic transmission.[388,389] The neuroactivity of L-leucine was recently demonstrated in the abdominal cord of the American cockroach, but its physiological role is not yet confirmed.[390]

Acetylcholine is synthesized in the nerve ending by the action of cholinacetylase (2.3.1.6) from choline and acetyl Co-A and stored in synaptic vesicles. The vesicles burst spontaneously to produce a miniature potential (1 to 2 mv) without giving any action to the associated cell. However, by the stimulation of the action current, the burst and the release of acetylcholine increase quickly by 100- to 1,000-fold. This causes synapse or end-plate potential (10 to 20 mv) and, consequently, the excitation of the postsynaptic membrane. The released acetylcholine is then rapidly hydrolyzed by acetylcholinesterase into inactive acetic acid and choline before the second impulse arrives. Most of acetylcholinesterase in the end-plates of mammals is localized near the receptors on the postsynaptic membrane.[391] As long as acetylcholine remains in the region of the synaptic cleft, the original state of the postsynaptic membrane cannot be reestablished. Therefore, the inhibition of acetylcholinesterase results in the disturbance of the nervous function which leads to severe and often lethal damage in the organism. This is the principal lethal lesion caused by organophosphorus and carbamate insecticides.

The usual cause of death of mammals is respiratory paralysis. Depending on the type of cholinesterase inhibitor, this lethal symptom may be caused by a peripheral paralysis owing to the blockage of neuromuscular transmission, by a disturbance of the function of the respiratory center in the medulla oblongata, or by a combination of both factors.[392] Histochemical investigations on the cholinesterase of insects treated with organophosphorus insecticides have shown that the peripheral area of the thoracic ganglia in houseflies and the brain and nerve cord in crickets are the vital sites of inhibition.[350] In fatally poisoned cockroaches, some synaptic pathways in abdominal ganglion are relatively normal, although function in thoracic ganglion is affected by cholinesterase inhibition.[393]

Certain insecticides are known to disturb the nerve function by blocking the acetylcholine receptors in the synapses. The insecticide Cartap (1,3-bis(carbamoylthio)-2-(N,N-dimethylamino)propane hydrochloride) is converted in vivo into an active principle, nereistoxin (4-(N,N-dimethylamino)-1,2-dithiolane),[394] which is a toxin isolated from the sea annelids *Lumbrineris* spp. and blocks synapses in the insect central nervous system by competing for receptor sites.[395] Nicotine noncompetitively blocks the synaptic transmission in insect ganglia.[396] The new insecticide chlorfenamidine (N',-(4-chloro-o-tolyl)-N,N-dimethylformamidine) interacts with the acetylcholine receptors on the neuromuscular junction of vertebrates.[397]

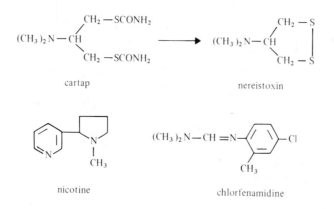

cartap

nereistoxin

nicotine

chlorfenamidine

Some organophosphates appear to block acetylcholine receptors reversibly at high concentrations.[398] However, it may happen rarely in normal poisoning, because the concentration needed for the receptor blockage is 100 to 1,000 times greater than that needed for acetylcholinesterase inhibition.

The nature of cholinesterases is well known,[399-401] but the acetylcholine receptor was rather hypothetical until the very recent investigations in the attempted isolation. The characteristics of the receptor are not identical, depending on the source and investigators. O'Brien studied the receptor in electroplax from the electric skate, *Torpedo marmorata*, using tritiated muscarone as the binding chemical, and suggested it to be a phospholipoprotein.[402] A receptor proteolipid having a high affinity for binding acetylcholine and cholinergic drugs was also isolated from electroplax of *Electrophorus electricus*.[403] However, the receptor molecule obtained as a complex with ³H-acetyl-α-bungarotoxin from the guinea pig cerebral cortex has the properties of a protein.[404] α-Bungarotoxin is a polypeptide isolated from the venom of the elapid snake, and is an irreversible binder of the cholinergic receptor.

b. Cholinesterases and Some Esterases

There are a variety of esterases which hydrolyze carboxylic esters. They may be classified into three groups (A, B, and C esterases) on the basis of their reaction with organophosphates such as DFP.[405] The esterases of A-type, named arylesterase (3.1.1.2), are not inhibited by organophosphates but hydrolyze them. They serve an important role in the detoxication mechanism of organophosphorus pesticides. B esterases are readily inhibited by organophosphates and involve carboxyesterase (3.1.1.1) or aliesterase and cholinesterases. C esterases, or acetylesterases (3.1.1.6), neither hydrolyze organophosphates nor are inhibited by them.

Cholinesterases are distinguished from other B esterases by their specific properties to hydrolyze choline esters in preference to other carboxylic esters and to be inhibited by physostigmine (eserine) at low concentrations (10^{-5} M). They are divided into two groups: acetylcholinesterase (3.1.1.7)(AChE) and cholinesterase (3.1.1.8)(BuChE). Table 14, shows the differences between them. AChE hydrolyzed most rapidly its natural substrate, acetylcholine, and plays an important role, described above, in the nervous system. It occurs mainly in the erythrocytes of most mammals, nervous tissues including central nervous system, ganglia and motor end-plate, and electric organs. On the other hand, BuChE is mainly found in serum and some organs, such as the pancreas, heart, and liver. It does not participate in nerve functions and its physiological function is still unknown. BuChE prefers butyryl and propionylcholines as substrate rather than acetylcholine and, with the exception of avian plasma cholinesterase,[406] does not split acetyl-β-methylcholine. No cholinesterases have an absolute specificity for choline esters. They split ordinary esters with more or less specificity patterns. Thus, certain noncholine esters may be utilized as convenient substrates in certain cases, as mentioned in Section III.G.1.d. Highly purified preparations of cholinesterases have been obtained from various sources, and some of them are now commercially available.[406] AChE from the electric organ of electric eels was obtained in a crystalline state.[407]

In connection with the mode of action of organophosphorus insecticides, cholinesterases of insects are most interesting to understand. The substrate specificities of the enzymes have been surveyed, using crude homogenates,[408,409] in many species of insects. Brain homogenates of adult insects are generally utilized as enzyme preparations. Housefly brain cholinesterase has been purified partially and only one type of cholinesterase has been recognized.[410] It may be classified as acetylcholinesterase: it prefers acetylcholine to any other choline esters of homologous higher acids as substrate; the esterase activity is inhibited by high concentrations of acetylcholine.[408] However, flyhead cholinesterase is distinctly

TABLE 14

Some Properties and Nomenclature of Cholinesterases

	Acetylcholinesterase	Cholinesterase
Systematic name	acetylcholine acetylhydrase (3.1.1.7)	acylcholine acylhydrase (3.1.1.8)
Alternative names	specific cholinesterase	nonspecific cholinesterase
	true cholinesterase	pseudocholinesterase
	erythrocyte cholinesterase	serum cholinesterase
	e-type cholinesterase	s-type cholinesterase
	(AChE)	butyrylcholinesterase (BuChE)
		propionylcholinesterase
Source	electric organ,	serum, pancreas, heart,
	nervous tissues,	liver
	erythrocytes, cobra venom	
Substrate specificity	ACh $\gg$ PrCh $>$ BuCh	BuCh, PrCh $>$ ACh
	D-acetyl-β-methylcholine	benzoylcholine,
	acetyl esters	butyryl and propionyl
		esters
Inhibited by:		
Excess substrates	+ ($>3 \times 10^{-3}$ M)	– (at 10^{-1} M)
Quaternary ammonium	++	+
Iso-OMPA*	+	++

*N,N'-diisopropylphosphorodiamidic

different from mammalian AChE because it hydrolyzes butyrylcholine at about half the rate of acetylcholine, whereas mammalian acetylcholinesterase activity to butyrylcholine is very low or not present. Another interesting property of the insect cholinesterase is that it is activated about 60% in the presence of 2 to 3% n-butanol and is protected by the organic solvent from inhibition by organophosphates.[410] The cholinesterase of pea aphids is rather similar to BuChE of mammals: it hydrolyzes acetylcholine and benzoylcholine, but not acetyl-β-methylcholine, and is not inhibited with excess acetylcholine.[409] Cholinesterase is distributed mainly in the central nervous system of insects, particularly in the membrane.[350]

c. Mechanism of Action of Acetylcholinesterase

It is generally accepted that the hydrolysis of choline esters (AX) by cholinesterases (EH) may be expressed in the following scheme:

$$EH + AX \underset{k_{-1}}{\overset{k_1}{\rightleftharpoons}} \underset{-HX}{EH \cdot AX \overset{k_2}{\longrightarrow}} \underset{+H_2O}{EA \overset{k_3}{\longrightarrow}} EH + AOH \tag{4}$$

where A and X are acyl group and choline, respectively. The enzyme and the substrate combine at first to form a Michaelis enzyme-substrate complex (EH·AX). The acyl group transfers to the esterase molecule to form acylated enzyme, which is then rapidly hydrolyzed and the active enzyme is recovered.

Cholinesterases neither contain nor require any specific prosthetic group or metal, though the activity is enhanced with various metals such as Ca^{++} and Mg^{++}. The catalytic action is, therefore, due to the structure of the enzyme protein itself. Folding the protein molecule, certain amino acid residues at distant sites of the chain are brought close to one another to form an active zone. The active zone contains two active sites: one fixes the substrate and is concerned chiefly with specificity (binding site or "anionic site"), and the other one catalyzes the hydrolytic process of the substrate (esteratic site or catalytic site).[399]

i. Binding Site

The substrate specificity of AChE for esters containing a cationic group suggests the presence of an anionic site in the active zone which attracts, binds, and orients the substrate by electrostatic forces, facilitating the attack of the esteratic site. In BuChE, such an electrostatic force appears to be unimportant.[400] Kinetic studies with reversible competitive inhibitors are more informative. The inhibitory action of choline and tetraethylammonium with a quaternary nitrogen decreases below pH 7, probably owing to the protonation of the anionic site whose pK_a value is about 6.2.[411] This value of pK_a differs considerably from those of carboxylic groups in proteins; the pK_a values of the ω-carboxylic group of glutamic acid and asparatic acid are 4.25 and 3.65, respectively. Krupka recently showed that the apparent pK value of about 6 of the anionic site is an artifact, and that the true value is 4.3.[412]

Tetraalkylammonium ion is, however, not bound by coulombic forces alone, but a considerable part of the affinity is from the alkyl group due to hydrophobic binding forces. Comparing the inhibitory potency of the series of methyl-substituted ammonium ions, Wilson computed an increase in the binding by a factor of 7 per methyl group, which corresponds to a free energy change of 1.2 kcal per methyl group (RTln 7).[399] This would indicate that the contribution of two methyl groups to the binding exceeds the contribution of electrostatic forces ($-\Delta F = 2$ kcal) between the negative charge of the anionic site and the cationic nitrogen atom.[405] In other words, the methyl groups of acetylcholine may contribute considerably to the binding of the substrate to the enzyme protein in terms of hydrophobic forces between the alkyl groups and nonpolar portions of the protein molecule. The nonpolar portions are evidently at the anionic site, so that the anionic site binds the cationic portion of the substrate by electrostatic and hydrophobic forces.[399] A weak interaction between the carbonyl group and the esteratic site may also participate in the fixation of the substrate.

As mentioned above, AChE also catalyzes the hydrolysis of some noncationic esters, such as phenyl acetate, indoxyl acetate, and indophenyl acetate. It appears that these substrates bind with AChE at different sites from that for acetylcholine binding. For example, an alkylating agent MCP (2-chloro-N-

(chloroethyl)-N-methyl-2-phenylethylamine) inhibits completely the activity of AChE for acetylcholine, probably due to alkylating the anionic site but, on the contrary, enhances the activity for indophenyl acetate.[353] The alkylation causes a moderate inhibition for phenyl acetate and indoxyl acetate. These effects are all due to changes in k_2 (acylation process) but not Km (formation of enzyme-substrate complex). O'Brien proposed, therefore, three different binding sites for AChE: α, β, and γ sites.[354] The α-site corresponds to the anionic site and binds with some alkylating agents such as MCP as well as cationic substances such as acetylcholine, tetraethylamine, choline, and amiton. These substances are called "α agents." The β-site is responsible for binding with organophosphates, carbamates, phenyl acetate, and indoxyl acetate. It is a hydrophobic portion, and the interaction with these inhibitors and substrates is weakened by α-agents. The γ-site combines with indophenyl acetate, acetyl fluoride, dimethylcarbamyl fluoride, and methylsulfonyl fluoride, and is postulated to shift into vicinity of the esteratic site by the configurational change induced by the action of α-agents.

ii. Esteratic Site

Bell shaped pH activity curves for cholinesterases, with a maximum at approximately pH 8, indicate the requirement of basic and acidic groups in the catalytic site of these enzymes because the charge of acetylcholine does not change in this pH range. The decrease in activity at lower pH may be due to inactivation of the basic group by protonation. The activity drop at the higher pH range, on the other hand, is attributable to the deprotonation of the acidic group. According to Krupka, two catalytically active basic groups are involved in AChE: one basic group (B_1) has a pK_a value of about 5.5 and the other (B_2) has a pK_a of 6.3.[412] The catalytically active acidic group (AH) has a pK_a value in the range of 9.2 and 10.4.[405,412] Referring to the pK values of functional groups involved in protein, the imidazole group of histidine is most probable as each of the two basic groups B_1 and B_2. This is also supported by the photochemical study of a similar enzyme; the destruction of histidine of chymotrypsin by selective photooxidation causes the loss of both enzyme activity and its reactivity with organophosphorus esters.[414]

In addition to this indirect evidence based on kinetic methods to establish knowledge on the structure of the active site of cholinesterases, more direct evidence on chemical analysis of components of the enzyme protein has been found. The latter method involves a specific reaction of the active site of an enzyme followed by the analysis of the groups involved. Cholinesterases, carboxyesterases, and so-called "serine proteases" such as chymotrypsin and trypsin react with organophosphorus esters such as diisopropyl phosphorofluoridate (DFP) to form an enzymatically inactive product. The reaction proceeds stoichiometrically: one mole of DFP phosphorylates one mole of the enzyme and inhibits completely the enzyme activity.[18]* The enzyme activity is recovered by removing the phosphoryl group, as discussed later. Furthermore, the substrate prevents the inhibition, indicating that the reaction occurs on a common active site, that is, an esteratic site. All such enzymes inhibited by an organophosphate labeled with ^{32}P give O-phosphoryl serine after hydrolysis of the protein. The amino acid sequences around the ^{32}P-labeled serine obtained from many inhibited enzymes are shown in Table 15.[405,417,418] They are very similar to each other. With the single exception of subtilisin, a bacterial proteinase, all hydrolases sensitive to organophosphorus inhibitors have a dibasic amino acid (glutamic acid in esterases and asparatic acid in proteases) preceding the phosphorylizable serine, which is followed by alanine (in esterases) or glycine (in proteases). These findings suggest that the hydroxyl group of serine in the amino acid sequence Glu-Ser-Ala or Asp-Ser-Gly plays an important part in the catalytic site of esterases or proteases. Although the substrate specificity of each enzyme may be determined by additional groups (binding site) in the enzyme molecule, cholinesterases have, thus, a great deal in common with other enzymes susceptible to organophosphates, and much information on the mode of action has come from serine proteases which can be obtained much more easily in pure state than esterases.

iii. The Catalytic Mechanism

As free serine neither catalyzes the hydrolysis of esters nor reacts with organophosphates, the hydroxyl

---

* Some enzyme proteins are phosphorylated without loss of enzyme activity. For example, stem bromelain is phosphorylated with DFP on the hydroxyl group of tyrosine residue, and the enzyme activity remains completely.[415,416]

TABLE 15

Amino Acid Sequences in the Active Sites of Organophosphate Sensitive Hydrolases[405,417,418]

Labeled enzyme	Sequence
AChE-DFP	P Glu-Ser-Ala
ChE-DFP	P Phe-Gly-Glu-Ser-Ala-Gly
Liver aliesterase-DFP	P Gly-Glu-Ser-Ala-Gly
Chymotrypsin-DFP	P Gly-Asp-Ser-Gly
Trypsin-DFP	P Gly-Asp-Ser-Gly-Pro-Val
Thrombin-DFP	P Gly-Asp-Ser-Gly
Elastase-DFP	P Asp-Ser-Gly
Subtilisin-DFP	P Thr-Ser-Met-Ala

group must be activated by other amino acid groups in the enzyme molecule. Porter et al. suggested that serine may be activated through the formation of an oxazoline ring by the interaction of the α-carbonyl of the neighboring dibasic amino acid.[419]

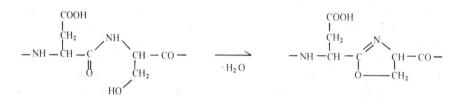

However, it looks more probable that the imidazole group of histidine is responsible for the activation of serine-hydroxyl,[420] if its possible participation in catalytic process is taken into consideration as mentioned above. The formation of a hydrogen bond between the doubly bonded nitrogen of the unprotonated imidazole ring and the hydroxyl group of serine residue may create a partial negative charge on the oxygen of the serine which may attack nucleophilically the carbonyl of the substrate (or the phosphoryl of inhibitors).

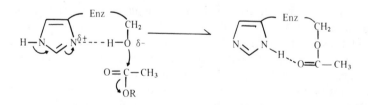

The acylated serine thus formed should be rapidly hydrolyzed. Krupka showed that one of the two active basic groups (B_2; pK_a 5.5) in the catalytic site of acetylcholinesterase functions in acetylation of serine hydroxyl and is located at 9 Å from the anionic site, and another basic group (B_1; pK_a 6.3) functions in deacetylation and is located within 5 Å of the anionic site.[412] The higher pK value of B_1 is probably due to the interaction of carboxylic acid near the base (within 5 Å). This imidazole group activates a water molecule to attack the acetyl serine by abstracting a proton in a similar way as the activation of serine-hydroxyl in the step of acetylation. The active zone of the enzyme must have a suitable steric configuration to permit the joint attack of the active groups, which are not adjacent like imidazoles and serine hydroxyl. According to the induced-fit theory proposed by Koshland, it is not a rigid template, but is flexible and induceable by the substrate to align catalytic groups properly.[421] Thus, the most probable mechanism is presented schematically after Krupka, as shown in Figure 11.[412] In this scheme, the active acidic group AH (pK_a 9.2 to 10.4) appears to be the phenolic hydroxyl group of tyrosine.

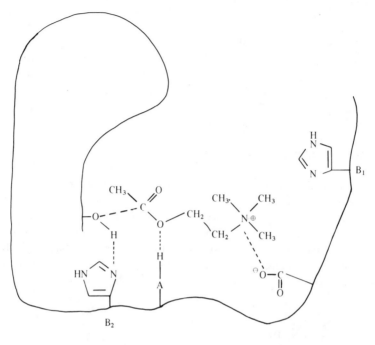

A

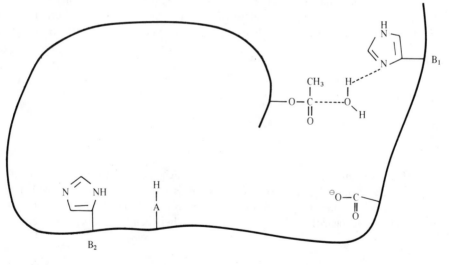

B

FIGURE 11.   Schematic mechanism of action of AChE, after Krupka. A) Enzyme–substrate complex in AChE. B) Deacetylation of acetyl-AChE.

It is also conceivable that the ω-carboxyl group of glutamic acid neighboring the active serine participates in the catalytic reaction of esterases, for instance, as the receptor of the leaving alcohol.[417] Another possibility of its participation is in the step of deacetylation. As demonstrated in the hydrolysis of acetylsalicylic acid, an intramolecular carboxylate ion in the sterically proper position relative to the ester bond accelerates the hydrolysis by nucleophilic attack to form an intermedial anhydride, which is then rapidly hydrolyzed.[422]

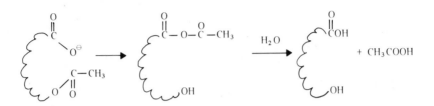

2. Reaction of Organophosphorus Esters with Acetylcholinesterase

a. Inhibition of Enzyme Activity

Unlike the reversible inhibition by quaternary ammonium compounds, the inhibition of acetylcholinesterase by organophosphate esters is irreversible and based on phosphorylation of the esteratic site, i.e., the serine hydroxyl group. The reaction basically corresponds to the acetylation of the site in the normal process for the enzymatic hydrolysis of the true substrate acetylcholine. However, in sharp contrast to the acetylated AChE, which is unstable and hydrolyzed very rapidly (the half-life is about 0.1 msec), the phosphorylated enzyme is much more stable by a factor of at least 10^7, as will be discussed later.[423] The reaction is progressive and temperature dependent.[424] It may be shown by Equation 5.

$$(RO)_2 PX + EOH \xrightarrow{k_i} (RO)_2 POE + HX \tag{5}$$

where k_i is the bimolecular rate constant.

Since the concentration of inhibitor (i) is much higher (at least ten times) than that of the enzyme (e) in most ordinary experiments, and consequently almost constant over the course of reaction, the bimolecular reaction shows first-order kinetics. Thus, the rate of phosphorylated enzyme (p) formation is given by

$$dp/dt = k_i(e-p)i \tag{6}$$

then the rate constant is calculated by:

$$k_i = \frac{2.303}{ti} \log \frac{100}{a} \tag{7}$$

where a is the residual enzyme activity (%) at time t. Although the inhibitory potency is more properly expressed by the rate constant k_i, it is very often expressed by the I_{50} value, which is the molar concentration of the inhibitor needed to cause 50% inhibition of the enzyme activity at a fixed time of incubation, or by its negative logarithm pI_{50}. The relationship between I_{50} and k_i is derived from Equation 7 for $a = 50$.

$$I_{50} = 0.695/tk_i \tag{8}$$

The pI_{50} values of organophosphorus inhibitors generally lie between 6 and 9. Actual organophosphorus insecticides, except phosphates and phosphorothiolates, are much less effective as inhibitors. They are activated in the body of the organism to manifest their biological activity, as will be discussed later.

The phosphorylation of the enzyme is similar to the alkaline hydrolysis of organophosphorus esters (S_N 2 reaction). In place of hydroxide ion, the serine hydroxyl group activated by imidazole serves as a nucleophilic agent to attack the phosphorus atom with a partial positive charge. However, organophosphates react much faster by several powers of ten with the enzyme than with hydroxide ion, suggesting the formation of an enzyme inhibitor complex as an intermediate which favors the subsequent reaction. This is also suggested by the competition of the substrate acetylcholine with organophosphates and the resemblance in structural requirements between suitable substrates and effective inhibitors.[425] The inhibition depends on pH in the same way as the normal enzymatic reaction. Therefore, organophosphates could be regarded as particularly poor substrates and Equation 5 should be amended as follows:

$$EH + PX \underset{k_{-1}}{\overset{k_1}{\rightleftharpoons}} EH{\cdot}PX \overset{k_2}{\longrightarrow} EP + HX \tag{9}$$

where EH is free enzyme, PX is an organophosphorus inhibitor with a leaving group X. The reaction scheme involves a reversible stage to form an enzyme inhibitor complex (EH·PX) analogous to a Michaelis complex with substrate and subsequent phosphorylation that results in an irreversible inhibition.

The first step depends on the affinity of the inhibitor for the active zone of the enzyme and is characterized by a dissociation constant (K_a), which is a measure of the affinity: the smaller the constant, the greater the affinity is.

$$K_a = k_{-1}/k_1 \tag{10}$$

The second step is due to the phosphorylating ability of the inhibitor, which is measured by the phosphorylation constant k_2, a monomolecular rate constant.

It was difficult to demonstrate the formation of the complex until Main recently succeeded in demonstrating this by introducing a simple kinetic treatment to evaluate separately the affinity and phosphorylation ability as follows:[426,427] When the concentration of an inhibitor (i) is much higher than that of the enzyme (e), the rate of phosphorylated enzyme formation is given by Equation 11.

$$\frac{dp}{dt} = k_2 c = \frac{i}{i + K_a} k_2 (e{-}p) \tag{11}$$

where c and p are the concentrations of the enzyme inhibitor complex and phosphorylated enzyme, respectively. Upon integration, Equation 11 gives Equation 12.

$$\frac{1}{i} = \frac{t}{2.3 \Delta \log v} \frac{k_2}{K_a} - \frac{1}{K_a} \tag{12}$$

where $\Delta \log v$ is the velocity change in logarithm of substrate hydrolysis at time interval (t). Thus, by plotting $1/i$ against $t/2.3 \Delta \log v$, K_a and k_2 will be obtained graphically.

The ratio k_2/K_a has the dimensions of the bimolecular rate constant $(M^{-1} min^{-1})$. Main defined $k_i = k_2/K_a$; then Equation 11 is modified as:

$$dp/dt = k_i (e{-}c{-}p)i \tag{13}$$

Thus, k_i is the measure of the overall inhibitory power of the inhibitor and is expressed by Equation 14, which is derived from Equation 12.

$$k_i = \frac{2.3 \Delta \log v}{t} (\frac{1}{i} + \frac{1}{K_a}) \tag{14}$$

The k_i was proposed to be called the bimolecular reaction constant, because it is not a simple rate constant but includes an equilibrium as well as a rate constant. Some examples of these constants are given in Table 16.[427-432]

By an analogy of the enzyme substrate interaction, the model proposed by Krupka may be applied to illustrate schematically the reaction of organophosphorus inhibitors with AChE (Figure 12).[433] In this scheme, S is the anionic site or the hydrophobic binding site (O'Brien's β-site).

It is reasonable to expect that inhibitors resembling the structure of acetylcholine, such as amiton (V), or having a positive charge at a proper distance from phosphoryl group as compound II, show a higher affinity to AChE than others. The affinity is mainly affected by the steric factor. The phosphorylation is essentially an $S_N 2$ reaction on phosphorus, replacing the most acidic group, and the phosphorylation ability is mainly affected by the electronic factor of the inhibitor. The source of enzyme influences, of course, these reaction constants. For example, DFP has 160-fold higher affinity for serum cholinesterase than for erythrocyte acetylcholinesterase.[427]

Although the measurements of K_a and k_2 require relatively high concentrations of inhibitors (10^{-3} to 10^{-4} M), the rates of inhibition are generally measured at lower inhibitior concentrations ($\sim 10^{-6}$ M), which are much less than the value of K_a (10^{-3} to 10^{-4}). Under such conditions, the concentration of the complex can be negligible and the Main equation (14) becomes identical with the classical Aldridge equation (7), which has been widely utilized for the evaluation of the inhibitory potency of organophosphates.

$$k_i = \frac{2.3 \, \Delta \log v}{ti}$$

(15)

b. Recovery of Enzyme Activity
1. Spontaneous Reactivation

Even if the reaction proceeds very slowly, the phosphorylated enzyme is spontaneously hydrolyzed by water, liberating phosphorus acid and the original active enzyme. The reaction may be expressed by Equation 16 and is called spontaneous reactivation or dephosphorylation.

$$EP + H_2O \xrightarrow{k_3} EH + POH$$

(16)

The rate of spontaneous reactivation does not, of course, depend on the leaving group of the original organophosphorus inhibitor, but on the remaining substituted groups on the phosphorus atom and the source of the enzyme. The rate of the reaction may be expressed in terms of the half-life or the catalytic center activity, which is the number of phosphoryl groups hydrolyzed off the enzyme per minute. In Table 17, the hydrolysis rates of phosphorylated enzymes are compared with those of carbamylated enzymes, which are the reaction products from carbamate inhibitors, and acetyl enzymes, the normal intermediates in the course of enzymatic hydrolysis of acetylcholine.

The hydrolysis rate of phosphorylated cholinesterases is lower by a factor of 10^7 to 10^9 than that for normal substrate and is also slower than that of carbamylated enzymes. Spontaneous reactivation does not occur at all in the diisopropoxyphosphinyl cholinesterases. The dimethoxyphosphinyl group is more readily removed than the diethoxyphosphinyl group.

However, the stability of phosphorylated AChE is comparable with that of organophosphorus inhibitors. For example, the half-life for hydrolysis of sarin is about 50 hr and that of diethoxyphosphinyl enzyme from electric eel about 48 hr. Therefore, AChE can be regarded as a relatively good leaving group.[435] Moreover, introducing a chlorine atom into the ethyl group makes the phosphorylated AChE much more unstable; the half-life decreases from 2.4 days to 23 min. This is, however, not the case in the cholinesterase of the parasitic nematode, *Haemonchus contortus*. The nematode cholinesterase inhibited by the anthelmintic Haloxon (bis-(2-chloroethyl) 3-chloro-4-methylcoumarin-7-yl phosphate) is not reactivated at all. This is regarded as a main reason for the selective toxicity of this anthelmintic agent.[436]

TABLE 16

Reaction Constants of Some Organophosphorus Compounds and AChE

Compound		Temp (°C)	K_a (M)	k_2 (min^{-1})	k_i (M^{-1} min^{-1})	Ref.
$(C_2H_5O)_2PO$—C$_6$H$_4$—NO$_2$ (para)	(I)	5	3.6×10^{-4}	42.7	1.2×10^5	428
$(C_2H_5O)_2PO$—C$_6$H$_4$—$\overset{\oplus}{N}(CH_3)_3$ (meta)	(II)	25	1.2×10^{-4}	5.2	4.2×10^4	429
$(CH_3O)_2\overset{O}{\overset{\|}{P}}$—SCHCO$_2C_2H_5$ $\mid$ CH$_2$CO$_2$C$_2$H$_5$	(III)	5	2.4×10^{-3}	67.0	2.8×10^4	428
$(CH_3O)_2\overset{O}{\overset{\|}{P}}$—SCH$_2\overset{O}{\overset{\|}{C}}$NHCH$_3$	(IV)	25	1.38×10^{-2}	6.2	4.5×10^2	430
$(C_2H_5O)_2\overset{O}{\overset{\|}{P}}SCH_2CH_2$N(CH$_3$)$_2$	(V)	5	1.8×10^{-4}	126	7.1×10^5	431
$(CH_3O)_2\overset{O}{\overset{\|}{P}}$O—$\overset{H}{\underset{CH_3}{\overset{\|}{C}}}$=CCO$_2CH_3$ mix.	(VI)	5	2.1×10^{-3}	42	2.0×10^4	432
$(i\text{-}C_3H_7O)_2\overset{O}{\overset{\|}{P}}F$	(VII)	5	1.58×10^{-3}	11.9	7.5×10^3	427

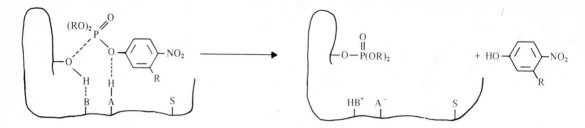

FIGURE 12. Schematic mechanism of reaction of organophosphate with AChE.

TABLE 17

Spontaneous Deacylation of Acylated Cholinesterases

Enzyme derivative	Enzyme source*	Catalytic center activity k_3 (min^{-1})	Half-life	Temp (°C)
$(CH_3O)_2P(O)-E$	rabbit e	0.0085	1.3 hr	37
	horse s	>0.0058	< 2 hr	37
	fly head	<0.0023	> 5 hr^a	37
$(C_2H_5O)_2P(O)-E$	human e	0.0002	2.4 days	37
	human s	0.000016	30 days	37
$(ClCH_2CH_2O)_2P(O)-E$	sheep e	0.03	23 min	36
$(i\text{-}C_3H_7O)_2P(O)-E$	guinea pig e	b	b	
$NH_2C(O)-E$	electric eel	0.35	1.9 min	25
	fly head	0.046	15 min	25
$CH_3NHC(O)-E$	bovine e	0.023	30 min	25
	fly head	0.0097	1.2 hr	25
$(CH_3)_2NC(O)-E$	bovine e	0.0123	56 min	25
	fly head	0.0029	4 hr	25
$CH_3C(O)-E$	bovine e	295,000	0.14 msec	37
	electric eel	610,000	0.07 msec	25

From References 423, 434
*e: erythrocyte, s: serum,
 a: no reactivation in 5 hr,
 b: no reactivation

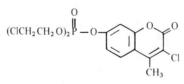

Haloxon

Insect cholinesterases are generally low in catalytic center activity for phosphoryl and carbamyl groups. This is not the case in in vivo. The cholinesterase level of a housefly intoxicated by organophosphorus compounds recovers at a considerable rate owing probably to the action of natural reactivators.[437,438]

As has been surveyed, the fundamental reaction of AChE with organophosphorus esters is essentially identical with that with acetylcholine; both the reactions can be expressed by the common Scheme 4. The most decisive variation is in k_3 and, as a result, the turnover number for diethyl phosphates is one billion times less than that of acetylcholine. Thus, organophosphates behave as inhibitors and acetylcholine as a substrate.

ii. Reactivators and Their Mode of Action

Since the dephosphorylation of phosphorylated enzyme or the reactivation of inhibited enzyme is essentially a nucleophilic displacement reaction on the phosphorus atom leaving the enzyme, some nucleophilic agents can promote the reactivation.

The nucleophilicity of the agent may be determined by some factors such as basicity, polarizability, and α-effect.[435] As mentioned previously (II.D.2 and III.B.1), hard bases, i.e., nonpolarizable bases, preferentially react with phosphoryl group. Thus, fluoride has a considerable reactivating activity.[439] Its efficiency of reactivation increases with decreasing pH and is negligible above pH 9 in contrast to reactivation by oximes. Under physiological conditions, fluoride is much less active than 2-PAM, a well-known oxime reactivator (see below).

As discussed in Section III.A.3.b, some bases having a lone pair of electrons on the adjacent atom show a higher nucleophilic reactivity (α-effect) than expected from their basicity, and catalyze the hydrolysis of phosphorus esters including phosphorylated enzymes. Thus, hydroxylamine and its derivatives, such as oximes and hydroxamic acids, have been investigated in order to find effective reactivators. The reactivation of phosphorylated AChE by such a reactivator, monoisonitrosoacetone (MINA), for example, is pH-dependent and has an optimum pH at about 7.8. The bell shaped reactivation rate pH curve indicates that only the anion of the reactivator is effective and a protonated form of the inhibited enzyme can be reactivated. Therefore, relating to the pK_a of a reactivator, two opposing effects become apparent. The pK_a value of the conjugate acid is a measure for both the supply and the basicity of the anion; a lower pK_a will guarantee a better supply of anion that is the actual reactive species, but will indicate a poorer nucleophilicity. The optimum pK_a is a function of pH, and is given by

$$pK_a \text{(optimum)} = -\log\left(\frac{1}{\beta} - 1\right) + pH,$$

where β is a constant for the same series of reactivator and is greater than zero and less than one. When the value of β is 0.7 and at a physiological pH, say, pH 7.2, the optimum pK_a is 7.6.[435] In the series of pyridinium oximes, the optimum pK_a value for the reactivator was found to be 7.8.[440]

On the other hand, notwithstanding the extremely low concentration of its conjugate base, choline is considerably active as a reactivator. It is reasonable to presume that choline may attach with its positive charge to the anionic site of the inhibited enzyme in a way to arrange the hydroxyl group near the phosphorylated esteratic site, and its nucleophilicity is enhanced by hydrogen bonding with some basic group such as imidazole. Thus, the introduction of a cationic center into the molecule of reactivator at an appropriate distance from the nucleophilic center will make it more active. Some examples are listed in Table 18.[435]

TABLE 18

Effect of Cationic Charge and Position of Nucleophilic Center of Pyridinealdoximes on Reactivation Rate of Acetylcholinesterase Inhibited by TEPP[435] As Second-order Rate Constant (1 mole⁻¹ min⁻¹)

R		
2-CH=NOH	1.8×10^{-1}	1.4×10^{4}
3-CH=NOH	1.6×10^{-1}	1.0×10^{-2}
4-CH=NOH	5.0×10^{-1}	3.0×10^{2}

Tertiary pyridinealdoximes reactivate the inhibited enzyme at a rate which would be expected from their nucleophilicity. However, the quaternary compound 2-PAM (pyridine-2-aldoxime methiodide; pralidoxime iodide) reactivates the inhibited enzyme 2×10^4 times more rapidly than would be expected from its pK_a (8.0). Introduction of a cationic center to the 4-aldoxime also causes high reactivity, but to less extent than in the 2-position isomer, while 3-PAM is less active than the corresponding tertiary compound.*

Thus, the ability of a reactivator (R) to reactivate the phosphorylated AChE (EP) depends on factors that influence the formation of an inhibited enzyme reactivator complex (EP·R), in addition to factors that influence the nucleophilic substitution reaction on the phosphorus. The reaction is described as follows:

$$EP + R \underset{k_{-1}}{\overset{k_1}{\rightleftharpoons}} EP{\cdot}R \overset{k_2}{\longrightarrow} E + PR \qquad (17)$$

$$\text{where } K_r = \frac{[EP][R]}{[EP{\cdot}R]} = \frac{k_{-1} + k_2}{k_1}$$

As shown in Table 19, the equilibrium constant K_r is much lower for the reactivator having a cationic center, such as 2-PAM, than for that having no cationic center, such as MINA.[443]

A further advance in reaching more potent reactivators was made by the introduction of a second cationic center into the molecule by linking two hydroxyiminomethylpyridinium groups together by a chain. In this case, 4-PAM derivatives are more effective than 2-PAM derivatives. Thus, 1,1'-trimethylene-(4-hydroxyiminomethylpyridinium) dibromide (TMB-4; trimedoxime bromide) is about 5 to 50 times more active than 2-PAM and 500 times more active than 4-PAM in reactivating diethoxyphosphinyl AChE.[435] However, trimedoxime is more toxic than 2-PAM (pralidoxime); the intravenous LD_{50} values (mg/kg) for mice are 57 and 94 for the chlorides of trimedoxime and pralidoxime, respectively.[444] This restricts the wide use of trimedoxime as an antidote for organophosphorus compounds. By replacing a methylene with an ethereal oxygen atom in the three-membered bridge, a less toxic and more effective reactivator, oxybis-(4-hydroxyiminomethylpyridinium-1-methyl) dichloride (obidoxime chloride; Toxogonin®) is

TABLE 19

Reaction Constants of Reactivators and Phosphorylated Acetylcholinesterase[443]
(pH 7.4, 25°C)

Reactivator	Phosphorylated enzyme	pK_a of oxime	Kr (mole/l)	k_2 (min^{-1})
MINA	$(C_2H_5O)_2 P(O)E$	8.3	2.0×10^{-2}	1.6×10^{-1}
2-PAM	$(C_2H_5O)_2 P(O)E$	8.0	1.4×10^{-4}	8.1×10^{-2}
2-PAM	$(i-C_3H_7O)_2 P(O)E$	8.0	8.0×10^{-4}	1.5×10^{-2}
4-PAM	$(C_2H_5O)_2 P(O)E$	8.3	3.1×10^{-3}	3.4×10^{-2}

* The steric features surrounding phosphorus in the esteratic site of the phosphorylated AChE were presumed to be best satisfied by the anti-form of 2-PAM and then by the anti-form of 4-PAM, but not by 3-PAM.[441] However, it is now known that the active 2-PAM used is usually syn-form and the syn-4-PAM is more active than the anti-isomer.[442]

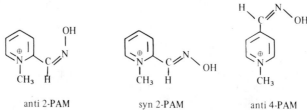

anti 2-PAM   syn 2-PAM   anti 4-PAM

obtained. Its LD_{50} value is 70.0. Obidoxime has been supposed to penetrate the blood-brain barrier and to reactivate inhibited AChE in the central nervous system,[444] though there are objections to this hypothesis.[445]

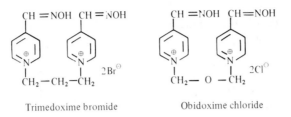

Trimedoxime bromide   Obidoxime chloride

These oximes are useful for therapy of organophosphate poisoning. They are used most desirably in combination with atropine, which blocks the synaptic transmission by replacing acetylcholine to bind with synaptic acetylcholine receptor. These two kinds of drugs are, therefore, complementary in action. Atropine is effective at the parasympathetic nerve terminals, central nervous synapses, and autonomic ganglia, but not at the neuromuscular junctions.[444] Oximes counteract organophosphorus poisons by reactivating inhibited synaptic AChE and also by decomposing organophosphates directly (see III.A.3.b).

The dephosphorylation of phosphorylated AChE by oximes reaches an equilibrium, indicating that the liberated enzyme is phosphorylated again (rephosphorylation) by the produced phosphorylated oxime.[446] The structure of phosphorylated oximes fits in with the P-XYZ system of phosphorylating agents (see Section III.B.2).[222]

$$R-CH=N-O^{\ominus} + \begin{matrix} RO \\ \diagdown \\ P \\ \diagup \\ RO \end{matrix} \begin{matrix} O \\ \diagup\diagup \\ \diagdown \\ O-Enz \end{matrix} \quad \rightleftharpoons \quad R-CH=N-O-\overset{\overset{O}{\parallel}}{P}\overset{OR}{\underset{OR}{\diagup}} + {}^{\ominus}O-Enz \qquad (18)$$

Actually, the phosphorylated products of pralidoxime, trimedoxime, and obidoxime are more active as anticholinesterase agents than original phosphorus inhibitors such as paraoxon and sarin.[446,447] This fact stimulated researchers to investigate for the purpose of finding new insecticides in a series of oxime phosphates and led to the finding of the new insecticide phoxim.

$$(C_2H_5O)_2\overset{\overset{S}{\parallel}}{P}-O-N=\overset{\overset{CN}{|}}{C}-\!\!\!\bigcirc$$

phoxim

In order to avoid the undesirable reverse reaction in the reactivation of inhibited AChE, oximes whose phosphorylated products decompose rapidly should be selected as reactivators. The decomposition rate constants of phosphorylated products of pralidoxime, trimedoxime, and obidoxime are in the range of 10^{-2} min^{-1}. The decomposition rate is related to the acidity of the methine-proton of the phosphorylated oxime.[440]

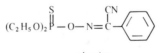

$$R'-\overset{H}{\underset{N-O}{C}}\overset{O}{\underset{}{\diagup\diagup}}\overset{OR}{\underset{OR}{P}} \quad \longrightarrow \quad R'-C=N + \begin{matrix} HO \\ \diagdown \\ P \\ \diagup \\ O \end{matrix} \begin{matrix} OR \\ \diagdown \\ \diagup \\ OR \end{matrix} \qquad (19)$$

The rate of reactivation induced by oximes is also affected by the substituted groups on phosphorus. Diethoxyphosphinyl AChE and isopropoxymethylphosphinyl AChE, which is produced by the reaction of sarin, are readily reactivated by oximes. However, AChE inhibited with DFP is only slowly reactivated. As shown in Table 19, the K_r value of 2-PAM with diisopropoxyphosphinyl AChE is much larger than that with diethyl homolog. This suggests that a bulky group hinders the access of the reactivator to the active zone of the phosphorylated enzyme. Complete prevention of the access of the cationic reactivator by the substituent group is observed in AChE inhibited by cholinyl methylphosphonofluoridate; the quaternary nitrogen of phosphoryl choline may occupy the anionic site of the inhibited enzyme to prohibit the binding of 2-PAM (Figure 13).

The inability of reactivation of AChE inhibited with tabun (VIII) or mipafox (XI) may be due to the lower electrophilicity of the phosphorylated enzyme that is phosphoramidate (see Section III.A.2.b).

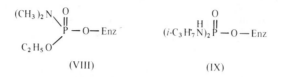

$$(CH_3)_2N \diagdown \quad \overset{O}{\underset{\|}{P}} - O - Enz^- \qquad\qquad (i\text{-}C_3H_7N)_2 \overset{H}{\underset{\|}{P}} \overset{O}{\underset{\|}{}} - O - Enz$$

$$C_2H_5O \diagup$$

(VIII)   (IX)

AChE inhibited by soman (pinacolyl methylphosphonofluoridate) is also very hard to be reactivated by oximes. This may be due to another reason, that is, a rapid "aging," which will be discussed in the following Section.

c. Aging of Phosphorylated Enzymes

As mentioned above, AChE inhibited by organophosphates may be reactivated by treatment with nucleophilic agents like some oximes. However, the inhibited cholinesterases change gradually into a nonreactivatable form on storage. This phenomenon is called "aging." It was assumed that the aging may be caused by a migration of the phosphoryl group from an initial position to form a more stable bond[448] or by the β-elimination of serine phosphate to lose the serine hydroxyl group (see Chapter III, p. 65).[2,449] However, it is now generally accepted that the aging is due to dealkylation of the dialkoxyphosphinyl enzyme (Equation 20).[405] There is a close quantitative parallelism between the degree of aging and the formation of alkoxyhydroxyphosphinyl enzyme in DRP-inhibited pseudocholinesterase.[417,450] The existence of the same mechanism for the aging of rat brain AChE inhibited in vivo with sarin was also confirmed.[451]

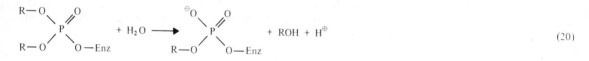

$$R-O \diagdown \quad \diagup O \qquad\qquad \overset{\ominus}{O} \diagdown \quad \diagup O$$
$$\qquad P \qquad + H_2O \longrightarrow \qquad P \qquad + ROH + H^\oplus \qquad\qquad (20)$$
$$R-O \diagup \quad \diagdown O-Enz \qquad\qquad R-O \diagup \quad \diagdown O-Enz$$

After dealkylation, the phosphorylated enzyme becomes stable and resists the nucleophilic attack of oximes because of its negative charge (see Section III.A.1).

The rate of aging is largely affected by the alkyl group, as shown in Table 20. Diethoxyphosphinyl AChE undergoes aging only very slowly, and its half-life at pH 7.4 and 37°C is 40 hr. However, methyl, secondary alkyl, and benzyl esters undergo aging much more readily. The aging of dimethoxyphosphinyl AChE proceeds 10% in 3 hr and 100% in 24 hr at pH 7.0 and 25°C. The half-lives of diisopropoxy- and isopropoxymethyl-phosphinyl AChEs are about 2.5 hr. That of AChE inhibited by soman is only 6 min.[452]

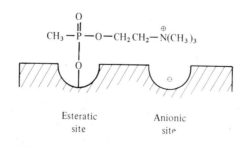

$$CH_3 - \overset{O}{\underset{\|}{P}} - O - CH_2CH_2 - \overset{\oplus}{N}(CH_3)_3$$

Esteratic   Anionic
site   site

FIGURE 13. AChE inhibited by cholinyl methylphosphonoflouridate.

TABLE 20

Aging Rates of Some Phosphorylated Esterases (k hr^{-1} at pH 7.4)

Enzyme derivative	Temp (°C)	AChE	BuChE	Chymotrypsin
$(C_2H_5O)_2P(O)Enz$	37	0.017	—	—
$(i\text{-}C_3H_7O)_2P(O)Enz$	37	0.26	0.28	0.007
$i\text{-}C_3H_7O$ / CH_3 P(O)Enz	37	0.23	0.02	0.000
$(CH_3)_3CCHO$ (CH₃) / CH_3 P(O)Enz	25	6.9	—	—
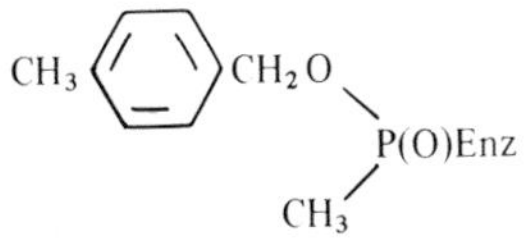 CH_3–phenyl–CH_2O / CH_3 P(O)Enz	25	>12.8	12.0	—
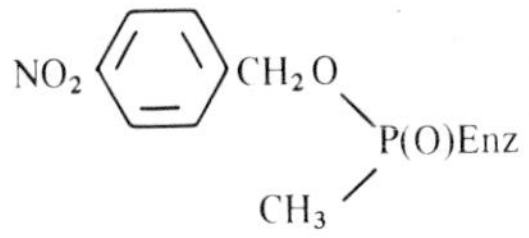 NO_2–phenyl–CH_2O / CH_3 P(O)Enz	25	<0.03	<0.02	—

Data from References 405, 448, 450, and 453.

Aging of p-methylbenzyloxymethylphosphinyl cholinesterases proceeds about 500 times as fast as that of the p-nitro analog.[453] There is a good correlation between the aging rate of substituted benzyloxymethylphosphinyl cholinesterases and the rates of unimolecular solvolysis of the corresponding tosylates.[453] Thus, the rate determining step in the aging reaction may be governed largely by a nonenzymatic chemical force that is the unimolecular fission of the C-O bond in the alkyl phosphate ester group. This indicates that aging occurs readily in the enzyme inhibited by an organophosphorus ester having a high alkylating ability (cf. III.C).

The aging rate is, however, influenced greatly by the chirality around the phosphorus atom.[454] For example, $(R)_P$-cyclopentoxymethylphosphinyl cholinesterase undergoes aging more than 1,000 times as fast as its enantiomer. In cholinesterase inhibited by soman, the configuration around the α-carbon atom in the alkoxy group also influences the rate of aging, but to a lesser extent. Thus, the aging rate of cholinesterase inhibited by $(R)_C(R)_P$-soman is only 7 times less than that of the enzyme inhibited by the isomer around the carbon atom, $(S)_C(R)_P$-soman, but is 1,000 times faster than that of the enzyme inhibited by the stereoisomer around the phosphorus atom, $(R)_C(S)_P$-soman.

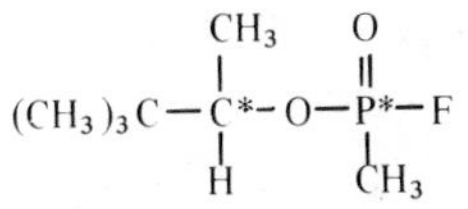

$$(CH_3)_3C-\underset{H}{\overset{CH_3}{C^*}}-O-\underset{CH_3}{\overset{O}{P^*}}-F$$

soman

The aging rate also differs with the enzyme concerned. Phosphorylated chymotrypsin undergoes aging very slightly as shown in Table 20. It appears likely that certain functional groups or structures of the enzyme molecule may be responsible for the aging process in some cases. Denatured DFP-inhibited ChE does not release isopropanol.[417]

In a special case, oximes can also act as the acceptors of the liberated alkyl group. For example, methoxysalicyloxyphosphinyl chymotrypsin cannot be reactivated with MINA, but yields salicyloxy-iminoacetone. Thus, in this case the oxime does not act to dephosphorylate, but to dealkylate from the phosphorylated enzyme; it acts not as a reactivator, but as an aging promoter.[251] A similar phenomenon has been reported for diethoxyphosphinyl chymotrypsin, which is dealkylated by the action of hydroxylamine.[455] As the possible reactivator of aged phosphorylated enzymes, the derivatives of benzonitrile oxide were proposed. The nitrile oxide may realkylate the aged enzyme, forming a neutral ester which has an oxime moiety close to the phosphoryl group.[456]

$$
\text{Ar}-\text{C}\overset{\oplus}{\equiv}\text{N}-\text{O}^{\ominus} + \underset{\underset{\text{HO}}{}{}}{\overset{\overset{\text{R}}{}}{\text{P}}}\underset{\text{O}-\text{Enz}}{\overset{\text{O}}{}} \longrightarrow \quad \longrightarrow \quad + \text{ HO}-\text{Enz} \tag{21}
$$

All the principles discussed in this chapter for the reactions of organophosphorus esters with acetylcholinesterase are also applicable to cholinesterase, other B esterases, and serine proteases in general. The overall reaction of an organophosphorus ester with such an organophosphate sensitive hydrolase may be written as the following scheme:

$$
\text{E} + \text{XPR} \underset{}{\overset{K_a}{\rightleftharpoons}} \text{E}\cdot\text{XPR} \overset{k_2}{\longrightarrow} \text{X}^{\ominus} + \text{EPR} \overset{k_3}{\longrightarrow} \text{E} + \text{PR} \tag{22}
$$

$$
\overset{k_4}{\searrow}
$$

$$
\text{EP}^{\ominus} + \text{R}^{\oplus}
$$

The rates of all the steps of the reaction will be different, depending on the source of the enzyme. A desirable organophosphorus insecticide must have properties to react differently with acetylcholinesterase of man and pest, as shown in Table 21.[423]

3. Structure-activity Relationship in Cholinesterase Inhibition

The relationship between chemical structure and anticholinesterase activity of organophosphorus compounds has been studied extensively. As discussed in the preceding Section 2, the inhibition of AChE by organophosphorus esters is due to the phosphorylation of the esteratic site, serine hydroxyl group, of AChE. Thus, the anticholinesterase activity depends largely on the phosphorylating ability of the ester. Many phosphorylating agents contain an anhydride linkage (Schrader's acyl rule) or a structure fitted with the P-XYZ system, as mentioned in Section III.B.2. In other words, organophosphorus esters containing a

TABLE 21

Reaction Constants with Acetylcholinesterase Required for Good Insecticide (Activated Form)[423]

Step of reaction		Acetylcholinesterase
Affinity for enzyme	K_a	pest > man
Phosphorylation	k_2	pest > man
Dephosphorylation	k_3	pest < man
Aging	k_4	pest > man

readily displaceable group are generally good inhibitors. In addition to the reactivity, the steric properties of the esters are also an important factor that determines the anticholinesterase activity.

a. General Considerations: Some Physicochemical Parameters

Aldridge and Davison first showed in 1952 a direct relationship between the inhibition constant k_i for erythrocyte AChE and the hydrolysis rates of diethyl substituted phenyl phosphates in phosphate buffer.[457] This is widely acceptable for many types of organophosphorus esterase inhibitors. Therefore, any structural factors relating to hydrolyzability, which have been discussed in a previous chapter (III.A.2), correlate more or less with anticholinesterase activity. More extended investigations by Fukuto and Metcalf demonstrated that the inhibition of fly-head AChE by diethyl substituted phenyl phosphates was related to the effect of the substituent on the lability of the P-O phenyl bond, as estimated by Hammett's σ constant, shifts in P-O-phenyl stretching frequencies, and hydrolysis rates.[197,458]

The effect of the substituent on the reactivity of the P-O phenyl bond can be expressed by the well-known Hammett's equation:

$$\log k/k_0 = \rho\sigma \tag{23}$$

where k and k_0 are rate or equilibrium constants for reactions of the substituted and unsubstituted compounds, respectively, ρ is the reaction constant which depends on the reaction and the conditions under which it takes place, and σ is the substituent constant which depends on the nature and position of the substituent.[459] The σ constant is a parameter that indicates the electron-withdrawing or -releasing property of the substituent: the positive σ constant means that the substituent is electron-withdrawing, and the negative constant indicates the substituent has electron-releasing properties.

Phenyl phosphate esters containing such a strongly electron attractive substituent as nitro or cyano group on the benzene ring are strong inhibitors of AChE. The more electron-withdrawing the substituent, the more active as anticholinesterase the phosphate ester is, as shown in Table 22:[197] the reaction constant ρ is positive, consistent with the nucleophilic displacement reaction supposed for the inhibition mechanism. The electron-withdrawing substituent will make the P-O phenyl bond deficient in electron density, and consequently make the ester readily undergo nucleophilic substitution on phosphorus.

Molecular orbital calculation indicated such active inhibitors have a dispositive bond between phosphorus and its connected atom (see Figure 5, III.A.2).[199] Fukui et al. corrected the electron density for the energy level of the molecular orbital to get "superdelocalizability" for a nucleophilic reaction on the phosphorus atom $(S_p^{(N)})$ of diethyl substituted phenyl phosphates, and found good correlation among superdelocalizability, alkali hydrolysis rate, and anticholinesterase activity (Table 22).[460] Excellent reviews for the application of molecular orbital theory to drug research have appeared recently.[461,462]

Table 22 also involves the insecticidal activity of the esters. There are some derivatives whose parameters are poorly correlated with the biological activity. 2,4-Dinitrophenyl phosphate is much less toxic to insects than expected from the anticholinesterase activity. This may be due to the extremely high $S_p^{(N)}$ and consequently to the extremely high reactivity which may result in the decomposition of the compound before its interaction with AChE starts in the insect body. In vivo activity is also influenced by many other factors, such as penetration and metabolism, as will be discussed in later sections. The absence of insecticidal activity in the p-formyl derivative, which should be expected to be considerably active from the parameters, including anticholinesterase activity, might be due to an in vivo metabolic mechanism for its rapid detoxication.

Some compounds which have a substituent on the m-position are much more effective as inhibitors or insecticides than predicted from the electronic parameters σ, $S_p^{(N)}$, and K_{hyd}. This may be due to the steric properties of m-substituents. It is presumed, moreover, that hydrophobic properties of the compounds may relate to conformational changes caused in enzymes, as well as to the distribution of the compounds between aqueous and hydrophobic phases in living tissues. Thus, change in biological activity induced by modification of chemical structure may be expressed as a function of changes in these three properties: electronic, steric and hydrophobic. The Hammett's σ and the Taft's E_s[463] or their modifications can be used as parameters related, respectively, to electronic and steric effects of the

TABLE 22

Physicochemical Parameters Related with Biological Activities of Diethyl Substituted Phenyl Phosphates

$$(C_2H_5O)_2P(O)O \text{—} \bigotimes \text{—} X$$

X	$S_P^{(N)*1}$	$\sigma*^2$	$K_{hyd}*^3$	$pI_{50}*^4$	$LD_{50}*^5$
2,4-di-NO$_2$	1.132	–	5.7×10^{-3}	8.52	155
o-NO$_2$	1.120	–	3×10^{-4}	7.30	7.0
p-NO$_2$	1.119	1.27	2.7×10^{-4}	7.59	0.5
p-CHO	1.106	1.13	–	6.82	>500
p-CN	1.106	1.00	–	6.88	3.5
2,4,5-tri-Cl	1.100	–	7.9×10^{-5}	8.22	8.0
2,4-di-Cl	1.100	–	4.8×10^{-5}	6.30	15.0
o-Cl	1.099	–	5.1×10^{-5}	4.70	250
p-Cl	1.099	0.23	3.2×10^{-5}	4.52	150
m-NO$_2$	1.097	0.71	9.8×10^{-5}	7.30	9.8
H	1.097	0.00	9.2×10^{-6}	<3.00	>500
p-CH$_3$	1.097	-0.17	–	<3.00	>500
p-tert-C$_4$H$_9$	–	-0.20	–	4.00	>500
m-tert-C$_4$H$_9$	–	-0.12	8.6×10^{-6}	6.05	500
m-OCH$_3$	1.096	0.12	8.9×10^{-6}	3.89	>500
p-OCH$_3$	1.096	-0.27	–	<3.00	>500
m-N(CH$_3$)$_2$	1.096	-0.21	1.9×10^{-6}	6.40	25
m-$\overset{+}{N}$(CH$_3$)$_3$	–	0.82	–	7.52	>500

*1. Superdelocalizability on phosphorus.[460]
*2. Substituent constant.[459]
*3. First-order hydrolysis constant at pH 9.5.[197]
*4. Negative logarithm of median inhibitory molar concentration for fly-head AChE in 15 min.[197]
*5. Topical median lethal dose for housefly $\mu g/g$.[197]

substituent. Hansch and Fujita introduced the partition coefficient of a compound between octanol and water as a measure of relative hydrophobic character, and defined the relative hydrophobic constant π of a substituent as the difference in the logarithm of the partition coefficient between the parent (P_0) and the derivative (P).[464]

$$\pi = \log P/P_0 \tag{24}$$

Assuming that the observed biological response is governed by one rate-limiting physical or chemical reaction rate or equilibrium constant (k), they got the following Equation 25 by the linear combination of these three energy-related physicochemical parameters.[464-466]

$$\log k/k_0 = a\pi + \rho\sigma + bE_s \tag{25}$$

where k_0 is the reaction rate or equilibrium constant of the parent compound, and a and b are constants. Biological activities are very often expressed by the concentration (C) to produce a definite biological response in a definite time interval, for example, like LD_{50} and I_{50}. Therefore, the above Equation 25 is further modified as follows:

$$\log 1/C = -a\pi^2 + b\pi + \rho\sigma + cEs + d \tag{26}$$

Using these substituent constants and regression analysis, the relative importance of electronic, steric, and hydrophobic effects of substituents can be separately examined. In certain cases, one or more members of the equation can be deleted to express the biological activity of a series of compounds. This may be because the contribution of the factor(s) is either constant or negligible. When all π values in a series of compounds are extremely larger or smaller than the optimum π value, a becomes zero.

A good correlation was obtained for anticholinesterase activity in a series of the para derivatives of diethyl phenyl phosphate, using σ alone;

$$pI_{50} = 3.451\sigma + 4.461 \qquad (r = 0.954) \tag{27}$$

A similar equation was established for the corresponding phosphonate series, and methyl p-substituted phenyl ethylphosphonates:[467]

$$pI_{50} = 2.46\sigma + 4.31 \qquad (r = 0.98) \tag{28}$$

On the other hand, in the meta substituted derivatives, the anticholinesterase activity was not correlated well by σ and π.[468] However, if a steric parameter E_s^{m*} was employed, a good correlation was found in the meta substituted derivatives as well as in the para substituted ones of diethyl phenyl phosphates by the following equation:[469]

$$pI_{50} = -0.97E_s^m + 2.29\sigma - 1.20X + 5.52 \quad \begin{array}{ccc} n & r & s \\ 13 & 0.980 & 0.313 \end{array} \tag{29}$$

where X is a position term which is 1 for meta derivatives and 0 for para derivatives, n is the number of points used, s is the standard deviation, and r is the correlation coefficient.

Equation 29 indicates that the steric role of the substituent on the meta position is of some importance, suggesting the meta derivatives might inhibit by binding to the anionic site of AChE. The hydrophobic role of the substituent is unimportant for the anticholinesterase activity of diethyl phenyl phosphates, although it is most important in N-methyl substituted phenyl carbamates.[469]

Compared with the effect of the leaving group, that of the alkyl ester group is less significant for reactivity. In a series of paraoxon homologs, unbranched chains of C_1 through C_4 are roughly equipotent for a phosphorylation constant, whereas the isopropyl group causes a considerable drop (about 20 times) in the activity.[428] Isopropyl group also causes a decrease in the affinity with AChE, but to a smaller extent (about 2 to 3 times), probably due to the steric effect of the branched chain. Similar phenomena were also observed in a series of malaoxon homologs. For selectivity in enzyme inhibition, the effect of the alkyl group is very important by virtue of the steric characteristics, as will be discussed later.

b. Effects of Some Groups
i. Thiono and Thiolo Groups

Many practical insecticides contain a thiophosphoryl group (P=S). They are generally almost inactive as the inhibitor of AChE in vitro, so that they must be transformed into oxo-analogs to manifest their biological activity. The poor electron-withdrawing ability of the sulfur atom may be the main reason for the inactivity of thiophosphoryl compounds. Actually, the superdelocalizability on the phosphorus atom of parathion is much less (1.056) in comparison with that of its oxo-analog, paraoxon (1.119).[460] However, the difference in reactivity (alkaline hydrolyzability) between them is only 10-fold, while that in anticholinesterase activity is greater than 10,000-fold. Moreover, the thiono effect on anticholinesterase activity varies greatly with the structure of the inhibitor. For example, the difference in anticholinesterase activity between cyclohexyl methylphosphonofluoridate and its thiono analog is more than 3,000-fold, whereas that between pinacolyl methylphosphonofluoridate and its thiono analog is only 3-fold.[354] These

* The steric parameter is directly related to van der Waals' radii and is given by the equation:[470]

$$E_s = -1.839r_v \text{ (av)} + 3.484 \tag{30}$$

where r_v (av) is the average of the maximum and minimum radii.

facts suggest that additional reasons, including the lack of hydrogen bonding between the sulfur atom and the enzyme active center, should be taken into account for the thiono effect.

Phosphorothiolate esters are generally more active as anticholinesterase agents for corresponding phosphate esters. The possible reasons for their high reactivity are discussed in the preceding Chapter (III.A.2.a). The enhancement of the anticholinesterase activity (the thiolo effect)[354] by replacing an oxygen atom of a phosphate ester linkage with sulfur atom is particulary outstanding in the cases when the original phosphate esters are poor inhibitors (Table 23).

Regardless of the nature of the substituent, any phosphorothiolate ester has considerably high anticholinesterase activity. In other words, the electronic properties of substituent do not affect the inhibitory activity so greatly as in phosphate esters. For example, increase in the activity by the introduction of a p-nitro group into diethyl phenyl phosphate is 20,000-fold, while that for corresponding thiolate analog is only 30-fold. Moreover, an electron releasing methyl group enhances the inhibitory activity more than ten-fold. The same phenomenon was also observed in corresponding ethylphosphono-thiolate. The thiolo effect is not observed in a phosphoramidothiolate.

Thermal or photocatalytic thiono-thiolo rearrangement reaction causes mainly the formation of S-alkyl thiolates during the manufacturing of thiono type pesticides and possibly in the field sprayed with the pesticides. Although such an S-alkyl group is not the leaving group, anticholinesterase activity is greatly enhanced by this reaction. Table 24 shows some examples.[472] The S-alkyl thiolates are 40 to 3,000 times more active than the original thionates, but somewhat less than the corresponding oxidatively desulfurized products.

TABLE 23

Thiolo Effect in Anticholinesterase Activity

| | I_{50} AChE (M) | | |
Compound	X = O	X = S	Ratio
$(C_2H_5O)_2\overset{O}{\overset{\|}{P}}-X-CH_2CH_2N(C_2H_5)_2$	$>10^{-3}$	4.9×10^{-8}	$>20,400$
$(C_2H_5O)_2\overset{O}{\overset{\|}{P}}-X-CH_2CH_3$	$>10^{-2}$	1.4×10^{-5}	>710
$(C_2H_5O)_2\overset{O}{\overset{\|}{P}}-X-CH_2CH_2CH_3$	4.5×10^{-5}	7.0×10^{-6}	6.4
$(C_2H_5O)_2\overset{O}{\overset{\|}{P}}-X-\langle\!\!\bigcirc\!\!\rangle$	$>10^{-3}$	2.8×10^{-7}	$>3,600$
$(C_2H_5O)_2\overset{O}{\overset{\|}{P}}-X-\langle\!\!\bigcirc\!\!\rangle-CH_3$	$>10^{-3}$	2.0×10^{-8}	$>50,000$
$(C_2H_5O)_2\overset{O}{\overset{\|}{P}}-X-\langle\!\!\bigcirc\!\!\rangle-NO_2$	5.5×10^{-8}	9.0×10^{-9}	6
$\overset{C_2H_5O}{\underset{H_2N}{>}}\overset{O}{\overset{\|}{P}}-X-C_2H_5$	1.7×10^{-5}	2.3×10^{-5}	0.7

From References 121, 198, 471.

TABLE 24

Anticholinesterase Activity of Some Thiono Type Insecticides and Their Rearranged and Desulfurized Products

I_{50} fly head AChE (M)

Compounds	X = S, Y = O	X = O, Y = S	X = O, Y = O
C_2H_5Y, C_2H_5O — P(=X) — O — (C₆H₄) — NO_2	2.4×10^{-5}	2.4×10^{-7}	1.1×10^{-7}
CH_3Y, CH_3O — P(=X) — S — CH($CO_2C_2H_5$) — $CH_2CO_2C_2H_5$	4.5×10^{-5}	1.2×10^{-6}	6.0×10^{-8}
CH_3Y, CH_3O — P(=X) — S — $CH_2CONHCH_3$	1.5×10^{-2}	5.6×10^{-6}	2.8×10^{-6}

ii. Amido Group

Since the amido nitrogen atom has a high capability to donate electrons, phosphoramides are relatively less active as anticholinesterases. This effect is stronger in the dialkylamino group than in the monoalkylamino group. Thus, for example, schradan (octamethylpyrophosphoramide) and dimefox (tetramethylphosphorodiamidic fluoride) are almost completely inactive in vitro. Their anticholinesterase activities are greatly intensified by oxidation or chlorination.[285,286,473] Their high toxicity may be due to biooxidative transformation, probably into either N-oxides or methylols. Tabun (ethyl N,N-dimethylphosphoramidocyanidate) is, however, so abnormally active in vitro that the pI_{50} value is 8.4, and is probably due to the high polarizability of the cyano group.[10]

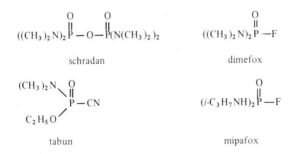

schradan   dimefox

tabun   mipafox

Monoalkylamides are usually more active than dialkylamides. Thus, saligenin cyclic N-methylphosphoramidate inhibits housefly AChE 86% at 10^{-7} M.[474] Mipafox (diisopropylphosphorodiamidic fluoride) inhibits 50% of erythrocyte AChE at $10^{-4.6}$ M, and plasma ChE at $10^{-7.3}$ M. In all the series of alkyl aryl phosphoramidates, alkyl S-aryl phosphoramidothiolates, and O,S-dialkyl phosphoramidothiolates, anticholinesterase activity decreases in the order: $NH_2 > NHR > NR_2$.

Hansch and Deutsch analyzed data obtained by Fukuto et al. from a series of methyl 2,4,5-trichlorophenyl N-alkylphosphoramidates (X) in order to clarify the effect of the N-alkyl group on the inhibitory activity for fly-head AChE.[468,475] They found that the logarithm of the bimolecular inhibition constant k_i is correlated excellently with Taft's steric constant E_s and polar constant σ^* of substituent;

$$\log k_i = 2.359E_s - 3.913\sigma^* + 4.948 \qquad \begin{array}{ccc} n & r & s \\ 8 & 0.939 & 0.438 \end{array} \qquad (31)$$

This equation shows the great importance of the steric effects: the bulky isopropyl and *tert*-butyl moieties decrease inhibition rates by steric interference. Moreover, the negative coefficient with σ^* means that electron releasing groups are favorable for the inhibition of fly-head AChE, if their steric effects are not significant. In this context, it is interesting to note that phosphoramidates are generally susceptible to acid hydrolysis due to the formation of a positive charge on the nitrogen atom by protonation (see III.A.2.b) and that the esteratic site of AChE involves an acid group (IV.A.1.c). Phosphoramidates carrying more electron releasing groups on nitrogen may be activated more readily by the acid group.

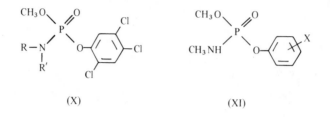

(X)   (XI)

On the other hand, the ring substituents of methyl phenyl *N*-methylphosphoramidates (XI) directly affect the anticholinesterase activity by virtue of the electronic and hydrophobic properties, as shown by the following equation.[476]

$$\log k_i = 1.17 \log K_{hyd} + 0.905\pi + 3.21 \qquad \begin{array}{ccc} n & r^2 & s \\ 9 & 0.92 & 0.315 \end{array} \tag{32}$$

However, in a series of ethyl *S*-(substituted)-phenyl phosphoramidothiolates, no correlation was observed between the rates of cholinesterase inhibition and hydrolysis rates or any of the free energy parameters for ring substituents.[215] Moreover, anticholinesterase activity of phosphoramidothiolates is not always correlated with their insecticidal activity.

Dialkyl *N*-acetyl-*N*-phenylphosphoramidates (XII) were recently found to be good inhibitors of housefly AChE.[306] The anticholinesterase activity decreases with the introduction of any ring substituent, regardless of its electronic characteristics.

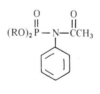

(XII)

iii. Enol Ester Group

Several enol phosphates have been developed as insecticides. They are very potent anticholinesterase agents. However, the anticholinesterase activity does not appear to be correlated with alkaline hydrolyzability. For example, the ethyl homolog of α-Bomyl (XIII) is hydrolyzed with alkali 575 times more rapidly than the ethyl homolog of α-mevinphos (XIV), whereas the former is 1.5 times less active in inhibitory activity than the latter.[219]

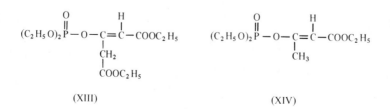

(XIII)   (XIV)

Since the phosphorylating ability of enol phosphates is promoted by protonation (see III.B.2),[477] it is presumable that the cholinesterase inhibition may be caused by the interaction of the acid group in the esteratic site. Such an interaction may be influenced by the steric factors. *cis*-Mevinphos is 10 to 20 times more active than the *trans*-isomer as an anticholinesterase agent. The activation entropy (90.3 eu) of the reaction of the *cis*-isomer with fly-head AChE is much greater than that for the *trans*-isomer (6.6 eu), suggesting that the carbomethoxy group *trans* to the dimethyl phosphoryl moiety is less likely to interfere sterically with the reaction between the esteratic site and the phosphorus.[214] Kinetic study by other investigators indicated, however, that the phosphorylation rate was more affected by the geometrical isomerism than the affinity for erythrocyte AChE.[432] There is little difference in anticholinesterase activity between *cis* and *trans* isomers of Bomyl.[478]

The high anticholinesterase activity of diazoxon (XV) and related imidoyl phosphates is also presumed to be due to protonation on the nitrogen atom by the acidic group of the enzyme. Their susceptibility to acid is well known (see III.A.2.c).

Oxime phosphates also have high anticholinesterase activity. The oxo-analog of the insecticide phoxim (XVI) is so selective to insect cholinesterase as to inhibit housefly AChE 270 times faster than erythrocyte AChE.[479]

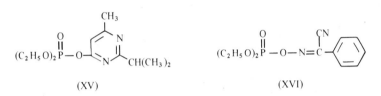

(XV)   (XVI)

The effect of substituents on the anticholinesterase activity of diethyl phosphate esters derived from *p*-substituted acetophenoximes was opposite to that for substituted phenyl phosphates: electron releasing groups enhanced the inhibitory activity.[480] This apparent abnormality was due to the fact that the oxime phosphate preparations contained highly reactive imidoyl phosphates which were readily produced from the oxime phosphates, particularly those having an electron releasing substituent, through the Beckmann-type rearrangement, by heat, during preparation.[306]

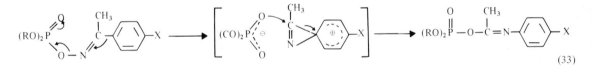

(33)

Actually, the pure preparation of diethyl *p*-methoxyacetophenoxime phosphate was 500-fold less effective in inhibiting fly-head AChE ($I_{50} = 3.45 \times 10^{-6}$ M) than the crude preparation containing the rearranged product ($I_{50} = 7.2 \times 10^{-9}$). There is also a possibility that the crude preparations contained tetraethyl pyrophosphate, which may be formed in the presence of small amounts of water from the imidoyl phosphate.

iv. P-C Bond

Phosphonates are generally more reactive than the corresponding phosphate esters because of the absence of $p\pi$-$d\pi$ contribution (see III.A.2.e). However, it is not always the case in anticholinesterase activity. The cholinesterase inhibition and insecticidal activity within a series of alkylphosphonate esters are generally less dependent on variations in the alkyl and alkoxy groups than they are in the corresponding phosphate series having two alkoxy groups.

When the alkyl group of ethyl *p*-nitrophenyl alkylphosphonate is increased in size from ethyl to *n*-hexyl, the inhibition rate (k_i) for housefly-head AChE drops to a greater extent than expected from the drop in hydrolyzability, suggesting that the larger alkyl groups prevent proper contact of the inhibitor with the esteratic site.[237] Multiple regression analysis showed that k_i was correlated well with the corrected Taft's steric substituent constant E_s^c as follows:[469]

$$\log k_i = 2.58\ E_s^c + 7.94 \qquad \begin{array}{ccc} n & r & s \\ 13 & 0.927 & 0.648 \end{array} \qquad (34)$$

Thus, the steric role of the alkyl group is most important for the anticholinesterase activity of the phosphonate series.

It may be interesting to discuss whether P-C bond cleavage occurs for the reaction with AChE. The P-C bond in the cyanidate tabun is undoubtedly active and responsible for the anticholinesterase activity. Trichlorfon (dimethyl α-hydroxytrichlorethylphosphonate) inhibits AChE, forming the inhibited enzyme whose reactivation rate is almost identical with that of dimethoxyphosphinyl AChE produced by dichlorvos (dimethyl dichlorovinyl phosphate).[481] The anticholinesterase activity of trichlorfon decreases by lowering the pH; housefly AChE was completely inhibited above pH 8 by $10^{-7}\ M$ trichlorfon, while only 11% at pH 5.4.[482] In contrast with trichlorfon, dichlorvos inhibited the enzyme completely over the entire pH range at $10^{-8}\ M$. As mentioned previously (III.A.2.e), trichlorfon is rather stable under acidic conditions but readily undergoes a nonenzymatic dehydrochlorination and rearrangement to form dichlorvos under alkaline conditions. Thus, the vinyl phosphate formed during the inhibition process may be responsible for the major, if not whole, part of the anticholinesterase activity of the phosphonate.

$$(CH_3O)_2\overset{O}{\underset{\underset{OH}{|}}{\overset{\|}{P}}}-CH-CCl_3 \qquad (CH_3O)_2\overset{O}{\overset{\|}{P}}-O-CH=CCl_2$$

trichlorfon   dichlorvos

On the other hand, Melnikov reported the high insecticidal and anticholinesterase activities of diethyl ethylthiocarbonylmethylphosphonate (XVII) and related compounds, suggesting that the inhibition of cholinesterase is probably not due to phosphorylation.[483]

$$(C_2H_5O)_2\overset{O}{\overset{\|}{P}}-CH_2\overset{O}{\overset{\|}{C}}SC_2H_5$$

(XVII)

v. Cyclic Ester Structure

Table 25 shows the anticholinesterase activity of several five- and six-membered cyclic phosphates and phosphoramidates. Although five-membered cyclic phosphate esters are very reactive, as mentioned in Section III.A.2.d, they are only poor inhibitors of AChE, with few exceptions. It has been supposed that they are too unstable to allow the phosphorylation reaction of the enzyme to proceed.[232,484] On the other hand, five-membered cyclic phosphoramidates have moderate anticholinesterase and insecticidal activities.[485] Some cyclic phosphorothionates derived from catechol have considerable inhibitory activity for plasma ChE, but no insecticidal activity.[486] Catechol cyclic phosphate diesters phosphorylate chymotrypsin stoichiometrically to inhibit the enzyme activity.[236]

Six-membered cyclic phosphate esters are not so unstable as five-membered esters, and the alkaline hydrolizability of their p-nitrophenyl ester derivatives (p-nitrophenoxy-1,3,2-dioxaphosphorinane-2-oxides; $K_{OH} = 0.23 \sim 1.56\ M^{-1}\ min^{-1}$) is comparable to that of the corresponding acyclic ester paraoxon (diethyl p-nitrophenyl phosphate; $K_{OH} = 0.94\ M^{-1}\ min^{-1}$). In spite of the proper reactivity, their anticholinesterase activity is only 1/100,000 that of paraoxon. Fukuto attributed their inactivity to the steric effects of the cyclic structure.[232] However, six-membered cyclic phosphates derived from o-hydroxybenzyl alcohol (saligenin), which are characterized by a hetero ring involving an enol and a benzyl ester linkage, show high anticholinesterase activity without any electron-withdrawing substituents on the benzene ring.[234,487]

TABLE 25

Anticholinesterase and Insecticidal Activities for Five- and Six-membered Ring Phosphate Esters

Compound	I_{50} fly AChE (M)	LD_{50} housefly (μg/g)
CH₃ structure with P(=O), O, O-phenyl-NO₂, CH₃	$>1.3 \times 10^{-3}$	>500
six-membered ring O,O,P(=O)-O-phenyl-NO₂	$>1.3 \times 10^{-3}$	>500
$(CH_3)_2CHCH_2$—NH,O ring P(=O)—OCH₃	3.0×10^{-6}	16
NH,O ring P(=O)—OCH₃	$>1 \times 10^{-3}$	>500
benzo O,O ring P(=S)—O-phenyl-NO₂	8.0×10^{-8} *	>500
benzo six-membered ring O,O,CH₂ P(=O)—OCH₃	7.6×10^{-8}	1.7
C_2H_5O, C_2H_5O P(=O)—O-phenyl-NO₂	2.6×10^{-8}	0.5

From References 232, 234, 485, 486
*For plasma ChE.

Studies on the reaction with chymotrypsin indicated that a saligenin cyclic phosphate ester inhibits the enzyme by stoichiometric phosphorylation accompanied by opening of the enol ester bond.[489] The inhibition of acetylcholinesterase may proceed similarly.

4. Steric Effects and Selectivity in Enzyme Inhibition
a. Effects of Bulky Groups
Some examples of steric effects on anticholinesterase activity were described in the preceding sections:

TABLE 26

Specificity in Housefly Esterase Inhibition of

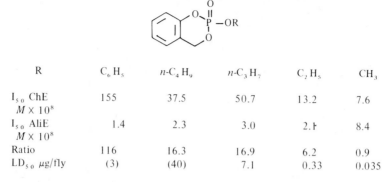

R	C_6H_5	$n\text{-}C_4H_9$	$n\text{-}C_3H_7$	C_2H_5	CH_3
I_{50} ChE $M \times 10^8$	155	37.5	50.7	13.2	7.6
I_{50} AliE $M \times 10^8$	1.4	2.3	3.0	2.1	8.4
Ratio	116	16.3	16.9	6.2	0.9
LD_{50} $\mu g/fly$	(3)	(40)	7.1	0.33	0.035

Figures in parentheses are mortality (%) at 10 $\mu g/fly$.

they included alkyl groups in dialkyl phenyl phosphates, alkylphosphonates, and *N*-alkylphosphoramidates, geometric isomers of vinyl phosphates, and meta substituents of dialkyl phenyl phosphates.

As enzyme action is greatly influenced by steric factors even a small modification of certain substituents often causes great selectivity in enzyme inhibition. Such examples will be given in this section.

When the alkyl group of ethyl *p*-nitrophenyl alkylphosphonates increases from C_3 to C_{10}, anticholinesterase activity decreases little between C_3 and C_5 (about ten-fold) and reaches almost constant activity above C_5, whereas their inhibitory potency against chymotrypsin and trypsin attains maxima at C_7 and C_6, respectively.[488]

The selectivity of saligenin cyclic phosphates to inhibit housefly AChE and aliesterase changes greatly by the size of exocyclic ester group, as shown in Table 26.[487] In spite of higher reactivity, the phenyl derivative is much less active than the alkyl derivatives in AChE inhibition. The structure-activity relation is reversed for the inhibition of aliesterase. The smaller the substituent, the more selective to AChE and the less to aliesterase the inhibitor is. There is, thus, a significant difference in behavior against insect esterases between the highly insecticidal methyl phosphate, salioxon, and the noninsecticidal phenyl phosphate. Tri-*o*-tolyl phosphate (TOCP) has been known as a specific inhibitor of aliesterase in vivo. It was demonstrated that TOCP is transformed in vivo into *o*-tolyl saligenin cyclic phosphate, the actual biologically active principle.[489] The cyclic TOCP-metabolite is about 130-fold more specific to aliesterase than to cholinesterase of housefly.[487]

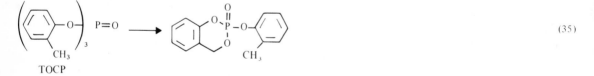

(35)

TOCP

Comparing with true AChE, pseudo ChE is preferably inhibited by organophosphorus compounds carrying bulky substituents like an isopropyl group.[425] Thus, iso-OMPA (*N,N'*-diisopropyl phosphorodiamidic anhydride) and mipafox (*N,N'*-diisopropyl phosphorodiamidic fluoride) are more than 1,000-fold more specific to pseudo ChE than to true AChE of horse or dog blood. The specificity varies according to the enzyme sources, and is only 56-fold specific for the enzymes from man. In a series of paraoxon homologs, the effect of the isopropyl group on the selectivity is not as great as in amide type inhibitors: the inhibitory ratio for pseudo ChE to AChe (I_{50} AChE/I_{50} ChE) of isopropyl paraoxon and methyl paraoxon is 10 and 0.7, respectively. It has been known that bee AChE is much less sensitive to isopropyl paraoxon than housefly AChE,[490,491] while both the enzymes are equally sensitive to ethyl paraoxon. This may be responsible for the high selective toxicity of isopropyl parathion; it is at least 100 times more toxic to

TABLE 27

Effect of Cationic or Bulky Substituents at β-Position of Alkyl Ester Group on the Anti-ChE Activity of Phosphorus Esters

Phosphorus ester	pI$_{50}$ AChE	Ref.
$(C_2H_5O)_2\overset{\text{O}}{\overset{\|}{P}}SCH_2CH_2\overset{+}{N}(CH_3)_3$	8.4 (human)	493
$(C_2H_5O)_2\overset{\text{O}}{\overset{\|}{P}}SCH_2CH_2\overset{+}{S}\overset{C_2H_5}{\underset{CH_3}{\big<}}$	8.3 (sheep)	257
$(C_2H_5O)_2\overset{\text{O}}{\overset{\|}{P}}SCH_2CH_2\overset{\overset{\uparrow}{O}}{\underset{\downarrow}{\underset{O}{S}}}C_2H_5$	6.2 (fly)	204
$(C_2H_5O)_2\overset{\text{O}}{\overset{\|}{P}}SCH_2CH_2SC_2H_5$	5.5 (fly)	204
$(C_2H_5O)_2\overset{\text{O}}{\overset{\|}{P}}SCH_2CH_2CH_2CH_3$	5.0 (erythrocyte)	472
$(C_2H_5O)_2\overset{\text{O}}{\overset{\|}{P}}OCH_2CH_2C(CH_3)_3$	6.5	458
$(C_2H_5O)_2\overset{\text{O}}{\overset{\|}{P}}OCH_2CH_2CH_2CH_3$	3.5 (erythrocyte)	472

houseflies than to bees. The selectivity in anticholinesterase activity of isopropyl paraoxon is caused by its poor affinity for bee AChE, being 10 times lower than that for fly AChE, and by the high phosphorylation reactivity for fly AChE, being 4 times more reactive than that for bee AChE.[214,492]

b. Choline-like Structure

Organophosphorus compounds which structurally resemble acetylcholine have much higher anticholinesterase activity than expected from the chemical reactivity (Table 27). Phosphorus esters having a cationic head, such as quaternary ammonium or sulfonium moiety which is separated from phosphorus by three atoms (about 5 Å distance), are typical examples. The cationic head may bind with the anionic site of AChE; consequently, the phosphoryl group may approach proper position to phosphorylate the esteratic site. Sulfone is also effective, though not so extensive as sulfonium. Moreover, even a noncharged tertiary butyl group exerts a considerable effect, indicating that such a hydrophobic bulky group can fit into a complementary hole in the enzyme and contribute to enzyme-inhibitor binding.

It has been mentioned already that the anticholinesterase activity towards housefly ChE of diethyl phenyl phosphate is greatly enhanced by the introduction of *t*-butyl or trimethylammonium group in the meta position (Table 22). In this context, it is interesting to discuss the high selective toxicity of some insecticides which have a methyl group on the meta position of the phenyl ester group, such as fenitrothion and fenthion.

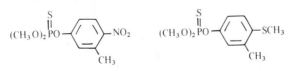

fenitrothion   fenthion

TABLE 28

Effects of 3-Methyl Group on Biological Activities of Dimethyl *p*-Nitrophenyl Phosphate and Phosphorothionate[494]

Compound	Toxicity (X = S) LD_{50} (mg/kg)		Anticholinesterase activity (X = O)					
			$k_i \times 10^{-5}$ $(M^{-1} min^{-1})$		$K_a \times 10^5$ (M)		k_p (min^{-1})	
	F	M	F	E	F	E	F	E
$(CH_3O)_2 \overset{X}{\underset{\parallel}{P}} O \!-\! \langle \rangle \!-\! NO_2$	1.2	23	2.9	5.2	3.7	1.3	10.6	6.6
$(CH_3O)_2 \overset{X}{\underset{\parallel}{P}} O \!-\! \langle \rangle \!-\! NO_2$ (CH₃)	3.1	1,250	7.6	0.73	1.1	6.7	8.3	5.0

F: housefly; M: mouse; E: Bovine erythrocyte.

Fenitrothion is 180 times less toxic to mammals than parathion-methyl, but is comparable to the latter in insecticidal activity. The anticholinesterase data of their oxo-analogs, shown in Table 28, indicate that introduction of a methyl group in the 3-position of the phenyl ring renders the compound more inhibitory to housefly-head AChE, while less to mammalian AChE.[494] Such an effect of the 3-methyl group appears general. For example, the selectivity ratio of *p*-acetylphenyl diethyl phosphate and its 3-methyl derivative for housefly AChE to human erythrocyte AChE is 60 and 1,500, respectively.[495] The 3-methyl effect is primarily due to change in affinity with the enzyme: the 3-methyl group causes the affinity to fly AChE to increase and that to mammalian AChE to decrease, as shown in Table 28. When an isopropyl group is introduced in 3-position, the effects are more significant. Phosphorylating activity decreases slightly for both enzymes by the methyl group, probably due to its electron-releasing effect and to the inhibition of resonance of the nitro group by steric interference. Affinity with the enzyme should be dependent largely on the steric features of the molecule. The formation of the enzyme-inhibitor complex with housefly AChE appears to be assisted by the interaction of the 3-alkyl group with the "anionic site" but the complex formation is hindered in the case of mammalian AChE. Hollingsworth et al. suggested, therefore, that the distance between the anionic site and the esteratic site differs between insect (5.0 to 5.5 Å) and mammalian (4.3 to 4.7 Å) cholinesterases, and that the 3-alkyl group, being 5.2 to 6.5 Å distant from the phosphorus atom, fits well on the anionic site of the insect enzyme but poorly on the mammalian enzyme, as illustrated in Figure 14.[494]

c. Effect of Asymmetry

Although it is well recognized that enzymes exhibit stereospecificity, in general, in their reaction with asymmetric carbon compounds, the effect of asymmetry in the phosphorus molecule on the inhibitory activity for AChE and other organophosphorus sensitive enzymes has been scarcely investigated. The first suggestion for the stereospecificity of AChE in the inhibition reaction with asymmetric organophosphorus compounds was presented in 1955 by Michel's observation that sarin (isopropyl methyl phosphono-fluoridate) reacts with AChE as if two components were present in equal amounts, one of which reacted rapidly, the other slowly.[495a] Because the sarin molecule is asymmetric, the existence of enantiomers may be responsible for the biphasic behavior of the inhibition reaction.

sarin

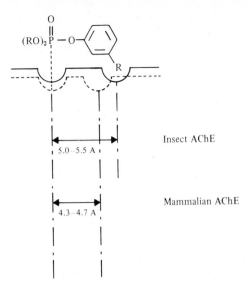

FIGURE 14. Interaction of 3-alkyl group with anionic site of acetylcholinesterases.

Since Aaron et al.[49] first succeeded, in 1958, in demonstrating that the levo-isomer of ethylphosphonate analog of demeton-S reacted 10 to 20 times faster with cholinesterases than the dextro-isomer, several enantiomers of biologically active organophosphorus compounds have been isolated and examined for their inhibitory activity on enzymes. The results which have been obtained are summarized in Table 29. It is interesting to note that AChE is highly stereospecific for asymmetric organophosphorus inhibitors, although its natural substrate, acetylcholine, has no asymmetric center. Acetylcholinesterase is generally more stereospecific than pseudocholinesterase.

Toxicity is also affected by asymmetry. The levo-isomer of *O*-ethyl *S*-2-(ethylthio)ethyl ethylphosphonothiolate is ten times more toxic to several insects than the dextro-isomer. The levo-isomer of *O,S*-dimethyl *p*-nitrophenyl phosphorothiolate is about five times more toxic to rats than the dextro-isomer. Recently, the toxicological properties of the four isomers of *O-sec*-butyl *S*-2-(ethylthio)ethyl ethylphosphonothiolate (XVIII) were reported.[214] They have two asymmetric centers, i.e., phosphorus and α-carbon of *sec*-butyl group.

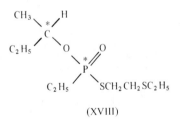

(XVIII)

The compound with both carbon and phosphorus atoms designated as levo (levo-C-levo-P) was the strongest inhibitor of AChE. The effect of asymmetry in the phosphorus moiety on anticholinesterase activity is much greater than that in the *sec*-butyl moiety. The latter is, however, also significant. Thus, the relative anticholinesterase activity of the four isomers for fly-head AChE was 1, 2, 1,270, and 1,440 for dextro-C-dextro-P, levo-C-dextro-P, dextro-C-levo-P, and levo-C-levo-P, respectively. Levo-P appears essential for the activity of XVIII. Similar stereospecificity was observed for bovine erythrocyte AChE. However, in the inhibition of horse serum ChE, stereospecificity was almost absent. Insecticidal activity of these isomers was consistent with the anticholinesterase activity. The two enantiomers containing levo-P showed the greatest insecticidal activity.

TABLE 29

Stereospecificity of Hydrolases in Reaction with Enantiomers of Asymmetric Organophosphorus Compounds

Asymmetric compound	Enzyme	Ratio of activity*	Ref.
CH₃O, O / CH₃S—P—O—⟨C₆H₄⟩—NO₂ (structure)	Rat-brain AChE	5.5	47
C₂H₅S, O / CH₃—P—O—⟨C₆H₄⟩—NO₂ (structure)	Bovine e-AChE	13.1	48
	Horse s-ChE	0.86	48
	Aliesterase	0.15	496
	Acetylesterase	49	496
	Chymotrypsin	0.29	496
	Trypsin	1.5	496
C₂H₅O, O / C₂H₅—P—SCH₂CH₂SC₂H₅ (structure)	Eel AChE	18	49
	Human e-AChE	9.6	49
	Bovine e-AChE	9.6	49
	Horse s-ChE	20	49
	Housefly AChE	11.3	50
i-C₃H₇O, O / CH₃—P—F (structure)	Bovine e-AChE	4,200	52
	Horse s-ChE	1	52

*k_i for levo-isomer/k_i for dextro-isomer, or I_{50} for dextro-isomer/I_{50} for levo-isomer.

Asymmetry in the succinate moiety, the leaving group, of malaoxon also affects the inhibitory activity toward esterases. *d*-Malaoxon was 4 and 8 times better than the *l*-isomer in the inhibition of AChE and carboxyesterase, respectively.[55]

$$(CH_3O)_2 \overset{\displaystyle O}{\overset{\displaystyle \|}{P}} - S - \overset{\displaystyle H}{\underset{\displaystyle CH_2COOC_2H_5}{\overset{\displaystyle |}{\underset{\displaystyle |}{C^*}}}} - COOC_2H_5$$

malaoxon

This may be responsible for the higher mammalian toxicity of *d*-malathion (the oral LD_{50} to mice 1,014 mg/kg) in comparison with the *l*-isomer (LD_{50} 2,357 mg/kg).

B. METABOLISM

Foreign compounds undergo a variety of metabolism in the body of organisms. Pesticides are usually less polar and are lipophilic in nature in order to penetrate readily into the pests through the skin, and the biotransformation generally proceeds in a direction in which the molecule becomes more polar. Such biological modifications of organophosphorus pesticides result in the great change in the toxicity of the compounds. The majority of organophosphorus insecticides, except phosphates and phosphorothiolates, inhibit little acetylcholinesterase in pure state unless they are activated. Their activity as effective anticholinesterase in vivo is the net result of competing biochemical processes: activation and detoxication. Activation means the metabolic conversion of intrinsically inactive compounds to active compounds or the conversion of active compounds to other active ones. Detoxication means the biotransformation of compounds to nontoxic substances. Detoxication or degradation is by far the major route of organophosphate ester metabolism, and activation itself can be regarded as a process resulting finally in degradation.

Metabolism often causes a compound to change to an extremely toxic product. Even if such a biotransformation occurred only to a very small extent, it is very important from the toxicological point of view.

The general types of biological reactions that are most prominent in the modification of organophosphorus compounds involve oxidation, hydrolysis, group transfer, reduction, and conjugation. The pattern of metabolism may differ by the chemical structure of a pesticide molecule and by the species of organisms. Effective pesticides should be readily activated in pests but rapidly degraded and excreted in man and other nontarget organisms.

Many excellent reviews for the metabolism of xenobiotics, including organophosphorus pesticides in animals and plants, have been published.[32,295,301,302,497-509]

1. Reactions Resulting in Activation

The biotransformation of essentially inactive organophosphorus compounds into active metabolites may be classified into the following five types of patterns:

1. Oxidative desulfuration of thiophosphoryl group: formation of phosphoryl compounds.

2. Oxidation of sulfide moiety: formation of sulfoxide and sulfone.

$$-S- \longrightarrow -SO- \longrightarrow -SO_2-$$

3. Oxidation of amides: methylphosphoramides may be converted to N-oxide or N-methylol, and N-alkylcarboxyamide may be dealkylated.

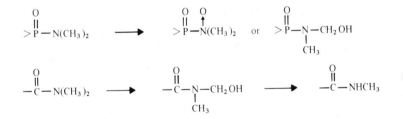

4. Hydroxylation of alkyl group followed by subsequent reactions: formation of cyclic phosphate esters or ketones.

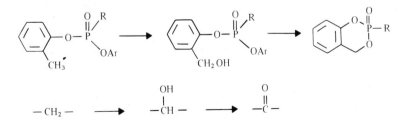

5. Miscellaneous nonoxidative reactions: rearrangement of α-hydroxy-β-haloalkylphosphonates, removal of protective group, and others.

Thus, the bioactivation of organophosphorus compounds is mainly due to oxidative reactions, and microsomal oxidases are responsible for these biological oxidations.[510,511]

a. Microsomal Mixed-function Oxidase System

When tissue homogenates are subjected to differential centrifugation, the microsomal fraction is obtained as the high speed pellet (from 100,000 to 250,000 X g for 60 to 120 min) from the supernatant fluid after the cell debris, nuclei, and mitochondrial fractions have been removed at 10,000 X g for 20 min. Microsomes are derived by homogenation from the endoplasmic reticulum, which is a tubular network of lipoprotein extending throughout the cytoplasm. There are two types of the endoplasmic reticulums: rough and smooth. The former is studded with ribosomes, which serve an important role in biosynthesis of proteins, but the latter is free of the particles. The smooth endoplasmic reticulum contains high enzyme activity to oxidize a great variety of lipophilic substrates including steroids, lipids, and foreign organic compounds.

Such oxidative reactions, catalyzed by the microsomal oxidases for the biotransformation of xenobiotics, are listed in Figure 15. Some of them relate to the activation of the compound, others to detoxication.

The reactions require molecular oxygen and reduced nicotinamide adenine dinucleotide phosphate (NADPH). Reduced nicotinamide adenine dinucleotide (NADH) may replace NADPH, though it is generally less effective. One of the atoms of molecular oxygen is incorporated into the substrate and the other is reduced to water. The reaction may be generalized as follows:

$$RH + NADPH + H^+ + O_2 \longrightarrow ROH + NADP^+ + H_2O \tag{36}$$

The enzymes are called "mixed function oxidases" (mfo),[512] "monooxygenases."[513] or "drug metabolizing enzymes."[514] These enzymes are abundant in the vertebrate liver and the fat body, Malpighian tubules and the digestive tract of insects. The activity differs by animal species and decreases in the following order: mammals > avian > fish. The mfo activity of female rat liver microsomes is significantly lower than that of the male rat. Ascaris appears to lack the microsomal oxidase activity.[515]

The microsomal mfo enzymes show poor substrate specificity and require only high lipophilicity of the substrate. A variety of substrates competitively inhibit the hydroxylation of a given substrate, suggesting that only one or a few enzymes are responsible for the hydroxylating activities of liver microsomes. 2-Diethylaminoethyl 2,2-diphenylvalerate hydrochloride (SKF 525A) and methylenedioxyphenyl compounds (1,3-benzodioxoles), which are the well-known inhibitors of the microsomal mixed-function oxidases, may exert their inhibitory action, at least in part, by serving as alternative substrates.[516]

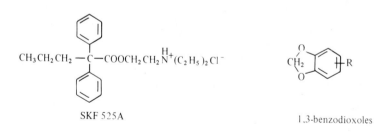

SKF 525A

1,3-benzodioxoles

The microsomal hydroxylation of foreign compounds depends on the participation of a special microsomal pigment, which is called cytochrome P-450 because an intense absorption at 450 nm appears by the reaction of the reduced pigment with carbon monoxide, resulting in the inactivation of the enzyme.[517] Reduced cytochrome P-450 reacts with molecular oxygen to form an "active oxygen" complex which oxidizes the substrate. The cofactor NADPH serves for the reduction of the oxidized cytochrome P-450, being mediated with a flavoprotein (fp), known as cytochrome c reductase, and with another unknown microsomal component (X). A probable scheme for the microsomal hydroxylation of foreign compounds is shown in Figure 16.

There is another electron transport system in liver microsomes. It is characterized by cytochrome b₅

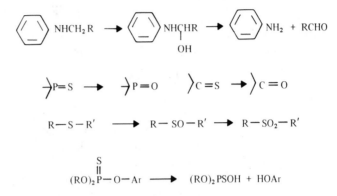

Aromatic hydroxylation	
Aliphatic hydroxylation	$RCH_3 \longrightarrow RCH_2OH$
O-Dealkylation	
N-Oxide formation	$R-N(CH_3)_2 \longrightarrow R-\overset{\overset{O}{\uparrow}}{N}(CH_3)_2$
N-Dealkylation	
Desulfuration	$\rangle P=S \longrightarrow \rangle P=O \qquad \rangle C=S \longrightarrow \rangle C=O$
Sulfide oxidation	$R-S-R' \longrightarrow R-SO-R' \longrightarrow R-SO_2-R'$
Deesterification	$(RO)_2\overset{\overset{S}{\|}}{P}-O-Ar \longrightarrow (RO)_2PSOH + HOAr$
Epoxidation	

FIGURE 15. Mixed-function oxidase catalyzed biotransformation of xenobiotics.

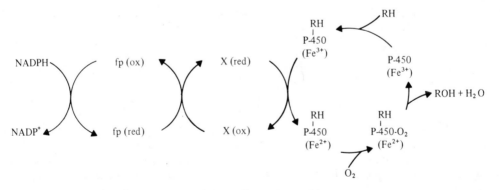

FIGURE 16. Schematic representation for microsomal hydroxylation of foreign compounds; fp: flavoprotein, X: unknown microsomal component, P-450: cytochrome P-450, RH: substrate.

which is reduced by electrons derived mainly from NADH, contrasting with the above mentioned cytochrome P-450 NADPH system. The function of the cytochrome b_5 system is not well established, but appears to interconnect with the cytochrome P-450 system in some way.[518-520]

The metabolism of foreign compounds is stimulated by exposure of animals to certain chemicals, due to the induction of the microsomal mixed-function oxidase by these chemicals. A number and variety of foreign organic compounds have such ability to induce the hepatic enzymes. They include drugs, food additives, and insecticides, particularly chlorinated hydrocarbons. 3-Methylcholanthrene (3MC), phenobarbital, and DDT are well-established examples.

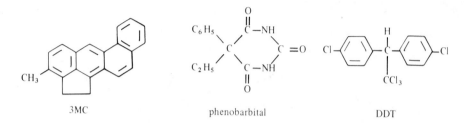

3MC   phenobarbital   DDT

Insect microsomes have similar mixed-function oxidase systems containing cytochrome P-450 which is sensitive to carbon monoxide.[521] Crude insect homogenates, however, often lack the ability to hydroxylate foreign compounds because of the presence of inhibitors of the mfo of both insects and mammals.[522,523] Strong inhibitors appear in the nuclei and debris fractions from the head and thorax homogenates of houseflies, and a weaker inhibitor in the soluble fraction from the abdomen homogenates. The effect of the inhibitor in the soluble fraction from the abdomen homogenates is greatly minimized by bovine serum albumin.[524] The inhibitor from housefly head appears to be the eye pigment, xanthommatin, and seems to function as an electron acceptor at the NADPH-cytochrome c reductase level.[525-527] It inhibits the housefly microsomal oxidase at concentrations as low as 5×10^{-7} M.

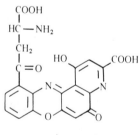

xanthommatin

b. Oxidative Desulfuration

In the pure state, phosphorothionate esters, such as parathion, and phosphorothiolothionate esters, such as malathion, are poor inhibitors of cholinesterases, whereas the corresponding oxo-analogs are highly potent anticholinesterase agents. The toxic action of those thiono compounds is attributable to their oxo-analogs formed in vivo by oxidative desulfuration of the thiophosphoryl group.[19] An electronic interpretation for the high activity of oxo-analogs has been given in the preceding Section (IV.A.3.b.i). Microsomal mixed-function oxidase systems are responsible for the conversion of thiono esters to oxo-analogs in animals. In plants, peroxidases may play a role in the thiono-oxon transformation,[528] although the same products may be formed photochemically on and in plants. The oxo-analogs hydrolyze with relative ease and usually do not accumulate in organisms; actually, it is often difficult to find oxons in tissues.

Oxidative desulfuration of thiophosphoryl sulfur atom has been demonstrated with a wide variety of insecticides, including phosphorothionates such as parathion (Equation 37), fenitrothion, and diazinon, phosphorothiolothionates such as malathion (Equation 38), dimethoate, and azinphosmethyl, and phosphonothiolothionates such as fonofos (Dyfonate) (Equation 39).

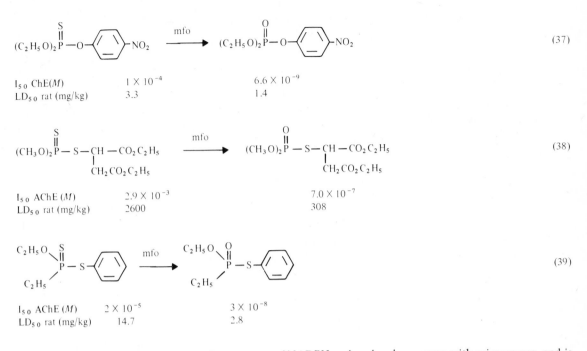

$$I_{50} \text{ ChE}(M) \quad\quad 1 \times 10^{-4} \quad\quad\quad\quad 6.6 \times 10^{-9}$$
$$\text{LD}_{50} \text{ rat (mg/kg)} \quad 3.3 \quad\quad\quad\quad\quad 1.4$$

$$I_{50} \text{ AChE}(M) \quad\quad 2.9 \times 10^{-3} \quad\quad\quad 7.0 \times 10^{-7}$$
$$\text{LD}_{50} \text{ rat (mg/kg)} \quad 2600 \quad\quad\quad\quad 308$$

$$I_{50} \text{ AChE}(M) \quad\quad 2 \times 10^{-5} \quad\quad\quad 3 \times 10^{-8}$$
$$\text{LD}_{50} \text{ rat (mg/kg)} \quad 14.7 \quad\quad\quad\quad 2.8$$

The formation of oxons requires the presence of NADPH and molecular oxygen with microsomes, and is inhibited by carbon monoxide, SKF 525A, and piperonyl butoxide, indicating that the reaction is catalyzed by the mixed-function oxidase systems. The eliminated sulfur is adsorbed on the microsomes in vitro and is excreted into urine in vivo.[529,530]

Isopropyl parathion appears to be a good substrate for desulfuration by the fly oxidase but a poor one for the bee enzyme; a higher level of isopropyl paraoxon was found in houseflies than in honey bees soon after treatment with isopropyl parathion.[531] This, with the greater sensitivity of housefly cholinesterase to isopropyl paraoxon,[492] contributes the greater toxicity of isopropyl parathion to houseflies than to bees (see IV.A.4.a).

Steric effect in the oxidative desulfuration was reported with respect to the metabolism of the thiono analogs of mevinphos: cis-thiono mevinphos, in contrast to the trans isomer, was selectively activated by the mouse liver mfo system.[532] Partial esters of phosphorothioate appear to be similarly desulfurized, at least in vivo: desmethyl fenchlorphos (Ronnel) fed to rats was excreted in the urine almost exclusively as the corresponding phosphate.[533]

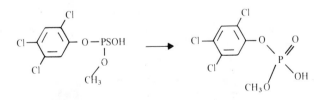

However, desulfuration was not observed in vivo with ethyl ethylphosphonothioic acid.[381] The oxidative desulfuration of partial thioate esters has not yet been confirmed by in vitro studies.

Hypothetical cyclic intermediates such as (XIX) and (XX) were postulated for microsomal oxidation of parathion and fonofos (Dyfonate), respectively, from studies with peracid oxidation models.[534,535]

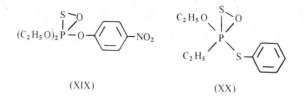

(XIX)   (XX)

c. Oxidation of Thioether Moiety

The oxidation of thioether moiety in certain organophosphorus esters has been observed in vivo in plants, mammals, and insects. Sulfide is converted into sulfoxide and then sulfone, as exemplified by the metabolic pathways of Abate $(O,O,O',O'$-tetramethyl O,O'-thiodi-p-phenylene phosphorothionate) in mosquito larvae,[536] and of phorate (diethyl S-ethylthiomethyl phosphorothiolothionate) in plants.[537]

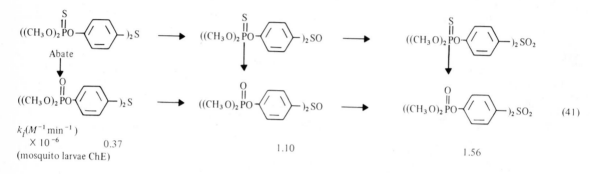

$$\tag{41}$$

$k_i(M^{-1} \min^{-1})$
$\times 10^{-6}$ 0.37   1.10   1.56
(mosquito larvae ChE)

The schemes show the sulfide oxidation as well as oxidative desulfuration and consequent change in anticholinesterase activity. In addition to these examples, oxidative metabolism of this type has been demonstrated with a variety of insecticides, including carbophenothion, demeton, disulfoton, fenthion, phenamiphos (Nemacur),[538] thiometon, and others, but not with the thiadiazole ring of methidathion (Supracide). (However, for the methylated derivative of the heterocycle, see 2.d.iv of this Chapter.)

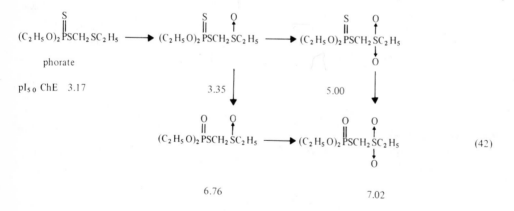

$$\tag{42}$$

6.76   7.02

The sulfide oxidation in plants is particularly important for systemic insecticides. The initial oxidation of sulfide into sulfoxide proceeds rapidly, whereas subsequent oxidation to sulfone and desulfuration go more slowly, though desulfuration occurs rapidly in animals.[205] With vamidothion, further oxidation of the sulfoxide does not appear to occur in plants. This delay factor allows the active compound to be translocated in lethal amounts in a relatively stable state before being destroyed by plant enzymes.[539]

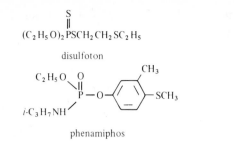

$$(C_2H_5O)_2\overset{\overset{\displaystyle S}{\|}}{P}SCH_2CH_2SC_2H_5$$

disulfoton

$$(CH_3O)_2\overset{\overset{\displaystyle S}{\|}}{P}SCH_2CH_2SC_2H_5$$

thiometon

phenamiphos

$$(CH_3O)_2\overset{\overset{\displaystyle O}{\|}}{P}SCH_2CH_2\overset{\overset{\displaystyle CH_3}{|}}{S}CHCONHCH_3$$

vamidothion

Similar transformations of sulfides also occur in soils,[540] probably by microorganisms living in soils. It has been reported that disulfoton sulfoxide is reduced to the original disulfoton in such special cases as in flooded soils which are in a reduced state.[541] It is well known that sulfoxide-oxygen atom is readily removed by the action of reducing agents.

Any enzyme systems responsible for the oxidation of the thioether moiety contained in organophosphorus molecules have not been confirmed by in vitro studies. However, from in vitro studies with other sulfur compounds,[542] it is reasonable to assume that the animal mixed-function oxidase systems are involved in the biooxidation of the thioether containing organophosphorus compounds, too. It appears, however, that the thioether oxidation in plants is in large measure due to photocatalysis, though thiophosphoryl oxidation is mainly due to enzymatic catalysis.[301]

The anticholinesterase activity generally increases in the order: sulfide < sulfoxide < sulfone. The activation is significant, but is not so great as that resulting from oxidative desulfuration. The oxidation of a thioether group attached to phenyl phosphorus esters, such as that in phenthion, causes a great increase in the electron-withdrawing capacity of the substituent: σ value changes from 0.047 to +0.567 and 1.049, according to the increasing oxidation state from CH_3S- to CH_3SO- and CH_3SO_2-, respectively. This should result in the enhancement of anticholinesterase activity. On the other hand, in phosphorus ester molecules containing a dialkylsulfide moiety, such as demeton, the inductive effect of the produced sulfoxide or sulfone group may be reduced greatly by the screening effect of the interposed methylenes. The inductive effect decreases, in general, by about 0.45 for each methylene. The anticholinesterase activity of the oxidation metabolites may be due to their structural suitability to fit with the binding site of acetylcholinesterase.

Toxicity of the oxidation metabolites is not necessarily higher than that of the original sulfide compounds. The sulfoxide and sulfone of demeton-S are almost equal in toxicity to the parent compound, probably because of in vivo metabolism. The insecticidal activity of metabolites of Abate decreases with increasing oxidation state because of lower lipophilicity, and consequently less penetrability, of the metabolites.[536] However, there are some cases in which the partially oxidized materials are more efficient as insecticides: the sulfoxides fensulfothion, oxydisulfoton, and Aphidan have been actually utilized as systemic insecticides.

fensulfothion

oxydisulfoton

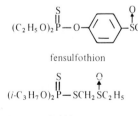

$$(i\text{-}C_3H_7O)_2\overset{\overset{\displaystyle S}{\|}}{P}SCH_2\overset{\overset{\displaystyle O}{\uparrow}}{S}C_2H_5$$

Aphidan

d. Oxidation of Amide Groups

The biooxidative activation of the phosphoramide insecticide, schradan, was noted in one of the earliest toxicological investigations of organophosphorus pesticides. Schradan, which lacks anticholinesterase activity, is oxidized biologically or chemically into a strong anticholinesterase agent, which is 10^5 times as active as the parent compound.[285,543] The biooxidation results in N-demethylation. Although the intermediates are not adequately characterized, the evidence suggests that schradan may be metabolized according to the following scheme:[285,543,544]

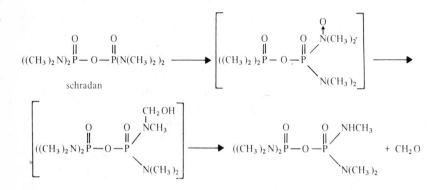

schradan

$$\text{(43)}$$

The presence of oxygen in the dimethylamido moiety should be responsible for the increased anticholinesterase activity of the metabolite. Spencer postulated that the methylol group allows an inductive effect on the phosphorus to increase its phosphorylating ability.[221]

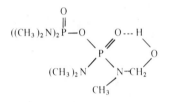

Cyclophosphamide, a known antitumor agent,[22] can be applied to the defleecing of sheep.[545] Although it is not effective in vitro against tumor cells at 400 μg/ml, it is activated by microsomal oxidation.[546] A carbonyl compound (XXII), a ring-opened carboxylic acid (XXIII), and some others were found as metabolites from urine of sheep and rabbits.[547,548] These metabolites were inactive as antitumor agents both in vivo and in vitro. However, the hydroxy compound (XXI), a possible metabolic intermediate, showed a high in vitro activity at 1 μg/ml.[547] The carbonyl compound (XXII) was considerably resistant to hydrolysis. Thus, the following metabolic pathway was postulated.

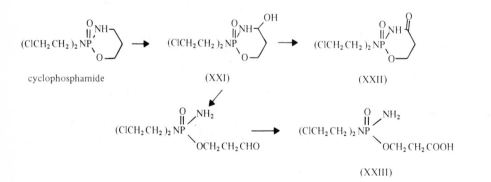

$$\text{(44)}$$

In the case of carboxyamides, the metabolic pathway is more clearly confirmed than that of phosphoramides. Dicrotophos (cis-2-dimethylcarbamoyl-1-methylvinyl dimethyl phosphate: Bidrin®) serves as a typical example of this class. Its metabolic pathway is shown by the following scheme,[549] where

R is $(CH_3O)_2 \overset{\overset{\text{O}}{\|}}{P}-O-\underset{\underset{CH_3}{|}}{C}=CHCO-$:

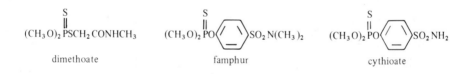

$$R-N(CH_3)_2 \longrightarrow R-N\overset{CH_2OH}{\underset{CH_3}{\diagup}} \longrightarrow R-N\overset{H}{\underset{CH_3}{\diagup}} \longrightarrow R-N\overset{H}{\underset{CH_2OH}{\diagup}} \longrightarrow R-NH_2 \qquad (45)$$

pI_{50} fly ChE (M)	7.2	7.0	6.8	6.9	6.5
LD_{50} fly (mg/kg)	38	14	6.4	30	1.0
LD_{50} mouse (mg/kg)	14	18	8	12	3

The N-methyl group is oxidized initially to the N-methylol group, which then undergoes either conjugation or elimination as formaldehyde. Thus, by the oxidative N-demethylation, dicrotophos is transformed to another insecticide, monocrotophos (Azodrin), which may be further metabolized to form finally the nonsubstituted amide. Including two N-methylol intermediates, at least four active metabolites from dicrotophos appeared in mammals, insects, and plants.[549,550] All of these metabolites, as well as the parent compound, are toxic; the metabolic transformation may be classified as an activation process. In the presence of Sesamex, an inhibitor for microsomal mixed-function oxidases, the oxidative N-demethylation of dicrotophos was blocked in houseflies. It is known that secondary amines are dealkylated by the mfo system to primary amines and aldehyde.

Similarly, N-demethylation has been observed not only with the N-methylcarboxyamide dimethoate,[551] but also with a sulfonamide like famphur; the N-demethylation of famphur appeared to occur significantly in insects but not in mice.[552] In both cases, N-demethylated derivatives show toxicity comparable to that of the parent compounds. N-Unsubstituted homolog of famphur has been developed as a systemic insecticide named cythioate.

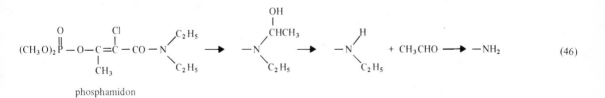

$(CH_3O)_2 \overset{\overset{\text{S}}{\|}}{P}SCH_2CONHCH_3$
dimethoate

$(CH_3O)_2 \overset{\overset{\text{S}}{\|}}{P}O\langle\!\!\!\!\bigcirc\!\!\!\!\rangle SO_2N(CH_3)_2$
famphur

$(CH_3O)_2 \overset{\overset{\text{S}}{\|}}{P}O\langle\!\!\!\!\bigcirc\!\!\!\!\rangle SO_2NH_2$
cythioate

The N-ethyl group of phosphamidon is also oxidatively eliminated as acetaldehyde through a hydroxylation on the α-carbon atom.[504] The oxidative N-dealkylation occurs faster with cis-phosphamidon than with the trans isomer. The des-N-ethyl metabolite is biologically more active than the parent insecticide.

$$(CH_3O)_2\overset{\overset{\text{O}}{\|}}{P}-O-\underset{\underset{CH_3}{|}}{C}=\overset{\overset{Cl}{|}}{C}-CO-N\overset{C_2H_5}{\underset{C_2H_5}{\diagup}} \longrightarrow -N\overset{\overset{\overset{OH}{|}}{CHCH_3}}{\underset{C_2H_5}{\diagup}} \longrightarrow -N\overset{H}{\underset{C_2H_5}{\diagup}} + CH_3CHO \longrightarrow -NH_2 \qquad (46)$$

phosphamidon

The N-formyl group of formothion is removed in vivo to form dimethoate. The reaction mechanism is not known. It is possible to postulate two mechanisms: an oxidative one, involving the conversion of the formyl group into the acid followed by decarboxylation, and a reductive one, involving a conversion into a methylol group.[295]

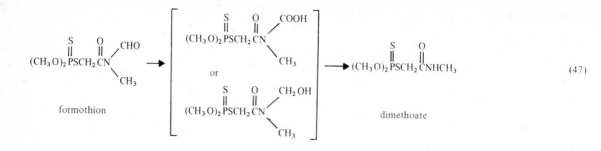

The *N*-methoxy group appears to be biologically more stable than the *N*-methyl group; at least, the former was not the initial site for the metabolism of C-2307 (dimethyl O-(1-*N*-methyl-*N*-methoxycarbamoyl-1-propen-2-yl phosphate), since the mono-*N*-methyl metabolite was not found in the urine of rats treated with the compound.[504]

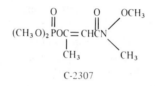

C-2307

e. Aliphatic Hydroxylation and Subsequent Reactions

The side chain of arylalkanes is oxidized by the action of an mfo system to form alcohols, which may be further transformed by subsequent reactions.

Tri-*o*-tolyl phosphate (known as TOCP), which is not a pesticide but a neurotoxic phosphate ester having a synergistic activity with malathion, is converted in vivo into the biologically active cyclic phosphate esters of saligenin.[489,553] The main active metabolite, *o*-tolyl saligenin cyclic phosphate, is a potent inhibitor of serum cholinesterase, being ten million times more active than TOCP.[489] Liver microsomal mfo system is effective in the activation of TOCP in vitro. Thus, the first step in TOCP activation is probably hydroxylation to yield di-*o*-tolyl *o*-α-hydroxytolyl phosphate. The conversion of this hydroxylated intermediate to the corresponding cyclic phosphate was greatly accelerated by plasma albumin, though spontaneous cyclization also took place more slowly.[554] The metabolic activation of TOCP or its analogs has been observed in mammals, birds, and insects, but not in plants.[555,556] For activation of related phosphate triesters, at least one *o*-alkylphenyl group is necessary; it may be hydroxylated on the α-carbon as the first step. Another aryl ester group is also required for the subsequent cyclization reaction; it may be liberated in the course of intramolecular transphosphorylation (Equation 48). Alkyl ester groups do not participate in the cyclization reaction.[554]

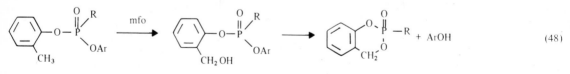

Metabolic activation of *o*-alkyl phenyl phosphates
R = alkoxy, aryloxy, or alkylamido group

Another substituted group remaining on phosphorus determines the physiological property of the parent compound: for example, diaryl *o*-tolyl phosphates are transformed to aryl saligenin cyclic phosphates and are neurotoxic and potentiate the toxicity of malathion; alkyl di-*o*-tolyl phosphates give alkyl saligenin cyclic phosphates, which are not neurotoxic but insecticidal; and di-*o*-tolyl *N*-methylphosphoramidate is converted into the corresponding cyclic *N*-methylphosphoramidate and shows systemic insecticidal activity.[487,556]

Further investigations on the structure-activity relationship of saligenin cyclic phosphorus esters have brought about the development of the insecticide Salithion.[233,556]

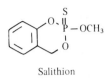

Salithion

Another example of aralkyl group hydroxylation related to the activation of phosphate esters was found in the metabolism of tri-*p*-ethylphenyl phosphate (TPEP), a neurotoxic substance. The *p*-ethyl group of TPEP was hydroxylated by the action of rat liver microsomal oxidase fortified with NADPH in the presence of air to form an α-hydroxyethyl group, which was then transformed to an acetyl group by the action of a soluble dehydrogenase.[557-559] This transformation causes a great change in the σ-value of the *p*-substituent group: from -0.151 for C_2H_5 to $+0.874$ for CH_3CO. As expected, the α-oxo metabolites of TPEP showed higher antiesterase activity and neurotoxicity, but no insecticidal activity. On the contrary, their homologous dialkyl esters are highly toxic to insects, but do not cause neurotoxicity in hens.[495]

This type of activation occurs generally in mammals, birds, and insects. Thus, the metabolic pathway of *p*-ethylphenyl phosphates may be summarized as follows:

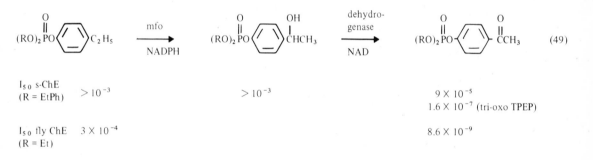

$$(49)$$

I_{50} s-ChE (R = EtPh)	$> 10^{-3}$		$> 10^{-3}$		9×10^{-5}
					1.6×10^{-7} (tri-oxo TPEP)
I_{50} fly ChE (R = Et)	3×10^{-4}				8.6×10^{-9}

The methyl group on *m*-position of fenitrothion is also oxidized, and 3-hydroxy-6-nitrobenzoic acid is finally excreted.[523] In this case, the alkyl group oxidation may not be related to activation but to detoxication.

The side-chain on a heteroaromatic ring is also hydroxylated by mfo systems. EDTA-treated microsomes from rat liver and cockroach fat body catalyzed the oxidation of diazinon to transform into the side-chain hydroxylated metabolites, as well as diazoxon and degradation products.[560] EDTA inhibited further hydrolysis of oxon products by the Ca^{++}-dependent esterase; consequently, the oxidation products accumulated.[561] The hydroxylation occurs on the tertiary carbon atom of diazinon and diazoxon. Hydroxydiazinon has been found in various organs from sheep dosed with diazinon[562] and also in field-sprayed kale.[299] It is not known that hydroxydiazinon in plants is produced either by an enzymatic or by a photochemical reaction. In addition to hydroxy-diazinon and -diazoxon, the 6-methylol derivative of diazinon (isohydroxydiazinon) and the dehydration product of hydroxydiazinon were recently found as active metabolites in mice and sheep.[563,564] All these oxidation products having intact ester linkages should be, of course, toxicologically active. The biotransformation of diazinon into the active metabolites is summarized in Equation 50.

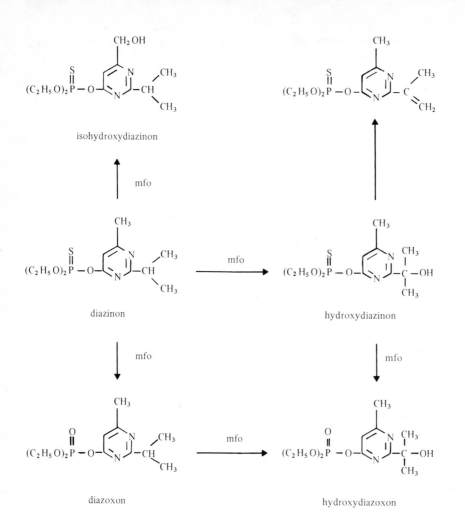

isohydroxydiazinon

diazinon

hydroxydiazinon

diazoxon

hydroxydiazoxon

Formation of toxic metabolites from diazinon (50)

An interesting hydroxylation and subsequent dehydrogenation on a heteroalicyclic ring was recently reported as the metabolic pathway of a new experimental systemic insecticide of phosphoramidothiolate esters, R-16,661, in the cotton plant and houseflies (Equation 51).[565] The keto metabolite has slightly less toxicological activity than the parent compound, which itself shows a high anticholinesterase activity (I_{50} fly ChE: 3×10^{-8} M). The mode of action of phosphoramidothiolates, including this compound and methamidophos (Monitor) (O-methyl S-methyl phosphoramidothiolate), is not well understood yet.

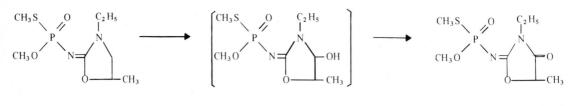

R-16,661 (51)

The hydroxylation of an aromatic ring is a very common biochemical reaction catalyzed by mixed-function oxidases. However, it is rarely observed with organophosphorus pesticides. Triphenyl phosphate, known as a malathion synergist, was metabolized in vivo in houseflies to form diphenyl p-hydroxyphenyl phosphate, which inhibited the SH-enzyme, yeast alcohol dehydrogenase.[566]

As a minor metabolite of the fungicide edifenphos (Hinosan; S,S-diphenyl ethyl phosphorodithiolate), its p-hydroxy derivative was produced in vivo in rats, cockroaches, and the fungus *Piricularia oryzae* (Equation 52).[302] On the other hand, with the S-benzyl thiolate fungicides Kitazin P (S-benzyl diiospropyl phosphorothiolate) and Inezin (S-benzyl ethyl phenylphosphonothiolate), hydroxylation occurred neither on para nor ortho position, but on m-position of the benzyl ester group by *P. oryzae* (Equation 53).[567,568] The m-hydroxylated Kitazin P was less active as a fungicide against the rice-blast fungus.

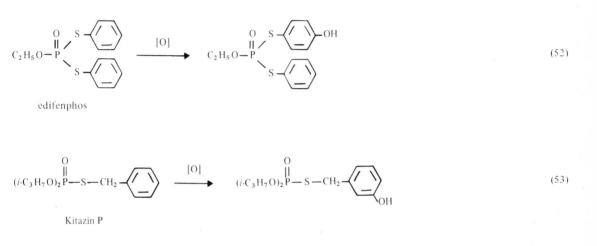

edifenphos

(52)

Kitazin P

(53)

f. Miscellaneous Nonoxidative Reactions

Besides oxidative reactions, only a few other reactions are concerned with activation of organophosphorus compounds. The rearrangement of the α-hydroxyethylphosphonate trichlorfon to the vinyl phosphate dichlorvos is considered an activation reaction. In the true sense, however, this reaction is not a metabolic reaction, but occurs spontaneously and rapidly at physiological pH.[482,569]

The insecticide butonate, which is the butyryl ester derivative of trichlorfon, is hydrolyzed by esterase action to yield trichlorfon, which is then transformed ultimately into the actual toxicant, dichlorvos. The conversion of butonate into trichlorfon is rapid, particularly in plants. Another related organophosphorus insecticide, naled, is the dibromo addition product of dichlorvos and it reacts with thiol groups or other natural constituents to debrominate, forming dichlorvos, which is probably the actual toxic agent.[500] For example, cysteine reacts with naled almost instantaneously.

$$(CH_3O)_2 \overset{\overset{O}{\|}}{P}-O-\overset{\overset{Br}{|}}{\underset{\underset{H}{|}}{C}}-\overset{\overset{Br}{|}}{\underset{\underset{Cl}{|}}{C}}-Cl + 2RS^- \longrightarrow (CH_3O)_2 \overset{\overset{O}{\|}}{P}-O-CH=CCl_2 + RSSR + 2Br^- \qquad (54)$$

The following scheme illustrates the relationship among these insecticides.

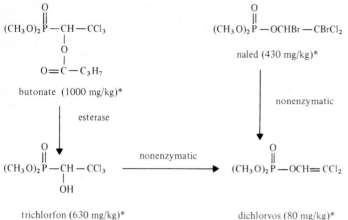

butonate (1000 mg/kg)*

naled (430 mg/kg)*

esterase

nonenzymatic

$$(CH_3O)_2\overset{O}{\overset{\|}{P}}-\underset{OH}{\overset{}{CH}}-CCl_3 \xrightarrow{\text{nonenzymatic}} (CH_3O)_2\overset{O}{\overset{\|}{P}}-OCH=CCl_2 \qquad (55)$$

trichlorfon (630 mg/kg)*   dichlorvos (80 mg/kg)*

* Figures in parentheses are LD_{50} for rats.

Demeton-O rearranges readily into the more toxic thiol isomer, as described in Chapter III.C.3. This reaction appears to occur in plants, too. Benjamini suggested that the thiono-thiolo rearrangement could occur enzymatically with fensulfothion in plant tissues,[570] though the reaction takes place nonenzymatically with heat or light under environmental conditions, as discussed in detail in the previous Chapter. Such rearrangement did not occur with the closely related sulfide fenthion, unless more vigorous thermal conditions were applied.

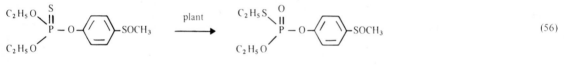

(56)

fensulfothion

It was recently reported that the displacement reaction of the vinyl chlorine atom by a hydroxyl group occurred readily in vivo, and consequently that phosphamidon was transformed in rats into the N-bisdesethylated hydroxy derivative.[571]

$$(CH_3O)_2\overset{O}{\overset{\|}{P}}-O-\underset{CH_3}{\overset{Cl}{\overset{|}{C}}}=C-CON(C_2H_5)_2 \xrightarrow{\text{rat}} (CH_3O)_2\overset{O}{\overset{\|}{P}}-O-\underset{CH_3}{\overset{OH}{\overset{|}{C}}}=C-CONH_2 \qquad (57)$$

phosphamidon

The formation of the reductively dechlorinated **metabolite** of phosphamidon was once reported erroneously,[572] but was corrected later.[571] Small amounts of a reductively dechlorinated active metabolite (potasan) were found in the feces of cows, which had been orally treated with the animal insecticide, coumaphos, for the control of housefly larvae in manure.[573] These metabolites were at least as toxic as the parent insecticides. Metabolic dehalogenation of dialkyl halophenyl phosphorothionate insecticides, such as fenchlorphos (ronnel), bromophos, and iodophos, is not known.

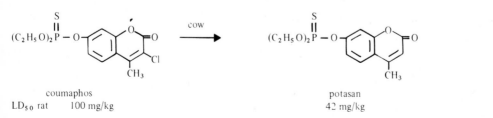

(58)

coumaphos   potasan
LD_{50} rat   100 mg/kg   42 mg/kg

2. Reactions Resulting in Detoxication

The metabolic detoxication of organophosphorus compounds is mainly due to the cleavage of a phosphorus ester bond which results in the formation of a negative charge in the molecule. The negative charge causes the phosphorus compound to be inactive as a phosphorylating agent or an anticholinesterase agent. Only in very special cases, anionic phosphorus esters like desethyl amiton have a considerable anticholinesterase activity, provided that they have a basic group which is properly located so as to form an internal salt and to neutralize the anionic charge.[354] Bis-(p-nitrophenyl) hydrogen phosphate is also active as a carboxyesterase inhibitor.

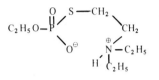

desethyl amiton

Although there are these exceptions of active partial esters, the cleavage of the phosphorus ester bond is generally accepted as detoxication because the resulting products are not only much less active as anticholinesterase agents, but are also much more soluble in water and can be readily excreted in urine. The leaving organic moiety usually possesses a hydroxy, amino, or thiol group and undergoes conjugation reactions, which make the compound generally more water soluble and excretable.

Two different types of ester bonds in organophosphorus pesticides can be cleaved: an anhydride bond and alkyl ester bonds. The phosphorus ester bond cleavages had been attributed exclusively to the action of hydrolytic enzymes. However, current research is disclosing that oxidative degradation mechanisms and group transfer mechanisms play important roles in the phosphorus ester cleavages and consequently in the detoxication of organophosphorus pesticides.

The biotransformation of nonphosphorus functional groups is also important for the detoxication of certain pesticides which contain a carboxyester, carboxyamide, or nitro group. Selective toxicity of organophosphorus pesticides is greatly due to detoxication mechanisms. Detoxication is responsible for many resistance mechanisms of insects, too.

a. Anhydride Bond Cleavage
i. Oxidative Dearylation

Knaak et al. in 1962 found that the cleavage of an ester bond in parathion took place oxidatively by the action of both an enzyme system and a nonenzymic model system and was closely related to the oxidative desulfuration reaction.[528] Five years later, it was reported from two laboratories that mammalian liver microsomes catalyze both the desulfuration of parathion to yield paraoxon and the cleavage of the aryl ester bond to yield nitrophenol and diethyl phosphorothioic acid.[529,574] Both reactions required the presence of NADPH as a cofactor and molecular oxygen, and were strongly inhibited by carbon monoxide, SKF 525A, and benzodioxoles, indicating clearly that these reactions were catalyzed by mixed-function oxidase systems.

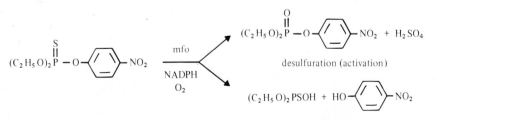

(59)

The oxidative dearylation by mfo systems from mammalian livers or housefly abdomens has been demonstrated with aryl phosphorothionates such as parathion, parathion-methyl, dicapthon, and fenitrothion;[575] with heteroaryl phosphorothionates such as diazinon;[576,577] phosphonothionates such as EPN;[575] and S-aryl phosphonothiolothionates such as fonofos (Dyfonate).[534] In the last case, a P-S (aryl) bond is disrupted. Moreover, a P-S (alkyl) bond in phosphorothiolothionates such as azinphosmethyl and malathion also appeared to be disrupted oxidatively by the mfo systems of mouse liver and resistant houseflies,[578,579] although the product of the nonphosphorus moiety was not identified.

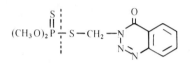

azinphosmethyl

It appears somewhat equivocal whether a thiophosphoryl group is required in the substrates for the oxidative dearylation catalyzed by the mfo systems.[580] Of eight homologous oxons examined, only n-propyl paraoxon was oxidatively decomposed to liberate nitrophenol by the mammalian liver mfo system.[575] Diazoxon could not serve as the substrate for the mfo system from rat liver, but was dearylated by the microsomal system from houseflies, although it was a less preferred substrate than its thiono analog, diazinon.[581] The oxygen analog of azinphosmethyl was slowly degraded by the mfo systems from liver and resistant houseflies. The anhydride cleavage of phosphate esters is mostly due to esterases.

The enzyme systems of microsomes for the oxidative desulfuration and the oxidative dearylation appear somewhat different from each other in nature. More activation than detoxication often occurs with dimethyl phosphorothionates, whereas the reverse is true with higher homologs. Variation of the ratio of desulfuration/dearylation among enzyme sources, especially with isopropyl parathion, indicated that these reactions may be catalyzed by separate enzyme systems.[575] Effects of some chemicals on these enzyme systems are also different.[582] Although the mixed-function oxidase inhibitor, SKF 525-A, inhibited both activities almost to the same extent, the chelating agent 8-hydroxyquinoline and the SH reagent PCMB (p-chloromercuribenzoate) inhibited the oxidative desulfuration of parathion more than the oxidative dearylation; some thiols such as mercaptoethanol affected the desulfuration little, but stimulated the ester cleavage, suggesting the participation of certain SH-enzymes in the latter reaction. The effects of electron-acceptors are complicated: riboflavin inhibited the ester cleavage more than the desulfuration, while menadione and methylene blue stimulated the ester cleavage but inhibited the desulfuration. Finally, of the inducers of liver drug metabolizing enzymes, phenobarbital stimulated both activities to the same degree, whereas another inducer benzo[a]pyrene induced the activity of the desulfuration more than that of the ester cleavage.

McBain et al. studied the reaction mechanisms, using oxygen-18 to trace the origin of each oxygen incorporated into the oxon and phosphonothioic acid which are derived from fonofos (Dyfonate) by the microsomal mfo system:[534] the oxygen of Dyfoxon came from molecular oxygen, whereas that of ethyl ethylphosphonothioic acid (ETP) came from water, indicating that a hydrolytic step was involved for the ester cleavage. In addition, ethyl ethylphosphonic acid (EOP) was produced from the oxon by the action of a microsomal esterase without NADPH. Thus, the mechanism for the oxidative dearylation and desulfuration by the mfo systems was postulated as follows:[276,534,535]

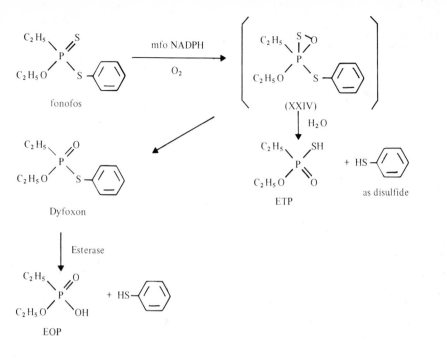

The postulated intermediate (XXIV), having a three-membered hetero ring, has not as yet been isolated. An oxgenated product obtained by an oxidation model system using *m*-chloroperbenzoic acid was erroneously postulated as the intermediate, but was recently corrected to be the oxidation by-product, phenyl phosphinyl disulfide (XXV).[277] However, it is interesting to note that the disulfide was sometimes produced when microsomes were incubated with fonofos in the presence – but not in the absence – of NADPH.[276]

(XXV)

ii. Enzymatic Hydrolysis

Certain esterases catalyze the hydrolysis of organophosphorus compounds.[424] They are called arylesterases, A-esterases, or phosphotriesterases, and are widely distributed in many mammalian tissues, particularly plasma, liver, and kidney. Similar esterases are found in insects and microorganisms. For details on A-esterases, see Aldridge's recent review.[424]

Besides organophosphorus compounds, these enzymes may hydrolyze some carbamic and carboxylic esters. A-esterases are regarded as enzymes different from usual phosphatases: phosphatases hydrolyze only partial esters of orthophosphoric acid, but A-esterases hydrolyze a variety of neutral phosphorus compounds at the bond between the phosphorus atom and the most acidic group that is the so-called leaving group; it is the same bond which is split when the compound reacts with B-esterases. Thus, A-esterases may cleave not only phosphorus ester bonds, such as $>\overset{O}{\overset{\|}{P}}-O$(aryl) and $>\overset{O}{\overset{\|}{P}}-S$, but also anhydride bonds, such as $>\overset{O}{\overset{\|}{P}}-O-\overset{O}{\overset{\|}{P}}<$, $>\overset{O}{\overset{\|}{P}}-F$, and $>\overset{O}{\overset{\|}{P}}-CN$.

$$(i\text{-}C_3H_7O)_2 \overset{\displaystyle O}{\overset{\|}{P}} - F + H_2O \xrightarrow{\text{A-Esterase}} (i\text{-}C_3H_7O)_2 \overset{\displaystyle O}{\overset{\|}{P}}OH + HF \tag{61}$$

$$(C_2H_5O)_2 \overset{\displaystyle O}{\overset{\|}{P}} - O - \text{⟨benzene⟩} - NO_2 + H_2O \xrightarrow{\text{A-Esterase}} (C_2H_5O)_2 \overset{\displaystyle O}{\overset{\|}{P}}OH + HO - \text{⟨benzene⟩} - NO_2 \tag{62}$$

$$(C_2H_5O)_2 \overset{\displaystyle O}{\overset{\|}{P}} - O - \text{⟨pyrimidine⟩} + H_2O \xrightarrow{\text{A-Esterase}} (C_2H_5O)_2 \overset{\displaystyle O}{\overset{\|}{P}}OH + HO - \text{⟨pyrimidine⟩} \tag{63}$$

It appears that thiophosphoryl compounds are not hydrolyzed by A-esterases, except by certain insect enzymes. The cleavage of methyl phosphate ester linkage in dichlorvos by the soluble enzyme of rat liver was reported.[583] However, it may not be due to a hydrolase, but probably to a glutathione S-alkyl transferase (see Section b.ii). The substrate specificity of A-esterases is complicated and not well established because it is rather difficult to obtain the enzyme in pure state. However, it has been established with some A-esterases by certain criteria that the same enzyme can hydrolyze different organophosphorus compounds, carbamates, and carboxylic esters. For example, an A-esterase in rabbit plasma hydrolyzes DFP, N-dimethylcarbamyl fluoride, and 4-nitrophenyl acetate;[584] a purified esterase from sheep plasma catalyzes the hydrolysis of paraoxon, DFP, tabun, and p-nitrophenyl acetate, but not TEPP and phenyl acetate.[585] An enzyme from hog kidney decomposes DFP and TEPP.[586] Plasma albumins from many animal species, but not ovalbumin, have esteratic activity and decompose various organophosphorus compounds as well as carboxylic esters.[587] The question arises whether albumin itself is catalytically active, or whether an enzyme is associated with albumin. An interesting reaction catalyzed by albumin is the intramolecular transesterification of o-α-hydroxytolyl phenyl phosphates to form cyclic esters (Equation 64), resulting in activation.[554] Only the aryl ester group can be displaced, and the alkyl ester group is never liberated by the reaction. This catalytic activity could not be separated from albumin and esterase activity.

$$\text{⟨benzene-CH}_2\text{OH⟩} - O - \overset{\displaystyle O}{\overset{\|}{P}} \overset{O-Ar}{\underset{O-R}{}} \xrightarrow{\text{albumin}} \text{⟨cyclic⟩} \overset{\displaystyle O}{\overset{\|}{P}} - OR + HOAr \tag{64}$$

The behavior of A-esterases to metallic ion activators may be useful for the classification of the enzymes:[505,588] 1) calcium activated and 2) manganese or cobalt activated. The calcium activated hydrolases are found in many mammalian tissues, especially in liver and serum, but not in insects.[581,588,589] This contributes in part to the selective toxicity of diazinon; both hydrolases in rat liver microsomes and serum hydrolyze diazoxon much more rapidly than paraoxon.[589] Heavy metal (Hg^{2+}, Cd^{2+}) and rare earth metal (Ce^{3+}, La^{3+}) ions cause a complete inhibition of the enzyme activity. EDTA also causes strong inhibition. The inhibition by SH-reagents, such as mercury compounds, is reversed by glutathione and L-cysteine, indicating the enzyme activity depends upon SH groups. The enzyme requires Ca^{2+} ion for its activity, and the removal of the calcium bond to the enzyme protein causes the complete loss of the activity.

Manganese activated hydrolases have been found in mammalian tissues,[583,586] insects,[590,591] and microorganisms.[586] The enzyme of hog kidney does not hydrolyze paraoxon, but hydrolyzes dichlorvos, DFP, and tabun.[586] Housefly enzyme can hydrolyze dichlorvos (Equation 65), TEPP, and DFP.[590]

$$(CH_3O)_2 \overset{\displaystyle O}{\overset{\|}{P}} - O - CH = CCl_2 + H_2O \longrightarrow (CH_3O)_2 \overset{\displaystyle O}{\overset{\|}{P}}OH + O = CH - CHCl_2 \tag{65}$$

dichlorvos

The mid-gut of lepidopterous caterpillars contains two kinds of phosphotriester hydrolases; one prefers phosphate esters as the substrate (P=O enzymes) and the other hydrolyzes preferably phosphorothionate esters (P=S enzymes).[591] Both the lepidopterous hydrolases are inhibited with heavy metal ions, but only the P=O enzymes are activated by SH compounds such as glutathione and cysteine. Several hydrolases of American cockroaches were partially purified recently.[592] They showed characteristics of A-type esterases to degrade organophosphate with certain substrate specificity. Their activities were stimulated by Mn^{2+} and glutathione.

In resistant strains of houseflies, there is a contradiction in the literature on the enzymatic degradation of parathion; Matsumura and Hogendijk reported its enzymatic hydrolysis,[593] but other researchers could not find such an enzyme activity in houseflies.[594,595] In certain organophosphorus resistant strains of houseflies, Oppenoorth and van Asperen found an enzyme activity which was absent in susceptible houseflies and catalyzed the degradation of organophosphorus insecticides.[596] Both the increased detoxifying ability and organophosphate resistance were associated with an abnormally low activity of aliesterase in these strains, and these characters were proved to be genetically inseparable.

Thus, these researchers proposed to call the degrading enzyme "mutant aliesterase," supposing that the enzyme may be produced by a mutant gene in the place of aliesterase that is present in susceptible strains.[597] The enzyme hydrolyzed paraoxon, but not parathion, producing diethyl phosphoric acid and p-nitrophenol. The activity was inhibited by the propyl homolog of paraoxon and p-chloromercuribenzoic acid but not by sesamex.[595] As the hydrolytic rate of organophosphate was very slow [V_{max} for paraoxon was 1.5×10^{-10} mole/fly/hr (=0.04 μg)], this enzyme may not be the major factor in the resistance of houseflies, but may contribute only in part to their resistance.[598] Oxidative dealkylation appears to play a more important role for the detoxication of paraoxon, particularly at high concentrations. For example, though the ratio of oxidation to hydrolysis of paraoxon in a resistant strain was 0.6 at 10^{-6} M, it was reversed to 6 at 10^{-3} M.[599] (For the oxidative dealkylation, see later Section b.i.)

Hydrolases which catalyze the hydrolysis of phosphodiesters, such as the desmethyl derivatives of dichlorvos and tetrachlorvinphos, into monomethyl phosphate, liberating the "leaving group," were found in rat liver soluble fraction.[583,600] The enzymes appear to be different from ordinary phosphodiesterases from snake venom and spleen; the phosphodiesterases do not hydrolyze the substrates mentioned above, but the RNA "core," to which the liver enzymes showed no activity.

iii. Glutathione S-Aryltransfer Reaction

Some reports describe that certain enzymic cleavages of an anhydride bond in organophosphorus compounds were promoted by glutathione.[530,581,591] In rat liver and cockroach fat body, Shishido and his co-workers found an enzymic activity, degrading diazinon, in the presence of glutathione (GSH), into diethyl phosphorothioic acid and S-(2-isopropyl-4-methyl-6-pyrimidinyl) glutathione:[601]

$$(C_2H_5O)_2\overset{S}{\overset{\|}{P}} - O\text{-pyrimidinyl} + GSH \xrightarrow[\text{transferase}]{\text{glutathione S-}} (C_2H_5O)_2PSOH + GS\text{-pyrimidinyl} \tag{66}$$

diazinon

Diazoxon and some higher alkyl homologs of diazinon were also degraded similarly by the transferase. Glutathione could not be replaced by other thiol compounds such as cysteine. Some SH reagents, such as phenylmercuric acetate, and oxidizing agents, such as N-bromosuccinimide, inhibited the enzyme activity. S,S,S-Tributyl phosphorotrithioate appeared to inhibit glutathione dependent dearylation of diazinon in resistant houseflies. A chelating agent, o-phenanthroline, inhibited specifically the enzyme of American cockroach, but had only a slight effect on the rat liver enzyme. The conjugation reactions of glutathione catalyzed by glutathione S-transferases are known with many substrates.[602] The relation between the enzyme catalyzing the reaction of diazinon with glutathione and glutathione S-aryltransferase, the typical substrates of which are halogenated nitrobenzenes, is uncertain.

b. Alkyl Ester Bond Cleavage

The importance of P-O-alkyl bond cleavage as a biodegradation pathway of organophosphorus compounds in animals was first found in 1958, by Plapp and Casida, who examined the metabolism of a variety of organophosphorus insecticides, including ronnel (fenchlorphos) and others.[201,533] This degradation reaction has been regarded as being catalyzed by hydrolases. However, recent studies elucidated that there are other biochemical mechanisms which are responsible for the formation of dealkylated metabolites. They are oxidative mechanisms and glutathione *S*-alkytransferase mechanisms.

i. Oxidative O-Dealkylation

Donninger et al. found that chlorfenvinphos (2-chloro-1-(2,4-dichlorophenyl)vinyl diethyl phosphate) was oxidatively deethylated by liver microsomes in the presence of NADPH and oxygen.[603] The products of the reaction were characterized as acetaldehyde and the diester. The mechanism was proposed as follows:

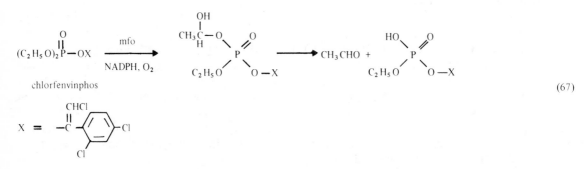

(67)

chlorfenvinphos

Besides chlorfenvinphos, a variety of vinyl, phenylvinyl, phenyl, and naphthyl phosphate triesters are substrates for this reaction. The enzyme is named phosphoric acid triester, oxygen:NADPH$_2$ oxidoreductase, and catalyzes dealkylation of dimethyl, diethyl, diisopropyl, and di-*n*-butyl ester groups. Ethyl esters appear to be the most suitable substrates; methyl esters may be more preferably degraded by glutathione *S*-alkyltransferase (see the following section). For example, in rabbit liver, oxidative dealkylation of chlorfenvinphos was three times faster than the transfer reaction, whereas for the closely related methyl ester, tetrachlorvinphos (Gardona), the dealkylation by the alkyltransfer mechanism was 2.7 times as great as by the oxidative mechanism.[505]

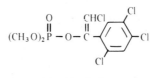

tetrachlorvinphos

There is a great variety in enzyme distribution from species to species of mammals: the liver microsomes of dogs contain the highest enzyme activity of examined animals, while those of rats contain only one-hundredth as much as the former. Thus, remarkable differences in the metabolism of chlorfenvinphos have been observed between rats and dogs. Whereas *O*-deethylation predominated in dogs (70% of the dose), esteratic cleavage followed by conjugation was the main pathway in rats. The difference in the metabolism is reflected, at least in part, on the selective toxicity of this insecticide: acute oral LD$_{50}$ for dog is 12,000 mg/kg, while for rat only 10 mg/kg. The microsomal oxidase system can be readily induced, especially in the rat, by the administration of phenobarbital or dieldrin. The enzyme activity in the liver of dieldrin treated rats was increased by 600-fold; the acute oral LD$_{50}$ was also increased by the pretreatment of rats with dieldrin.[603]

Similar dealkylation occurs with diazoxon and paraoxon (but not with their thiono analogs) by a microsomal preparation from resistant strains of the housefly in the presence of NADPH and oxygen.[599],[604] The oxidative O-dealkylation appears to be an important route to detoxication with phosphate triesters, particularly those having ethyl ester linkages, but not with phosphorothionate esters.[295]

ii. Glutathione S-Alkyltransfer Reaction

Demethylation of dichlorvos with a soluble enzyme fraction from rat liver was demonstrated by Hodgson and Casida.[583] Fukami and his co-workers found that the enzyme activity catalyzing the demethylation of parathion-methyl and fenitrothion in the supernatant fraction of rat liver homogenates was greatly increased by the addition of glutathione (GSH), and they suggested that glutathione may act as a methyl group acceptor.[605-607] It has been confirmed subsequently with several organophosphorus methyl esters, including mevinphos,[608] tetrachlorvinphos (Gardona),[603] bromophos,[609] fenitrothion, and parathion-methyl,[610] that the products of the glutathione dependent enzymic reactions are S-methylglutathione and the mono-desmethyl derivative of the corresponding organophosphorus ester. The methyl group is transfered directly to glutathione (Equation 68). This enzyme may be identical with glutathione S-alkyltransferase which is responsible for the degradation of a variety of alkyl halides.[602]

$$(CH_3O)_2 \overset{\overset{\displaystyle S(O)}{\|}}{P} -X \ + \ GSH \quad \xrightarrow[\text{alkyltransferase}]{\text{Glutathione } S\text{-}} \quad \overset{\overset{\displaystyle S(O)}{\|}}{\underset{CH_3O}{\overset{HO}{\diagdown}}} P -X \ + \ GS-CH_3 \tag{68}$$

Plapp and Casida indicated that ronnel was detoxified in rats mainly by demethylation, whereas in houseflies by slow dearylation.[533] The alkyl transferase activity is greatest in the soluble fraction of mammalian liver (Table 30). Less activity was found in the mid-gut and fat body of insects. No demethylation activity was found in sucking insects and mites; this might be responsible for the effectiveness of methyl esters towards sucking insects.[2] On the contrary, with methyl phosphonate esters, such as trichlorfon and its dichloro analog, demethylation occurs more readily in sucking insects, e.g., green rice leafhoppers and black rice bugs, as compared to other insects.[611] Mammalian transferase is highly specific to methyl esters, in comparison with ethyl and isopropyl esters (Table 30).[607],[610] Insect enzymes

TABLE 30

Enzymatic Degradation of Parathion and Parathion-methyl by Soluble Fractions of Various Tissues with or without Addition of Glutathione[607]

Tissue	μg Insecticide decomposed/g of equiv. fresh wt/120 min	
	Parathion-methyl	Parathion
Rat		
Liver	165.3	3.92
Liver + GSH	243.8	4.73
Brain	5.3	0.27
Brain + GSH	24.9	1.88
Kidney	4.5	0.14
Kidney + GSH	10.0	1.60
Blood	3.9	—
Blood + GSH	4.0	—
Horn beetle larvae		
Mid-gut	11.9	0.12
Mid-gut + GSH	44.8	0.17
Fat body	2.8	—
Fat body + GSH	10.8	—
Haemolymph	1.4	—
Haemolymph + GSH	2.3	—

TABLE 31

Effect of Alkyl Group on Mammalian Toxicity

Oral toxicity to mouse
LD_{50} mg/kg

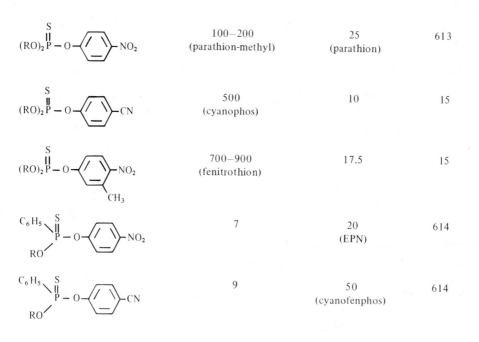

	R = CH_3	C_2H_5	Ref.
	100–200 (parathion-methyl)	25 (parathion)	613
	500 (cyanophos)	10	15
	700–900 (fenitrothion)	17.5	15
	7	20 (EPN)	614
	9	50 (cyanofenphos)	614

of resistant strains appear to be more specific to diethyl phosphorus esters: the glutathione dependent *O*-deethylation of diazinon, diazoxon, and parathion occurs in certain organophosphorus resistant strains of houseflies which show greater resistance to diethoxy phosphorus insecticides than to dimethoxy insecticides.[604,612] The substrate specificity of the mammalian enzyme may be responsible, at least in part, for the fact that methyl phosphorothionate insecticides are generally less toxic to mammals than corresponding ethyl homologs (see Table 31).[613] This generalization cannot be applied to phosphonate type insecticides: the ethyl esters EPN and cyanofenphos are less toxic to mammals than corresponding methyl esters.[614] It is not certain whether phosphonate methyl esters can be degraded by the transferase or not.

Both phosphorothionates and phosphate esters can be substrates. Only one of two *O*-methyl groups in the molecule is removed by this reaction: the transferase is inactive towards the produced mono-desmethyl derivatives. The formation of bis-desmethyl bromophos had been reported, but was later denied by the same author.[615] Of two geometric isomers of mevinphos, only *cis* isomer is readily degraded by the mouse liver glutathione *S*-alkyltransferase.[608] On the contrary, the *trans* isomer of mevinphos and both *trans* and *cis* isomers of the related insecticide Bomyl are not demethylated (but are degraded by esterases splitting the vinyl ester linkage), perhaps for steric reasons:[221] the dimethoxyphosphinyl group is close to a carboxy ester group in these molecules.

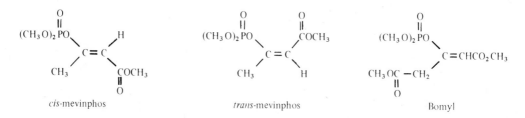

cis-mevinphos   trans-mevinphos   Bomyl

The transferase catalyzed reaction appears to be stereospecific: *O,S*-dimethyl *O*-1-naphthyl phosphorothiolate, obtained by the methylation of the enzymatically demethylated product from *O,O*-dimethyl *O*-1-naphthyl phosphorothionate, was predominant in the (-)-enantiomorph.[603]

It is reasonable to assume that the low toxicity of fenitrothion depends, at least in part, on the *O*-demethylation mechanism in mammals. The demethylated derivatives were prominent detoxified metabolites in mice, particularly at higher dosage level of the insecticide: 75.5% and 83.2% of the urinary metabolites of fenitrothion, which were excreted for 24 hr from mice administered with the insecticide 200 mg/kg and 850 mg/kg, respectively, were the *O*-demethylated metabolites, whereas with parathion-methyl, only 42% of the metabolites were the demethylated ones at the 17 mg/kg dose.[616] Methylglutathione was not excreted in the urine, probably due to its rapid degradation in vivo.

As the glutathione dependent demethylation plays an important role as the first step of detoxication of the methyl ester series of organophosphorus pesticides, an attempt to use glutathione with oxime antidote for the therapy of organophosphate poisoning was undertaken successfully.[617] On the other hand, certain alkylating agents, such as methyl iodide, greatly potentiated the toxicity of fenitrothion to mice (11.5-fold), but not the ethyl homolog of the insecticide (1.4-fold),[610] showing again the great importance of glutathione for the detoxication of methyl phosphorus esters, but little importance for that of ethyl derivatives. Methyl iodide may alkylate glutathione before it can accept the methyl group from the organophosphorus compound.

iii. Miscellaneous Phosphorus Bond Splitting Reactions

Hydrolytic cleavage of alkyl phosphate ester bond — Kojima and O'Brien reported that, in addition to three different Ca^{2+}-activated liver enzymes which hydrolyze paraoxon to diethyl phosphate and *p*-nitrophenol, a fourth enzyme, responsible for degrading paraoxon to desethyl paraoxon and ethanol, was present in the soluble fraction of rat liver.[588] Recently, similar enzyme activity was found in vivo in houseflies.[618] Since labeled ethanol was formed, rather than acetaldehyde or *S*-ethylglutathione, from ^{3}H-paraoxon, this reaction had to be catalyzed by a hydrolase (Equation 69). Deethylation is a major degradation pathway in a susceptible housefly strain.

$$\underset{C_2H_5O}{\overset{C_2H_5O}{\diagdown}} \overset{O}{\underset{\parallel}{P}} - O - \langle \rangle - NO_2 \longrightarrow \underset{HO}{\overset{C_2H_5O}{\diagdown}} \overset{O}{\underset{\parallel}{P}} - O - \langle \rangle - NO_2 + C_2H_5OH \qquad (69)$$

Cleavage of *S*-alkyl phosphorothiolate bond — The metabolic cleavage of P-S-aryl linkages takes place, apparently, at P-S with an oxidative mechanism for thiolothionates or with a hydrolytic mechanism for thiolate esters, as demonstrated in the metabolism of fonofos (Equation 60). However, the destruction of P-S-alkyl linkages is not certain and depends on the structure of the compound and the species of organisms if P-S or S-C bond is attacked, i.e., either the P-OH or P-SH product is formed. Though the detoxication of malathion is due mainly to carboxyester hydrolysis and then to demethylation (5%), the thiolothionate ester linkage was disrupted to a small extent at both P-S (1%) and S-C bond (0.5%) by human liver homogenate.[619] Different mechanisms may be involved for the cleavages of these bonds: the S-C bond splitting, as well as carboxyester hydrolysis by wheat germ or human serum, was prevented by triphenyl phosphate, but the P-S bond cleavage was not.[508] The degradation of phenthoate on the surface of plants proceeded chiefly by P-S bond cleavage, yielding bis α-(carboethoxy)benzyl disulfide, probably owing to photocatalysis, whereas within the plant interior, carboxyester hydrolysis occurred first.[620]

$$\underset{}{\overset{S}{\underset{\parallel}{(CH_3O)_2 P}}} - S - \underset{\underset{CH_2CO_2C_2H_5}{|}}{CH} - CO_2C_2H_5$$

malathion

$$\underset{}{\overset{S}{\underset{\parallel}{(CH_3O)_2 P}}} - S - CH - CO_2C_2H_5$$

phenthoate

Dimethoate may be degraded by 1) amide hdyrolysis, 2) *O*-demethylation, 3) P-S cleavage, and 4) S-C cleavage. The contribution of each reaction to dimethoate degradation by the homogenates of rat liver, houseflies, and rice leaf is shown in Table 32.[621] It is interesting to note that *O*-demethylation is preponderant in rice plants and houseflies, but the thiolate bond cleavage plays an important role for detoxication in rat liver. The S-C bond splitting appears to be very specific in rats. This is true in the metabolism of vamidothion, too. It is uncertain how P-S and S-C bonds of the dithioates are split. Since the glucuronide of *N*-methylhydroxyacetamide was found,[622] as the metabolite of dimethoate, the S-C bond cleavage should be due to a hydrolytic reaction.

Menazon was mainly disrupted at the P-S bond in rats; the main excretion metabolite was 2-methylsulfinylmethyl-4,6-diamino-*s*-triazine, which might be· produced by methylation of the liberated mercaptan followed by oxidation of the resultant sulfide (see d.iv).[623] Dioxathion gave urinary metabolites owing to the cleavage of each bond of thiolophosphorus and thiolocarbon.[624]

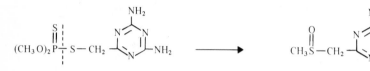

menazon

(70)

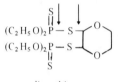

dioxathion

Metabolic degradation of phorate proceeded mainly by P-S bond cleavage, but *O,O*-diethyl phosphorodithioic acid was also found in plants and insects.[205]

Among dithiophosphates with a thiomethylene bridge between the phosphorus and a heterocyclic nitrogen atom, phosmet (Imidan) appeared to be disrupted at the S-C bond in rats, giving *N*-hydroxymethylphthalimide, which was then decomposed rapidly into phthalamic acid, the main metabolite, and further to phthalic acid and others.[625] Similar metabolites were also found in insects and plants.

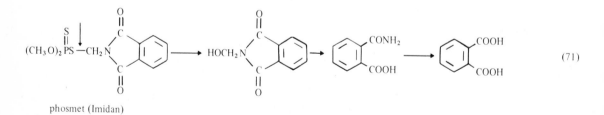

phosmet (Imidan)

(71)

On the other hand, methidathion (Supracide) was degraded mainly by P-S bond splitting in rats, being excreted as *S*-methylated metabolites (see Section 2.d.iv of this Chapter). Moreover, phosalone was degraded at C-S bond in plants and soils, forming diethyl phosphorodithioic acid, chlorobenzoxazolone (XXVI), and formaldehyde, whereas either of the C-S and P-S bonds was split in rats.[210] Chlorobenzoxazolone-3-yl-methanethiol (XXVII) produced by the P-S bond cleavage was methylated and then oxidized, giving finally the sulfone (XXVIII) in rats. Chlorobenzoxazolone (XXVI) produced by the C-S bond cleavage accompanied by leaving of formaldehyde was further transformed in plants into *N*-glucoside (XXIX), whereas in rats and soils it was biotransformed into phenoxazinone (XXX). The metabolic pathways of phosalone are summarized in Equation 72.

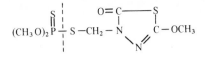

methidathion (Supracide)

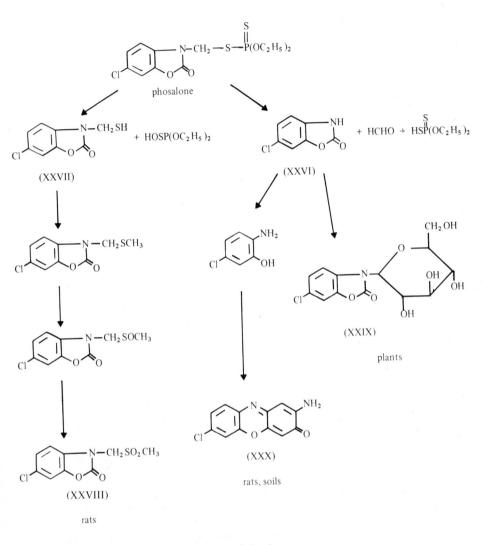

Degradation pathways of phosalone

The fungicides Kitazin and Inezin are *S*-benzyl phosphorothiolate and phosphonothiolate esters, respectively. Cleavage of S-C linkage occurred remarkably in metabolism of these fungicides by rats, cockroaches, rice plants, and the fungus *Piricularia oryzae,* and in soils, yielding *O,O*-diisopropyl phosphorothioic acid or *O*-ethyl phenylphosphonothioic acid.[302] The P-S bond cleavage was also found with Kitazin P in rice plants, but to only a small extent in rats, insects, and fungi. In contrast, the P-S bond cleavage was a main degradation pathway for Inezin in the fungus.[302]

(72)

TABLE 32

Cleavage Sites of Dimethoate by Homogenates of Different Organisms (%)

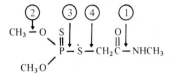

Cleavage site	Rat liver	Houseflies	Rice leaf
1 C-N	2.5	27.3	0
2 C-O	11.6	63.1	100
3 P-S	58.6	6.9	0
4 S-C	25.8	0	0

Kitazin P   Inezin

Phosphoramide bond cleavage – Only a little information is available about phosphoramide bond cleavage in biological systems. Crufomate (Ruelene) was first disrupted at the P-N bond in sheep: desamino crufomate was the main metabolite excreted in urine for the first 9 hr after oral administration.[626]

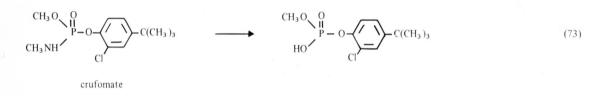

$$(73)$$

crufomate

The mechanism for the metabolic cleavage of the P-N bond is not known, though the acid catalyzed hydrolytic mechanism could be a possible model (see Section III.A.2.b).

P-C bond cleavage – The cleavage of the P-C bond in biological systems will be discussed briefly here. Tabun is enzymatically hydrolyzed at the P-CN bond by A-type esterases. β-Chloroethylphosphonate is also highly reactive and is disrupted at the P-C bond (see Section V.E.3). Although trichlorfon is readily transformed nonenzymatically into a vinyl phosphate, dichlorvos, which may be the actual toxicant, certain species of animals and plants do not appear to involve this pathway in their detoxication process; trichlorfon and its acetyl derivative hydrolyze more quickly than dichlorvos.[481] The nonphosphorus portion of trichlorfon injected intravenously into a dog was excreted in the urine as trichloroethyl glucuronide, indicating the disruption of the P-C bonding. However, in the urine of a rabbit administered with the insecticide, another glucuronide consisting of desmethyl dichlorvos was observed but not the glucuronide of trichloroethanol.[627]

184   *Organophosphorus Pesticides: Organic and Biological Chemistry*

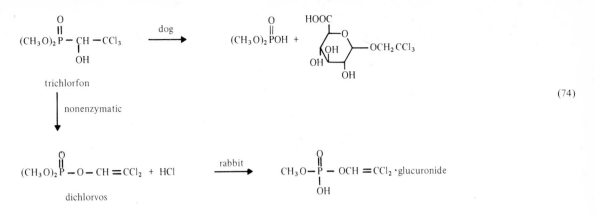

(74)

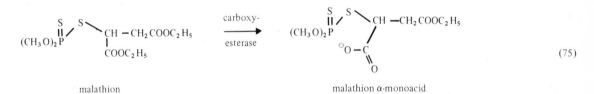

(75)

Many of the reactions described in Section iii are not necessarily certain from the chemical or biochemical point of view: all enzymes concerned have not been studied thoroughly, and some products have not been fully characterized. Further investigations with modern techniques may be necessary for confirmation or correction, if erroneous.

c. Biotransformation of Nonphosphorus Functional Groups
i. Hydrolysis of Carboxyester Linkage

A number of organophosphorus insecticides containing a carboxyester linkage in the molecule, such as malathion and acethion, owe their low toxicity to mammalian carboxyesterase (EC 3.1.1.1). The hydrolysis of the ester linkage results in detoxication. An anionic charge on the carboxyl group may be brought up close to the phosphorus so that the electrophilicity of the phosphorus is reduced by the field effect; consequently, malathion monoacid is inactive.

Extensive studies on the substrate-specificity of a rat liver carboxyesterase were carried out by Dauterman and his co-workers (reviewed in Reference).[504] Their conclusions are as follows: nonphosphorus mono- and di-carboxylic esters serve as substrates; long-chained carbalkoxy compounds are better substrates; the α-carboxyester linkage of malathion is preferably hydrolyzed to yield O,O-dimethyl S-(1-carboxy-2-carbethoxy)ethyl phosphorodithioate (malathion α-monoacid)*;[628,629] of the optical isomers of malathion, the d-isomer binds better to the enzyme than the l-isomer; the substrate must be unionized at physiological pH. Thus, the second carbethoxy group of malathion is not hydrolyzed by rat liver carboxyesterase. However, malathion diacid was found as an in vivo metabolite; it may be produced under the influence of an unknown factor.

Carboxyesterases are also called ali-esterases or B-esterases, and are widely distributed in mammals and have been found in the liver, kidney, serum, lung, spleen, and ileum. The carboxyesterase activity is low in susceptible insects.[631] This is responsible, at least in part, for the high selective toxicity of malathion. The metabolism of malathion in insects and mammals is qualitatively similar, but greatly different quantitatively. The degradation to nontoxic ionic products is much more extensive in mammals than in insects: the level of the toxic metabolite malaoxon in the American cockroach was about 20-fold of that in

* Welling and Blackmeer reported recently that horse liver aliesterase and rat liver microsomal esterase produced mainly the β-monoacid (ratio α/β = 0.1), but housefly homogenate gave the α-monoacid more than the β-isomer (ratio α/β = 3.5 to 5.0).[630]

TABLE 33

Toxicity of Some Organophosphorus Compounds Having a Carboxy Ester or an Amide Group

Compound	Structure	LD$_{50}$ (mg/kg)		Ratio
		Mouse	Housefly	
Malathion	$(CH_3O)_2\overset{\overset{S}{\|}}{P}SCHCO_2C_2H_5$ $\quad\quad\quad\quad\|$ $\quad\quad CH_2CO_2C_2H_5$	815	30	27
Malaoxon	$-P{=}O$	75	15	5
Acethion	$(C_2H_5O)_2\overset{\overset{S}{\|}}{P}SCH_2CO_2C_2H_5$	1,280	9.4	130
Acetoxon	$-P{=}O$	214	3.4	63
Mevinphos	$(CH_3O)_2\overset{\overset{O}{\|}}{P}OC{=}CHCO_2CH_3$ $\quad\quad\quad\quad\|$ $\quad\quad\quad CH_3$	5	1	5
Dimethoate	$(CH_3O)_2\overset{\overset{S}{\|}}{P}SCH_2CONHCH_3$	140	0.4	350
Dimethoxon	$-P{=}O$	55	0.1	550
Phosphamidon	$(CH_3O)_2\overset{\overset{O}{\|}}{P}OC{=}CCON(C_2H_5)_2$ $\quad\quad\quad\quad\| \quad\quad Cl$ $\quad\quad\quad CH_3$	11	19.5	0.6

the mouse 2 hr after injection.[699] The degradation of malathion by carboxyesterase hydrolysis predominates in mammals.

Certain strains of arthropods which are resistant to malathion show the higher activity of carboxyesterase which is responsible, at least partly, for the malathion resistance of mosquitoes, houseflies, leafhoppers, and citrus red mites.[632-635] In a leafhopper (*Nephotettix cincticeps*) and a planthopper (*Laodelphax striatellus*), resistance to malathion was associated with a great increase in an esterase activity towards β-naphthyl acetate.[636] Similar increase of esterase activity was observed in certain resistant strains of houseflies. However, the esterases of the housefly, detected on agar gel-electrophoregram by the use of β-naphthyl acetate as substrate, played only a minor role for degrading malathion. Another esterase was found, however, which was more specific to hydrolyze malathion carboxyester but was not well detected by β-naphthyl acetate.[637] The carboxyesterases are inhibited by synergists, such as *n*-propyl paraoxon, the oxo-analog of EPN,[633] and saligenin cyclic phenyl phosphate or phenylphosphonate,[635,637] equivalent to the active metabolite of TOCP. Malaoxon plays a role as substrate and inhibitor for the carboxyesterase.

The carboxyester group in organophosphorus insecticides does not always contribute to a decrease in the toxicity; mevinphos is a much more toxic insecticide than expected from its structure, which contains a carbethoxy group (see Table 33). For the detoxication of mevinphos, carboxyesterases play an insignificant part, if any; carboxyester hydrolyzed metabolites could not be found in the incubation mixtures with liver homogenates, though the carboxyester linkage was hydrolyzed nonenzymatically.[221]

Lactone ring opening was postulated for the metabolic breakdown of coumaphos (Co-Ral) in cow and goat as judged from the partition behavior of the metabolites in urine.[638]

$$(76)$$

ii. Hydrolysis of Carboxyamide Linkage

Another hydrolyzable nonphosphorus functional group is a carboxyamide linkage found, for example, in dimethoate and related insecticides.

$$
\underset{\text{dimethoate}}{(CH_3O)_2\overset{S}{\overset{\|}{P}}SCH_2\overset{O}{\overset{\|}{C}}NHCH_3} \longrightarrow (CH_3O)_2\overset{S}{\overset{\|}{P}}SCH_2\overset{O}{\overset{\|}{C}}OH
$$

$$(77)$$

The importance of the cleavage of the amide bond for selective toxicity has been demonstrated; there is a correlation between in vitro degradation and in vivo toxicity.[639] Carboxyesterase, mentioned in the preceding section, does not catalyze the hydrolysis of dimethoate carboxyamide.[640] The enzymic carboxyamide cleavage of dimethoate in vertebrates occurs almost exclusively in the liver. Sheep liver is a good source of the carboxyamidase, which is primarily associated with the microsomal fraction. The sheep liver amidase was solubilized from microsomes and purified 50-fold. Various divalent cations and nucleotides do not affect the enzymatic activity.[640] The amidase cannot hydrolyze N-methylamides of propionic acid and shorter homologs. The best substrate in this series is N-methylcaproamide.[504] With N,N-dialkylcaproamides, only the dimethyl compound was hydrolyzed. In a series of N-monoalkyl-caproamides, methylamide binds best with the enzyme, and the binding decreases with an increase in the chain length of the alkyl group. In the homologous series of dimethoate, N-n-propyl homolog was the best substrate. The oxygen analog of dimethoate is not hydrolyzed by the amidase, but inhibits the enzyme; only the thiono analog can serve as substrate. The dimethoate amidases from different sources differ in their sensitivity to inhibition by EPN in vivo: those of houseflies and milkweed bugs were insensitive to EPN, but those of mammals were highly sensitive.[641]

Metabolic amide cleavage is not necessarily significant with all organophosphorus pesticides containing a carboxyamide moiety nor in all organisms. For example, although the degradation of dimethoate in sheep liver homogenate was almost completely due to amidase, that in guinea pig liver was predominantly due to the cleavage of the S-C bond to form dimethyl phosphorodithioic acid.[642] Moreover, the amide group of dimethoate was hydrolyzed nonenzymatically on the leaf surface after oxidative desulfuration.[643] Amide bond cleavage was observed with dicrotophos in rats, whereas the metabolic formation of phosphamidon acid from phosphamidon was not observed in animals and plants.[644]

Decarboxylation of dimethoate carboxylic acid, the product of amidase activity, was suggested to occur by the action of enzymes of the germ and aleurone layer of wheat forming trimethyl phosphoro-dithioate.[508]

dicrotophos   phosphamidon

iii. Reduction

The nitro group of parathion, EPN, and similar compounds is reduced enzymatically to an amino group (Equation 78). The resulting amino derivatives are essentially inactive as insecticides or as anticholinesterase agents because the amino group is electron releasing ($\sigma = -0.66$) in contrast to the nitro group which is a strong electron-withdrawing group ($\sigma = +1.27$).

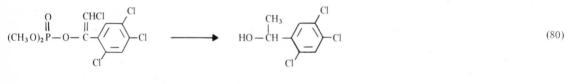

$$(C_2H_5O)_2\overset{\overset{\text{S}}{\|}}{P}-O-\underset{}{\underset{}{\bigcirc}}-NO_2 \xrightarrow[\text{NADPH}]{\text{reductase}} (C_2H_5O)_2\overset{\overset{\text{S}}{\|}}{P}-O-\underset{}{\underset{}{\bigcirc}}-NH_2 \qquad (78)$$

The reduction occurs readily in rumen juice and microorganisms:[645,646] amino-parathion is a major metabolite in ruminants. Although oxidative systems play a much more important role for parathion degradation in insects and vertebrates, nitroreductases which are effective for the detoxication of parathion and paraoxon have been found in housefly abdomen and vertebrate livers.[647,648] The reductases require the cofactor NADPH.

Reduction of sulfoxides to sulfides was recently observed in soils and plants. The sulfide of fensulfothion represented 10 to 20% of the chloroform soluble metabolites of fensulfothion in the bean plant.[301] Disulfoton sulfoxide was transformed partly into disulfoton in flooded soils.[541]

$$(C_2H_5O)_2\overset{\overset{\text{S}}{\|}}{P}-O-\underset{}{\underset{}{\bigcirc}}-\overset{\overset{\text{O}}{\uparrow}}{S}CH_3 \xrightarrow{\text{plant}} (C_2H_5O)_2\overset{\overset{\text{S}}{\|}}{P}-O-\underset{}{\underset{}{\bigcirc}}-SCH_3 \qquad (79)$$

fensulfothion

The reduction of sulfoxides cannot be regarded as detoxication.

Reductive dechlorination is also known. For example, chlorfenvinphos and tetrachlorvinphos are degraded into dichloro- and trichloro-phenylethanol, respectively, in mammals, plants, and soils.[649,650] The reaction occurs in vitro in beef liver homogenate.[651]

$$(CH_3O)_2\overset{\overset{\text{O}}{\|}}{P}-O-\overset{\overset{\text{CHCl}}{\|}}{C}-\underset{}{\underset{}{\bigcirc}}\overset{\text{Cl}}{\underset{\text{Cl}}{-Cl}} \longrightarrow HO-\overset{\overset{\text{CH}_3}{|}}{C}H-\underset{}{\underset{}{\bigcirc}}\overset{\text{Cl}}{\underset{\text{Cl}}{-Cl}} \qquad (80)$$

tetrachlorvinphos

Reductive dechlorination of phosphamidon amide was once suggested from the erroneous identification of small quantities of dechlorophosphamidon amide in rat urine and goat milk.[572] It was, however, corrected later by the same author (see Equation 57).[571]

d. Conjugation

Conjugation reactions are biosynthetic pathways by which foreign compounds and their metabolites containing certain functional groups are linked to endogenous substrates leading generally to less toxic compounds. As discussed in the preceding sections, organophosphorus compounds undergo a variety of biological modifications that may result in the introduction of a functional group on the nonphosphorus organic moiety for conjugation. Thus, conjugation reactions serve for the further metabolism of disrupted leaving groups. Conjugation generally enhances the excretion of the xenobiotics from the body by 1) promoting the glomerular filtration process in the kidneys, and 2) preventing the storage of originally lipophilic substances in body lipid. Important conjugation reactions that may be involved in the metabolism of organophosphorus compounds are the formation of glucuronides, glycosides, sulfates, and methylated derivatives.

i. Glucuronide Formation

Alcohols, phenols, carboxylic acids, amines, and thiols can form glucuronic acid conjugates by the action of glucuronyl transferases in all mammals and in most vertebrates other than fish.[497] The formation of O- and N-glucuronides appears to be catalyzed by different enzymes. The enzyme activities are located in the microsomal fraction of liver cells. Uridine-5′-diphospho-D-glucuronic acid (UDPGA) functions as the donor of the D-glucuronopyranosyl group to a variety of substrates to form glucuronides. The linkage between glucuronic acid and phosphoric acid in UDPGA cannot be hydrolyzed by β-glucuronidase, suggesting that it is of α-glycoside configuration. On the other hand, all the produced glucuronic acid conjugates have a β-glycoside linkage, without exception. Therefore, the transfer reaction is accompanied by the Walden conversion.

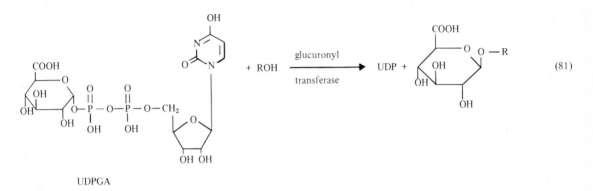

UDPGA

(81)

Glucuronide formation has been observed in in vivo metabolism of organophosphorus insecticides, as in the following examples: dichloroethyl glucuronide from dichlorvos in rats (Equation 82);[654] 1-(2,4-dichlorophenyl)ethyl glucuronide from chlorfenvinphos in rats and dogs (Equation 83);[649] methyl- and dimethyl-sulfamoylphenyl glucuronides from famphur in the calf (Equation 84);[652] and p-aminophenyl glucuronide from parathion in the cow (Equation 85).[653]

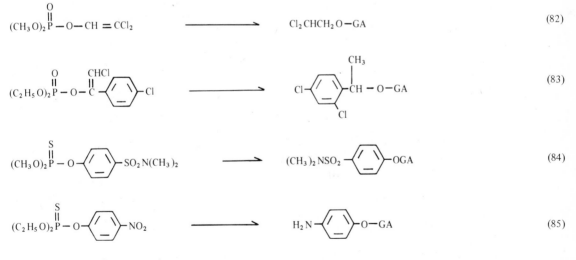

GA = glucuronic acid

ii. Glycoside Formation

Unlike vertebrates, plants and insects can utilize the D-glucopyranosyl group transferred from uridine-5′-diphospho-D-glucose (UDPG) in the formation of conjugates. For example, the active metabolites of an experimental insecticide GC 6506, which are the sulfoxide and sulfone, were rapidly detoxified to form the glucuronides and sulfates of the corresponding phenols in rats, whereas in insects and plants, the β-glucosides of the phenols were produced (Equation 86).[655,656] The conjugated metabolites in animals were rapidly excreted from the body, but the glucosides in plants remained in the treated foliage without translocation and were slowly transformed into secondary metabolites, probably gentiobiosides. Under certain conditions, sugars other than glucose may be involved in the glycoside conjugation.[650] N-glucoside formation is also known; the metabolite of phosalone in plants has been identified as N-glucoside of benzoxazolone (Equation 72).

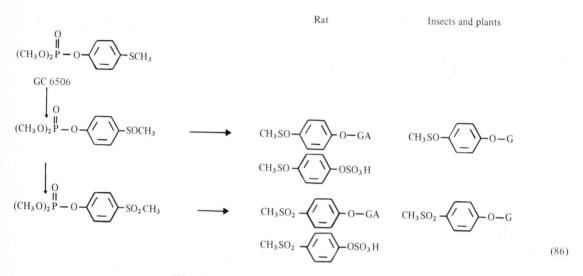

(86)

(GA = glucuronic acid; G = glucose)

iii. Sulfate Formation

Sulfate conjugation is an important metabolic pathway for phenols and alcohols. The insecticide cyanophos (dimethyl p-cyanophenyl phosphorothionate), for example, was degraded in rats and p-cyanophenyl sulfate was excreted in the urine as a main metabolite.[657]

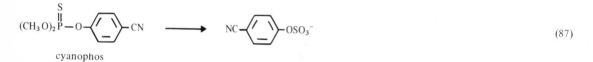

(87)

Aromatic amines also undergo sulfate conjugation, forming sulfamates. The sulfate group is transferred from 3'-phosphoadenosine-5'-phosphosulfate (PAPS) by the catalytic action of an enzyme named sulfotransferase (or sulfokinase) in the soluble fraction of liver and other cells.

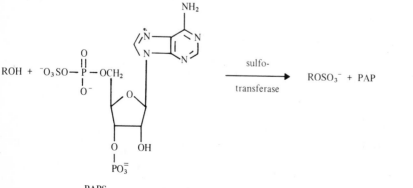

(88)

iv. Methylation

Methylation is a relatively minor pathway of xenobiotics metabolism. However, organophosphorus compounds containing a thiolate ester bond which can be disrupted biochemically may yield S-methylated metabolites. For example, the insecticide methidathion (Supracide) decomposes rapidly in rats, forming the 4-methylsulfinylmethyl and 4-methylsulfonylmethyl derivatives of 2-methoxy-1,3,4-thiadiazol-5-one heterocycle as the main metabolites.[658] The metabolites may originate by the methylation and subsequent oxidation of the mercaptomethyl derivative liberated by the cleavage of the P-S bond. Phosalone yields similar metabolites in rats (see Equation 72).[210] A similar methylated sulfoxide metabolite was found as one of the major degradation metabolites excreted in the urine of rats which had been administered with menazon (see Equation 70).[623]

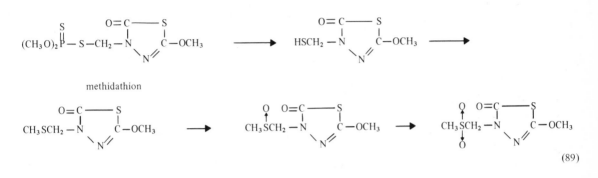

(89)

These methylated metabolites have no anticholinesterase activity and are 15 to 50 times less toxic than the original insecticides.[659]

The soil insecticide fonofos (Dyfonate) undergoes similar biotransformation: the thiophenol moiety, released by the oxidative P-S cleavage of fonofos or by the hydrolysis of the oxon, is rapidly methylated and subsequently oxidized to sulfoxide and sulfone in both rats and potato plants.[381,660,661] These reactions are a real detoxication process, as indicated by the LD_{50} values of the metabolites in Equation 90. Moreover, trace amounts of neutral esters which appeared to be the methylated products of O-ethyl ethylphosphonothioic acid and its oxon were discovered, though the evidence for metabolic methylation of these acids was not conclusive.

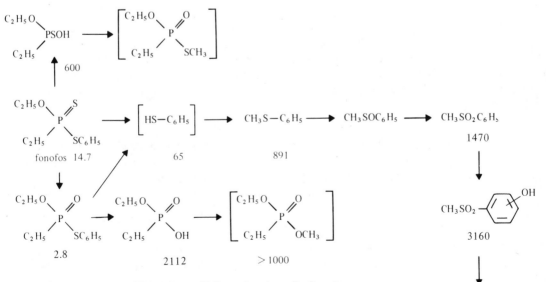

Figures are oral LD_{50} values in mg/kg for rats.

(90)

The methyl group is transferred by the action of methyltransferases from "active methionine," that is S-adenosylmethionine biosynthesized from methionine and ATP.

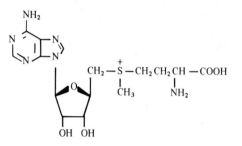

S-adenosylmethionine

v. Glutathione Conjugation

Some alkyl or aryl halides undergo a reaction with glutathione by transferases to form corresponding S-alkyl (or aryl) glutathiones, which may be finally converted in vivo into mercapturic acids (S-substituted N-acetylcysteine). The important role of glutathione conjugation for the detoxication mechanisms of organophosphorus compounds has been discussed in the preceding sections (2.a.iii and 2.b.ii).

2. Fate in the Environment

Residual pesticides were generally preferred in the past because they permitted the control of pests for a long time by a single treatment. However, some chlorinated hydrocarbon insecticides such as DDT and dieldrin persist amazingly long in soils, water, plants, and animal bodies and have been used in a huge amount over the world, so that the contamination of the environment has caused serious problems in this decade. Being lipophilic even after biotransformation into noneffective compounds, the chlorinated hydrocarbons are readily stored in fatty tissue with little excretion and are ecologically magnified in organisms through food chains.

On the contrary, organophosphorus pesticides are generally short-lived in the environment and biological systems; they are commonly degraded through hydrolysis and other reactions into nontoxic and water soluble compounds which are rapidly excreted from animal bodies. In Chapter III and preceding sections, a variety of chemical and biochemical reactions of organophosphorus compounds have been described. The reactions may contribute to the transformation and disappearance of organophosphorus pesticides under environmental conditions too.

When pesticides are sprayed on plants, materials not adhering tenaciously to the plant surface slough off immediately and the remaining portion of the initial deposit becomes an effective deposit, which then begins to be absorbed into the plant. The dissipation of the pesticide deposit on foliage may proceed primarily through codistillation with plant respirants, volatilization, photodecomposition by sunlight, and mechanical removal by rain and wind.

After penetration into the plant, metabolism may be principally operative. Although the enzyme activities are generally lower in plants than in animals and the contribution of each reaction to the whole metabolism may differ more or less among species, the metabolic patterns are not essentially different;

oxidative desulfuration of thiophosphoryl group, oxidation of thioethers, oxidative *N*-demethylation, *O*-demethylation, and ester bond cleavage by esterases or oxidases commonly occur in both organisms. Only one distinctive difference concerns excretion: the animal excretes the greater part of an applied phosphoric ester, in changed or unchanged form, within a few hours or days, whereas the plant tends to store chemicals. In animals the major part of the liberated leaving group or pesticide modified by hydroxylation may be excreted through conjugation with sulfuric acid or glucuronic acid, while glycoside conjugation occurs in plants (and insects). The glycoside can be cleaved to liberate again the aglycon.

In addition to the enzymatic transformation mechanism, nonenzymatic reactions also occur on plants to produce toxic compounds: oxidation of thiophosphoryl and thioether groups and thiono-thiolo rearrangement are promoted by sunlight and/or heat. The toxicity of such products transformed from parathion, for example, is compared in Equation 91. LD_{50} values in mg/kg for mice (M) by subcutaneous treatment and for houseflies (F) by topical application are shown.

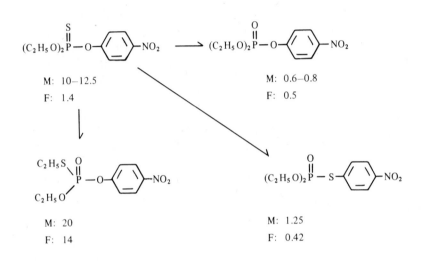

M: 10—12.5
F: 1.4

M: 0.6—0.8
F: 0.5

M: 20
F: 14

M: 1.25
F: 0.42

(91)

The activated organophosphorus pesticides do not represent a serious environmental persistence problem, since they tend to hydrolyze readily in plants, with the exception of some compounds containing the oxidized thioether group. Although hydrolysis products do not have the typical toxicity of organophosphorus, i.e., anticholinesterase, toxicological investigations are still necessary in certain cases, particularly for heterocyclic compounds.

The rate of chemical and biochemical transformation may depend on a number of factors involving the nature of the compound, application method, the kind and growing stage of the plant, and climate. Physicochemical properties of the compound, for example, stability to sunlight and vapor pressure, may greatly influence the residue. The residue is also affected by the formulation. Deposit of pesticides generally decreases in the following order: emulsion > wettable powder > dusts. The physical properties of the plant surface are one of the main factors in deciding the amount of pesticides adsorbed on the plant. The thickness of the wax layer in the subsurface influences the absorption of organophosphorus pesticides and the transfer of the compound into the plant. Organophosphorus pesticides absorbed into the wax layer are rather stable for a long time. For example, 15 to 30 days after treatment, the residues of azinphosmethyl, dicrotophos, dimethoate, dioxathion, ethion, and malathion in the rind of oranges were 2.6 to 7 ppm and, in contrast, those in the juice were only 0.03 to less than 0.2 ppm.[662]

Soils are contaminated with pesticides falling out of aerial sprays on crops or by direct application. Factors influencing the disappearance of pesticides from soils may involve physicochemical characteristics of the compound and the soil, moisture or water content, and microflora in the soil. Pesticides on the surface may dissipate by volatilization, photodecomposition, and mechanical removal. Once the pesticide is incorporated, chemical decomposition catalyzed by metallic ions and other components of soils, adsorption to soil colloids, microbial metabolism, and possibly plant uptake will operate as important mechanisms for the disappearance of the pesticide.

Temperature and pH affect, of course, the rate of degradation. For example, diazinon is most stable at the neutral pH region; the half-life at 20° in water was 155 days at pH 7.4, whereas that at pH 3.1 and 10.4 was 0.5 day and 6 days, respectively. (Diazinon is more susceptible to acid than alkali, contrary to ordinary organophosphorus esters.)[223] The half-life of parathion dropped from 1,000 days at 10° and pH 1 to 5, to 15 days at 50°.[663]

The presence of soils influences greatly the rate of pesticide decomposition. In spite of stability at neutral pH, diazinon hydrolysis occurs rapidly in a silty clay at pH 7.2. The soil type is one of the most important factors influencing the persistence of organophosphorus pesticides. It is often more influential than temperature. For example, the disappearance rate of disulfoton and phorate in a loamy sand in winter was greater than that in a silt loam in summer.[664] Some soil components apparently catalyze the hydrolysis.

Soil colloids play an important role in the fate of pesticides in soils. Adsorption on soil colloids may result in inactivation of the pesticide, contribute to its persistence, or in contrast, catalyze the chemical change of the pesticide molecules. Some systemic organophosphorus insecticides, including mevinphos, phorate, schradan, and demeton, were more effective to control aphids over a long time period when applied to sand or sandy soil than when applied to silt loam, clay loam, or muck.[502] Furthermore, the amount of mevinphos bound by soils correlated positively with the organic matter content, and it appeared that organic matter in soils was a major factor in restricting absorption of phorate by plants from the soil.[665] The adsorption of disulfoton and carbophenothion to soils also correlated with the content of organic matter rather than the content of clay. Organic matter content is considered to be the most important single factor affecting the persistence of insecticide residues.

Metallic ions in soils interact with organophosphorus pesticides. Malathion is quickly incorporated into the interlayer region of some metallic montmorillonite and adsorbed through hydrogen bonding interaction between the carbonyl oxygen and a water molecule in the primary hydration shell of the metallic ion. In the dehydrated state, the carbonyl group interacts directly with the metallic ion, and the intensity of the ion-dipole interaction increases with the cationic valence of the metal.[666] In either case, adsorption was sufficiently strong that no degradation was observed. Mortland and Raman found that cupric ion was the most effective catalyst for degradation of certain organophosphorus esters.[244] Diazinon and chloropyrifos (Dursban) were decomposed rapidly by contact with Cu(II)-montmorillonite; their half-lives at 20° were 4 and 0.9 hours, respectively (see III.A.3.a). However, Ca- and Mg-montmorillonites, vermiculite, and beidellite hardly catalyzed the degradation.

Moisture is important by virtue of competition with pesticides for the adsorption sites on the soil particles, resulting in desorption of the pesticide molecules. Thus, diazinon, parathion, trichlorfon, and mevinphos in moist mineral soils are 135-fold, 28-fold, 20-fold and 1.4-fold, respectively, as active as in dry soils.[667] However, organophosphorus pesticides do not move freely in soils with water, and the loss by leaching does not appear to be appreciable. For example, when ^{32}P-labeled phorate was applied on the surface of various soils and the soil watered every 2 days with 10 to 30 mm of water with a total of 250 mm for 24 days, more than 80% of ^{32}P remained near the surface of silt loam and muck, although more leaching was observed in sandy soil (50%) and quartz sand (76%).[668] Parathion adsorption to soil from aqueous solutions is mainly affected by the organic matter and the expanding-lattice clays.[669]

It is rather difficult to distinguish the effect on pesticide degradation of microflora and chemical action, because sterilization of soils may often be imperfect, autoclaving may cause the change of the physicochemical properties of the soil and the destruction of heat-labile soil components which accelerate the degradation of organophosphorus pesticides, and residues of chemicals applied for soil sterilization may react directly with the pesticide. Gamma irradiation is considered as the best method for soil sterilization.[670] Diazinon decomposed about 90% after 16 weeks in a silt loam containing 20% of water at 25° and about 70% in the sterilized soil, suggesting the initial degradation of diazinon in the soil was mainly due to nonbiochemical reactions.[671] However, another report indicates that sterilization affects most extensively diazinon degradation in loam and sandy loam.[672] When sodium azide was used for soil sterilization, the degradation of parathion was simply retarded, while the reagent catalyzed the degradation of diazinon into diethyl phosphorothioic acid, 2-isopropyl-4-methyl-6-pyrimidinol and 3 other compounds.[673] Fumigation of soils with propylene oxide did not inhibit the hydrolysis of diazinon but inhibited the evolution of carbon dioxide from the pyrimidinol. This indicates that the hydrolysis of diazinon in soils proceeded chemically, and the degradation of the produced pyrimidinol was due to the action of soil microorganisms.[674] Actually the predominant microbial population arising in diazinon treated soils, *Pseudomonas, Arthrobacer,* and *Streptomyces,* attack the chemically hydrolyzed products, i.e., the pyrimidinol and diethyl phosphorothioic acid, rather than intact diazinon.[675] The soil microbes first oxidize the thiophosphoryl group and then degrade the resultant diethyl hydrogen phosphate. The degradation of malathion, demeton, and mevinphos in soils also appears to be initiated by chemical hydrolysis.[675] However, in flooded paddy soils, soil microflora appear to participate much more in the degradation of the parent compound: diazinon disappeared more rapidly in a nonsterilized submerged soil than in a sterilized one, transforming mainly into 2-isopropyl-4-methyl-6-pyrimidinol.[676] The pyrimidinol did not decompose further, probably owing to anaerobic conditions in the submerged soil, in contrast with the above-mentioned nonflooded soils. No diazoxon was formed under this condition.

When parathion was added to a silt loam and incubated for 6 days at 30°, the degradation rate dropped from 46% to 14% by sterilization. Addition of glucose to the nonsterilized soil increased the degradation, forming the reduced product aminoparathion besides the hydrolysis product.[677] The reduction of nitro group to amino group was primarily attributed to yeast in the soil. In addition, *Rhizobium* reduces parathion rapidly into aminoparathion.[678] Aminoparathion has a negligible toxicity compared with parathion. *Bacillus subtilis* converted fenitrothion rapidly into the nontoxic metabolites such as the amino derivative and desmethyl derivatives, but not into the toxic oxon metabolite.[679]

Only a few reports deal with the effect of organophosphorus insecticides on nitrification by soil bacteria. Although parathion as well as some chlorinated hydrocarbon insecticides such as aldrin completely inhibited nitrification by *Nitrobacter agilis* in pure culture at 10 μg/ml, malathion caused only delayed nitrification at 1,000 μg/ml.[680] In soils, organophosphorus insecticides had little effect on nitrification, whereas chlorinated hydrocarbons, carbamates, and ureas markedly inhibited nitrification.[681] Moreover, an increase of nitrate-N in volcanic ash soils resulted from treatment with the systemic organophosphorus insecticides, Aphidan and disulfoton granules; this caused an increase in the yield of potatoes.[682]

It is generally confirmed that organophosphorus pesticides are much less persistent in soils than some chlorinated hydrocarbons, and that with some exceptions they dissipate within a few weeks after application.[683] Pesticides may be divided into three groups by their persistency in soils: 1) highly residual or highly persistent, 2) moderately residual or intermediate, and 3) short residual or rapidly decomposing.[684-686] Group 1 involves DDT and dieldrin as representatives which showed no significant decrease in biological activity in sandy loam over 48 weeks. Many organophosphorus insecticides belong to group 3, losing insecticidal activity within 2 to 4 weeks; they include azinphosmethyl, bromophos, crotoxyphos, diazinon, dichlorvos, dimethoate, disulfoton, chloropyrifos (Dursban), malathion, mevinphos, mecarbam, parathion, parathion-methyl, phorate, and others. This does not mean, however, that these compounds were completely degraded within this period in soils. For example, phorate disappeared rapidly, but its sulfoxide and sulfone derivatives persisted beyond 16 weeks in a silt loam, although the biological activity was lost by strong adsorption to soil constituents.[540] The formulation also affects pesticide persistency; 50% of azinphosmethyl sprayed as an emulsion on soil surface dissipated within 12 days, while that in granules dissipated 50% within 28 days, and 13% of the applied dosage remained as original form and breakdown products after one year.[677] Some detergents such as ABS and LAS prolonged the persistence of parathion and diazinon in soils and also had a synergistic effect on both insecticides in soils.[677]

Fensulfothion, Mocap, and trichloronate are moderately persistent in soils and belong to group 2. Persisting over 36 weeks in soils, chlorfenvinphos, phosfolan (Cyolane), dichlorfenthion, and oxydisulfoton belong to group 1. These organophosphorus insecticides belonging to group 1 or 2 may be useful for soil insect control.

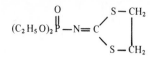

phosfolan
(Cyolane)

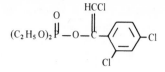

chlorfenvinphos
(Birlane)

$$(C_2H_5O)_2\overset{\overset{\text{S}}{\|}}{P}-O-\text{(aryl, Cl, Cl)}$$

dichlorfenthion
(Nemacide)

$$(C_2H_5O)_2\overset{\overset{\text{S}}{\|}}{P}SCH_2CH_2SOC_2H_5$$

oxydisulfoton

$$\overset{\displaystyle C_2H_5O}{\underset{\displaystyle C_2H_5}{\diagdown}}\overset{\overset{\text{S}}{\|}}{P}-O-\text{(aryl, Cl, Cl, Cl)}$$

trichloronate
(Agritox)

$$(C_2H_5O)_2\overset{\overset{\text{S}}{\|}}{P}-O-\text{(aryl)}-SOCH_3$$

fensulfothion
(Dasanit)

C. SELECTIVE TOXICITY AND RESISTANCE

The toxicity of pesticides differs depending upon both the nature of the chemicals and the species of organisms involved. Selectivity is undoubtedly the most important property which pesticides should have in order to yield maximum effects and to avoid any hazard and unfavorable changes in the ecological environment into which the pesticides are introduced. The selective toxicity may be expressed by the ratio of toxicity indexes between different species in question, for example LD_{50} mouse/LD_{50} housefly. Desirable insecticides should have a selectivity higher than 50-fold for the target pest compared to mammals. The toxicity of some insecticides to mice and some species of insects is compared in Table 34.[631] The high selectivity of malathion for green rice leafhoppers and fenitrothion for rice stem-borers is quite evident.

Difference in toxicity between different strains of the same species due to pesticide resistance is presented as a special case in selective toxicity. Pesticide resistance is a result of selection by the extensive use of pesticides; the pesticide does not appear to induce any heritable changes. A resistant strain developed by exposure to one pesticide often shows resistance to other pesticides too (cross resistance), probably owing to a common resistance mechanism. For example, a resistant strain of houseflies, Fc, selected by diazinon shows high resistance not only to diazinon but also to DDT and other insecticides, including carbamates and pyrethroids, as shown in Table 35.[687,688] The resistance mechanism will be mentioned later.

Only a brief discussion will be made on these subjects in this book. Readers may refer for details to O'Brien's excellent reviews on selective toxicity and many fine reviews on insecticide resistance.[31,32,597,689-696]

When a toxicant acts on an organism, it may undergo the several following steps: 1) contact with the organism, 2) penetration into the organism, 3) metabolism, 4) storage at inert sites and excretion, 5) transport to the target, and 6) interaction with the target. Any qualitative or quantitative difference between organisms in these steps may have an effect on the selective toxicity. On the basis of ecological differences among different species, it is possible that a pest species is made to come into contact with the toxicant in preference to other friendly species. This may be called ecological selectivity. Systemic

TABLE 34

Selective Toxicity of Some Organophosphorus Insecticides[631]

Insecticide	Mouse oral LD_{50} mg/kg	Toxicity ratio (LD_{50} mouse/LD_{50} insect)		
		Housefly	Rice stem-borer	Green rice leafhopper
Selective insecticides				
Cyanophos	860	302	–	–
Dimethoate	56	280	–	–
Fenitrothion	870	153	870	100
Dichlorvos	75	94	72	–
Chlorthion	270	86	85	–
Salithion	91	61	114	3
Trichlorfon	390	57	107	5
Acethion	800	52	–	–
Fenthion	88	27	88	7
Parathion-methyl	32	25	64	3
Diazinon	48	25	20	21
Cyanofenphos	50	24	–	–
Malathion	347	20	386	434
Phenkapton	296	15	–	–
EPN	24	9	27	5
Nonselective insecticides				
Mevinphos	5	5	4	–
Parathion	5	5	1	1.3
Paraoxon	3	3	0.5	–
Thionazin	5	1.5	2.6	–
Azinphosmethyl	8	0.6	8	–
Phosphamidon	11	0.6	6	–
Demeton-methyl	26	0.6	1	–
Phorate	9	0.4	0.9	–
TEPP	2	0.4	0.2	–
Demeton	6	0.3	0.3	–
Oxydemeton-methyl	10	0.1	–	–
Selective mammalicide				
Schradan	29	0.007	0.005	0.2

insecticides which are absorbed into plants and distributed throughout the plants are really selective ecologically and have a great advantage in reducing hazards, because only pest insects will come into contact with the chemicals. This type of insecticide will be described in Section V.A.2. Selectivity occurring after contact is due to physiological differences and may be called physiological selectivity.

1. Penetration

Toxicants may enter into the body through mouth, skin, insect cuticle, or spiracle against some impedance. O'Brien proposed the permeability factor P for a rough measure of the impedance, defining it as LD_{50} (topical)/LD_{50} (injection) for insects and LD_{50} (cutaneous)/LD_{50} (oral) for mammals.[31] If the penetration is not the rate limiting step in the toxic action of an insecticide, the factor will be unity. The permeability factor P of some organophosphorus insecticides for rats and rice stem-borers (*Chilo supressalis* larvae) is shown in Table 36.[631] It appears that there is no distinct correlation between selective toxicity and permeability of organophosphorus insecticides. However, in certain resistant strains of houseflies, a factor that delays the penetration of organophosphorus insecticides contributes somewhat to the resistance (see also Section 6 of this Chapter).[697,698]

TABLE 35

Cross Resistance in Strain Fc Housefly[688]

	LC$_{50}$ (μg/jar)		
	Susceptible strain	Fc strain	Resistance factor
Insecticide	(A)	(B)	(B/A)
Organophosphates			
Diazinon	0.3	25	83.3
Malathion	20	100	5
Carbamates			
Propoxur	6	60	10
Chlorinated hydrocarbons			
Aldrin	0.45	4.5	10
DDT	4.5	4,000	889
Pyrethroids			
Allethrin	6	40	6.7

TABLE 36

Permeability Factor P of Some Organophosphorus Insecticides for Rats and Rice Stem-borers[631]

	Permeability factor P	
Insecticide	Rat[a]	Rice stem-borer[b]
Phosphates		
Dichlorvos	1.3	5.5
Mevinphos	0.8	1.8
Phosphorothionates		
Fenthion	1.5	—
Chlorthion	1.7	2.0
Demeton	2.3	—
Diazinon	8.3	4.8
Dicapthon	2.0	2.3
Parathion-methyl	4.5	1.7
Parathion	1.6	5.7
Phosphorothiolothionates		
Dioxathion	5.5	—
Azinphosmethyl	16.9	0.7
Malathion	3.2	1.4
Phorate	2.7	—
Carbophenothion	1.8	11.5
Phosmet	>17.8	—
Phosphonothionates		
EPN	6.4	2.1
Phosphonates		
Trichlorfon	>3.2	3.6
Pyrophosphoramides		
Schradan	1.6	—

a. LD$_{50}$ (cutaneous)/LD$_{50}$ (oral).
b. LD$_{50}$ (topical)/LD$_{50}$ (injection).

2. Metabolism

Differences in metabolic detoxication of pesticides in different organisms are perhaps most important for selectivity and resistance mechanisms and have been extensively investigated.

The high selective toxicity of malathion has been attributed to high carboxyesterase activity in mammals, in contrast with its low activity in susceptible insects.[699] It is interesting to note that mammalian liver carboxyesterase was about 2,500 times as insensitive as the insect enzyme towards inhibition by malaoxon.[631] The carboxyester function was regarded as a "selectophore" and the thiophosphoryl group was called the "opportunity factor" because it would give time for detoxication to occur.[689] Based on this hypothesis, O'Brien designed and obtained the selective insecticide acethion $(C_2H_5O)_2P(S)SCH_2CO_2C_2H_5$ (see also IV.B.2.c).[700]

In contrast with diazinon resistant insects, malathion resistant strains of insects, such as *Blattella, Chrysomya, Culex,* and *Musca,* are very specific and, with few exceptions,[696] have no cross-resistance to other insecticides, indicating that resistance depends on carboxyesterase degradation. High carboxyesterase activity has been actually demonstrated in some resistant insects[630,632-634] and mites.[701] Carboxyesterase inhibitors act as the synergists of malathion to almost abolish the resistance (see IV.D). The carboxyesterase of the resistant *Culex* prefers to hydrolyze the carbethoxy group and carbo-*n*-propoxy group but cannot hydrolyze a carbomethoxy group. Thus, the carbomethoxy homolog of malathion is effective as an insecticide towards the resistant as well as the susceptible strain.[702] This is not necessarily the case for other insect species: the definite malathion resistant strains of *Musca, Chrysomya* and *Cimex* are resistant against the malathion methyl homolog.[703] It is interesting to note that the modification of the $(RO)_2P$ group decreases the resistance level more distinctively. For example, the resistance level of the malathion resistant strain of *Culex tarsalis* ($\times$ 100), *Musca domestica* ($\times$ 200), and *Chrysomya putoria* ($\times$ 260) decreased into 2 to 5 by replacing the methyl group of malathion with the isopropyl group.

For dimethoate, unlike malathion, there is a great variety in sensitivity in both vertebrates and insects. Guinea pigs (LD_{50} 400 mg/kg) and rats (240 mg/kg) are insensitive but hens are highly sensitive (LD_{50} 30 mg/kg). The toxicity to vertebrates is well correlated with the breakdown rate of the amide and phosphorus ester bonds by liver homogenate.[639] In insects, however, the species specific toxicity of dimethoate is not attributed simply to the degradation rate, but to the sum of every factor, such as penetration rate, activation rate, and sensitivity of acetylcholinesterase to dimethoxon.[704] For example, dimethoate was degraded more rapidly in the American cockroach, a sensitive insect (LD_{50} 7 μg/g), than in the milkweed bug, an insensitive insect (LD_{50} 205), whereas cockroach cholinesterase was 40 times more sensitive (I_{50} 2 $\times$ 10^{-7} M) than that of the bug (I_{50} 5 $\times$ 10^{-5} M). In houseflies, the most sensitive insect (LD_{50} 0.53 μg/g), the degradation proceeded most slowly and cholinesterase was most susceptible to the oxon, whose formation occurred most rapidly in the fly.

Species difference in hydrolase activity may confer a selective toxicity on some pesticides containing a hydrolyzable protective group. Butonate, the butyrate ester of trichlorfon, is a typical example. The toxicity ratio (LD_{50} rat/LD_{50} housefly) increased about 8 times by the acylation of the hydroxyl group of trichlorfon.[177] The metabolic hydrolysis of the carboxyester group results in activation.

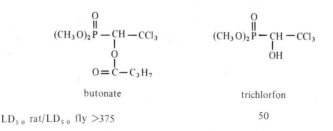

butonate

LD_{50} rat/LD_{50} fly >375

trichlorfon

50

Similarly, *N*-acylation of carbamate insecticides increases the selective toxicity of the insecticides.[705] This idea was extended to *O,S*-dialkyl phosphoramidothiolate insecticides. Methamidophos (Monitor) is modified by *N*-acetylation to give acephate which is about 40 times less toxic to mice than the parent compound, while the insecticidal activity does not change by this modification.[706]

acephate   methamidophos

Furthermore, the carbonylation of phosphoramidothionates yields new low toxic insecticides.[141] For example, *m*-isopropylphenyl *N*-dimethoxyphosphinothioyl-*N*-methylcarbamate (XXXI) is 40 times less toxic to mice than the parent carbamate (XXXII) and more active as an insecticide, particularly for carbamate resistant insects. Fukuto supposed that in mammals the carbamate ester C-O bond would be disrupted to form nontoxic metabolites on one hand, but in insects the phosphoramidate N-P bond would be cleaved to form insecticidal parent carbamate.[707]

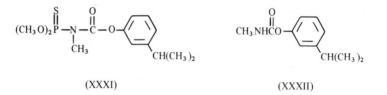

(XXXI)   (XXXII)

N-Alkoxycarbonyl phosphoramidates are known as selective systemic insecticides with low mammalian toxicity.[34] Avenin and demuphos are typical examples;[137,708-710] the LD_{50} of the former for experimental animals is more than 5,000 mg/kg.

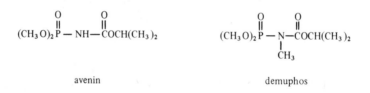

avenin   demuphos

The substrate specificity of mammalian glutathione *S*-alkyltransferase. which prefers methyl phosphorus esters to ethyl homologs, contributes to the selective toxicity of organophosphorus insecticides containing the methyl ester group (compare parathion-methyl with parathion in Table 34). The extremely low toxicity of fenitrothion for mammals in comparison with parathion-methyl is partly due to its higher susceptibility towards demethylation, particularly at higher dosages.[616] Insects generally have poor transferase activity. However, high activity has been reported in certain resistant strains of arthropods such as the diazinon resistant SKA strain of houseflies,[604] an organophosphorus resistant strain of tobacco budworms,[711] and an azinphosmethyl resistant strain of mites.[712] The glutathione *S*-alkyltransferase of the diazinon resistant houseflies (strain SKA) prefers ethyl esters rather than methyl esters. This may be a reason why SKA flies are more resistant to diethyl organophosphorus insecticides than the dimethyl insecticides.[604,713]

A calcium ion dependent phosphotriesterase which has a high substrate specificity towards diazoxon was found in rat liver and serum but not in insects; the selective toxicity of diazinon was attributed to this enzyme.[561,589] Although the rate of glutathione-conjugative degradation was somewhat higher in American cockroach fat body than that in rat liver, the degrading activity of the liver was approximately 200-fold higher than that of the roach fat body, owing to the great difference in the hydrolase activity.

Organophosphorus insecticide resistance in several strains of houseflies is associated with unusually low levels of aliesterase activity, and this property is controlled by a single gene *a*. A Dutch group assumed that a different allele produces modified enzyme proteins, which have little aliesterase activity but degrade the organophosphate ester bond.[714] Though the "mutant aliesterase" is now considered to play a smaller role than was originally thought, this mechanism still appears to contribute significantly to the resistance of the housefly strain E_1, at least at low concentrations of organophosphates (Table 38).[599] A similar reduction in aliesterase activity was also found to occur in an organophosphate resistant strain of blowflies *Chrysomya putoria*.[696] In contrast, in resistant leafhoppers *Nephottetix cincticeps* and aphids *Myzus persicae*, organophosphate resistance is associated with a great increase in esterase activity towards naphthyl acetates.[636,715]

Microsomal mixed-function oxidases (mfo) play a great role in the metabolism of organophosphorus insecticides in both activation and degradation. The ability of liver microsomes for oxidative deethylation of chlorfenvinphos differs remarkably from species, as shown in Table 37 (see also Section IV.B.2.b.i).[603] The difference causes the high selective toxicity of the insecticide. The activity level of the mfo system in insects varies greatly from one strain to another and relates to insecticide resistance. Microsome preparations from the diazinon resistant Rutgers strain of houseflies were about twice as active as that of the susceptible strain in NADPH oxidation, oxygen uptake, and cytochrome P_{450} content.[716] The organophosphate and DDT resistance of housefly Fc strain (Table 35) appears to be due to the microsomal oxidation mechanism. The responsible gene controls the level of NADPH dependent microsomal oxidase activity which detoxifies certain chlorinated hydrocarbons, pyrethroids, carbamates, and organophosphorus insecticides.[688] The metabolic degradation of these insecticides was inhibited by the mfo inhibitor piperonyl butoxide, except for that of malathion, which was rather stimulated because the inhibitor suppressed the formation of malaoxon, the inhibitor of carboxyesterase which should be responsible for malathion degradation. Paraoxon was oxidatively deethylated by the strain Fc microsomal enzyme, and the reaction was inhibited by another microsomal oxidase inhibitor, sesamex.[599] The inhibitor abolished the resistance of Fc houseflies but reduced only in part that of E_1 strain (Table 38), indicating that the resistance of the former is mainly but the latter is only partly due to the microsomal oxidation mechanism.[598,599]

3. Storage and Excretion

Organophosphorus pesticides are generally biotransformed into water soluble metabolites with relative ease in the animal body. The water soluble metabolites are rapidly excreted into urine. For example, in

TABLE 37

Species Distribution of Microsomal Phosphoric Acid Triester Oxygen: NADPH Oxidoreductase and Selective Toxicity of Chlorfenvinphos[603]

Species	Rate of deethylation (nmol/mg protein/min)	LD_{50} mg/kg oral
Rat	0.02	10
Mouse	0.65	100
Dog	2.00	>12,000

TABLE 38

Toxicity of Paraoxon and Its Microsomal Oxidation in Some Strains of Houseflies[598,599]

	LD_{50} μg/fly ♂			Resistance factor with sesamex	Microsomal oxidation	
Strain	Without sesamex	With sesamex	Synergism ratio		$K_m \times 10^6$ (M)	V_{max} (μg/abdomen/hr)
S	0.06	0.006	10	–	c. 11	0.04
Fc	0.20	0.010	20	1.7	5	0.40
$E_1{}^a$	0.5	0.10	5	17	8	0.17

a. K_m and V_{max} for hydrolysis of paraoxon in strain E_1 are 4×10^{-9} M and 0.04 μg/fly/hr.

mice Salithion was decomposed and excreted so rapidly that chloroform soluble ^{32}P in the body was only 2.4% of the administered ^{32}P-labeled Salithion 3 hr after oral treatment.[717] On the other hand, Salithion persisted in the housefly for a relatively long time. More than 4% of the applied or 10% of the absorbed Salithion was found in the chloroform soluble fraction of houseflies 24 hr after topical treatment. Lipid soluble organophosphates can also be excreted from the vertebrate body with bile into feces. With a few exceptions such as Narlene,[718] there is no evidence of accumulation of organophosphorus insecticides in mammalian tissues. The relatively low chronic toxicity of organophosphorus pesticides may be attributable to this general property.

The absorption of the insecticides into body fat appears to contribute to the insecticide tolerance of insects. For example, the female cockroach (*Periplaneta americana*) is twice as tolerant to schradan as the male insect, probably due, at least in part, to the fact that the female body absorbs about five times as much of the insecticide as the male.[719]

Besides reacting with acetylcholinesterase, organophosphorus anticholinesterase agents may react with noncritical B-esterases and certain proteins, leaving the phosphorus moiety bound to the protein. This may be regarded as one of the defense mechanisms, at least with respect to acute toxicity. For selectivity in delayed neurotoxicity, which appears to be due to inhibition of certain organophosphate sensitive enzymes other than acetylcholinesterase, see Chapter IV.E.1.

4. Transport to Target

In a series of 2-chlorovinyl phosphate insecticides, an interesting structure-activity relationship was found. Chlorfenvinphos has a relatively high mammalian toxicity except for dogs, where the oxidative deethylation mechanism may be responsible for their insensitivity (see Table 37). However, the introduction of another chlorine atom on the phenyl ring at the 5-position caused a great reduction in mammalian toxicity (about 30-fold reduction in toxicity for rats). The dimethyl ester homolog tetrachlorvinphos (Gardona) has such low toxicity to mammals that the oral LD_{50} for rats is 4,000 to 5,000 mg/kg, while the insecticidal activity is comparable to chlorfenvinphos. The low toxicity of tetrachlorvinphos was attributed to its poor solubility in both water and organic solvents and consequently slow penetration and transport from the site of application to the target area.[720]

chlorfenvinphos   tetrachlorvinphos

The low mammalian toxicity of fenitrothion was partly attributed to the low penetrability of the oxon analog into the mammalian brain.[721]

In spite of high toxicity towards vertebrates, acetylcholine and ionized anticholinesterase agents such as prostigmine are nontoxic towards insects. This may be explained by the difference in peripheral nerve systems between vertebrates and insects: the neuromuscular junction of insects is not cholinergic, but those in motor nerve and parasympathetic nerve systems of vertebrates are. Cholinergic junctions in insects are in ganglia or central nervous systems, which are protected from ions with a lipoidal "ion barrier." An ionizable organophosphorus insecticide, amiton $(C_2H_5O)_2P(O)SCH_2CH_2N(C_2H_5)_2$, shows extremely high toxicity towards mammals, Hemipterous insects, aphids, mites, and scale insects, but is not effective to many other insects. Similar selectivity was observed with schradan. Saito found by electron microscopy that the thoracic ganglia of schradan insensitive insects such as rice stem-borer larvae (LD_{50} 26,000 $\mu g/g$), American cockroaches (LD_{50} 2,170 $\mu g/g$), and houseflies (LD_{50} 1,932 $\mu g/g$) were covered by a thick sheath, whereas those of susceptible insects such as green rice leafhoppers (LD_{50} 160 $\mu g/g$), black rice bugs (LD_{50} 92 $\mu g/g$), and rice bugs (LD_{50} 23 $\mu g/g$) were surrounded by a thin membrane.[722] After treating insects with ^{32}P-labeled schradan, he found a much higher concentration of schradan in the ganglia of susceptible insects than in housefly ganglion. Thus, he suggested that the structure of the nerve sheath which acts as a barrier against the penetration of schradan may be the most important factor responsible for the selective toxicity of the insecticide. O'Brien, however, found that there is no distinct difference in permeability of ganglia between aphids and American cockroaches, and suggested the possibility that the neuromuscular junction of aphids may be cholinergic.[32] For detailed discussion on penetration of insecticides to the target, see O'Brien's reviews.[31,32,689]

5. Nature of Target

Difference in the nature of the target is undoubtedly an important factor not only for selective toxicity between species but also for the resistance mechanism. In the previous Section (IV.A.4), some examples of structure-selectivity relationship of anticholinesterase agents were presented. The remarkable selective toxicity of isopropyl parathion towards houseflies in comparison to honey bees is greatly due to the high sensitivity of fly cholinesterase and the insensitivity of the bee enzyme towards the oxon analog, though other factors involving difference in the metabolism may also contribute to the selectivity.[490-492,531,689] A part of the selective toxicity of fenitrothion may be attributed to the difference of affinity between mammal and insect cholinesterases (see IV.A.4, Table 28). Several phenyl phosphorus esters having a substituent group in the meta-position of the phenyl ring, such as fenthion, Chlorthion, ronnel, bromophos, and iodophos, enjoy low mammalian toxicity, probably, at least in part, for the same reason. These phosphorothionate insecticides with low mammalian toxicity are all methyl esters. For corresponding ethyl esters, the effect of *m*-substitution is not so distinctive as for the methyl esters (see V.A.1. Table 54). Toxicity reflects the net result of various effects. The important role of the methyl ester group in selective toxicity was mentioned above (Section 2). For *p*-cyanophenyl dimethyl phosphorothionate (Cyanophos) with very low mammalian toxicity, the introduction of a *m*-methyl group no longer yields a toxicity decreasing effect.

The recovery rate of phosphorylated cholinesterase also contributes to selective toxicity; mammalian acetylcholinesterase inhibited with the anthelmintic Haloxon recovers rapidly, whereas that of parasitic nematodes does not recover (see IV.A.2.b.).[436]

In certain resistant strains of spider mites (*Tetranychus urticae*) and cattle ticks (*Boophilus microplus*), the low sensitivity of the cholinesterase to anticholinesterases is regarded as the main factor of resistance.[694,723,724] For example, cholinesterase from resistant mites is 400 and 150 times less sensitive towards paraoxon and diazoxon, respectively, than that from susceptible mites. Smissaert et al. suggested that the essential alteration in the resistant enzyme might be a slight shift in position of a imidazole residue relative to the serine hydroxyl of the active site.[725] Nolan et al. found that five isozymes of cattle tick acetylcholinesterase were separable by electrophoresis.[726] Two of them exhibited similar sensitivity to the anticholinesterase agent in both the susceptible and resistant strains, and they represent the susceptible acetylcholinesterase components. The other 3 isozymes represent the major part of the acetylcholinesterase activity in each strain and differ 500-fold in sensitivity to the inhibitor coroxon between the strains.

6. Interaction of Factors

As the toxicity of a toxicant to an organism is the net result of integrated interactions at every step mentioned above which the toxicant may undergo in order to manifest the biological effect, selective toxicity cannot often be attributed to a sole factor. A great difference between species at one step may be compensated by other big differences in the reverse direction at another step, and even a small difference may cause a large effect by combination with other factors.

For example, in spite of the remarkable difference in degradation rate of the insecticide famphur between mice (90% degradation in 1 hr) and milkweed bugs (10% degradation in 1 hr), the toxicity is almost the same for both species (LD$_{50}$ mg/kg for mice 11.6; for the bug, 8.0). This discrepancy is explained by the observation that the mouse acetylcholinesterase is 32 times more sensitive to the oxon analog of the insecticide than the bug enzyme.[552] Another example is the selectivity of dimethoate among insect species as mentioned above (Section 2).

On the other hand, the effect of resistant genes in combination is not additive, but multiplicative. Thus, although the organotin resistance factor, *tin*, which was demonstrated to be identical with the penetration delaying factor, *Pen,* confers only two- to three-fold resistance on houseflies, it acts as a multiplier when combined with another resistant gene, for example, the a_{para} gene that bestows 30-fold resistance to parathion, giving 100-fold resistance.[695] A more complicated example is the housefly SKA strain, which was obtained by a cross between diazinon resistant strains (Danish and Italian) followed by continuous selection with diazinon.[604,713] The strain is very resistant to both diazinon and parathion. It has resistance factors on chromosomes II, V (*Ses*), and III (*Pen*). The resistant factors on chromosome II confer 13-fold resistance to diazinon due to phosphatase activity in the microsomal fraction accompanied with low aliesterase level (gene *a*) and glutathione dependent *S*-alkyltransferase activity in the supernatant fluid. The factor *Ses* bestows nine-fold resistance to diazinon, owing to sesamex sensitive microsomal oxidase activity which does not promote dearylation but gives unknown products from diazinon and diazoxon. As oxidative dearylation occurs both in susceptible strains and strains with SKA's single chromosomes to some extent, this reaction may not relate to the resistance mechanism. The factor *Pen* is the penetration delaying factor and increases resistance to organophosphorus insecticides two to three times. Thus, the combination of these factors confers upon the SKA strain 300-fold resistance to diazinon.

D. INTERACTION WITH OTHER CHEMICALS

The effectiveness or toxicity of organophosphorus pesticides is often remarkably influenced by the action of other chemicals involving solvents for formulations, impurities in technical products, and other pesticides mixed for simultaneous control of several pests. Admixing synergists may be possible but is not practical for organophosphorus pesticides at the present time. Pre-exposure of animals to certain chemicals involving organochlorine insecticides also causes alteration of the toxicity of organophosphorus pesticides. These interactions of chemicals may be classified into three categories: 1) direct interactions and indirect interactions through 2) the inhibition of or 3) the induction of pesticide metabolizing enzymes.

The joint action exhibited by the combination of two or more chemicals, in which the toxicity or biological effect is much greater than what would be expected from the simple summation of the effects caused by the individual components, is called synergism or potentiation. Thus, synergism gives a cotoxicity coefficient of a mixture (= actual toxicity index of a mixture/expected toxicity index of a mixture) significantly above one, while a coefficient less than one indicates antagonistic action. As the toxicity of the synergist itself is usually insignificant in comparison with the insecticide component, the cotoxicity coefficient may be simplified as follows:[727-729]

$$\text{cotoxicity coefficient} = \frac{LD_{50} \text{ of toxicant alone}}{LD_{50} \text{ of toxicant in mixture}} = \text{synergistic ratio (SR)}$$

Several excellent reviews on insecticide synergism and toxicity potentiation in mammals have appeared.[516,729-733]

1. Direct Interactions

Reactions of organophosphorus compounds have been discussed in detail in Chapter III. Many of these reactions bring about modifications in the toxicological properties of these pesticides. For example, hydrolysis causes the loss of pesticidal activity and reactions like oxidation, isomerization, and transalkylation may cause an increase in the toxicity.

As phosphorus esters are generally susceptible to alkali, mixing them with an alkaline pesticide such as Bordeaux mixture should be avoided or the mixture should be used as soon as possible after preparation. Direct correlation was found between the degradation rate of organophosphorus dust formulations and the base exchange capacity and iron content of formulation materials.[734] Certain organophosphorus pesticides which are susceptible to acid are stabilized effectively by the addition of bases to formulations (see III.A.2.c).

Polar solvents sometimes cause the occurrence of thiono-thiolo rearrangement in phosphorothionate esters to yield the thiolate isomers having a high anticholinesterase activity. Dimethylformamide is such a typical solvent (see III.C.3).[262] Casida and Sanderson found that dimethoate formulated with methyl "cellosolve" (2-methoxyethanol) was converted to highly toxic compounds during storage; the rat oral LD_{50} decreased from the initial value of 150 to 250 mg/kg to 15 mg/kg after 9 months' storage.[735] Such a deteriorated sample contained at least six phosphorus containing ionic and seven neutral phosphorus compounds other than the original insecticide. The most toxic compound in the stored preparation was probably 2-methoxyethyl methyl S-(N-methylcarbamoylmethyl) phosphorothiolate; oxidative desulfuration and the displacement of the O-methyl ester group with methoxyethanol had proceeded during the storage.[735,736]

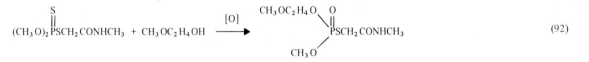

$$(CH_3O)_2 \overset{\overset{\displaystyle S}{\|}}{P}SCH_2CONHCH_3 \ + \ CH_3OC_2H_4OH \ \xrightarrow{[O]} \tag{92}$$

Similar elevation in mammalian toxicity was also observed with other methyl phosphorothionate type insecticides such as malathion, ronnel, and methyl 2,4,5-trichlorophenyl phosphoramidothionate after incubation with methyl "cellosolve" at 70° for 8 days.

Dimethyl sulfoxide (DMSO) appeared to alter cellular permeability and to affect drug absorption by acting as a "penetrant carrier." Thus, in the rabbit the dermal absorption of dicrotophos, dichlorvos, and mevinphos was increased when the compounds were formulated with DMSO.[737] However, the solvent showed no effect on the root absorption by cotton plants or on toxicity towards spider mites of some organophosphorus systemic insecticides such as oxydemeton-methyl, disulfoton, and demeton with cotton plants growing in nutrient solutions.[738]

The alkylbenzene sulfonate detergents often increase the toxicity of some organophosphorus insecticides. For example, the toxicity of parathion, ronnel, and carbophenothion towards the fathead minnow (*Pimephales promelas*) was synergized, but EPN and dicapthon were not synergized.[739]

2. Inhibition of Pesticide-metabolizing Enzymes

a. Inhibitors of Esterases and Amidases

Since Frawley et al. in 1957 found that EPN potentiated markedly the mammalian toxicity of malathion,[740] many insecticide combinations have been examined for toxicity potentiation.[730] When one half of the LD_{50} of each compound is given simultaneously to animals, 50% mortality will be expected if the effects are simply additive. If the mortality is much less than 50%, the effects are antagonistic. On the other hand, synergistic combinations will give a much higher mortality than 50%. DuBois reported that 4 out of about 50 tested combinations of organophosphorus insecticides were synergistic toward female rats; the synergistic pairs were malathion–coumaphos, malathion–EPN, malathion–trichlorfon, and azinphos-methyl–trichlorfon (Table 39).[730] Three of the synergistic combinations involve malathion as a component. The fourth combination, which consists of azinphosmethyl and trichlorfon, showed a lower degree of potentiation (cotoxicity coefficient was 1.5) in comparison with other combinations involving malathion.

Of the synergistic pairs, the effect of EPN on the metabolism of malathion was well studied. It strongly inhibited the enzymic detoxification of malathion by carboxyester hydrolysis in vivo.[730] As malathion owes its low toxicity to the degradation by carboxyesterase, the inhibition of this enzyme causes the increase of the toxicity. EPN may be converted in vivo into its oxon which is a good inhibitor of carboxyesterase rather than cholinesterase.[741]

Bearing in mind that any compounds which selectively inhibit carboxyesterases might potentiate malathion, Murphy et al. tested the potent esterase inhibitor tri-o-tolyl phosphate (TOCP), which was not an insecticide but had other wide industrial uses including employment as a plasticizer, and found its synergistic activity.[731] TOCP showed high activity provided that it had been administered to rats 24 hr before treatment with malathion, suggesting that it might take a considerable time for its conversion into active metabolites. Casida undertook to examine more than 150 organophosphorus compounds for potentiation of malathion to mice and found many potentiators.[742,743] The most active one among the tested triaryl phosphate and diaryl arylphosphonate esters was di-o-tolyl p-tolyl phosphate, and most of

TABLE 39

Effects of Organophosphate Combinations to Mammalian Toxicity

	Combination	Mortality (%)	
		Rats[a]	Mice[b]
Synergistic Effect			
Malathion	+ EPN	100	100
	+ Coumaphos	100	—
	+ Trichlorfon	100	—
	+ TOCP	—	100
Acethion	+ EPN	—	80
	+ TOCP	—	100
Dimethoate	+ EPN	—	100
	+ TOCP	—	100
Azinphosmethyl	+ Trichlorfon	100	—
Narlene	+ EPN	—	100
	+ TOCP	—	100
Antagonistic Effect			
Malathion	+ Parathion	10	—
	+ Azinphosmethyl	10	—
EPN	+ *cis* Mevinphos	—	0
	+ Trichlorfon	—	0
Carbophenothion	+ Trichlorfon	10	—
Additive Effect			
EPN	+ Parathion	45	—
	+ Demeton	45	—
	+ Azinphosmethyl	60	—
Coumaphos	+ Demeton	45	—
Disulfoton	+ Malathion	55	—

a. Half the LD_{50} of each component was injected simultaneously.[730]

b. Half the LD_{10} of each component was injected simultaneously.[744]

the outstanding potentiators in the series contained at least one *o*-alkyl substituent on the phenyl ester group. They were poor in vitro inhibitors of esterases but inhibited strongly in vivo the esterases hydrolyzing malathion. TOCP was converted in vivo into *o*-tolyl saligenin cyclic phosphate (M-1), which is a selective carboxyesterase inhibitor (pI_{50} for mouse plasma malathion esterase = 7.2) (see A.4.a; B.1.e).[489] TOCP itself caused 4-fold potentiation of malathion toxicity by 100 mg/kg administration to mice, whereas M-1 resulted in a 100-fold potentiation by a dose of 20 mg/kg with simultaneous administration.[743]

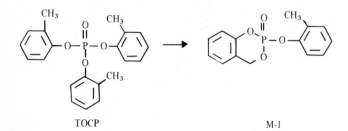

TOCP   M-1

(93)

Related aryl saligenin cyclic phosphates also showed high activity but alkyl homologs were much less active as potentiatiors.

Other triaryl phosphates having the *o*-alkyl group may be similarly activated by metabolic conversion into the corresponding cyclic aryl phosphates. One exceptional triaryl phosphate, which potentiated malathion remarkably without *o*-alkyl substituent, was tri-*p*-ethylphenyl phosphate (TPEP). It was demonstrated that the *p*-ethyl group was biotransformed into the acetyl group and that tri-*p*-acetylphenyl phosphate (trioxo TPEP) was a high liver esterase inhibitor: the I_{50} values for TPEP and trioxo TPEP were 2×10^{-4} M and 1.1×10^{-7} M, respectively.[558]

In addition to the above, some malathion potentiation in mice was observed with the insecticides chlorthion and Phostex, the defoliant DEF (*S,S,S*-tributyl phosphorotrithiolate) and the corresponding phosphite (merphos), and some diaryl *N*-methylphosphoramidates.[743] All of these compounds appear to be inhibitors, at least in vivo, of malathion detoxication by carboxyesterase. The highest synergistic activity of DEF was obtained when it was given to mice 6 hr before treatment with malathion; certain metabolic transformation might be involved in the manifestation of the activity.

$$(C_2H_5O)_2 \overset{\overset{S}{\|}}{P} - S - S - \overset{\overset{S}{\|}}{P} \overset{OC_2H_5}{\underset{OC_3H_7(i)}{}} \qquad (n\text{-}C_4H_9S)_3P{=}O$$

Phostex   DEF

The mammalian toxicity of another carbethoxyl group containing insecticide, acethion, was similarly potentiated in mice by EPN and TOCP, probably due to the same mechanism as in the case of malathion.[744] Although mevinphos contains a carboxy ester group in the molecule, it was not potentiated by EPN because its main detoxication mechanism does not depend on carboxyesterases (see B.2.c.i). The situation of dimethoate is rather complicated. The degradation mechanisms are different species by species. EPN markedly potentiated the toxicity of dimethoate in mice, but did little in guinea pigs because the detoxication mechanism in the former is due to carboxyamidase which is sensitive to the organophosphate inhibitor, but in the latter animal it is due to another enzyme catalyzing the S-C bond cleavage which is insensitive to the organophosphate.[642]

Impurities in technical preparations of insecticides may also contribute to the potentiation of the toxicity of certain organophosphorus insecticides containing a carboxylic acid derived moiety in the molecule (Table 40).[736,745] In addition to the oxon analog and the *S*-methyl isomer, the technical preparation of phenthoate contained *O,O,S*-trimethyl phosphorothionothiolate, *O,O,S*-trimethyl phosphorodithiolate, and O,O,S-trimethyl phosphorothiolate.[745] All of these impurities had an ability to potentiate the toxicity of phenthoate in mammals. The three trimethyl esters were commonly discovered in all technical preparations of malathion, phenthoate, and the carboisopropyl homolog of phenthoate, M 1703. These impurities probably originated from the by-products of *O,O*-dimethyl phosphorodithioic acid which is obtained by the reaction of phosphorus pentasulfide and methanol (see II.B.2). The dithiolate impurity is the most active potentiator for all 3 insecticides: it caused a twofold elevation of the toxicity to rats at concentrations less than 0.05% (Table 41). In contrast, the impurities caused a decrease in the insecticidal activity (Table 40). Although there was no experimental evidence, it is reasonable to presume, in analogy with the homologous series of DEF, that the mechanism of potentiation in rats by the trimethyl esters might be similar to that of other carboxyesterase inhibitors. The phosphoryl type impurities could be detoxified by treating the technical preparations with acyl halides.[745]

As in insects, particularly susceptible strains, carboxyester hydrolysis is not important for the biodetoxication of malathion, so the malathion potentiators for mammals are generally ineffective as synergists towards susceptible insects. However, against malathion resistant strains of houseflies and *Culex tarsalis* mosquitoes, some excellent organophosphorus synergists of malathion were found by Plapp et al. who tested more than 60 organophosphorus compounds.[746,747] TOCP was not effective for the resistant houseflies by simultaneous application with malathion, but was effective in the mosquito. It may be due to the slower metabolism rate of TOCP in houseflies; previous administration of TOCP is necessary for its

TABLE 40

Effects of Purity of Insecticides on Toxicity[736,745]

Insecticide		Purity (%)	LD_{50} rat oral (mg/kg)	Relative insecticidal activity[a]	
				Housefly	Mosquito larvae
Malathion	Purified	98.2	8,000	100	100
	Technical	92.2	1,580	116	81.9
Phenthoate	Purified	98.5	4,728	100	100
	Technical	90.5	242.5	93.9	90.5
	Technical	78.7	118	88.9	88.6
M 1703	Purified	98	2,750	100	100
	Technical	83.2	205	86.5	106.9
Dimethoate	Purified		648		
	Technical	77.8	336		

a. activity of purified insecticide = 100

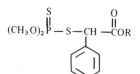

Phenthoate	R = C_2H_5
M 1703	R = i-C_3H_7

TABLE 41

Toxicity and Potentiation Activity of Organophosphorus Impurities[745]

Impurities	Toxicity rat oral LD_{50} mg/kg	% Impurity necessary to potentiate twice the toxicity of the insecticide		
		Malathion	Phenthoate	M 1703
$(MeS)_2P(O)OMe$	96	0.045	0.003	0.001
$(MeO)_2P(O)SMe$	47	0.18	0.0055	0.01
$(MeO)_2P(S)SMe$	450	3.5	0.25	0.12

effective synergism of malathion. The most active synergists in the tested esters were triphenyl phosphate, tributyl phosphorotrithiolate (DEF), its propyl homolog, and tributyl phosphorotrithioite (merphos). Trialkyl phosphates were ineffective. Similar results were obtained with the malathion resistant strain of *Chrysomya putoria*.[748] Some noninsecticidal carbamates were also highly effective as malathion synergists to resistant houseflies; they were dibutylcarbamate esters of *o*-chlorophenol, *m*-chlorophenol, and thiophenol.[749] These synergists reversed the resistance factor almost completely, but were not so effective against susceptible insects, as shown in Table 42. The synergism was first attributed to the inhibitory activity against "mutant aliesterase" which had been proposed by Dutch scientists.[596,746] However, the contribution of "mutant aliesterase" to organophosphate resistance has been criticized and is now believed to be much less than first thought.[750] The increased malathion carboxyesterase activity is apparently responsible for a major part of malathion resistance in the housefly as well as *Culex*.[630,632,633] Matsumura et al. performed a partial purification to characterize the carboxyesterases from the resistant strains of *Culex* and houseflies.[632,633] The esterases were inhibited 50% by propyl paraoxon and the oxon

TABLE 42

Toxicity of 1:1 Combinations of Malathion and Synergist to Houseflies and Mosquitoes (*Culex tarsalis*)[746,747,749]

Synergist	Houseflies Susceptible		Houseflies Resistant		Mosquitoes Susceptible		Mosquitoes Resistant	
	LD_{50}[a]	SR	LD_{50}[a]	SR	LC_{50}[b]	SR	LC_{50}[b]	SR
None	17	–	1,800	–	0.025	–	2.4	–
TOCP	–	–	800	2.3	0.014	1.8	0.045	53
$(C_6H_5O)_3P = O$	17	1	80	23	0.016	1.5	0.024	100
$(C_4H_9O)_3P = O$	–	–	900	2	–	–	>1.0	–
$(C_4H_9S)_3P = O$	9	2	25	72	0.014	1.8	0.03	80
$(C_4H_9S)_3P$	12	1.4	40	45	0.014	1.8	0.025	96
$(C_3H_7S)_3P = O$	–	–	20	90	–	–	0.1	24
o-Cl-$C_6H_4OCON(C_4H_9)_2$	–	–	25[c]	120[c]	–	–	–	–
$C_6H_5SCON(C_4H_9)_2$	–	–	30[c]	100[c]	–	–	–	–

a. 24 hr LD_{50} in μg malathion/jar
b. 24 hr LC_{50} in ppm malathion
c. LD_{50} for malathion alone was 3,000 μg/jar.

analog of EPN at concentrations as low as 10^{-8} *M*. Thus, EPN and propyl paraoxon showed considerable synergistic activity in combination with malathion. Aryl saligenin cyclic phosphates that are homologs of TOCP metabolite were also effective as malathion synergists toward resistant houseflies.[751] Triphenyl phosphate is a rather nonselective and poor inhibitor of housefly esterases in vitro.[752] However, it greatly reduced the carboxyesterase metabolites of malathion in vivo, for example, in the resistant strain of flour beetles *Tribolium castaneum*.[753]

Although the organophosphate resistance of the Leverkusen R-strain of spider mites, *Tetranychus urticae*, and of the resistant strain of ticks, *Boophilus microplus*, is attributed to decreased sensitivity of the acetylcholinesterase to anticholinesterase agents,[694,723] a high malathion carboxyesterase activity was found in Blauvelt R-strain of *Tetranychus telarius*.[701] Resistance in a strain of red citrus mites, *Panonychus citri*, resembles that in the *T. telarius* Blauvelt strain. For the citrus mite strain, saligenin cyclic phenylphosphonate was the most effective malathion synergist of 46 tested compounds, and retarded both the in vivo and in vitro production of carboxyesterase metabolites.[635,754] Many compounds synergistic to malathion towards resistant arthropods potentiate the toxicity of malathion to mammals, too, and have another deleterious property; they produce a delayed neurotoxicity, as will be discussed in the following Chapter. For this reason, their practical use as synergists must be carefully controlled.

In plants similar interactions are known between the anilide herbicide propanil (3',4'-dichloropropionanilide) and esterase inhibitors such as organophosphate and carbamate insecticides.[755,756] Propanil has a high selective toxicity towards rice plants and weeds, especially barnyard grass: 50% growth inhibition was attained at concentrations as low as 0.07% of liquid cultures for the weed whereas at 2.5% for rice plants. This is attributed to the high hydrolyzing activity for the anilide in rice plants and the lack of the enzyme activity in the weed. The propanil degrading enzyme in rice plants is highly sensitive to certain organophosphate or carbamate antiesterase agents; for example, trichlorfon and the carbamate insecticide carbaryl inhibited almost completely the enzyme activity at 10 ppm.[755] Thus, these insecticides cause severe plant toxicity to rice plants in combination with the herbicide. In certain cases, such a combination is useful to control propanil tolerant weeds. Actually, a herbicide formulation containing propanil and carbaryl named WYDAC is practically employed for crabgrass control in citrus orchards.

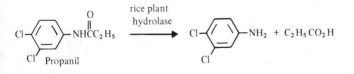

$$ (94) $$

Other interactions between several organophosphate insecticides and herbicides have been reported.[756] The metabolism of 3,6-dichloro-o-anisic acid (dicamba), isopropyl m-chlorocarbanilate (chlorpropham), and 3-(3,4-dichlorophenyl)-1-methoxy-1-methylurea (linuron) in wheat, bean, and plantain, respectively, was commonly inhibited by organophosphate insecticides. Moreover, linuron and propanil inhibited the metabolism of fonofos (Dyfonate) and malathion in bean tissue.

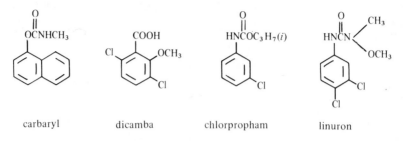

carbaryl   dicamba   chlorpropham   linuron

Organophosphate synergism is not restricted to combinations involving malathion or compounds containing a portion derived from carboxylic acid in the molecule as one of the components. Such synergistic combinations of organophosphorus insecticides towards mice involve EPN with Narlene (Dowco 109),[744] chlorthion with Phostex,[757] and ronnel (fenchlorphos) with EPN, parathion, azinphosmethyl, or demeton.[758] The potentiation of fenitrothion toxicity by phosphamidon in male rats was also reported.[759] It has been reported that DEF and EPN acted as synergists not only for dicrotophos and dimethoate, but also for phorate against boll weevils, *Anthonomus grandis*.[760]

In animal tissues there are a variety of B-type esterases and proteins which are able to bind with organophosphate esters. These enzymes or protein molecules compete with acetylcholinesterase in reaction with organophosphate anticholinesterase agents. Such reactions are not critical for the life of the organism, in contrast to acetylcholinesterase. Thus, these organophosphate bindable proteins might play a role of defense mechanism. If a component of the organophosphate combinations reacts more readily with such proteins than another insecticidal component, then the latter survives to penetrate tissues near the target enzyme acetylcholinesterase.[731]

Moreover, against certain resistant insect strains, DEF synergized to some extent paraoxon, azinphosmethyl, the carbamate insecticides, and DDT ("cross-synergism").[693,749,761,762] These facts suggest that the synergists are able to inhibit some pesticide degrading enzymes, probably including mixed-function oxidase as well as carboxyesterase.

b. Inhibitors of Mixed-function Oxidases

Recently another plausible explanation has appeared for the synergism of organophosphorus insecticides which contain no carboxyester or related function in the molecule. As discussed in the preceding chapter, organophosphorus compounds undergo a variety of the mixed-function oxidase catalyzed metabolic reactions: oxidative disulfuration results in the activation, whereas dearylation and dealkylation also proceed oxidatively, leading to detoxication. Thus, in certain cases, thiophosphoryl compounds may interfere with the detoxication of organophosphate anticholinesterase agents. Oppenoorth found that parathion, diazinon, and nontoxic diethyl phenyl phosphorothionate (SV_1) effectively inhibited the microsomal oxidative dealkylation of paraoxon in in vitro preparations from some resistant strains of houseflies: their I_{50} values (μg/ml) for the microsomal enzyme of the Fc strain flies were about 0.05, 0.05, and 0.012, respectively.[599] The thiophosphoryl compound SV_1 synergizes not only organophosphorus insecticides, including paraoxon, parathion, and diazinon, but also the carbamate insecticide Talcord (S-(2-cyanoethyl)-N-[(methylcarbamoyl)oxy] thioacetimidate) as shown in Table 43.[599]

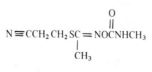

Talcord

TABLE 43

TABLE 43

Synergistic Activity of Sesamex and SV_1 Towards Resistant Housefly Strain Fc[598]

	Sesamex		SV_1	
Insecticide	LD_{50} (ng/fly)	SR	LD_{50} (ng/fly)	SR
Paraoxon	15(10)	13	3 (10)	65
Parathion	25 (1)	6	8 (1)	19
Parathion	13 (10)	12	4 (10)	38
Diazinon	50 (20)	30	6 (10)	250
Talcord	150 (10)	28	15 (10)	300

Figures in parentheses indicate the amounts of synergist in μg.

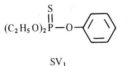

SV_1

TABLE 44

Effect of Some Mixed-function Oxidase Inhibitors on Parathion Metabolism by Rabbit Liver Microsomes[529]

	% Inhibition of metabolite formation			
	Dearylation		Desulfuration	
Inhibitor[a]	Acidic metabolite	p-nitrophenol	Detached sulfur atom	Paraoxon
Piperonyl butoxide	85	76	64	87
Propyl isome	61	68	56	60
Sesamex	30	–	35	36
Sulfoxide	74	44	57	76
SKF 525-A	64	53	62	55
WARF antiresistant	48	45	45	40
MGK 264	56	66	67	54

a. Inhibitor: 4×10^{-4} M, parathion: 2.5×10^{-5} M. For structure, see Figure 17.

The best established mixed-function oxidase (mfo) inhibitors are methylenedioxyphenyl compounds (MDP), which are also called 1,3-benzodioxoles. MDPs were originally developed as the synergists of pyrethroids, but then their synergistic activity with carbamate insecticides was found. Although the detoxication of these insecticides is almost exclusively due to the mfo systems, in the metabolism of organophosphorus pesticides the microsomal mfo systems are concerned with both the activation and degradation. Thus, the effects of microsomal enzyme inhibitors on the toxicity of organophosphorus compounds are complicated and are often difficult to interpret. For example, MDP and some other mfo inhibitors reduce both the desulfuration and dearylation of parathion in houseflies and mammalian liver microsomes (see Table 44).[529,594]

Table 45 shows the effect of sesamex on the toxicity of various organophosphorus compounds against housefly NAIDM strain.[728,763,764] Some insecticides are synergized and some others are antagonized

TABLE 45

Effect of Sesamex on the Insecticidal Activity of Organophosphorus Insecticides Towards Housefly NAIDM Strain[728,763,764]

Insecticide	LC$_{50}$ of toxicant (%)		
	Alone	With 1% sesamex	SR
Dicrotophos	0.059	0.003	19.7
SD 5656	0.0425	0.0014	30.4
Monocrotophos	0.014	0.0021	6.7
SD 11319	0.0037	0.00197	1.9
Phosphamidon	0.024	0.00165	14.6
Dichlorvos	0.0055	0.00245	2.2
Mevinphos	0.0054	0.0022	2.5
Malathion	0.00845	0.0725	1.1
Azinphosmethyl	0.020	0.0113	1.8
Paraoxon-methyl	0.0039	0.0022	1.8
Parathion-methyl	0.0053	0.013	0.41
Parathion	0.0041	0.0065	0.63
Chlorthion	0.058	0.198	0.30
EPN	0.0103	0.0147	0.70
Diazinon	0.053	0.0137	3.9
Amiton-oxalate	0.77	0.0265	29.1
Phorate	0.043	0.0052	8.3
Demeton	0.13	0.0036	36.1
Disulfoton	0.13	0.0036	36.1
Carbophenothion	0.080	0.014	5.7
Fenthion	0.014	0.0071	2.0
Ethion	0.11	0.0096	11.4
Dioxathion	0.23	0.0245	9.4
Schradan	3.1	1.95	1.9

SD 5656: *N,N*-diallyl homolog of dicrotophos; SD 11319: *N*-demethyl monocrotophos.

with sesamex. No significant effect of sesamex is observed with some organophosphorus insecticides. Whether any particular organophosphorus insecticide is synergized or not by MDP may depend on a metabolic balance between the critical pathways responsible for activation and degradation of the compound catalyzed by the mfo systems of the organism involved, and on the degree to which each of these pathways is inhibited by the MDP. It relates to the structure of the insecticide, the species involved, the development of resistance, and other factors. The anatagonistic action of sesamex on the toxicity of phosphorothionate or phosphonothionate insecticides such as parathion, parathion-methyl, chlorthion, and EPN in susceptible housefly NAIDM strain is probably due to the inhibition of oxidative desulfuration. Parathion was synergized with sesamex in the resistant strain Fc of houseflies (as shown in Table 43), in which oxidative deethylation was probably the more important metabolic pathway. Some diethyl phosphorus esters, such as diazinon, phorate, demeton, disulfoton, and phosphamidon, are actually synergized in NAIDM, too. However, high synergism specific to the phosphorus esters containing a thioether group is rather difficult to interpret, because thioether oxidation is an important metabolic activation pathway, although it has not been confirmed yet that such thioether oxidation is actually catalyzed by housefly mfo systems. It is supposed that the sulfur oxidized products undergo biodegradation rapidly or the thioether moiety assists oxidative degradation at some bond around the phosphorus atom.

The metabolism of the phosphate insecticides dichlorvos, mevinphos, and paraoxon-methyl appears to be not so dependent on the mfo systems as the mfo inhibitor affects the toxicity significantly. It is interesting to note that in a series of dicrotophos the degree of sesamex synergism depends on the *N*-alkyl group: the diallylamide homolog SD 5656 is synergized 30-fold, the dimethylamide dicrotophos 20-fold, but the monomethylamide monocrotophos and the unsubstituted amide SD 11319 are synergized only 7- and 2-fold, respectively. It has been demonstrated that dicrotophos is oxidatively demethylated successively into monocrotophos and finally the unsubstituted amide via *N*-methylol derivatives (see B.1.d; Equation 45).[549] The oxidative *N*-demethylation was greatly retarded in sesamex treated houseflies.[549] As the *N*-demethylation causes the increase of the insecticidal activity, so the treatment with sesamex would be expected to result in antagonism rather than synergism of dicrotophos. In order to explain this discrepancy, Wilkinson presented two possibilities: 1) the dealkylated products might be more susceptible to enzymatic degradation or 2) the mfo system could attack some sites other than the *N*-methyl group in the molecule and more lipophilic dialkyl derivatives would be more susceptible to the attack.[733] Another possible interpretation concerns the hypothesis that *N*-methylol derivatives would be the key metabolites for detoxication. The *N*-dealkylation of dicrotophos does not appear to be a necessary step in the poisoning process; dicrotophos is only 20 times less toxic to houseflies than the unsubstituted amide derivative, it is intrinsically more active than the latter as an anticholinesterase agent (pI_{50} for fly cholinesterase is 7.2, 6.8, and 6.5 for dicrotophos, monocrotophos, and the unsubstituted amide, respectively), and the unsubstituted amide is only a minor metabolite (0.1% of administered dicrotophos).[549] On the other hand, 90% of administered dicrotophos was found as hydrolysis products or chloroform unextractable products, which were reduced to 85% by treatment with sesamex. Of the hydrolysis products, the *O*-desmethyl derivative (XXXIV) and the deamidated carboxylic acid (XXXV) are quantitatively significant. It was suggested that the acid is more probably formed from the *N*-methylol derivative (XXXIII) than from dicrotophos or the *N*-demethylated derivatives.[765] Detoxication by conjugation of the methylol may be also possible as demonstrated in plants.[766] Thus, the oxidative *N*-demethylation pathway leads to toxic metabolites on the one hand, but can be regarded as an important detoxication mechanism on the other hand.

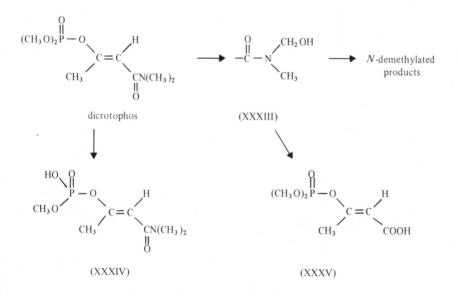

dicrotophos

(XXXIII)

N-demethylated products

(XXXIV)

(XXXV)

(95)

Mixed-function oxidase inhibitors generally stabilize xenobiotics in biological systems. In addition to the above mentioned examples, the biochemical stabilization of organophosphorus insecticides by some MDP compounds has been reported: that of parathion by sesamex in the fat body microsomes of cockroaches;[529], phosphamidon by sesamex in boll weevil (synergistic);[767] diazinon by piperonyl butoxide in resistant fly abdomens (synergistic);[688] coumaphos and its oxon by piperonyl butoxide in mammals and insects (synergistic);[768] the chemosterilant hexamethylphosphortriamide (hempa) by Tropital in houseflies.[769] The inhibitors of mfo activity are not restricted to MDP compounds. They include alkylamine derivatives such as SKF 525-A, Lilly 18947, MGK 264, and WARF antiresistant, propynol derivatives such as RO 5-8019 and NIA 16824, and other miscellaneous compounds including benzothiadiazoles and thiocyanates. Their structures are shown in Figure 17. SKF 525-A inhibited the conversion of thiophosphoryl into phosphoryl by mammalian microsomes and cockroach guts.[529,770,771] MGK 264 and WARF antiresistant also showed a similar effect.[529,772] SKF 525-A protected mice against poisoning by schradan, azinphosmethyl, and dimethoate, but substantially increased the toxicity of phorate and amiton.[770] In houseflies, however, SKF 525-A synergized azinphosmethyl. It stabilizes paraoxon in vivo in mice and cockroaches, probably due to the inhibition of oxidative deethylation.

The mode of action of these mixed-function oxidase inhibitors is not yet fully understood, but the following possible mechanisms have been proposed. Hennessy suggested that the benzodioxole oxidatively loses a hydride ion (H^-) from the methylene group to form an electrophilic benzodioxolium ion (XXXVI), which as an aromatic system having 10 π-electrons may form a π-bonded complex with iron,[773] possibly displacing a ligand or adding at the P-450 hemochrome. This hybride transfer mechanism was supported by a molecular orbital study.[774]

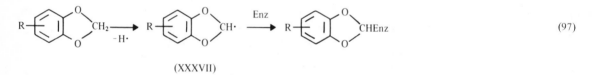

$$(96)$$

(XXXVI)

For the propynyloxy compounds and thiocyanates, a similar mechanism involving the formation of coordination complexes with the iron component of the metabolizing enzymes was also proposed.[773]

Based on a study of structure-activity relationships employing substituent constants and regression analysis, Hansch proposed another possible mechanism.[775] It involves homolytic hydrogen abstraction from the methylene group of the benzodioxole by the action of a microsomal enzyme to generate a free radical (XXXVII) which may play a role in the inhibition:

$$(97)$$

(XXXVII)

Studies with nonenzymatic oxidation model systems demonstrated that the benzodioxoles are able to inhibit the oxidation by acting as free-radical acceptors. For example, parathion activation by a Udenfriend model system consisting of H_2O_2, Fe^{2+}, ascorbic acid, and EDTA was inhibited by some MDP compounds and WARF antiresistant.[772] By a similar oxidation system, a modified Fenton's reagent consisting of H_2O_2, Fe^{2+}, EDTA, and bovine serum albumin, but not by a peracid system, benzodioxols themselves are oxidatively degraded (Equation 98). The degradation rate of the substituted benzodioxoles correlated well (r = 0.98) with their inhibitory activity with respect to aldrin epoxidation by the Fenton's system.[776] In another type of mfo inhibitor, benzothiadiazoles, a similar correlation was obtained.[776] It is believed that in these model systems, hydrogen peroxide reacts with ferrous ion to form hydroxy radical. The radical may cause the homolysis of the benzodioxole.

$$H_2O_2 + Fe^{2+} + H^+ \longrightarrow H_2O + \cdot OH + Fe^{3+}$$

$$(98)$$

Although these nonenzymatic model systems cannot adequately reproduce the unique characteristics of the microsomal mfo systems, the hydroxyl radical is regarded as one possible form of active oxygen at cytochrome P-450.

Methylenedioxyphenyl Compounds

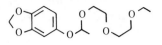

sesamex

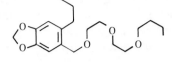

piperonyl butoxide

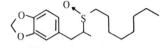

sulfoxide

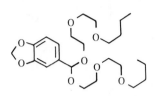

Tropital

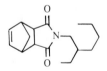

propyl isome

Alkyl Amines and Amides

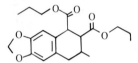

SKF 525-A

Lilly 18947

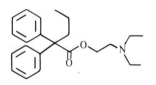

MGK 264

WARF antiresistant

Propynol Derivatives

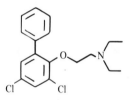

RO 5-8019

NIA 16824

Miscellaneous Compounds

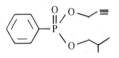

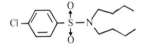

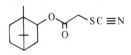

Thanite

WL 19,255

FIGURE 17.   Structures of selected inhibitors of mixed function oxidases.

Casida et al. suggested the MDP compounds and some other mfo inhibitors may act as alterntive substrates for the mfo systems, saving the insecticide from the mfo dependent biotransformation.[516],[777] Using the MDP labeled with ^{14}C in the methylene group, they confirmed that the dioxole ring was cleaved by mammalian liver and insect microsomal mfo systems to yield formic acid and the catechol, probably via a 2-hydroxybenzodioxole intermediate (XXXVIII):[777-779]

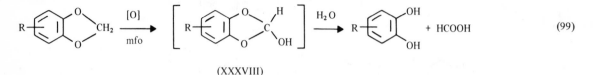

$$(99)$$

(XXXVIII)

The hydroxybenzodioxole can be regarded as a common intermediate for all the mentioned possible pathways. It is the pseudo base of the benzodioxolium ion (XXXVI). On the other hand, the benzodioxole radical (XXXVII) could react with the hydroxy radical to give the hydroxybenzodioxol (XXXVIII) as an intermediate in pyrocatechol formation by the microsomal system. The possibility that the produced catechol might act as the mfo inhibitor is hardly supported because of the inability of catechols to inhibit the enzyme activity on addition to the microsomal system.[780]

The mfo inhibitors other than the MDP compounds are also able to serve as alternative substrates for the mfo system. Alkylamines are oxidatively dealkylated by the action of the mfo system (Equation 100). Propynyl ether bond is also cleaved oxidatively (Equation 101).[516]

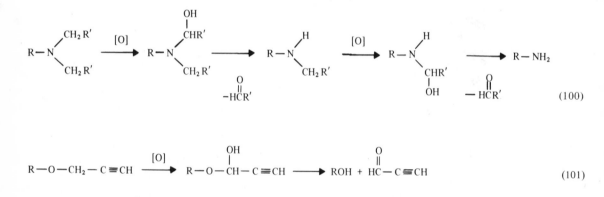

$$(100)$$

$$R—O—CH_2—C≡CH \xrightarrow{[O]} R—O—\underset{\underset{OH}{|}}{CH}—C≡CH \longrightarrow ROH + HC—C≡CH$$

$$(101)$$

Exceptionally, organic thiocyanates such as Thanite undergo a glutathione dependent reaction catalyzed by glutathione S-transferase rather than the mfo catalyzed oxidation.[781]

In order to act optimally as the inhibitor, the "alternative substrate" must have a lower turnover and also a higher affinity (lower K_m) with the mfo system than the substrate insecticide. It was found, however, that no relationship exists between enzyme affinity and the inhibition of aniline hydroxylation by nitrobenzodioxole and other related compounds.[780] Based on this and studies on the difference spectra of the P-450, which differs by substrates and inhibitors, Kuwatsuka postulated that the enzyme has different binding sites for different substrates, and that allosteric effects may be involved for the interaction between the enzyme and MDP inhibitors.[780]

The mfo enzyme of insect microsomes is more sensitive to inhibition by MDP inhibitors than that of mammalian microsomes. For example, the I_{50} values for 4,5,6,7-tetrachlorobenzo-[d]-1,3-dioxole towards aldrin epoxidation in the mouse and army-worm microsomal preparations are 7.4×10^{-5} M and 7.7×10^{-7} M, respectively.[779]

c. Inhibition of Glutathione S-transfer Reaction

The glutathione S-alkyltransferase reaction is now recognized as an important detoxication mechanism for dimethyl phosphorothionate insecticides (see IV.B.2.b.ii).

Treatment with methyl iodide (135 mg/kg) caused a depression in the liver glutathione level of mice for several hours and potentiated the toxicity of fenitrothion more than tenfold.[610] It may compete with the phosphorus methyl ester as a methyl donor to form S-methylglutathione. Glutathione appears to be a limiting factor for the detoxication of fenitrothion in mice by glutathione S-alkyltransferase. Diethyl maleate also reduced liver glutathione and strongly synergized fenitrothion toxicity.[610] The ethyl homolog of fenitrothion was not synergized by methyl iodide.

On the other hand, organic thiocyanates such as the insecticide Lethane 384® (β-butoxy-β'-thiocyanodiethyl ether) are activated by the action of glutathione S-transferases in the presence of glutathione to liberate hydrogen cyanide (Equation 102).[781] Thus, fenitrothion acts as an antagonist for some organothiocyanates in mice, probably owing to reduction in the available glutathione level by formation of S-methylglutathione on in vivo reaction of fenitrothion with glutathione.[781]

$$RCH_2 SCN \xrightarrow[\substack{- HCN}]{\substack{GSH\ S\text{-transferase} \\ + GSH}} RCH_2 S - SG \xrightarrow{\substack{+ GSH}} RCH_2 SH + GS - SG \qquad (102)$$

3. Effect of Enzyme Induction

It was reported by Ball et al. in 1954 that rats were protected from normally lethal doses of parathion by preliminary treatment with aldrin, an organochlorine insecticide; parathion toxicity to female rats decreased by about sevenfold several days after administration of aldrin.[782] Thereafter, many papers describing similar observations have appeared.[731,765,783,784] DuBois and Kinoshita found that the toxicity of almost all examined organophosphorus insecticides to female rats and male mice decreased by pretreatment with phenobarbital (see Table 46). Exceptions were schradan and dimethoate: the toxicity of schradan was increased about twice in rats but not in mice;[784] that of dimethoate was also enhanced 1.5- to 3-fold in mice.[765] In addition, other organochlorine insecticides including dieldrin, chlordane, and DDT, and other compounds like chlorcyclizine, nikethamide, 3,4-benzpyrene, and several steroids yield similar

TABLE 46

Effect of Phenobarbital Pretreatment on the Toxicity of Organophosphorus Insecticides[765,784]

Organophosphorus insecticides	$\dfrac{\text{Pretreated LD}_{50}}{\text{Control LD}_{50}}$	
	Mice	Rats
Mevinphos	1.0	2.0
Dicrotophos	1.9	—
Monocrotophos	2.0	—
Phosphamidon	1.8	—
Parathion	1.8	2.9
Parathion-methyl	1.5	1.1
Ronnel	1.2	1.1
Coumaphos	1.0	1.8
EPN	2.7	10.3
Demeton	1.2	4.2
Disulfoton	2.4	8.1
Carbophenothion	1.6	6.6
Azinphosmethyl	1.2	1.3
Malathion	1.2	1.5
Dimethoate	0.3	—
Dioxathion	2.6	7.0
Ethion	1.2	11.7
Schradan	1.0	0.5

TABLE 47

Effect of Enzyme-inducing Agents on the Formation of
Parathion Metabolites by Liver Microsomes[574,790]

Enzyme inducing agent	Ratio, pretreated/control			
	$(EtO)_2 PSOH$		Paraoxon[a]	
	Rat	Mouse	Rat	Mouse
3,4-Benzpyrene[b]	1.47	–	2.3	–
Chlordane[c]	2.36	2.49	2.67	2.28
DDT[c]	3.11	2.60	3.06	2.12
3-Methylcholanthrene[d]	1.83	0.53	2.82	0.71[f]
Phenobarbital[e]	1.72	–	1.71	–

a. As the sum of paraoxon and diethyl phosphoric acid.
b. 20 mg/kg/day × 3.
c. 100 mg/kg/day × 3.
d. 20 mg/kg/day × 4.
e. 80 mg/kg/day × 4.
f. Statistically not significant.

effects. These chemicals are now understood to induce drug metabolizing enzymes. Several reviews on the induction phenomena and the mechanism in connection with pesticides were recently compiled in a book.[785]

Aldrin pretreatment caused the enhancement of A-esterase activity in the liver of mammals.[786,787] Based on the finding that pretreatment of mice daily for several days with chlorcyclizine, phenobarbital, or SKF 525-A produced marked protection against the toxicity of some organophosphorus insecticides and increased certain B-type esterases rather than A-esterases in liver and plasma, Welch and Coon suggested that the decrease of the toxicity might be at least partly attributed to the increased esterases which could combine with organophosphates to spare acetylcholinesterase.[783]

The phenobarbital induced protection mechanism against malathion in rats apparently differs from that against other phosphorus esters. It took much more time to achieve the maximum protection for malathion (5 days) than for EPN (2 days). The administration of TOCP, a selective aliesterase inhibitor, did not affect the protective effect of phenobarbital against EPN, but abolished that against malathion, indicating that the induction of liver aliesterases plays a major role in the protection against malathion.[788]

Recently, it was confirmed that phosphorothionate insecticides undergo oxidative degradation as well as activation, and both reactions are catalyzed by the microsomal mixed-function oxidases as discussed in the preceding Chapter. The mfo system contains two important enzyme proteins, i.e., cytochrome P-450, which is responsible for the activation of oxygen molecules, and a flavoprotein which transfers the necessary electrons derived from NADPH to the cytochrome. Both the enzymes are inducible by such lipid soluble compounds as those mentioned above.[789] Actually, pretreatment of rats with the enzyme inducing agents increased the production of all three parathion metabolites, i.e., paraoxon, diethyl phosphoric acid, and diethyl phosphorothioic acid, by liver microsomes.[574,790] The degree of the effect on the different types of reaction, desulfuration and dearylation, appears to depend on the inducing agent employed and the animal species involved. Thus, as shown in Table 47, chlordane in mice enhanced both reactions of parathion, whereas DDT preferentially stimulated the dearylation, and 3-methylcholanthrene repressed the dearylation. In rats, chlordane, DDT, and phenobarbital equally enhanced both reactions, but benzpyrene and 3-methylcholanthrene preferentially enhanced the pathway leading to paraoxon formation.[790] Since a variety of metabolic pathways of organophosphorus compounds are catalyzed by microsomal oxidase systems (see IV.B), the effect of enzyme inducing agents on the toxicity of organophosphorus insecticides is complicated. Generally speaking, for phosphate esters the pretreatment with inducers results in the greater production of hydrolytic metabolites, because the increased oxidative activity probably leads to the production of more hydrolyzable metabolites. For phosphorothionate esters, the increased oxidative activity results in the increase of the more hydrolyzable but much more active oxon analogs. Dimethoate, for example, has two oxidizable sites in the molecule, i.e., thiophosphoryl group and N-methyl-carboxyamide group. The former has to be oxidized for the insecticide to be active. Oxidation of the N-alkyl group gives products which are still toxic but are more susceptible to hydrolysis and thereby aid in the detoxication process.[765] The observed toxicity change caused by the enzyme inducer is the net result of increased oxidation to effect activation and detoxication.

Nearly all organic lipid soluble compounds which can be bound to cytochrome P-450 protein are potential inducers. Many mfo inhibitors have, therefore, the inducing ability: piperonyl butoxide, NIA 16824, and WL 19255 inhibited hepatic microsomal mfo activity for several hours after a single intraperitoneal dose to mice, but after 24 to 72 hr, increased mfo activity and P-450 content were induced.[791] In order to induce the microsomal enzymes, the agent must interact with the microsomal enzymes for at least several hours. Thus, short acting biodegradable chemicals do not have a significant inducing capacity because they are oxidized rapidly and lose their binding to the enzyme.[789]

After the administration of phenobarbital (80 mg/kg) the synthetase of δ-aminolevulinic acid, a precursor of heme porphyrin, achieves its maximal activity during the first few hours, then cytochrome P-450 increases to about threefold during 2 days.[789] DDT induces rather slowly, and the maximum content of the cytochrome is achieved 6 days after the treatment and lasts for several weeks. The induction of the mfo enzymes by chlorinated hydrocarbon insecticides and other enzyme inducing agents is observed in insects too.[792,793] Pretreatment of 6th instar larvae of the wax moth (*Galleria mellonella*) with chlorcyclizine decreased the mortality due to parathion.[794]

As a tentative conclusion, the effect of enzyme inducing chemicals on the toxicity of organophosphorus pesticides may be due to 1) increased liver A-esterases, 2) increased noncritical binding sites such as B-esterases, 3) increased rate of formation of oxon analogs to provide a substrate for A-esterases, and/or 4) increased rate of direct cleavage of the parent (P=S) compounds, and the effect differs depending upon the pretreatment chemical, the insecticide in question, and the animal involved.[731]

E. SIDE EFFECTS

Certain toxic effects of organophosphorus compounds which do not appear to be directly related to acetylcholinesterase inhibition in the nervous systems are known. This Section deals with some of such biological effects.

1. Delayed Neurotoxicity

Many organophosphorus compounds show a special chronic neurotoxicity, which is not due to acetylcholinesterase inhibition but to another unknown mechanism, and is accompanied by irreversible demyelination both in the central and peripheral nervous systems of certain species of vertebrates. The best known neurotoxic organophosphate compound is tri-*o*-tolyl phosphate or tri-*o*-cresyl phosphate (TOCP), which is not a pesticide but a component isomer of technical tritolyl phosphate preparations usable as a plasticizer, oil additive, lubricant and solvent. About ten outbreaks of TOCP poisoning have been reported. The biggest one occurred in 1930 in the USA: more than 10,000 people were stricken with a flaccid paralysis (so-called "ginger paralysis") of the lower limbs about 10 days after drinking a contaminated extract of Jamaica ginger.[795] The more recent outbreak took place in Morocco in 1959 from cooking oil contaminated with the lubricating oil of turbo-jet aircraft engines.[796]

Many animal species, including man, monkeys, dogs, cats, cows, sheep, rats, hens, dogs, and pheasants, are more or less susceptible to the delayed neurotoxic poisoning. Rats are rather resistant and need repeated doses of the compounds to be attacked with an ataxia. Cats and hens (young chicks are insensitive before a critical age of 55 to 70 days) are most sensitive and suitable for testing compounds for the neurotoxicity. Many investigations have been undertaken with adult hens. The onset of the first symptom, leg weakness, is usually delayed 8 to 14 days after dosing with a neurotoxic compound. If poisoning is severe, the legs do not recover from paralysis. Histological studies show that the long axons in the spinal cord and in the periphery, such as the sciatic nerve, sustain degenerative lesions which always occur at the distal first.[558,797,798]

a. Structure-activity Relationship

Extensive studies have been done in order to find the relationship between the structure and the neurotoxicity of triaryl phosphates.[799-803] In this class, all neurotoxic compounds, with the exceptions of tri-*p*-ethylphenyl phosphate and diphenyl *p*-ethylphenyl phosphate, contain at least one *o*-alkylphenyl ester group as shown in Table 48. It has already been mentioned (Section IV.B.1.e) that TOCP was biotransformed into an active phosphorylating agent, saligenin cyclic *o*-tolyl phosphate,[489,553] which caused ataxia in the hen by a single dose of such a small amount as 2 to 5 mg/kg.[743] The clinical and neuropathological signs produced by the cyclic metabolite were essentially identical with those by TOCP.[798] These were confirmed later in cats too.[804,805] Other neurotoxic *o*-alkylphenyl phosphates are presumably metabolized in the body into the corresponding saligenin cyclic phosphate through hydroxylation on the α-carbon of the side chain and intramolecular transphosphorylation. The neurotoxicity of several synthesized saligenin cyclic phosphate esters is shown in Table 49. Meta or para alkyl-substituted triphenyl phosphates cannot, of course, give active metabolites of the cyclic phosphate type. A few compounds having the *o*-alkylphenyl ester group, such as tri-*o*-propyl phosphate, *o*-propylphenyl bis(3,5-dimethylphenyl) phosphate, tris(2,5-dimethylphenyl) phosphate, tris(2,6-dimethylphenyl) phosphate, and tris(2-methyl-4-ethylphenyl) phosphate, were not able to produce ataxia in hens. The lack of neurotoxicity in these compounds appears not to be absolute; although the non-neurotoxicity of tri-*o*-ethylphenyl phosphate had been repeatedly reported,[800,802,803] it was found recently that higher or repeated doses lead to ataxia in hens.[806] The reason for their non-neurotoxicity may possibly be either that their metabolic activation occurs too slowly or that the structure requirements to produce neurotoxicity are not suitably fulfilled by the metabolites, probably due to rather high inhibitor specificity of the "neurotoxic site," the phosphorylation of which may cause the neurotoxicity.[806] The cyclization reaction of the second step in the metabolic activation of *o*-alkylphenyl phosphates involves liberation of the phenol but not of the alcohol component.[554] Moreover, saligenin cyclic alkyl phosphates do not yield ataxia, as shown in Table 49.[743] Thus, di-*o*-tolyl methyl phosphate and diethyl *o*-tolyl phosphate are not neurotoxic.[487]

In tri-*p*-ethylphenyl phosphate (TPEP), an exceptional neurotic triaryl phosphate which has no *o*-alkyl substituent, the ethyl group was biotransformed into the acetyl group, a highly electron attractive group, through α-hydroxylation followed by dehydrogenation (see IV.B.1.e).[557-559] The metabolite *p*-acetylphenyl di-*p*-ethylphenyl phosphate should be active as a phosphorylating agent and yielded ataxia in hens by a single oral dose of 100 mg/kg. Tri-*p*-acetylphenyl phosphate is chemically more active and might be the actual neurotoxic principle.[558] Although it was not discovered as an in vivo metabolite, tri-*p*-acetylphenyl phosphate was produced enzymatically in vitro from tri-*p*-α-hydroxyethylphenyl phosphate which was actually found as one of the in vivo metabolites of TPEP, and caused ataxia in the hen at a dose of 50 mg/kg if a small amount of TPEP (50 mg/kg) was combined as a solvent, and was active enough to inhibit "neurotoxic esterase."

In the TPEP series, however, the situation is more complicated than that in the TOCP series; di-*p*-ethylphenyl phenyl phosphate, di-*p*-ethylphenyl *p*-tolyl phosphate, and *p*-ethylphenyl di-*p*-tolyl phosphate are all non-neurotoxic. On the other hand, *p*-ethylphenyl diphenyl phosphate produced ataxia in the hen by a single dose of 1 g/kg, but its possible metabolite *p*-acetylphenyl diphenyl phosphate did not by the same dosage.[558,743] Dialkyl *p*-acetylphenyl phosphates do not cause ataxia in hens. It is very interesting to note the analogy in the relation of chemical structure to selective toxicity between a series related to TOCP or its active form, i.e., saligenin cyclic phosphates, and another series related to TPEP or its active form, i.e., *p*-acetylphenyl phosphates. In both series, bulky aryl esters are neurotoxic, but not insecticidal. In contrast, small alkyl esters are not neurotoxic, but insecticidal (Table 50).[495,807] The cyclic ester structure and the electron attractive substituent group contribute to the chemical reactivity as phosphorylating agents, whereas the steric property of the nonleaving group appears very important for the selectivity (see also IV.A.4).

TABLE 48

Structure and Neurotoxicity of Triaryl Phosphates

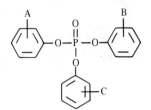

A	B	C	Dose mg/kg	Neurotoxicity
H	H	H	1,000	–
2-Me	H	H	50	+
2-Me	2-Me	2-Me	25 – 200	+
2-Me	2-Me	3-Me	250	+
2-Me	2-Me	4-Me	25	+
2-Me	3-Me	3-Me	100	+
2-Me	3-Me	4-Me	100	+
2-Me	4-Me	4-Me	50	+
2-Me	3,5-diMe	3,5-diMe	1,000	+
2-Et	H	H	1,000	+
2-Et	4-Me	4-Me	200	+
2-Et	2-Et	H	250	+
2-Et	2-Et	4-Me	500	+
2-Et	2-Et	2-Et	500 × 5	+
2-Et	3,5-diMe	3,5-diMe	500	+
2-n-Pr	4-Et	4-Et	200	+
2-n-Pr	2-n-Pr	4-Me	1,000	+
2-n-Pr	2-n-Pr	2-n-Pr	1,000	–
2-n-Pr	3,5-diMe	3,5-diMe	1,000	–
2,3-diMe	2,3-diMe	2,3-diMe	4,250	+
2,4-diMe	2,4-diMe	2,4-diMe	2,000	+
2,4-diMe	2,4-diMe	3,5-diMe	10,400	+
2,5-diMe	2,5-diMe	2,5-diMe	1,000	–
2,6-diMe	2,6-diMe	2,6-diMe	1,000	–
2,6-diMe	3,5-diMe	3,5-diMe	24,000	+
2-Me-4-Et	2-Me-4-Et	2-Me-4-Et	500	–
3-Me	3-Me	3-Me	5,000	–
3-Et	3-Et	3-Et	1,000	–
4-Me	4-Me	4-Me	5,000	–
4-Et	4-Et	H	2,000	–
4-Et	4-Et	4-Et	200 – 1,000	+
4-Et	4-Et	4-Me	500	–
4-Et	4-Me	4-Me	500 × 5	–
4-Et	H	H	1,000	+
2-Cl	H	H	1,000	–
2-Cl	2-Cl	2-Cl	1,000	–

Data from References 743, 800-803, 806.

TABLE 49

The Neurotoxicity of TOCP-metabolite and Related Compounds for Hens[743]

R	Minimum ataxic dose* (mg/kg)
OPh	1.5−2
OPh-2-Me (TOCP-metabolite)	2−5
OPh-3-Me	1−2
OPh-4-Me	0.25−0.5
OPh-3,5-diMe	4−8
OPh-2-Cl	12−25
OMe	n.a.
Ph	200
Et	n.a.
CH_2Cl	n.a.
$N(CH_3)_2$	n.a.

*n.a. = No ataxia signs evident with any sublethal dosages.

TABLE 50

Analogy in the Relation of Chemical Structure to Selective Toxicity Between Two Different Series of Organophosphorus Esters[495]

		Neurotoxicity MAD*[1] (mg/kg)	Insecticidal activity LD_{50} μg/fly
Active metabolite of TOCP		2−5	>10*[3]
Modified insecticide		>80*[2]	0.05
Active metabolite of TPEP	$(C_2H_5 \langle \rangle O)_2 PO \langle \rangle COCH_3$	100	>10*[3]
Modified insecticide	$(CH_3O)_2 PO \langle \rangle COCH_3$	>1,000*[2]	0.17

*1. Minimum ataxic dose in the hen.
*2. No ataxia was induced by the indicated dose.
*3. No insecticidal activity was observed at a dose of 10 μg/fly.

Besides TOCP- and TPEP-related compounds, a great variety of organophosphorus compounds have been found to yield delayed neurotoxicity. They contain an alkyl ester group or alkylamide group in the molecule. The anthelmintic agent Haloxon and related di-2-chloroethyl phosphates, the insecticide EPN and related phosphonate esters, the defoliant DEF and related trialkyl phosphorotrithiolates and trithioites, the herbicide DMPA and related alky laryl N-isopropylphosphoramidates, and phosphorofluorine compounds such as DFP, sarin, mipafox and other related compounds are included. Table 51 shows the structures of the neurotoxic compounds and related non-neurotoxic compounds for the convenience of comparison. Although the structure-activity relationship is too complicated to be generalized, the neurotoxicity appears again to be rather closely associated with the structure of the nonleaving group than that of the leaving group. For example, many fluorides are neurotoxic, but dialkylphosphinic fluorides are not at all;[808] dichlorovinyl phosphates containing ethyl or larger alkyl ester groups are neurotoxic but the dimethyl homolog (dichlorvos) is not.[809] The most interesting fact is that no dimethyl ester is involved in the neurotoxic compounds. One exception is dimethyl phosphorofluoridate, which has the weakest neurotoxicity in the homologous series and is 10 to 100 times less active than other homologous members.[806]

Several other organophosphorus insecticides causing paralytic effect in hens have been reported.[810] They include abate, azinphosmethyl, coumaphos, crotoxyphos, crufomate, dicapthon, dioxathion, disulfoton, chloropyrifos, ethion, fenthion, malathion, menazon, parathion-methyl, methyl carbophenothion, phorate, and ronnel. However, they differ in the clinical condition from the above mentioned typical neurotoxic compounds: the former produced the onset of leg weakness within 24 hr after treatment and the paralyzed hens recovered completely within about 1 month. On the other hand, some nonphosphorus containing organic compounds such as p-bromophenylacetylurea were recently found to produce an "organophosphorus pattern" of degeneration in the nervous systems.[811]

b. Mechanism

No mechanism of the delayed neurotoxic action of organophosphorus compounds has been established as yet. However, recent studies clarified some aspects of the mode of action.

It was postulated that liberated fluorine may cause a biochemical lesion leading to the production of the delayed neurotoxic effects.[808] This appears, however, to be unconvincing because of the lack of neurotoxicity of dialkylphosphinic fluorides and the high neurotoxicity of many nonfluorine containing organophosphorus compounds.

On the other hand, it is natural to suppose that the neurotoxic lesion may arise from the phosphorylation of a protein or the inhibition of an esterase in the nervous system. All the neurotoxic compounds are esterase inhibitors with the phosphorylating property, at least after metabolic activation in vivo. It is a problem how to find the esterase or protein involved in the genesis of delayed neurotoxicity among a great variety of those which are sensitive to organophosphorus antiesterase agents. The inactivation of pseudocholinesterase was at first suggested to be the prelude to demyelination.[812] Actually, many neurotoxic agents are specific inhibitors of the cholinesterase rather than acetylcholinesterase. However, the reverse is not always the case: the selective inhibitors of the enzyme involve some non-neurotoxic organophosphorus compounds, such as iso-OMPA, tetraisopropyl pyrophosphate, and isopropyl paraoxon.

The relation of aliesterase inhibition with neurotoxicity has been studied extensively.[742,803,806] DFP and the active cyclic metabolite of TOCP were found to be the selective inhibitors of spinal cord esterases which hydrolyzed some butyryl esters.[813] In this context, it is interesting to note that most synergists of malathion, which are essentially inhibitors of aliesterases, show high neurotoxicity; TOCP and its related compounds, the oxon analog of EPN, tributyl phosphorotrithiolate (DEF), and corresponding trithioite (merphos) are typical examples. The structure-activity relationships in malathion synergism and neurotoxicity are similar, in general, but not identical.[742] Both types of biological activity may relate to esterase inhibition, but the structure requirement for the inhibition is somewhat different, indicating that the esterases involved are different. If the inhibition of an esterase is an essential prerequisite for the genesis of delayed neurotoxicity, the active site of the esterase must be phosphorylated by DFP or any other neurotoxic agent, but may resist non-neurotoxic organophosphorus compounds such as TEPP. Thus,

TABLE 51

Structures of Neurotoxic Phosphorus Esters and Related Non-neurotoxic Compounds[743,799,806,808,809]

Neurotoxic compound	Non-neurotoxic compound

Phosphates and phosphorothioates

$(RO)_2 P(O)OCH=CCl_2$

$R = C_2H_5, C_3H_7, C_5H_{11}$

$(CH_3O)_2 P(O)OCH=CCl_2$

(dichlorvos)

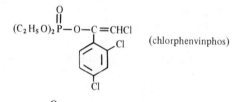

$(CH_3O)_2 \overset{O}{\overset{\|}{P}}OC=CHCOOCH_3$ (mevinphos)
$\overset{|}{CH_3}$

$(ClCH_2CH_2O)_2 \overset{O}{\overset{\|}{P}}OCH=CCl_2$

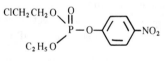

(chlorphenvinphos)

$(CH_3O)_2 \overset{O}{\overset{\|}{P}}-OC=CHCON(CH_3)_2$ (dicrotophos)
$\overset{|}{CH_3}$

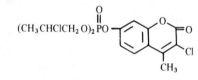

(Haloxon)

$(ClCH_2CH_2O)_2 \overset{O}{\overset{\|}{P}}-O-\!\!\!\!\bigcirc\!\!\!\!-NO_2$

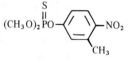

$(CH_3O)_2 \overset{S}{\overset{\|}{P}}O-\!\!\!\!\bigcirc\!\!\!\!-NO_2$
$\quad CH_3$

(fenitrothion)

$(C_2H_5O)_2 \overset{S}{\overset{\|}{P}}SCH_2S-\!\!\!\!\bigcirc\!\!\!\!-Cl$

(carbophenothion)

$(CH_3O)_2 \overset{O}{\overset{\|}{P}}SCH_2CH_2SOC_2H_5$

(oxydemetonmethyl)

TABLE 51 (continued)

Structures of Neurotoxic Phosphorus Esters and Related Non-neurotoxic Compounds[743,799,808,809]

Phosphonates

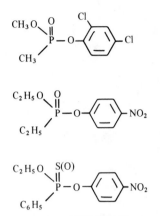

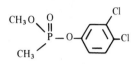

(paraoxon)

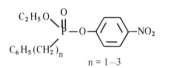
(EPN, EPN-O)

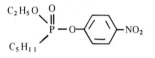

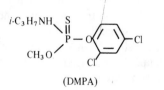
n = 1—3

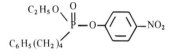

Trialkyl phosphorotrithiolates and trithioites

$(C_4H_9S)_3P=O$ (DEF) $(C_3H_5S)_3P=O$

$(C_3H_7S)_3P=O$ $(C_5H_{11}S)_3P=O$

$(C_4H_9S)_3P$ (merphos) $(C_6H_{13}S)_3P=O$

Phosphoramidothionates

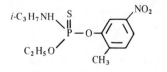
(DMPA)

i-C_3H_7NH, C_2H_5O — P(=S) — O — (ring) — CH_3, NO_2

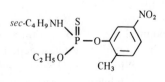
(Cremart)

TABLE 51 (continued)

Structures of Neurotoxic Phosphorus Esters and Related Non-neurotoxic Compounds[743,799,808,809]

Phosphorofluorine compounds

$$
\begin{array}{c}
O \\
\parallel \\
(RO)_2 PF
\end{array}
$$

$$
R = CH_3, C_2H_5, C_3H_7,
$$
$$
i\text{-}C_3H_7 \text{ (DFP)}, C_4H_9, C_5H_{11}
$$

$$
\begin{array}{c}
R \quad O \\
\diagdown \; \parallel \\
P - F \\
\diagup \\
R
\end{array}
\qquad R = C_2H_5 \sim C_4H_9
$$

$$
\begin{array}{c}
C_2H_5O \quad O \\
\diagdown \quad \parallel \\
P - F \\
\diagup \\
(CH_3)_2 N
\end{array}
$$

$$
\begin{array}{c}
O \\
\parallel \\
(RNH)_2 P - F
\end{array}
\qquad R = CH_3, C_2H_5, C_3H_7
$$

$$
i\text{-}C_3H_7 \text{ (mipafox)}, C_4H_9
$$

$$
\begin{array}{c}
RO \quad O \\
\diagdown \quad \parallel \\
P - F \\
\diagup \\
R'
\end{array}
\qquad R' = CH_3 \sim C_3H_7
$$

$$
\text{(sarin; } R = i\text{-}C_3H_7, R' = CH_3)
$$

Johnson and Aldridge developed an analytical method for such a protein ("neurotoxic protein") using ^{32}P-labeled DFP, TEPP, and mipafox.[424,799,814,815] If a nerve tissue homogenate is incubated previously with TEPP before treatment with ^{32}P-DFP, proteins phosphorylated with TEPP will prevent ^{32}P-labeling by ^{32}P-DFP, provided the proteins have only one phosphorylation site in each molecule as B-type esterases, and the mixture of proteins labeled (A) by ^{32}P-DFP treatment may contain the "neurotoxic protein." If the homogenate is similarly preincubated with TEPP and mipafox followed by ^{32}P-DFP treatment, labeled proteins (B) may not involve the "neurotoxic protein," since the protein should be phosphorylated previously by mipafox, a neurotoxic agent. Thus, the difference of the radioactive labelings of A and B may give a measure of the neurotoxic protein or the available phosphorylation site for neurotoxic organophosphorus compounds. The brains of hens which had been treated in vivo with neurotoxic or non-neurotoxic compounds were analyzed by this technique, and it was found that neurotoxic compounds phosphorylated the "neurotoxic protein" resulting in the great reduction (about 70% or more) of the available "phosphorylation sites," whereas non-neurotoxic compounds had little effect on the value, as shown in Table 52.[806] Similar phenomena were also observed in the spinal cord and sciatic nerve, where histological lesions occur markedly by treatment with neurotoxic agents.[816]

The "neurotoxic protein" was demonstrated to have an esterase activity to hydrolyze phenyl phenylacetate, used as a substrate because of structural similarity with the neurotoxic TOCP metabolite, saligenin cyclic o-tolyl phosphate.[424,816,817] Since only a small fraction (less than 4%) of the total hydrolysis of this substrate by hen brain homogenate is attributable to the "neurotoxic esterase," its activity can be assayed only after selective inhibition of the irrelevant esterases. The "neurotoxic esterase" is highly sensitive to in vivo inhibition by neurotoxic organophosphorus compounds but is resistant to non-neurotoxic compounds. A good correlation among the delayed neurotoxicity, the phosphorylation of the protein, and the inhibition of the esterase activity is shown in Table 52.

Some carbamates and sulfonyl fluorides inhibited the "neurotoxic esterase" but did not cause ataxia in hens. Although the esterase inhibited by carbamates recovered rather rapidly, the sulfonylated enzyme was much more stable. Moreover, these agents protected the hen from neurotoxicity caused by neurotoxic organophosphorus compounds, probably due to the blocking of the phosphorylation site.[806,816] These

TABLE 52

Phosphorylation and Inhibition of "Neurotoxic Esterase" by Some Organophosphorus Compounds (%)[806]

Compound	Phosphorylation	Inhibition
Non-neurotoxic		
Bromophos	3	0
Dimefox	13	1
Paraoxon	8–38	12
TEPP	22	25
Ethyl p-nitrophenyl 4-phenylbutylphosphonate	38	25
Dichlorvos	46	36
Neurotoxic		
Ethyl p-nitrophenyl 2-phenylethylphosphonate	65	86
Di-2-chloroethyl 4-nitrophenyl phosphate	73	92
4-Bromo-2,5-dichlorophenyl methyl ethylphosphonothionate	89	86
Saligenin cyclic phenyl phosphate	76–95	100
Diphenyl o-tolyl phosphate	87–95	93
Mipafox	89–92	100
DFP	92	90–95

facts indicate that the inhibition of the "neurotoxic esterase" is not a requisite for delayed neurotoxicity, but the nature of the acyl group bound to the esterase active site determines the toxic response. Among organophosphates, dimethyl esters have no or, at most, poor neurotoxicity. Dichlorvos cannot produce ataxia in hens even if 70% or more of the "neurotoxic esterase" were inhibited in vivo by a large dose such as 100 mg/kg.[806] Its higher alkyl homologs have high activities in both ataxia production and esterase inhibition. Neurotoxic esterase generally appears to be resistant to methyl phosphate esters. The reason that carbamoylation, sulfonylation, and methylphosphorylation of the enzyme hardly lead to delayed neurotoxicity remains to be solved. The "neurotoxic esterase" probably may be the primary site of action of the neurotoxic organophosphorus compounds, but how the phosphorylation of the enzyme causes demyelination in nerve tissues is unknown. The enzyme may play an important role for metabolism in nerve tissues: the phosphorylation may cause the block of unknown functions other than esterase activity leading to the neurotoxic lesion.

Significant decrease in protein synthesis and biochemical defects in phospholipid metabolism were found in some nerve tissues of hens administered neurotoxic organophosphorus compounds.[818] The relationship between these effects and "neurotoxic esterase" phosphorylation is not known.

In spite of extensive investigations, no effective therapy for delayed neuropathy caused by TOCP or other organophosphorus compounds has been found:[813] oximes and atropine which are effective for acute poisoning by organophosphorus compounds are completely ineffective against the delayed neurotoxicity. Some nicotinic acid derivatives can delay, but not prevent, the onset of TOCP induced ataxia in hens.[819] It was postulated that the nicotinic acid derivatives may have protective action on a particular site. Previous treatment with phenyl benzylcarbamate or phenylmethanesulfonyl fluoride protected hens from the delayed neurotoxicity caused by DFP and some other organosphosphorus compounds.[816] The protective effect of the carbamate lasted only several hours after dosing, while that of the sulfonyl flouride lasted for several days. If these agents, however, were given 1 hr after DFP treatment, no protection was observed. The agents may prevent phosphorylation of the active site of "neurotoxic protein."

2. Teratogenicity

Certain organophosphorus compounds show a teratogenic effect when injected into hen eggs before or within the early period of incubation.[819-822] Dicrotophos cis-isomer was the most active compound to produce teratogenic signs at a dosage of 0.03 mg/egg, while the trans-isomer lacked teratogenicity, producing no sign even at so high a dose as 10 mg/egg.[821] The teratogenic signs include lack of feathers, parrot beak, shortening and deformation of the legs and spine, wry neck, edema, and more rarely, syndactylia and visceral hernia. Some dicrotophos homologs, including N-diethyl, O-diethyl, N-monomethyl (monocrotophos), and N-unsubstituted ones, also showed high teratogenic activity comparable with that of dicrotophos cis-isomer. The latter two homologs may be produced in vivo as the metabolites from dicrotophos. The teratogenic effect was less marked with phosphamidon (3 mg/egg), mevinphos (10 mg), dichlorvos (3 mg), DEF (5 mg) and its phosphate analog (5 mg), and parathion (1 mg). No teratogenic effect was observed by treatment with malathion (1 mg), azinphosmethyl (1 mg), EPN (3 mg), dioxathion (1 mg), and TOCP (10 mg).

The inhibition of acetylcholinesterase activity caused by the injection of dicrotophos and other organophosphate esters did not correlate directly with the degree of teratogenesis. Although eserine, generally regarded as a specific cholinesterase inhibitor, produced teratogenic signs, neither acetylcholine nor choline was teratogenic and they did not potentiate or alleviate the action of dicrotophos. It appears possible that only dicrotophos- and eserine-sensitive esterases are involved in teratogenesis, while esterases inhibited by administration of EPN, for example, are not.

Nicotinic acid, nicotinamide, and certain related compounds alleviated dicrotophos induced teratogenesis. These agents yielded embryos normal in appearance at 21 days, although hatching was unsuccessful. Casida et al. supposed that the effect of nicotinic acid derivatives against dicrotophos induced teratogenesis may result from mechanisms different from those involving altered rates of metabolism of the teratogen or the alleviating agents.[821]

Formothion (30 mg/kg) and thiometon (5 mg) given daily by stomach tube to rabbits on days 6 to 18 of pregnancy had no teratogenic or embryotoxic effects.[823] However, fetal malformations, resorptions, or decreased fetal and placental weights resulted from intraperitoneal treatment of maternal rats at the 11th day of pregnancy with parathion (3 mg), dichlorvos (15 mg), diazinon (100 to 200 mg), apholate (10 mg), and tepa (5 to 10 mg). On the other hand, malathion at toxic dosages (900 mg) neither affected fetal weight nor produced malformations.[824] Chronic poisoning with disulfoton at a dosage level of 10 ppm in the diet reduced the number of pregnancies of rats.[825]

Deformation of puparia of the housefly was induced by treatment of 3rd stage larvae with some organophosphorus insecticides such as dichlorvos, trichlorfon, and malathion.[826]

The safety evaluation of pesticides may be conveniently performed by cell cultures. For He La cells, diazinon (10 ppm) showed a deleterious effect on DNA synthesis and disulfoton affected protein synthesis.[827] Human liver cells are susceptible to malathion (ID_{50} = 370 μg/ml) while mouse liver cells are resistant (1,804 μg/ml) because of their inability for malathion—malaoxon conversion.[828] Trichlorfon was most toxic of the tested organosphosphorus insecticides towards both cell cultures. The radiomimetic effect of dichlorvos was suggested by examinations using the onion root tip technique.[829]

3. Effects on Enzymes Other Than Esterases

It has been mentioned in Section IV.A that, like cholinesterases, other B-esterases and so-called "serine proteases" are susceptible to organophosphate inhibitors. There is some evidence that certain carboxy-amidases are also inhibited by organophosphorus compounds, which show synergistic activity with some insecticides and herbicides containing an amide group in the molecule like dimethoate and propanil (see Section IV.D.2.a). A rat liver arylamidase is more sensitive to some organophosphorus compounds than cholinesterases.[829a]

A few enzymes other than these hydrolases are also susceptible to certain organophosphorus esters. It was reported that some organophosphorus insecticides, including coumaphos, crufomate, and malathion, inhibited the beef liver glutamate dehydrogenase activity, although parathion and some organochlorine insecticides had no appreciable effects.[830] Certain saligenin cyclic phosphorus esters, such as S-methyl thiolate isomer of Salithion (2-methylthio-4H-1,3,2-benzodioxaphosphorin-2-oxide) showed high inhibitory activity towards the SH-enzymes yeast alcohol dehydrogenase (I_{50}: 4.5 $\times$ 10^{-5} M) and papain. They react with some SH-compounds such as cysteine to give corresponding S-o-hydroxybenzyl sulfides. A correlation between the inhibitory activity of the SH-enzymes and the reactivity with the SH-compounds was found in the series of saligenin cyclic phosphorus esters (see III.C).[251] Not only the cyclic triesters but also the ring-opened *o*-hydroxybenzyl esters are active as the SH-enzyme inhibitor and alkylating agent towards SH-compounds. Thus, the inhibition of the SH-enzymes is attributable to the alkylating property of the partially hydrolyzed benzyl esters:[255,256]

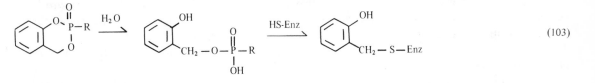

(103)

Alkylation of protein molecules by the insecticide trichlorfon was observed in vivo and in vitro: most of the ^{14}C-methyl group of trichlorfon administered to rats was covalently bound to liver proteins, particularly the albumin fraction.[831] Albumin has an ability to protect SH-enzymes from inactivation by certain alkylating agents.

Aminoparathion (diethyl *p*-aminophenyl phosphorothionate) has no antiesterase activity but a considerable activity to inhibit in vitro milk peroxidase activity; 50% inhibition was obtained at 10^{-5} to 10^{-6} *M*.[832] Horseradish peroxidase, however, was resistant to aminoparathion but was readily inhibited by isopropyl parathion (I_{50} = c. 5×10^{-5} *M*), which had no effect on the milk enzyme at 10^{-2} *M*. On the other hand, milk xanthine oxidase was not inhibited by various organophosphorus insecticides.[833]

Mitochondrial oxidative phosphorylation in rat liver was inhibited with chlorfenvinphos and ronnel (fenchlorphos) at concentrations of 0.1 to 1.0 m*M*, whereas malathion, malaoxon, and trichlorfon did not have any effect.[834] The effect of the former two compounds might be due to the presence of the halogenated aromatic ring moiety. Several chlorinated hydrocarbons such as aldrin, heptachlor, and γ-BHC, but not DDT, inhibited oxidative phosphorylation in beetle tissues, while the organophosphorus insecticides parathion and avenin and the carbamate insecticide carbaryl were not inhibitory.[835]

Acute poisoning by parathion produced a drastic depletion of myocardial phosphorylase activity in rats, but cytochrome oxidase and succinic dehydrogenase activity were not diminished.[836] Malathion induced greater depression in sulfate uptake by cartilage mucopolysaccharide than in brain acetylcholinesterase activity in rats.[837] Incorporation of glucosamine into the fungus mycelia cell wall was greatly reduced by the fungicide Kitazin (*S*-benzyl diethyl phosphorothiolate).[838] The high fungicidal activity and low mammalian toxicity of this fungicide are attributed to this interesting biological activity.

Recently, it was found that certain bicyclic phosphorus esters exerted an extremely high toxicity to mice without inhibition of acetylcholinesterase.[838a] For example, 4-isopropyl-2,6,7-trioxa-1-phospha-bicyclo[2.2.2]octane 1-oxide (XXXIX), which is not a pesticide, is more toxic than DFP and parathion, having an intraperitoneal LD_{50} for mice of 0.15 mg/kg. The mode of action is not known.

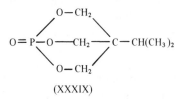

(XXXIX)

In certain cases of organophosphate poisoning, liberation of cellular enzymes or complex changes of the molecular architecture of membrane bound enzymes may be induced. Malathion released arylsulfatase from liver lysosome in vivo and in vitro. Malathion may interact with a structural component of the lysosome membrane and alter the permeability.[839] The central nervous system tissue became more permeable due to chronic poisoning with disulfoton.[840] Proteinase activity of some nervous systems in cats increased after administration of organophosphorus, particularly delayed neurotoxic compounds.[818] Daily administration of parathion increases β-glucuronidase in serum but not in liver of rats.[841] Enzyme secretion from the intestine of dogs increased by dosing of trichlorfon in the diet.[842]

4. Release of Physiologically Active Substances

It is known that a common physiological response to acute stress is an increased secretory activity of the stress sensitive endocrine glands, such as the pituitary and adrenal glands. This is the case in acute sublethal poisoning by organophosphorus pesticides: it results in an increased level of plasma corticosterone in rats and consequently of glucocorticoid responsive enzymes in liver such as alkaline phosphatase, tyrosine-α-ketoglutarate transaminase, and tryptophan pyrolase activities.[843] The magnitude of the increase in the enzyme activities depends not only upon the plasma corticosterone level, but also upon the rapidity and duration of this response. Thus, the short acting insecticide azinphosmethyl induced little alkaline phosphatase, a slowly inducible enzyme.

Organophosphorus poisoning also causes the release of catecholamines: excretion of adrenaline, noradrenaline, and their main metabolite 4-hydroxy-3-methoxymandelic acid was increased in the urine of rats which had been administered parathion, disulfoton, or dichlorvos.[844,845] Injection of parathion (5 mg/egg) into the allantoic sac of chick embryos on the 16th day of incubation caused release of catecholamines, resulting in almost complete absence of noradrenaline in the adrenal cells.[846]

In insects, similar release of pharmacologically active substances is induced by certain forms of stress, including some organophosphorus insecticides and other chemicals.[847] Sternburg and Kearns first found the release of a neuroactive substance from nervous tissue into the blood of American cockroaches and crayfish during the course of DDT poisoning.[848] A similar phenomenon was observed by poisoning with TEPP and nicotine and even by electrical stimulation.[848,849] Tashiro et al. isolated a neuroactive substance from the blood of silkworms prostrated with DDT treatment and identified it as L-leucine on the basis of chemical and physiological evidence.[390,850] L-Leucine causes a great increase of spontaneous discharge at 10^{-5} M on the isolated nerve cord of the cockroach. No other amino acids examined showed such activity, except L-alanine which appeared to have weak neuroactivity. Labeled L-leucine which had been incorporated into the nerve cord of the roach was released in vitro by the application of DDT or electrical stimulation. Dahm presumed that stress in chemical poisoning originates at the primary target of the insecticide, and the abnormal release of pharmacologically active substances may be a secondary effect.[851]

Casida and Maddrell found that treatment of fifth stage larvae of *Rhodnius prolixus* with a variety of insecticides, including organophosphorus compounds, chlorinated hydrocarbons, carbamate esters, and pyrethroids, induced diuresis due to the liberation of a diuretic factor, probably the diuretic hormone, from the central nervous system into the hemolymph at the paralytic stage of poisoning.[852] The resultant disturbances in water distribution and cation balance may contribute to the eventual death of the insects. The symptoms of poisoning and the volume of rectal fluid were correlated. Of the organophosphorus insecticides tested, diazinon, dicrotophos, and monocrotophos were most active in both biological effects. EPN and schradan were moderate in activity, but malathion did not show any activity by topical application at 600 μg/g. The structure of the diuretic factor has not been identified but was presumed to be an indolealkylamine resembling 5-hydroxytryptamine, which has an activity to induce diuresis but differs from the diuretic hormone in organ selectivity. Adenosine 3′,5′-monophosphate (cyclic AMP) showed similar diuretic acitivity and was considered to mediate the action of the diuretic hormone which probably stimulates the activity of adenyl cyclase.

5. Effects on Plant Metabolism

Since insecticides are commonly applied to plants for their protection, nonphytotoxicity is one of the most important properties which the insecticide should have. Very little is known of the interaction of organophosphorus pesticides with normal plant metabolism, which may be disturbed by phytotoxic compounds.

Catalase activity in *Lepidium sativum* germs was considerably inhibited by demeton-methyl, diazinon, and parathion but weakly by phenkapton.[853] Spraying a 0.1% emulsion of demeton-methyl on cotton plants during budding and blooming caused an increase in the activity of dehydrases and peroxidases, and a decrease in the activity of ascorbic oxidase and polyphenol oxidases, but had no effect on catalase activity. The increase of peroxidase activity was supposed to be a protective reaction of the plant against the insecticide.[854] Photosynthesis of cabbage leaves was apparently inhibited by 2-dichlorovinyl N,N,N',N'-

tetramethylphosphorodiamidate at 0.1 M.[855] Fenthion is a more potent photosynthesis inhibitor in estuarine phytoplanktons than abate, DDT, and the carbamate insecticide Baygon.[856]

Coloration of apples is disturbed by certain organophosphorus insecticides. The coloration relates closely to the content of anthocyanins and chlorophyll in the peel: the more anthocyanins and the less chlorophyll in the peel, the better the coloration is. Phenthoate was the most potent inhibitor of anthocyanin development in the peel.[857] This is also the case for the coloration of strawberries. Besides phenthoate, several organophosphorus methyl esters including Amiphos (S-2-(acetamido)ethyl dimethyl phosphoro-thiolothionate), menazon, phenitrothion, Salithion, fenthion, phosmet, and leptophos inhibited the coloration of apple peel cuttings in vitro. On the other hand, many ethyl esters, including EPN, diazinon, parathion, and phosalone, did not affect the coloration. In addition, some fungicides such as Bordeaux mixture and the mixture of zinc dimethyldithiocarbamate and N,N'-bis(dimethyldithiocarbamoyl) ethylenediamine also inhibit the coloration of apples. These fungicides retarded the chlorophyll disappearance and the formation of ethylene, and consequently ripening of the fruit. On the other hand, phenthoate did not affect the ethylene formation of apples. Ethylene is regarded as a ripening hormone.[858,859]

Ethephon (ethrel) liberates ethylene at physiological pH (Equation 104); consequently, maturation and ripening of fruits are accelerated by treatment with ethephon.[860] It also promotes flowering and defoliation.[859]

$$ClCH_2CH_2 - \overset{\overset{\displaystyle O}{\|}}{\underset{\underset{\displaystyle O^-}{|}}{P}} - OH + OH^- \longrightarrow CH_2{=}CH_2 + H_2PO_4^- + Cl^- \qquad (104)$$

ethephon

The mode of action of the recently developed herbicide glyphosate (N-phosphonomethylglycine) was attributed to its interference with the biosynthesis of phenylalanine, probably owing to the inhibition or repression of chorismate mutase and/or prephenate dehydratase of plants.[861]

Chapter V

INDIVIDUAL PESTICIDES

Organophosphorus compounds yield a variety of biological effects. Although the great majority of organophosphorus pesticides are employed as insecticides, their uses are not limited to this application, but have been extended to use as acaricides, anthelmintics, nematocides, chemosterilants, and rodenticides. Moreover, a number of organophosphorus fungicides and herbicides including plant growth regulators have been recently developed. A few compounds are useful as human medicines too. For example, paraoxon (diethyl p-nitrophenyl phosphate: Mintacol) and its phosphonate analog (ethyl p-nitrophenyl ethylphosphonate: Armine) are used against glaucoma, and cyclophosphamide 2-[bis(2-chloroethyl)amino] tetra-hydro-2H-1,3,2-oxazaphosphorin-2-oxide: Endoxan) is useful against malignant tumors.[22] Furthermore, cyclophosphamide was recently found to be useful as a defleecing agent of sheep.[545]

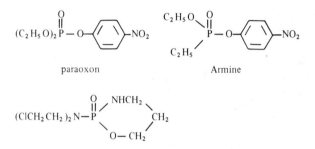

paraoxon

Armine

cyclophosphamide

Minor chemical modifications of an organophosphorus compound often cause a great change in toxicity from species to species.[862] Thus, compounds which are chemically very similar are often used for different purposes. The phosphorus esters of polyhalophenols present the most conspicuous examples, as shown in Table 53.[863]

Most of the biological activities are essentially due to esterase inhibition. However, the biological action of chemosterilants, fungicides, and herbicides is apparently due to different mechanisms.

A. INSECTICIDES AND ACARICIDES

A variety of neutral esters or amides derived from phosphorus oxyacids, including the thio analogs and the anhydrides too, are utilized for insecticides. They are principally contact poisons, but some compounds are useful as gut poisons and fumigants too. Although some compounds such as TEPP display very short residual action, some others such as azinphos have a prolonged residual action. Some insecticides such as parathion have a broad spectrum activity, but some others such as schradan have a very specific activity. Some organophosphorus insecticides are absorbed into the plant through foliage or roots and distributed in the whole plant to protect it from insects feeding thereon. These are called systemic insecticides (see V.A.2). Some others are usable on livestock to kill parasites by administration to the host animals (see Section V.A.3). In contrast to nonphosphorus containing insecticides, which are generally inactive to acarids, many organophosphorus insecticides, particularly plant systemics, have high acaricidal activity.

1. Contact Insecticides
a. Phosphates

The majority of commercialized phosphate type insecticides are vinyl ester derivatives which generally have an electron withdrawing group such as halogens and a carboxylic ester or amide group on the β-position. They are most conveniently produced by the Perkow reaction (see Section II.A.4.a) of trialkyl phosphites with appropriate α-halocarbonyl compounds such as chloral, haloacetophenones, and α-halo-acetoacetates. Some of the technical products consist of the geometrical isomers which differ from each

TABLE 53

Variety in Biological Activities of Some Closely Related Phosphorus Esters

Structure	Name	Use	LD_{50} rat, oral (mg/kg)
$(CH_3O)_2\overset{\overset{S}{\|\|}}{P}$—O—(aryl, Cl, Cl, Br)	Bromophos	Insecticide	4000
$(CH_3O)_2\overset{\overset{S}{\|\|}}{P}$—O—(aryl, Cl, Cl, Cl)	Ronnel	Animal systemic	1500
$(C_2H_5O)_2\overset{\overset{S}{\|\|}}{P}$—O—(aryl, Cl, Cl)	Dichlofenthion	Nematocide	270
$\underset{(CH_3)_2CHNH}{\overset{CH_3O}{\diagdown}}\overset{\overset{S}{\|\|}}{P}$—O—(aryl, Cl, Cl)	DMPA	Herbicide	270
$\underset{NH_2}{\overset{CH_3O}{\diagdown}}\overset{\overset{S}{\|\|}}{P}$—O—(aryl, Cl, Cl, Cl)	Dow ET-15	Insecticide	710
$\underset{C_2H_5}{\overset{C_2H_5O}{\diagdown}}\overset{\overset{S}{\|\|}}{P}$—O—(aryl, Cl, Cl, Cl)	Trichloronate	Soil insecticide	35
$\underset{\text{(phenyl)}}{\overset{C_2H_5O}{\diagdown}}\overset{\overset{O}{\|\|}}{P}$—O—(aryl, Cl, Cl, Cl)	H 6034	Fungicide	
$\underset{\text{(phenyl)}}{\overset{CH_3O}{\diagdown}}\overset{\overset{S}{\|\|}}{P}$—O—(aryl, Cl, Cl, Br)	Leptophos	Insecticide	46

other in chemical and biological properties. As dialkyl substituted phenyl phosphates are generally much more toxic to mammals and less effective as insecticides than the corresponding thiono analogs, only a few compounds of this class are utilized as insecticides.

Dichlorvos, DDVP, Vapona®

2,2-Dichlorovinyl dimethyl phosphate

$$\underset{\substack{\|\\O}}{(CH_3O)_2 P} - OCH = CCl_2$$

Dichlorvos was first found as a highly insecticidal impurity of trichlorfon in 1955.[864] Actually, trichlorfon is rapidly converted to the dichlorovinyl phosphate above pH 6 (see Section II.A.3; Equation 19). The manufacturing is done by the Perkow reaction of trimethyl phosphite and chloral (see Section II.A.4; Equation 23). Dichlorvos is a colorless liquid; bp 74°C at 1 mmHg; d_4^{25} 1.415; n_D^{25} 1.4523; soluble in water to about 1%; and miscible with most organic solvents. The half-life in water at pH 7.0 is about 8 hr.

Dichlorvos is a contact and stomach insecticide with fumigant action and low residual activity. It has rapid knockdown activity against flies, mosquitoes, moths, etc. It is also utilized as an anthelmintic for worms in pigs. The acute oral LD_{50} for male rats is 80 mg/kg. Dichlorvos is rapidly degraded in mammals through the initial splitting of either the vinyl or the methyl ester linkage.[583,654]

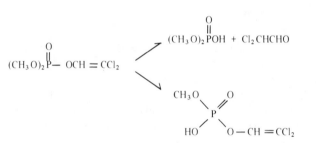

$$(1)$$

Many related compounds have been introduced as commercial or experimental insecticides. The calcium salt of the desmethyl dichlorvos makes a complex compound with 2 moles of dichlorvos and the complex is utilized as an insecticide under the name calvinphos in Japan.

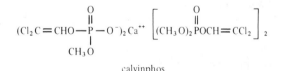

calvinphos

Bromine adds to dichlorvos to form the insecticide naled (Dibrom®) (see the next Section). Some selected insecticidal 2-halovinyl phosphate esters are shown as follows:

Structure	Remark	Ref.
$\underset{\substack{\|\\O}}{(CH_3O)_2 POCH} = CClF$	Low toxicity to fish	865
$\underset{\substack{\|\\O}}{(CH_3O)_2 POCH} = CHCl$	LD_{50} rat 50 mg/kg	13

Structure	Remark	Ref.

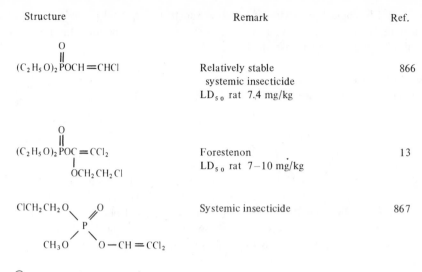

	Relatively stable systemic insecticide LD_{50} rat 7.4 mg/kg	866
	Forestenon LD_{50} rat 7–10 mg/kg	13
	Systemic insecticide	867

Naled, Dibrom®
1,2-Dibromo-2,2-dichloroethyl dimethyl phosphate

$$(CH_3O)_2\overset{O}{\underset{\|}{P}}-OCHBrCBrCl_2$$

Naled was introduced in 1956 by California Spray Chemical Corporation as an experimental insecticide, and is manufactured by the bromination of dichlorvos. The pure material melts at 26°C; bp 110°C at 0.5 mmHg. It is insoluble in water and readily soluble in organic solvents except aliphatic hydrocarbons. Although naled is somewhat more stable than dichlorvos, it is almost completely hydrolyzed in water within 2 days at room temperature. The mammalian toxicity is relatively low; the oral LD_{50} to rats is 430 mg/kg. Thus, naled is effective as a short-lived insecticide for use on vegetable crops and for fly and mosquito control. In order to prevent the rapid hydrolysis in aqueous formulations, poly(oxyethylene) partial esters of phosphoric acid are effective as stabilizers.

Naled is readily degraded to hydrolysis products in mammals.[654] By the reaction with natural thiol compounds, naled transforms readily into dichlorvos, which may be the actual insecticidal principle. The degradation metabolites are similar to those of dichlorvos (see IV.B.1.f). In addition, dichlorobromo-acetaldehyde is produced by the direct hydrolysis of naled.

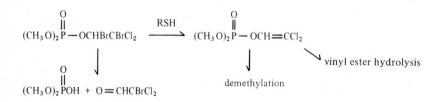

(2)

Chlorfenvinphos, Birlane®, Supona®
2-Chloro-1-(2,4-dichlorophenyl)vinyl diethyl phosphate

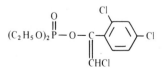

The insecticidal activity of chlorfenvinphos was first reported in 1961 for possible use in the control of livestock insects.[868] It is synthesized by the reaction of triethyl phosphite with 2,4-α,α-tetra-chloroacetophenone.

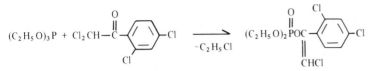

$$(3)$$

Technical chlorfenvinphos is a mixture of the geometrical isomers, of which the bulk is the *trans* one. It is an amber liquid; bp 168 to 170°C at 0.5 mmHg; sparingly soluble in water (145 mg/l); and miscible with acetone and aromatic hydrocarbons. The toxicity is particularly high to rats (oral LD_{50} 10 to 40 mg/kg), but is much lower to other mammals (oral LD_{50} for mice, 117 to 200 mg/kg; for rabbits, 500 mg/kg; for dogs, more than 12,000 mg/kg). This selective toxicity to rats is attributed to the poor degradative activity of rat liver microsomal enzymes. In other mammals, a liver microsomal enzyme catalyzes to degrade the insecticide through oxidative deethylation, which contributes greatly to the detoxication of the insecticide (see IV.B.2.b.i).[603]

In spite of being a phosphate triester, chlorfenvinphos is relatively stable against hydrolysis (see Table 5 in Chapter III); it persists a rather long time in soils and can be useful for the control of soil insects. In soils it is gradually decomposed to form α-dichlorophenylethanol, dichloroacetophenone, diethyl hydrogen phosphate, and desethyl chlorfenvinphos.[303,869]

The methyl homolog of chlorfenvinphos (SD-8280) and its 2′,5′-dichloro isomer (SD-8211) are very low toxic insecticides. They are effective to control flies on feces of hens by feeding.[870]

Tetrachlorvinphos, Gardona®, Rabon®
2-Chloro-1-(2′,4′,5′-trichlorophenyl)vinyl dimethyl phosphate

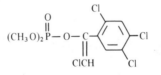

Tetrachlorvinphos was introduced in 1966 by the Shell Development Company.[720] It is synthesized by the Perkow reaction with trimethyl phosphite and 2,4,5-α,α-pentachloroacetophenone. By this reaction, the α- and β-isomers are produced in the ratio of 1 to 9, and the former is removed by crystallization. The mp of pure α-isomer and β-isomer is 62°C and 98°C, respectively. Technical Gardona is a white crystalline solid, mp 97 to 98°C, and contains 98% β-isomer, in which the phosphate group and chlorine atom are *cis* on the double bond. The solubility in water at 20°C is 11 ppm. It is soluble in most organic solvents except hydrocarbons. It is slowly hydrolyzed in acid media and more rapidly in alkaline media; the half-life at pH 3 and pH 10.5 is 54 and 3.3 days, respectively.

Tetrachlorvinphos is effective to control lepidoptera, diptera, and coleoptera and has very low mammalian toxicity; the oral LD_{50} to rats is 4,000 mg/kg and rats were safely fed up to a maximum of 125 ppm for 2 years.[720] Thus, it may be useful against pests of vegetables, fruit crops, and stored products. It is used in dairies and livestock barns and is also promising for making manure toxic to fly larvae by feeding.[871]

The metabolic breakdown of tetrachlorvinphos appears to take place as shown below.[650,872] On foliage, isomerization from β to α-form was observed.[650]

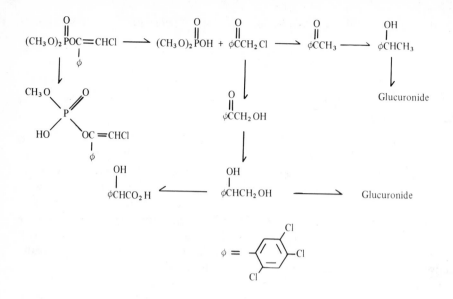

$$\phi = \text{(2,4,5-trichlorophenyl)}$$

Of the related compounds, the ethyl homolog (SD-8448) is a very effective agent against the cattle tick.[873] The bromine analogs (I, II) of tetrachlorvinphos and chlorfenvinphos are also known to have insecticidal activity.[867] For the thiono analogs, see Section b.

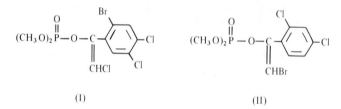

(I)   (II)

Bomyl®
1,3-Di(methoxycarbonyl)-1-propen-2-yl dimethyl phosphate

$$(CH_3O)_2\overset{\displaystyle O}{\overset{\|}{P}}-O-C{=}CHCO_2CH_3$$
$$\underset{\displaystyle CH_2CO_2CH_3}{|}$$

This insecticide was introduced in 1959 by the Allied Chemical Corporation under the code number GC-3707. It is a liquid; bp 155 to 164°C at 2 mmHg; soluble in most organic solvents except aliphatic hydrocarbons; and insoluble in water. It is effective for contact-residual treatment of insects and mites. Contrasting with many other vinyl phosphate insecticides, the two geometrical isomers of Bomyl resemble each other in biological activities.[874]

In spite of the existence of the carboxylic ester groups, Bomyl is quite toxic to mammals; the acute oral LD_{50} for rats is 32 mg/kg. The main process in the detoxication mechanism of Bomyl is the hydrolytic cleavage of the vinyl ester linkage, whereas the carboxylic ester hydrolysis and the dealkylation of the methyl phosphate ester group play a much less significant role in biodegradation.[221]

Propaphos, Kayaphos®
Di-n-propyl p-methylthiophenyl phosphate

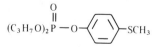

Propaphos was introduced in 1968 by Nippon Kayaku Company as an experimental insecticide. It is one of the rare commercial organophosphorus insecticides which have alkyl ester groups other than methyl or ethyl in the molecule. The insecticidal activity of propaphos was found in the course of examination of dipropyl aryl phosphates as synergists of organophosphorus insecticides against resistant insects. The pure propaphos is a colorless and odorless liquid; bp 176°C at 0.85 mmHg; and n_D^{25} 1.5106. It is scarcely soluble in water, but readily soluble in most organic solvents.

Propaphos is a highly selective insecticide effective against both rice stem-borers and green rice leafhoppers, two most serious pest of rice plants. The acute oral LD_{50} to rats is 70 mg/kg.

The dimethyl homolog (GC-6506) is an insecticide and acaricide for use in cotton. It has a higher insecticidal activity but is much more toxic to mammals in comparison with the propyl homolog; the acute oral LD_{50} to rats is 7 mg/kg. The diethyl homolog is also more toxic to mammals than propaphos.

b. Phosphorothionates

Dialkyl aryl phosphorothionates compose one of the most important classes of organophosphorus insecticides. Since the discovery of parathion in 1944 by Schrader, many compounds in this class have been developed into commercial insecticides. The alkyl group is methyl or ethyl with few exceptions, whereas the aryl group varies remarkably by the substituent groups. The substituent group is usually in the *p*-position and must be electron attractive, at least after metabolic modification. Additional substituent groups can be introduced to raise selective toxicity. The variation in structure increases enormously by the introduction of aromatic heterocycles in place of the aryl group.

The introduction of a second substituent in the benzene ring such as chlorine atom or the methyl group markedly reduces the mammalian toxicity (except the *p*-cyanophenyl ester) without any distinctive decrease in the insecticidal activity as shown in Table 54.[14,15,875] Meta substitution is particularly effective. Ortho substitution of a methyl group often causes a depression in the insecticidal activity.[495] The mammalian toxicity decreasing effect of the meta substitution is very conspicuous in the dimethyl esters, but much less obvious in the diethyl and isopropyl methyl esters (see Table 54).

The methyl esters are, in general, chemically less stable and toxic to mammals than the corresponding ethyl esters, but are often as effective as the ethyl esters as insecticides. The most remarkable difference in selective toxicity between the methyl and ethyl esters is shown in a *p*-cyanophenyl phosphorothionate series (see Table 54). This is partly due to the physical property of the methyl esters, which are less lipophilic and consequently slower to penetrate the mammalian skin than the ethyl esters.[862] The more important factor for the lower mammalian toxicity of the methyl esters is the detoxication mechanism via dealkylation; the activity of glutathione *S*-alkyltransferase, which has the substrate specificity to the methyl esters, is much higher in mammals than in insects (see IV.B.2.b.ii).

Diisopropyl aryl phosphorothionates show a high selectivity among insect species; isopropyl parathion is virtually nontoxic to the honeybee although it is highly toxic to the housefly (see Table 55). The tetraisopropyl ester of dithiopyrophosphoric acids shows a similar selectivity.[875] However, asymmetric dialkyl aryl phosphorothionates having an isopropyl ester group have a high toxicity to warm-blooded animals as well as a high insecticidal activity.[862]

Since phosphorothionates have generally higher vapor pressure, higher lipophilicity, and greater hydrolytic stability in aqueous conditions than the corresponding phosphate analogs, phosphorothionates are usually more active as insecticides than the phosphate analogs, although they have no anticholinesterase activity unless they are transformed into the latter in the body. Moreover, the thiophosphoryl group plays an important role for selective toxicity as a delaying factor to give time for detoxication.

Owing to readily occurring thermal thiono-thiolo rearrangement, ordinary preparations of phosphorothionate insecticides contain the thiolate isomer as an unwanted impurity, which is often less effective as the insecticide and more toxic to mammals than the original thionate.

TABLE 54

Effect of the Alkyl Ester Groups and the Second Substituent Group on the Selective Toxicity of Dialkyl Aryl Phosphorothionates

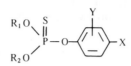

R_1	R_2	X	Y	Name	LD_{50} rat mg/kg	LD_{50} house fly $\mu g/g$	Mortality of insect %/conc. %
CH_3	CH_3	NO_2	H	Parathion-methyl	15	1.3	D 100/0.0008
C_2H_5	C_2H_5	NO_2	H	Parathion	6	0.9	
CH_3	CH_3	NO_2	2-Cl	Dicapthon	400	1.6	
CH_3	CH_3	NO_2	3-Cl	Chlorthion	880	11.5	D 100/0.001
C_2H_5	C_2H_5	NO_2	3-Cl		50		D 100/0.001
CH_3	$i\text{-}C_3H_7$	NO_2	3-Cl		50		D 100/0.004
							M 100/0.02
CH_3	CH_3	NO_2	$3\text{-}CF_3$	Fluorothion®	250		
							D 100/0.004
CH_3	CH_3	NO_2	$3\text{-}CH_3$	Fenitrothion	740	2.6	M 100/0.004
							C 100/0.006
C_2H_5	C_2H_5	NO_2	$3\text{-}CH_3$		10		D 100/0.02
CH_3	$i\text{-}C_3H_7$	NO_2	$3\text{-}CH_3$		25		D 40/0.0008
CH_3	CH_3	SCH_3	H		10	2.0	D 100/0.01
CH_3	CH_3	SCH_3	$3\text{-}CH_3$	Fenthion	500	2.3	D 70/0.001
C_2H_5	C_2H_5	SCH_3	$3\text{-}CH_3$		25		D 60/0.004
CH_3	$i\text{-}C_3H_7$	SCH_3	$3\text{-}CH_3$		17.5		D 90/0.02
CH_3	CH_3	CN	H	Cyanophos	500 (m)		C 97/0.006
CH_3	CH_3	CN	$3\text{-}CH_3$		500 (m)		C 77/0.012
C_2H_5	C_2H_5	CN	H		10 (m)		C 100/0.012
C_2H_5	C_2H_5	CN	$3\text{-}CH_3$		7.5 (m)		C 62/0.025

C: *Chilo suppressalis*, D: *Doralis fabae*, M: *Myzus persicae*, m: mouse.

TABLE 55

Selective Toxicity of Isopropyl Parathion

	Topical LD_{50} $(\mu g/g)$	
Insect	parathion	isopropyl parathion
Musca domestica	0.9	4.2
Apis mellifera	3.5	>1000
Dacus dorsalis	1.2	3.5
Opius longicaudatus	1.0	100
Opius persulcatus	2.0	100

Parathion, Folidol®, Thiophos®, E605®
Diethyl p-nitrophenyl phosphorothionate

$$(C_2H_5O)_2 \overset{\overset{\text{S}}{\|}}{P} - O - \langle\underline{}\rangle - NO_2$$

Parathion was discovered by Schrader in 1944. Parathion and the methyl homolog, parathion-methyl, have been the most widely used organophosphorus insecticides, although their use is now prohibited in some countries because of their extremely high mammalian toxicity.

Parathion is manufactured by the condensation of diethyl phosphorochloridothionate with sodium p-nitrophenate. The purity of the product may change by the preparation method of the phosphorus intermediate. As the intermediate preparations obtained from phosphorus thiochloride and sodium ethoxide usually contain ethyl phosphorodichloridothionate, crude parathion will be up to 80% pure and contains ethyl bis-(p-nitrophenyl) phosphorothioate. A purer intermediate may be obtained by chlorination of diethyl phosphorodithioate (see II.B.2; Equations 48 and 49); hence, parathion of 95% purity can be produced.

Pure parathion is a pale yellow liquid; bp 113°C at 0.05 mmHg; mp 6°C; d_4^{25} 1.265; and n_D^{25} 1.53668. It is highly soluble in most organic solvents except alkanes. The solubility in water is 24 ppm. Parathion is relatively stable in neutral or mild acid solutions (see III.A; Table 5), but rapidly hydrolyzed in alkaline solutions. The thermal isomerization to O,S-diethyl p-nitrophenyl phosphorothiolate takes place slowly at temperatures above 130°C, with about 1% isomerization in 1 hr at 130°C. The thiolo isomer is more reactive and more potent as the anticholinesterase agent, but less effective as the insecticide than the thiono isomer. In the environment, parathion appears to be much more persistent than previously believed; it remains as the intact molecule in plant tissues for many days. For example, the half-life of parathion sprayed on grape vines as an emulsifiable concentrate at the rate of 2 lb/acre was 2.2 days, but the residue was stabilized in the bark at the level of 25 ppm, being maintained for at least 70 days.[876]

Parathion shows a high insecticidal activity with a wide spectrum, and may be regarded as a measure of effectiveness for the evaluation of insecticides. It also has a high mammalian toxicity (oral LD_{50} to rats is 7 mg/kg), which is a limiting factor in its use.

For the biological activity, the presence of a nitro group is important; diethyl phenyl phosphorothionate lacks the insecticidal activity. Both the insecticidal activity and the mammalian toxicity of the positional isomers of parathion decrease in the following order: para > ortho > meta. The biological activity is remarkably affected by the alteration of the alkyl group. The toxicity of the mono-methyl homolog (ethyl methyl parathion) is almost the same as that of parathion, but the dimethyl homolog, i.e., the insecticide parathion-methyl, is considerably less toxic to mammals (see IV.B.2.b.ii). The isopropyl homolog (isopropyl parathion) is a highly selective insecticide;[490] it is at least 100 times less toxic to honeybees than to house flies, though parathion is only 4 times less toxic to the bee. The selectivity is the net result of the inability of the bee to convert the diisopropyl phosphorothionate into the oxon and of the lower affinity of the oxon for the bee cholinesterase.[492,531] Isopropyl parathion also shows a distinct selective toxicity between the oriental fruit fly *Dacus dorsalis* and its Brancoid parasites *Opius persulcatus* and *O. longicaudatus* (Table 55).[875]

Paraoxon (diethyl p-nitrophenyl phosphate) is the active form of parathion. It is a strong cholinesterase inhibitor and is too poisonous to mammals (oral LD_{50} rats = 2.5 mg/kg) to be employed practically for plant protection, though it has a high insecticidal activity. The structure-activity relationship in paraoxon series is presented in Table 22 in Section IV.A.3. Dipropyl paraoxon is known as the synergist of malathion (see IV.D.2.a).

The main metabolic pathways for parathion and the homologous compounds in animals are as follows:

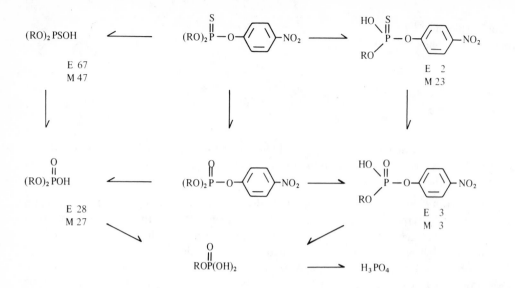

$$(5)$$

The level of the oxon formed in animals by biotransformation of parathion is generally low compared to other metabolic degradation products, although it is the actual active principle. Only several percent of the LD_{50} dose administered to animals may be converted to paraoxon. The figures in the metabolic pathways indicate the percentage of each metabolite of parathion (E) and parathion-methyl (M) in American cockroaches.[201] The hydrolysis of the nitrophenyl ester bond takes place extensively, but the cleavage of the alkyl ester linkage plays an important role only in the degradation of the methyl homolog.

Parathion-methyl, methyl parathion
Dimethyl p-nitrophenyl phosphorothionate

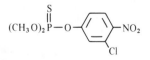

Parathion-methyl was introduced in 1949 by Bayer AG. It is a white crystalline powder; mp 36°C; bp 109°C at 0.05 mmHg; d_4^{20} 1.358; n_D^{35} 1.5515; soluble in most organic solvents except alkanes; and slightly soluble in water (55 mg/l). It is less stable than the ethyl homolog, parathion. The hydrolysis rate by alkali is 4.3 times higher than that of parathion. Dealkylation occurs readily; it is a good methylating agent. The thermal rearrangement into the S-alkyl phosphorothiolate isomer takes place more easily than with parathion.

The insecticidal activity is similar to that of parathion, but the mammalian toxicity is somewhat lower; the acute oral LD_{50} to rats is 25 to 50 mg/kg. This may be due to the great activity of mammalian liver enzyme to catalyze the glutathione S-methyl transfer reaction (see IV.B.2.b.ii).[606] The technical products are more toxic due to impurities.

Chlorthion®
3-Chloro-4-nitrophenyl dimethyl phosphorothionate

$$(CH_3O)_2 \overset{S}{\underset{\parallel}{P}} - O - \text{\LARGE{⟨⟩}} - NO_2$$
Cl

Chlorthion was introduced in 1952 by Bayer AG as a contact insecticide with low mammalian toxicity. The pure material is a crystalline powder; mp 21°C; and d_4^{20} 1.437. The solubility in water is 40 ppm. It is highly soluble in aromatic hydrocarbons. The half-life in aqueous solutions between pH 1 and 5 is 138 days. Chlorthion is effective for the control of aphids, caterpillars, and beetles. The acute oral LD_{50} for rats is 880 mg/kg. No mortality was observed in rats fed 50 mg/kg/day for 60 days.

The detoxication of chlorthion is due to both aryl ester hydrolysis and methyl ester bond cleavage in mammals, whereas the latter occurs only to a small extent in insects.[201] For example, the contribution of demethylation to the detoxication of chlorthion is 38% in the rats, but only 1% in the American cockroach. The difference in detoxication mechanisms among animal species may contribute to the selective toxicity of the insecticide.

Dicapthon, Di-Captan®
2-Chloro-4-nitrophenyl dimethyl phosphorothionate

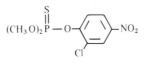

Dicapthon was introduced in 1954 by American Cyanamid Company as an experimental insecticide. The insecticidal activity was first described by Davich and Apple in 1951.[877] Dicapthon is a crystalline powder, mp 52 to 53°C. It is highly soluble in aromatic hydrocarbons, halogenated hydrocarbons, esters, and ketones, but only slightly soluble in alkanes. The solubility in water is 35 ppm.

Dicapthon is effective for controlling household insects, aphids, and boll weevils. The acute oral LD_{50} for male rats is 400 mg/kg. Rats fed a 25 ppm diet for 1 year showed no retardance in growth.

The positional isomer phosnichlor (Isochlorthion®) also has low mammalian toxicity; the acute oral LD_{50} to rats is 500 mg/kg.

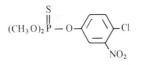

phosnichlor

Fenitrothion, Sumithion®, Folithion®, Metathion, MEP
Dimethyl 3-methyl-4-nitrophenyl phosphorothionate

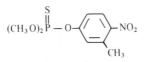

This compound was first prepared in Czechoslovakia in 1957 and then independently in Japan and Germany. It was introduced in 1959 by Sumitomo Chemical Company as an experimental insecticide. The high insecticidal activity and the low mammalian toxicity of fenitrothion were first reported by Nishizawa.[15] It is a liquid; bp 95°C at 0.01 mmHg; d_4^{25} 1.3227; and n_D^{25} 1.5528. Fenitrothion is highly soluble in alcohols, ethers, and aromatic hydrocarbons. The water solubility is 30 ppm. Fenitrothion is somewhat more stable than parathion-methyl: the half-life in 0.01N NaOH at 30°C is 272 min, whereas that of parathion-methyl is 210 min. It is isomerized partially into S-methyl isomer by distillation under reduced pressure (140 to 145°C at 0.1 mmHg).

Fenitrothion is synthesized by the reaction of dimethyl phosphorochloridothionate with 3-methyl-4-nitrophenol (see II.B.1; Equation 35). The direct nitration product of m-cresol containing 4-nitro and 6-nitro isomers is applied for the preparation of the final phosphorus ester in the USSR. Thus, the product called methylnitrophos in the USSR is a mixture of 3-methyl-4-nitro and 3-methyl-6-nitro isomers of fenitrothion. The 6-nitro isomer is about 200 times less effective than the 4-nitro isomer as an insecticide, but has synergistic activity with fenitrothion and parathion.[878]

Fenitrothion is as effective as parathion-methyl to control rice-stem borers, but the insecticide Clorthion (i.e., the chlorine analog of fenitrothion) is not practically effective against this pest. Fenitrothion is also useful to control flies and mosquitoes in public health programs. The acute oral LD_{50} to various mammals is between 142 and 1,000 mg/kg. No rats died after being fed 250 ppm for 90 days.

Although the main metabolic pathways of fenitrothion are essentially the same as those of parathion-methyl,[721] the rate of demethylation in mammals is greater with fenitrothion, particularly at high doses, than with parathion-methyl.[616] The high selective toxicity of fenitrothion between insects and mammals is attributed to the net difference in rates of activation, detoxication,[616] and transfer to the target[721] and in the sensitivity of the target enzyme.[494]

Cyanophos, CYAP, Cyanox®
p-Cyanophenyl dimethyl phosphorothionate

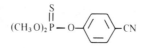

Cyanophos was introduced in 1962 and commercialized in 1966 by Sumitomo Chemical Company.[879] The pure preparation is an amber liquid; mp 14°C; d_4^{25} 1.260; and soluble in alcohols, ketones, and aromatic hydrocarbons. Cyanophos has a low mammalian toxicity and is effective in controlling rice stem borers and also useful for vegetables, fruits, and ornamentals. The acute oral LD_{50} to rats and mice is 580 and 1,020 mg/kg, respectively.

The main biodegradation pathway in mammals is aryl ester bond cleavage; the liberated cyanophenol is excreted in the form of the sulfate ester by conjugation.[657]

The introduction of a second substituent such as methyl, methoxy, or chlorine on the benzene ring does not decrease the mammalian toxicity. The diethyl homolog is about 50 times more toxic to mammals.

Bromophos, Nexion®, Brofene®
4-Bromo-2,5-dichlorophenyl dimethyl phosphorothionate

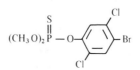

Bromophos was introduced in 1961 by Boehringer Sohn and Cela Landwirtschaftliche Chemie, Germany. It is a colorless crystalline solid; mp 54°C; and soluble in ketones, aromatic hydrocarbons, and halogenated hydrocarbons. The water solubility is 40 ppm. It is stable in media of pH up to 9.0 and is compatible with sulfur and organometallic fungicides.

Bromophos is a nonsystemic and persistent insecticide and acaricide for general use on fruits and vegetables, particularly effective against diptera. It is also useful as a public health insecticide and for ectoparasite control in big livestock. The mammalian toxicity is very low; the acute oral LD_{50} to rats is 3,750 to 6,100 mg/kg. It is safe for fish at 50 times the concentration toxic to mosquito larvae. Bromophos administered to mammals is eliminated almost completely within 24 hr.[880] The main metabolite is desmethyl bromophos.[881]

The ethyl homolog, bromophos-ethyl, is useful as an insecticide and acaricide, but is more toxic to mammals than bromophos; the oral LD_{50} to rats is 238 mg/kg. The chlorine analog of bromophos is the insecticide ronnel, which is useful as an animal systemic (see V.A.3).

Iodofenphos, Alfacron®, Nuvanol N®

2,5-Dichloro-4-iodophenyl dimethyl phosphorothionate

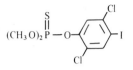

Iodofenphos was introduced in 1966 by Ciba Ltd. as a broad spectrum insecticide with low mammalian toxicity. Iodofenphos is a colorless crystalline solid; mp 76°C; and highly soluble in ketones, aromatic hydrocarbons, and methylene chloride. The solubility in water is less than 2 ppm. It is a nonsystemic contact and stomach insecticide and acaricide effective against a wide range of insect pests: for example, flies, mosquitoes, fleas, bedbugs, beetles, moths, roaches, and ticks. Owing to the low mammalian toxicity and relatively low persistence, it may be useful not only for plant protection, but also for public hygiene and the protection of stored products.[882] The acute oral LD_{50} to rats is 2,100 mg/kg.

The ethyl homolog is more than ten times as toxic to mammals as iodofenphos.

Fenthion, Baytex®, Lebaycid®

Dimethyl 3-methyl-4-methylthiophenyl phosphorothionate

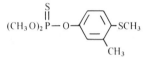

Fenthion was developed in 1958 by Bayer AG as an insecticide. It is produced by the condensation of dimethyl phosphorochloridothionate and 3-methyl-4-methylthiophenol. The intermediate phenol may be synthesized from *m*-cresol by the reaction with dimethylsulfoxide and hydrogen chloride or with methyldisulfide and sulfuryl chloride.

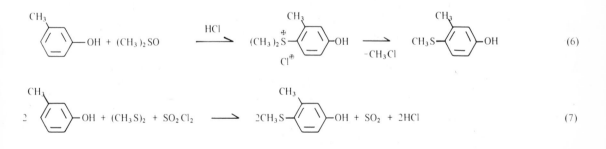

Pure fenthion is a colorless liquid; bp 87°C at 0.01 mmHg; and d_{20}^{20} 1.245. It is highly soluble in most organic solvents except for aliphatic hydrocarbons. The solubility in water is 54 ppm.

Fenthion is much more stable to acid and alkaline hydrolysis and heat in comparison with parathion-methyl; consequently, it is a highly persistent insecticide. The oxidation of the thioether group into sulfoxide and sulfone proceeds in plants, animals, and by sunlight. In plants, the oxidative desulfuration takes place mainly by the action of enzyme, while the thioether oxidation is chiefly due to the influence of light. The oxidized products are also insecticidal. The *S*-methyl isomers of fenthion and of the oxidation products are also found in plants. This is due to the thermal rearrangement reaction. The reaction sequences giving rise to insecticidal metabolites are given below (see also IV.B.1.c).[883,884]

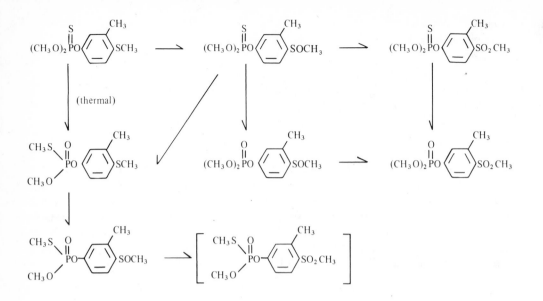

(8)

Fenthion is a general purpose insecticide with systemic action. It is particularly effective for fly and mosquito control. The acute oral LD_{50} to male and female rats is 215 and 615 mg/kg, respectively.

Of the related compounds, the *O*-ethyl homolog (Lucijet®) is an animal systemic insecticide (see V.A.3) and its sulfoxide is effective as a nematocide and soil insecticide (see V.B).

Abate®, Biothion®
O,O,O',O'-Tetramethyl *O,O'*-thiodi-*p*-phenylene phosphorothionate

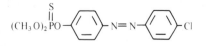

Abate was introduced in 1965 by American Cyanamid Company as a mosquito larvicide. The pure material is a white solid; mp 30°C; insoluble in water and aliphatic hydrocarbons; and soluble in ether, aromatic hydrocarbons, and chlorinated hydrocarbons. Abate has a very low mammalian toxicity (oral LD_{50} for rats 2,000 to 4,000 mg/kg), yet it is extremely effective in concentrations as low as 0.005 ppm for control of mosquito larvae. The same larvicidal activity is obtained with 0.008 ppm of parathion. No toxicological effect was observed with human volunteers administered 64 mg/day for 4 weeks.

Abate is transformed biologically by insects into active metabolites, the sulfoxide and sulfone phosphates (see IV.B.1.c).[536]

The disulfide analog of Abate has a higher larvicidal activity; 100% lethal concentration is $10^{-11}\%$. However, the mammalian toxicity is also very high; the oral LD_{50} to rats is 10 to 25 mg/kg.

Azothoate, Alamos®, Slam®
p-(*p*-Chlorophenylazo)phenyl dimethyl phosphorothionate

$$(CH_3O)_2 PO - \langle \rangle - N = N - \langle \rangle - Cl$$

Azothoate was introduced in 1956 as an experimental insecticide and acaricide by Montecatini. It is practically nontoxic to mammals.

Akton®
2-Chloro-1-(2',5'-dichlorophenyl)vinyl diethyl phosphorothionate

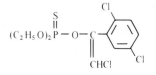

The active ingredient of Akton, which was introduced by Shell Development Company, is a brown liquid; bp 145°C at 0.005 mmHg; and soluble in most organic solvents. The solubility in water at 23°C is 1.39 ppm. The acute oral LD_{50} to rats is 146 mg/kg. It is a nonsystemic insecticide, effective particularly for soil insects such as lawn chinch bugs.

The methyl homolog and the 2',4',5'-trichloro analog of Akton are effective as housefly larvicides.[885]

Although the related halophenyl vinyl phosphates, as chlorfenvinphos, are generally synthesized by the Perkow reaction, the vinyl phosphorothionates are synthesized by the condensation of an appropriate acetophenone with a dialkyl phosphorochloridothionate.

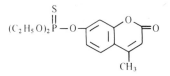

$$(9)$$

Potasan®
Diethyl 4-methyl-7-coumarinyl phosphorothionate

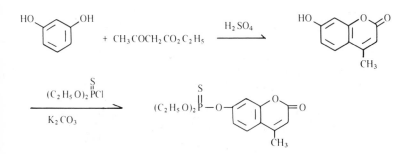

The active ingredient of Potasan was first synthesized by Schrader in 1947 from diethyl phosphorochloridothionate and 7-hydroxy-4-methylcoumarine, which was obtained by the reaction of ethyl acetoacetate with resorcinol.[13]

$$(10)$$

The pure compound is a colorless crystalline solid, mp 38°C. It is highly soluble in most organic solvents, but practically insoluble in water. It is stable in pH region 5 to 8, but the hydrolytic cleavage of the ethyl ester bond occurs slowly by heating with water for a long time.

Potasan is a stomach insecticide with weak contact activity. It is particularly effective against Colorado potato beetles, but not against aphids. The mammalian toxicity is relatively high; the acute oral LD_{50} for male and female rats is 42 and 19 mg/kg, respectively.

The 6-coumarinyl isomer has very low toxicity to mammals (LD_{50} rats 1,000 mg/kg), but is practically inactive as a pesticide. On the other hand, the 3-chloro derivative (coumaphos) has low mammalian toxicity and is effective for the control of parasites of animals (see V.A.3).

Chloropyrifos, Dursban®
Diethyl 3,5,6-trichloro-2-pyridyl phosphorothionate

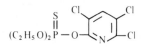

Chloropyrifos was discovered by Dow Chemical Company in 1965.[886] It is a white crystalline solid; mp 42.5 to 43°C; and highly soluble in most organic solvents, but not in water. It is stable except under strong alkaline or acid conditions. The hydrolysis rate is greatly enhanced by the catalytic action of cupric ion (see III.A.3.a).

Chloropyrifos is a moderately persistent insecticide, effective to control mosquito and fly larvae, sucking and chewing plant pests, and soil inhabitant plant pests. It is also useful to control ticks on livestock. The acute oral LD_{50} for rats is 163 mg/kg.

Chloropyrifos is rapidly metabolized in rats and excreted mainly in the urine (90%). The degradation appears to be mainly due to dealkylation; about 80% of the metabolites in the urine consist of trichloropyridyl phosphate.[887] The absorption of chloropyrifos by plants through roots and leaves is insignificant.

The methyl homolog (Dowco® 214) is much less toxic to mammals (oral LD_{50} rats 1,500 mg/kg), and more effective against adult mosquitoes but less effective against the larvae than chloropyrifos. It is much more unstable to hydrolysis than the ethyl homolog; at pH 5 the methyl ester is decomposed 110 times as fast as the ethyl ester.[888] The methyl oxon analog (Dowco® 217) also has low toxicity to mammals; the acute oral LD_{50} for rats is 869 mg/kg. The structurally related compound, 5-cyano-2-pyridyl diethyl phosphorothionate (III) is also effective against various insects and mites.

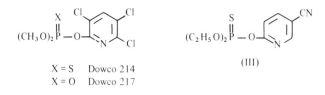

X = S   Dowco 214
X = O   Dowco 217

(III)

Diazinon, Basudin®, Srolex®
Diethyl 2-isopropyl-6-methyl-4-pyrimidinyl phosphorothionate

Diazinon was discovered by Gysin in 1952.[889] It is synthesized from diethyl phosphorochloridothionate and 2-isopropyl-6-methyl-4-pyridinol, which is produced by the condensation of acetoacetate ester with isobutylamidine according to the following synthetic pathway.

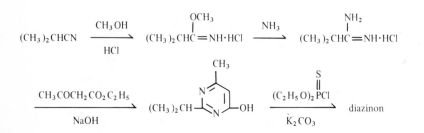

$$(11)$$

Pure diazinon is a colorless liquid; bp 89°C at 0.1 mmHg; d_4^{20} 1.115; and n_D^{20} 1.4978 to 1.4981. It is highly soluble in most organic solvents, but only 40 ppm in water at 20°C. Diazinon is a weak base whose pK value is 2.6, and is relatively unstable in acid in comparison with ordinary dialkyl aryl phosphorothionate insecticides such as parathion (see III.A.2.c).

Diazinon shows, however, a relatively long residual action and is effective as a control of soil, fruit, vegetable, and rice insects. When diazinon is applied to paddy water, it is absorbed and translocated in the leaf sheath and blade of rice plants. Diazinon is also useful for the control of pests in the household and on livestock.

Although the toxicity of pure diazinon is relatively low, that of technical products varies considerably by the batch, probably owing to impurities. The acute oral LD_{50} for rats is 108 to 250 mg/kg. The low toxicity to mammals is attributed in part to the hydrolase activity in the liver and plasma which hydrolyzes diazoxon.[589] The metabolism of diazinon in mammals, insects, and plants was precisely investigated in recent years.[560-564,576,577,581,890] The principal metabolites in mammals are the pyrimidinol, the side chain hydroxylated pyrimidinols, and desethyl diazinon. The pyrimidinols may be liberated from diazinon and hydroxydiazinon by the action of microsomal oxidase and from the corresponding oxons by the action of hydrolase. The latter mechanism is specific in mammals and contributes to the selective toxicity of diazinon.[589] For the metabolic pathways, see Equations 50, 63, and 66 in Chapter IV.

The n-propyl isomer of diazinon (Pyrazinon) is also effective as an insecticide. The acute oral LD_{50} for rats is 261 mg/kg.

Pirimiphos-ethyl
Diethyl 2-dimethylamino-4-methylpyrimidin-6-yl phosphorothionate

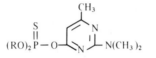

R = C_2H_5 pirimiphos-ethyl
R = CH_3 pirimiphos-methyl

Pirimiphos-ethyl was introduced by Imperial Chemical Industries in 1972. It is synthesized from diethylguanidine, ethyl acetoacetate, and diethyl phosphorochloridothionate. It is a straw colored liquid; d^{20} 1.14; n_D^{25} 1.520; and highly soluble in most organic solvents. The acute oral LD_{50} for rats is 140 to 200 mg/kg. Pirimiphos-ethyl has a broad insecticidal spectrum and is particularly recommended for the control of diptera and coleoptera. It has a fungicidal activity too.[891]

The dimethyl ester homolog (pirimiphos-methyl) is a liquid; d^{20} 1.157; and n_D^{25} 1.527. It is an insecticide and acaricide with contact and fumigant activity; the oral LD_{50} for rats is 2,050 mg/kg. Owing to the low mammalian toxicity, it is recommended for use with stored products and for public health.

Pyridafenthion, Ofnack®
Diethyl 3-oxo-2-phenyl-2H-pyridazin-6-yl phosphorothionate

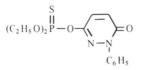

Pyridafenthion was commercialized in 1973 by Mitsui-Toatsu Chemicals. It was also known as American Cyanamid 12,503. Pyridafenthion is effective in the control of rice stem borers. The acute intraperitoneal LD_{50} for mice is 64 to 256 mg/kg. The metabolism in animals was recently reported.[892] More than 70% of the administered dose was excreted in the urine. The main metabolites excreted were phenyl maleic hydrazide (IV) and desethyl pyridafenthionloxon (V).

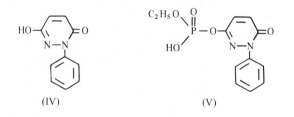

(IV)   (V)

Diethquinalphion, quinalphos, Bayrusil®
Diethyl 2-quinoxalyl phosphorothionate

$$(C_2H_5O)_2\overset{\overset{S}{\|}}{P}-O-$$

Diethquinalphion was discovered in 1965 by Bayer AG.[893] The pure material is a white crystalline powder with a melting point of 35 to 36°C and is readily soluble in most organic solvents; d_4^{20} 1.230; and n_D^{25} 1.5624. The solubility in water is about 20 ppm at 20°C. It is rather unstable, not only in alkaline solution, but also under acid conditions (see III.A.2.c).

Diethquinalphion is a nonpersistent insecticide effective against both biting and sucking pests on vegetables, especially against the diamond-back moth. It is also useful in the control of mosquitoes and mites. It has an excellent initial effect, but has short term action, with contact and stomach activity. The oral LD_{50} for rats is about 66 mg/kg.

Introduction of any substituent on the ring causes a decrease in the insecticidal activity.

Triazophos, HOE 2960, Hostathion®
1-Phenyl-3-(diethoxyphosphinothioyloxy)-1,2,4-triazole

$$(C_2H_5O)_2\overset{\overset{S}{\|}}{P}-O-$$

Triazophos was discovered in 1967 by Farbwerke Hoechst.[894] It is a yellow liquid; d 1.433; and readily soluble in most organic solvents. The solubility in water at 23°C is 39 ppm. Triazophos is a nonsystemic insecticide and acaricide, acting as a contact and stomach poison. The acute oral LD_{50} for rats is 82 mg/kg.

Isoxathion, Karphos®
Diethyl 5-phenyl-3-isoxazolyl phosphorothionate

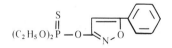

Isoxathion was discovered in 1965 and commercialized in 1972 by Sankyo Company, Japan. Isoxathion is synthesized from diethyl phosphorochloridothionate and 3-hydroxy-5-phenylisoxazole (see Equation 36 in Chapter II), which may be produced by the reaction of hydroxylamine with a β-phenyl-α,β-dihalopropionate ester in the presence of a large excess of alkali.[895,896]

$$\text{C}_6\text{H}_5-\text{CHX}-\text{CHX}-\text{CO}_2\text{R} \xrightarrow[\text{OH}^-]{\text{NH}_2\text{OH}} \quad \text{(isoxazole-OH)} \xrightarrow[\text{Na}_2\text{CO}_3]{(\text{C}_2\text{H}_5\text{O})_2\overset{\text{S}}{\underset{\|}{\text{P}}}\text{Cl}} \quad \text{isoxathion} \qquad (12)$$

The pure material is a slightly yellow liquid, bp 160°C at 0.15 mmHg. Isoxathion is a broad spectrum insecticide effective for a variety of insects, including scale insects and soil insects. The acute oral LD_{50} to rats is 112 mg/kg.

The dimethyl homolog (Dimex®) is much less toxic to mammals (LD_{50} mice = 593 to 697 mg/kg) and is less stable chemically than isoxathion. The diethyl phosphorothionate ester of nonsubstituted 3-hydroxyisoxazole appears to be more effective against insects than the 5-phenyl derivative, isoxathion, but is too toxic to mammals (LD_{50} mice = 5 to 10 mg/kg) to be utilized as an insecticide.

Some nonesterified 5-substituted 3-hydroxyisoxazoles (VI) are known to have interesting biological activities; for example, the 5-methyl derivative is a soil fungicide, and the amino acid having the isoxazole skeleton, ibotenic acid, is a housefly poison isolated from the fungus *Amanita strobiliformis* (ibotengu-dake).[897]

$$\begin{array}{ll} \text{(VI)} & R = CH_3 \quad \text{fungicide} \\ & R = CH(NH_2)COOH \quad \text{ibotenic acid} \end{array}$$

Phoxim, Valexon®, Baythion®
(Diethoxyphosphinothioyloxyimino)phenylacetonitrile

$$(\text{C}_2\text{H}_5\text{O})_2\overset{\text{S}}{\underset{\|}{\text{P}}}-\text{O}-\text{N}=\overset{\overset{\text{C}\equiv\text{N}}{|}}{\text{C}}-\text{C}_6\text{H}_5$$

Phoxim was discovered in 1965 by Bayer AG.[898] It is produced by the reaction of diethyl phosphorochloridothionate with α-cyanobenzaldoxime (see Equation 37 in Chapter II). Phoxim is a light yellow liquid; mp 5 to 6°C; bp 102°C at 0.01 mmHg; n_D^{20} 1.5405; and d_4^{20} 1.176. The solubility in water at 20°C is 7 ppm.

Phoxim has a broad spectrum insecticidal activity as a contact and stomach poison. It is effective to control leafhoppers, aphids, cutworms, black beetles, and insects affecting stored products and public health. The mammalian toxicity is very low; the acute oral LD_{50} for rats is larger than 2,000 mg/kg.

The high selective toxicity of phoxim may be due, in part, to the specificity of the oxon to inhibit insect cholinesterase and to the rapid degradation in mammals.[479] Phoxim is metabolized in vivo into the oxon, which inhibits housefly cholinesterase 270 times as fast as bovine erythrocyte acetylcholinesterase. The oxon produced in mammals may be hydrolyzed immediately into diethyl phosphate. In addition to this main metabolic pathway in mammals, the cleavage of the oxime ester bond without desulfuration, the hydrolytic transformation of the nitrile group into carboxyl, and deethylation also appear to contribute to the low mammalian toxicity. In susceptible insects, large amounts of phoxim and the oxon remain in the body.

On plant foliage, phoxim disappears rapidly; only 20% and less than 1% remained intact, respectively, 1 and 5 days after application to cotton leaves.[898]

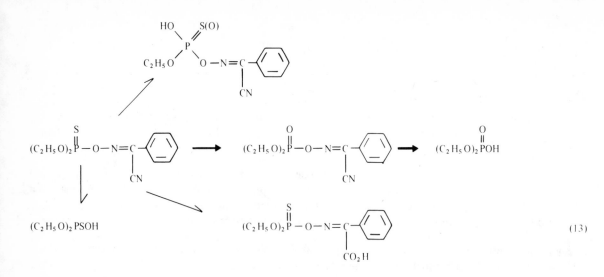

$$(13)$$

This is mainly attributed to photoreactions. By UV illumination, three neutral products were formed besides the hydrolysis products. The major product was the photoisomer (VII), which reached a maximum 53% of the applied amount of the insecticide after 1 day. Two other products were pyrophosphate derivatives: tetraethyl pyrophosphate (TEPP) (VIII) and its monothio analog (IX). The amounts of TEPP and monothio-TEPP were 10 and 1%, respectively, after 2 days. These three photoproducts were found to a smaller extent on cotton leaves treated with phoxim.

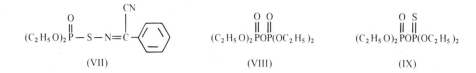

| (VII) | (VIII) | (IX) |

The phosphonate analog (X) also has high insecticidal activity and is more effective than phoxim against organophosphate resistant strains of houseflies. However, the phosphinate analog (XI) is almost inactive as an insecticide.

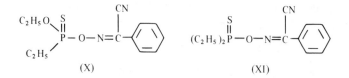

| (X) | (XI) |

The nitrile group appears to be essential for the insecticidal activity of oxime phosphates. No insecticidal compounds were found in the series of phosphorus esters of ring-substituted acetophenoximes and α-substituted benzaldoximes.[480] Some substituents on the benzene ring of phoxim fortify the insecticidal activity: the o-chloro derivative (chlorphoxim) is active enough to control an organophosphate resistant strain of houseflies to which phoxim is not effective.[899] The o-cyano derivative also has a high insecticidal activity.[82]

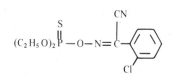

chlorphoxim

Salithion
2-Methoxy-4*H*-1,3,2-benzodioxaphosphorin-2-sulfide

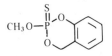

Salithion was discovered in 1963 at the Laboratory of Pesticide Chemistry, Kyushu University, Japan,[233] and developed into commercialization in 1968 by Sumitomo Chemical Company. It is synthesized by the reaction of *o*-hydroxybenzyl alcohol (saligenin) with methyl phosphorodichlorido-thionate in the presence of an aqueous caustic alkali solution (see Equation 38 in Chapter II).[84]

The pure compound is a white crystalline powder, mp 55.5 to 56°C. It is soluble in most organic solvents. Salithion is relatively unstable in storage and is stabilized in the presence of some secondary amines, such as carbazole and *N*-phenyl-α-naphthylamine.[900] Salithion is a short-lived insecticide particularly effective for protection of fruits and vegetables. The acute oral LD_{50} for rats is 102 mg/kg.

This unique structure of salithion comes from the active metabolite of tri-*o*-tolyl phosphate (TOCP) (see IV.B.1.e).[489] In spite of the high neurotoxicity of the TOCP-metabolite, salithion causes no such toxicity (see IV.E.1).

Salithion is only a weak cholinesterase inhibitor in vitro, but is converted in vivo to the oxon analog which is a strong anticholinesterase agent. Salithion-^{32}P administered orally to mice was rapidly decomposed and excreted; chloroform soluble ^{32}P in the body was only 2.4% of the administered ^{32}P at 3 hr after the treatment.[717] The degradation of salithion in mammals is greatly due to demethylation.

The introduction of any type of substituent at any position of the benzene ring and the carbon atom of the hetero ring of salithion decreases the insecticidal activity.[901] An outstanding contrast in the effect of *p*-substitution between the saligenin cyclic phosphorothionates (salithion series) and dialkyl phenyl phosphorothionates (parathion series) is noteworthy. Table 56 shows the effect of the electronic character of the substituent in the *p*-position of the phenolic ester group upon the insecticidal activity. In the series

TABLE 56

Effect of Para-substitution on Insecticidal Activity of Saligenin Cyclic Methyl Phosphorothionate (XII) and Diethyl Phenyl Phosphorothionate (XIII)

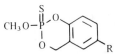

(XII)

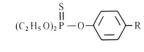

(XIII)

		Relative insecticidal activity*	
R	σ	**XII**	**XIII**
OCH$_3$	−0.268	9.2	0.1>
CH$_3$	−0.170	2.6	0.1>
H	0.000	100.0	0.1>
C$_6$H$_5$	+0.009	12.8	—
Cl	+0.226	3.0	0.33
COCH$_3$	+0.87	2.0	2.5
NO$_2$	+1.27	1.7	100.0

*Percentage of the most active compound in each series.

of diethyl phenyl phosphorothionates, the toxicity to insects is progressively increased by *p*-substitution of the phenyl ring in the order of increasing electron withdrawing ability, whereas neither an electron withdrawing nor releasing group increases the activity of salithion.

The ethyl and higher alkyl homologs of salithion are much less active as insecticides. Of the analogs of salithion, alkylphosphonates, alkylphosphonothionates, *S*-alkyl phosphorothiolates, and *S*-alkyl phosphorodithioates are less effective than salithion in insecticidal activity. In each series the most active one is the methyl derivative except in the alkylphosphonate series, in which the ethyl homolog is the best. *N*-Methyl phosphoramidate (saliamide) and its thiono analog are as active as salithion as a contact poison and have a systemic activity too.[556] *S*-Alkyl phosphorothiolate analogs have not only insecticidal but also fungicidal activity.[902] The biological activities of salithion and related compounds have been reviewed in detail.[487,556,903]

Cyclic esters derived from alkanediols lack insecticidal properties, even if they contain an "acyl group" such as the *p*-nitrophenyl ester group in the molecule.[232,484] Exceptions are the phosphorochloridates. *P*-Chloro-2,4-dioxa-5-methyl-*P*-thiono-3-phosphabicyclo [4.4.0]-decane (UC 8305) is effective to control the aphid infesting tobacco.[904] It was synthesized in 1960 and introduced as an experimental insecticide by Union Carbide Corporation.[905] It is a liquid; bp 78°C at 0.2 mmHg; and d_{15}^{30} 1.237. The solubility in water is 100 ppm. The acute oral LD_{50} for rats is 120 mg/kg.

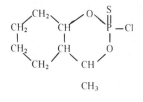

UC 8305

c. Phosphorothiolothionates

Phosphorothiolothionates are usually named phosphorodithioates for simplicity. In almost all useful insecticides in this class, the thiolo sulfur atom attaches an alkyl group which has an electron attractive group on the α-position, at least after metabolic activation. Many phosphorothiolothionate esters having a carboxyamide group or a sulfide group in the molecule are effective as systemic insecticides. They will be described separately in Section V.A.2. Dialkyl *S*-aryl phosphorothiolothionates analogous to parathion have only a poor insecticidal activity.

Malathion, Cythion®, Karbofos®
S-[1,2-Di(ethoxycarbonyl)ethyl] dimethyl phosphorothiolothionate

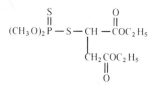

Malathion was introduced in 1950 by American Cyanamid Company and has been acknowledged as the first organophosphorus insecticide with high selective toxicity. It is synthesized by the addition of dimethyl hydrogen phosphorodithioate to diethyl maleate (see Equation 64 in Chapter II). Malathion is a yellow liquid; bp 120°C at 0.2 mmHg; d_4^{25} 1.23; and n_D^{25} 1.4985. The solubility in water is 145 ppm at 20°C. It is highly soluble in most organic solvents except alkanes. Malathion is rapidly hydrolyzed in aqueous solutions above pH 7.0 and below pH 5.0. Heavy metals, particularly iron, catalyze the decomposition.

Malathion is a safe general purpose insecticide suited for the control of sucking and chewing insects on fruits and vegetables, and for the control of mosquitoes and flies. Owing to its low mammalian toxicity and high insecticidal activity, malathion is used on a large scale by the WHO for the eradication of Anopheles. The acute oral LD_{50} to rats is 1,375 mg/kg. Rats fed with 1,000 ppm of technical malathion for 92 weeks showed normal growth.

The high selective toxicity of malathion is attributed to the presence of the carboxy ester group which is readily hydrolyzed by mammalian carboxyesterase (see IV.B.2.c.i and Table 33). The esterase activity is low in susceptible insects.[631] On the other hand, bioactivation, i.e., the formation of the oxon, occurs faster in insects than mammals. Cleavage of the P-S-C linkage occurs preferentially in insects. It may take place by phosphatase and/or microsomal oxidase. Demethylation takes place, but to a small extent. In human liver, for example, malathion is degraded mainly by the action of a carboxyesterase, whereas the demethylation accounts for about 5%, the P-S cleavage about 1%, and the S-C bond cleavage less than 0.5%.[619]

As the detoxication mechanism in mammals is due to the carboxyesterase activity, esterase inhibitors potentiate the toxicity of malathion. Typical examples of potentiators are tri-*o*-tolyl phosphate (TOCP) and EPN (see IV.D.2.a). Some impurities in technical malathion, particularly *O,S,S*-trimethyl phosphorodithiolate (a by-product formed in the course of manufacturing the intermediate *O,O*-dimethyl hydrogen phosphorodithioate from phosphorus pentasulfide and methanol), increase the mammalian toxicity of malathion (see IV.D.2.a; Tables 40 and 41).

Several other organophosphorus insecticides with low mammalian toxicity having a carboxy ester functional group have been developed, as exemplified by the following compounds.

Phenthoate, dimephenthoate, Cidial®, Elsan®, Papthion®
Dimethyl *S*-[(α-ethoxycarbonyl)benzyl] phosphorothiolothionate

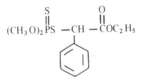

Phenthoate was introduced in 1964 by Soc. Montecatini, Italy. It is synthesized by the reaction of sodium dimethyl phosphorodithioate with ethyl α-bromophenylacetate. The pure product is a liquid with bp 70 to 80°C at 2×10^{-5} to 5×10^{-5} mmHg and n_D^{25} 1.5512. It is soluble in aromatic hydrocarbons, ketones, alcohols, and chlorinated hydrocarbons. Phenthoate is relatively unstable to heat, decomposing 50% at 120°C for 110 hr.

Phenthoate has a broad spectrum of insecticidal and acaricidal activities and is particularly effective against codling moth and scale insects. It is useful for the protection of vegetables, citrus, tea, and rice. The mammalian toxicity depends greatly upon the purity, as shown in Table 40 in Chapter IV. The purified product has a low mammalian toxicity, oral LD_{50} to rats 4,700 mg/kg, while a technical preparation of low purity (78.7%) is 40 times more toxic than the purified one.[745]

Several related compounds have been tested as experimental insecticides and acaricides. The mammalian toxicity appears to be remarkably influenced by the alkyl group of the carboxy ester; the fluoroethyl ester (M 1788) is as toxic as parathion.

TABLE 57

Mammalian Toxicity of Phenthoate Analogs

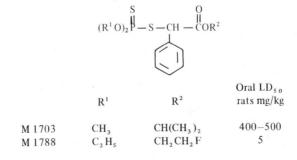

	R^1	R^2	Oral LD_{50} rats mg/kg
M 1703	CH_3	$CH(CH_3)_2$	400–500
M 1788	C_2H_5	CH_2CH_2F	5

Acethion

Diethyl *S*-(ethoxycarbonylmethyl) phosphorothiolothionate

$$(C_2H_5O)_2\overset{\overset{S}{\|}}{P}-SCH_2\overset{\overset{O}{\|}}{C}OC_2H_5$$

Acethion is a yellowish liquid; bp, 92°C at 0.01 mmHg; d_4^{20} 1.176; and n_D^{20} 1.4992. It is produced by the reaction of sodium *O,O*-diethyl phosphorodithioate with ethyl chloroacetate (see Equation 54 in Chapter II). Acethion is a selective insecticide with a low mammalian toxicity; the oral LD_{50} for rats is 1,100 mg/kg (see Table 33 in Chapter IV). It is, however, not used practically.

The carbomethoxy homolog is also an insecticide, named azethion. Some related insecticidal compounds are listed below.

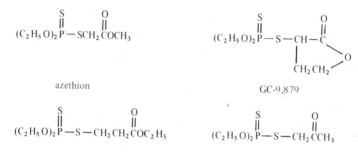

$$(C_2H_5O)_2\overset{\overset{S}{\|}}{P}-SCH_2\overset{\overset{O}{\|}}{C}OCH_3$$

azethion

$$(C_2H_5O)_2\overset{\overset{S}{\|}}{P}-S-CH_2CH_2\overset{\overset{O}{\|}}{C}OC_2H_5$$

propathion

GC-9,879

$$(C_2H_5O)_2\overset{\overset{S}{\|}}{P}-S-CH_2\overset{\overset{O}{\|}}{C}CH_3$$

ketothion

Chlormephos

S-Chloromethyl diethyl phosphorothiolothionate

$$(C_2H_5O)_2\overset{\overset{S}{\|}}{P}-S-CH_2Cl$$

Chlormephos was introduced in 1968 by Murphy Chemical as a soil insecticide. It is produced by the reaction of bromochloromethane with the alkali salt of diethyl phosphorodithioic acid. It is a colorless liquid; bp 81 to 85°C at 0.1 mmHg; n_D^{20} 1.5244; and d^{20} 1.260. Chlormephos is miscible with most organic solvents. It dissolves in water to 60 ppm and is stable in neutral solutions at room temperature.

Chlormephos is a nonsystemic insecticide effective to control soil insects, particularly larvae of Coleoptera (wireworms). It has a high mammalian toxicity; the oral LD_{50} for rats is 7 mg/kg.

Carbophenothion, Trithion®, Akarithion®, Garrathion®

S-(*p*-Chlorophenylthiomethyl) diethyl phosphorothiolothionate

$$(C_2H_5O)_2\overset{\overset{S}{\|}}{P}-S-CH_2-S-\!\!\left\langle\!\!\bigcirc\!\!\right\rangle\!\!-Cl$$

Carbophenothion was introduced in 1955 by Stauffer Chemical Company as an insecticide and acaricide. It is synthesized by the reaction of *p*-chlorophenyl chloromethyl sulfide with sodium diethyl phosphorodithioate (see Equation 59 in Chapter II). The technical product of 95% purity is a light amber colored liquid; bp 82°C at 0.01 mmHg; and d_{20}^{20} 1.590 to 1.597. It is miscible with most organic solvents.

Carbophenothion has a long residual action in controlling sucking plant pests, particularly mites. It is also useful as a dip for cattle tick. The mammalian toxicity is rather high; the acute oral LD_{50} for rats is 32 mg/kg.

The thioether oxidation into the sulfoxide and then the sulfone is observed on field growing vegetables.[906]

The methyl homolog of carbophenothion (methyl carbophenothion; Methyl Trithion®) was introduced in 1958 by Stauffer Chemical Company. It is a light yellow liquid; bp 125°C at 0.01 mmHg; d_{20}^{20} 1.360; n_D^{30} 1.6130; and miscible with most organic solvents. The solubility in water is 1 ppm at room temperature. It is effective to control a variety of insects and mites, particularly the cotton boll weevil.

Methyl carbophenothion is much less toxic to mammals than the ethyl homolog, carbophenothion; the acute oral LD_{50} for rats is 150 mg/kg.

Phenkapton, Phencapton®
S-(2,5,-Dichlorophenylthiomethyl) diethyl phosphorothiolothionate

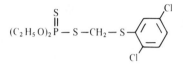

Phenkapton was introduced in 1956 by Geigy AG as an acaricide. It is synthesized by the reaction of ammonium diethyl phosphorodithioate with chloromethyl 2,5-dichlorophenyl sulfide. Phenkapton is a yellowish liquid; bp 120°C at 0.001 mmHg; d_4^{21} 1.3507; and n_D^{21} 1.6007. It is nearly insoluble in water, but miscible with most organic solvents.

Phenkapton is a selective acaricide with prolonged action, and effective against all development stages including eggs. It has only a low insecticidal activity, with little hazard for bees and no systemic property. The mammalian toxicity is much lower in comparison with that of carbophenthion; the acute oral LD_{50} for rats is about 200 mg/kg.

The methyl homolog (methyl phenkapton) is useful as an acaricide and insecticide. The acute oral LD_{50} is 375 mg/kg. The 3,4-dichloro isomer of methyl phenkapton (G-30493) is an insecticide and may be useful as a fly larvicide in manure from chicks by oral administration.[907]

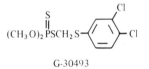

G-30493

Azothion
Diethyl S-(N,N-dimethylthiocarbamylthiomethyl) phosphorothiolothionate

$$\underset{(C_2H_5O)_2P-SCH_2S-CN(CH_3)_2}{\overset{S\qquad\quad S}{\overset{\|\qquad\quad\|}{}}}$$

Azothion was introduced in 1957 by Farbwerke Hoechst as an insecticide and acaricide. It is synthesized by the reaction of sodium dimethyldithiocarbamate with diethyl S-chloromethyl phosphorothiolothionate, which is produced from bromochloromethane and alkali salt of diethyl phosphorodithioate:

$$\underset{(C_2H_5O)_2PSK}{\overset{S}{\overset{\|}{}}} \xrightarrow{BrCH_2Cl} \underset{(C_2H_5O)_2PSCH_2Cl}{\overset{S}{\overset{\|}{}}} \xrightarrow{\overset{S}{\overset{\|}{NaSCN(CH_3)_2}}} Azothion \qquad (14)$$

Azothion has good insecticidal and acaricidal activity. The oral LD_{50} to rats is 150 mg/kg.

Ethion, Nialate®

Tetraethyl S,S'-methylene bis(phosphorothiolothionate)

$$\underset{(C_2H_5O)_2PSCH_2\,SP(OC_2H_5)_2}{\overset{S\quad\;\; S}{\underset{\|\quad\;\; \|}{}}}$$

Ethion was developed in 1956 by Food Machinery and Chemical Corporation. It is synthesized by the reaction of diethyl phosphorodithioic acid (2 mole) with bromochloromethane or dibromomethane (1 mole) in the presence of alkali. The product is a yellow liquid; bp 164 to 165°C at 0.3 mmHg; d_4^{20} 1.2277; and n_D^{20} 1.5490. Ethion is practically insoluble in water and highly soluble in aromatic hydrocarbons. It is slowly oxidized in air.

Ethion is useful for the control of aphids, scales, and mites. The acute oral LD_{50} to rats is 208 mg/kg for the purified grade. The technical preparations have higher toxicity.

The tetramethyl homolog is effective to bean leaf-lice as well as *Tetranychus telarius*. The benzylidene derivative (SD 7,438; tetramethyl S,S'-benzylidene bis (phosphorothiolothionate)) is an insecticide with relatively low mammalian toxicity; the acute oral LD_{50} to rats is 280 mg/kg.

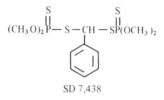

SD 7,438

Dioxathion, Delnav®, **Navadel**®

1,4-Dioxan-2,3-ylidene bis(O,O-diethyl phosphorothiolothionate)

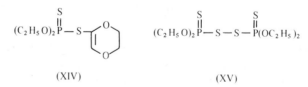

Dioxathion was introduced in 1954 by Hercules Incorporated as a contact insecticide and acaricide. It is synthesized by the condensation of 2,3-dichloro-*p*-dioxane with O,O-diethyl phosphorodithioic acid with the catalytic action of zinc chloride or pyridine or the addition of bis(diethoxyphosphinothioyl) disulfide to *p*-dioxene (see Equation 68 in Chapter II).[96] The product is a mixture of *cis*- and *trans*-isomers (1:1.5-2 ratio). Technical preparations contain about 30% of by-products such as the dioxene derivative (XIV), the disulfide (XV), and others.[96,624]

$$\underset{(C_2H_5O)_2P-S-}{\overset{S}{\underset{\|}{}}}\;\;\;\underset{}{\overset{O}{\bigcirc}}O \qquad\qquad \underset{(C_2H_5O)_2P-S-S-P(OC_2H_5)_2}{\overset{S\qquad\quad S}{\underset{\|\qquad\quad\|}{}}}$$

(XIV)   (XV)

Trans dioxathion is more unstable to heat than the *cis*-isomer; thermal decomposition gives the elimination product XIV (see Section III.F).[96]

The acute oral LD_{50} for rats of the technical dioxathion is about 175 mg/kg. The *cis*-isomer is more toxic to both insects and mammals than the *trans*-isomer; the acute subcutaneous LD_{50} to rats is 65 and 240 mg/kg for the *cis*- and the *trans*-isomer, respectively.[624] The by-product XIV also shows a high insecticidal activity and mammalian toxicity (LD_{50} rats = 90 mg/kg). The compound XV has much lower toxicity, but probably has some synergistic activity with organophosphorus insecticides.[908]

Dioxathion is an acaricide and insecticide with a long residual effect and is particularly useful to control mites on cotton and fruits and ticks, lice, and horn fly on cattle.

The methyl ester homolog also has a high insecticidal activity but lower mammalian toxicity (oral LD_{50} rat = 300 mg/kg).

Phostex®
Diethoxyphosphinothioyl ethoxyisopropoxyphosphinothioyl disulfide

$$(C_2H_5O)_2 \overset{\overset{S}{\|}}{P} S - S \overset{\overset{S}{\|}}{P} \overset{OC_2H_5}{\underset{OCH(CH_3)_2}{}}$$

Phostex was developed by Food Machinery and Chemical Corporation in 1954. It is prepared by the oxidation of dialkyl hydrogen phosphorodithioates produced by the reaction of phosphorus pentasulfide with a mixture of ethanol and isopropanol (3:1) (see Equation 153 in Section II.G). Therefore, diethoxyphosphinothioyl ethoxyisopropopoxyphosphinothioyl disulfide is not the sole component of Phostex. Phostex is slightly soluble in water and miscible with most organic solvents.

Phostex is a weak contact insecticide but a highly effective miticide with a good ovicidal activity. The mammalian toxicity is very low; the acute oral LD_{50} for rats is 2,500 mg/kg.

The homologous bis(dialkoxyphosphinothioyl) disulfides synergize the insecticidal activities of many organophosphorus insecticides.[908]

Phosmet, phthalophos, Imidan®, Prolate®
Dimethyl S-phthalimidomethyl phosphorothiolothionate

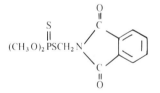

Phosmet was introduced in 1966 by the Stauffer Chemical Company as an insecticide and acaricide. It is synthesized by the reaction of sodium dimethyl phosphorodithioate and N-chloromethylphthalimide (see Equation 58 in Chapter II). Phosmet is an off-white crystalline solid with an offensive odor; mp 72°C; and soluble in most organic solvents except aliphatic hydrocarbons. The solubility in water at 25°C is 25 ppm. The half-life in aqueous solutions at pH 4.5, 7.0, and 8.3 is about 13 days, 12 hr, and 4 hr, respectively. It varies 3 to 19 days in soils according to the moisture and microbial population.[909]

Phosmet is a broad spectrum nonsystemic insecticide effective to both sucking and chewing insects. It may also be useful for cattle grub control.[910] The acute oral LD_{50} to male rats is 230 mg/kg.

Carbonyl-C^{14} labeled phosmet administered orally in rats was rapidly degraded and excreted; 79% of the dose was eliminated via the urine and 19% via feces.[911] Hydrolysis to water soluble products is the predominant metabolic pathway. The main metabolic route was shown in Chapter IV.B, Equation 71.

Dialifor, Torak®
S-(2-Chloro-1-phthalimidoethyl) diethyl phosphorothiolothionate

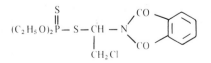

Dialifor was introduced in 1965 by Hercules Incorporated as an insecticide and acaricide. It is synthesized by chlorination of *N*-vinylphthalimide followed by the reaction with diethyl phosphorodithioic acid. Dialifor is a colorless crystalline solid, mp 67 to 69°C. It is highly soluble in acetone, chloroform, xylene, and ether. The solubility in water, ethanol, and hexane is less than 1%.

Dialifor is effective to control many insects and mites on apples, citrus, grapes, and vegetables. The acute oral LD_{50} for mammals varies greatly, depending on species and sex, from 5 to 97 mg/kg.

The dimethyl homolog (Herc. 14,504), as well as dialifor, is effective to control alfalfa weevil.[9][12] Succinimide derivatives such as XVI (Herc. 13,462) are also effective as insecticides.

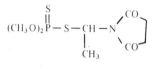

(XVI)

Phosalone, Zolone®

S-(6-Chlorobenzoxazolone-3-ylmethyl) diethyl phosphorothiolothionate

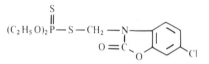

Phosalone was discovered in 1961 and commercialized in 1964 by Rhone-Poulanc, France. It is synthesized by the condensation of 3-chloromethyl-6-chlorobenzoxazolone with sodium diethyl phosphorodithioate.

$$(C_2H_5O)_2 \overset{\overset{\text{S}}{\|}}{P} SNa + ClCH_2 \underset{\underset{\text{Cl}}{\overset{\|}{O=C}}}{-N} \longrightarrow (C_2H_5O)_2 \overset{\overset{\text{S}}{\|}}{P} SCH_2 N \underset{\underset{\text{Cl}}{\overset{\|}{O=C}}}{} \qquad (15)$$

Phosalone is a colorless crystalline solid with a slight garlic odor; mp 47 to 48°C; and soluble in most organic solvents except aliphatic hydrocarbons. It is readily hydrolyzed by alkali to form 6-chlorobenzoxazolone, formaldehyde, and diethyl phosphorodithioate (see Equation 17 in Chapter III).

Phosalone is a broad spectrum insecticide and acaricide useful for the control of caterpillars, aphids, and the active stages of mites on apples, pears, and field crops.[9][13] The acute oral LD_{50} for male rats is 120 mg/kg. For the metabolic degradation pathways in rats, plants, and soils, see Equation 72 in Chapter IV.B.

Azinphosmethyl, Guthion®, Gusathion®

Dimethyl *S*-(4-oxobenzotriazino-3-methyl) phosphorothiolothionate

Azinphosethyl, Ethyl guthion®, Gusathion A®

Diethyl *S*-(4-oxobenzotriazino-3-methyl) phosphorothiolothionate

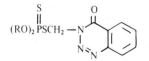

R = CH₃   azinphosmethyl

R = C₂H₅   azinphosethyl

These compounds were discovered in 1953 by Bayer AG.[13] They are synthesized by condensation of corresponding dialkyl phosphorodithioate with *N*-chloromethylbenzazimide (see Equations 60 and 61 in Chapter II).

Azinphosmethyl is a white crystalline substance, mp 73 to 74°C. It is soluble in most organic solvents. The water solubility at 25°C is 29 ppm. At elevated temperatures (above 200°C), it decomposes with gas evolution. The hydrolysis takes place under alkali and acid conditions (see Equation 16 in Chapter III). Under natural conditions, however, both homologs of azinphos have a long residual activity. Azinphosethyl forms colorless needles, mp 53°C.

Azinphosmethyl and -ethyl are nonsystemic insecticides and acaricides for use in field crops, fruits, and cottons. They have high mammalian toxicity; the acute oral LD_{50} for rats of methyl and ethyl homologs is 15 and 17.5 mg/kg, respectively.

Methidathion, Supracide®, Ustracide®
Dimethyl *S*-(2-methoxy-1,3,4-thiadiazol-5-(4*H*)-onyl-4-methyl) phosphorothiolothionate

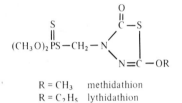

R = CH₃   methidathion
R = C₂H₅   lythidathion

Methidathion was introduced in 1963 by Geigy AG. It is synthesized from 2-methoxy-1,3,4-thiadiazol-5-(4*H*)-one, formaldehyde, and dimethyl phosphorodithioate (see Section II.C.1).[92] It is a colorless crystalline substance; mp 39 to 40°C; and readily soluble in acetone, benzene, and methanol. The solubility in water at 25°C is 240 ppm. It is stable in neutral and weakly acid media, but is readily hydrolyzed in alkali through P-S bond fission (see Equation 18 in Chapter III).

Methidathion is a nonsystemic insecticide and acaricide with a wide spectrum, particularly effective to control lepidopterous larvae, such as codling moth, and scales on fruit trees and cotton. It is inactive against eelworms and slugs. The mammalian toxicity is rather high; the acute oral LD_{50} to rats is 25 to 48 mg/kg. It is rapidly degraded in plants, soils, insects, and mammals.[659] The main pathway is the hydrolysis of the P-S ester bond followed by ring cleavage and oxidation to carbon dioxide. In addition, the hydrolysis product 4-mercaptomethylthiadiazolone is methylated, followed by oxidation to yield the methylsulfoxide and methylsulfone derivatives of the heterocycle in mammals (see Equation 89 in Section IV.B.2.d.iv).

In contrast to the high mammalian toxicity of methidathion, its 2-ethoxy homolog (lythidathion) (mp 49 to 50°C) has a much lower toxicity; the acute oral LD_{50} for rats is 268 to 443 mg/kg. It is active especially against lepidopterous and dipterous larvae and Orthoptera.

Another homolog, diethyl *S*-(2-isopropoxy-1,3,4,-thiadiazol-5-(4*H*)-onyl-4-methyl) phosphorothiolo-thionate (prothidathion), is also useful as an insecticide.

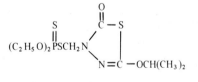

prothidathion

d. Phosphorothiolates
Although phosphorothiolates are the active form of phosphorothiolothionate insecticides and, in certain cases, of phosphorothionate esters such as demeton which readily undergo the thiono-thiolo isomerization, only a small number of this type of insecticides are known. Some of them are systemic insecticides (see

Section V.A.2.c). Many agriculturally useful phosphorothiolate esters are employed as fungicides, as will be described in Section V.D.

In general, phosphorothiolate esters are more active chemically and biochemically than their corresponding phosphate esters (see III.A.2.a and IV.A.3.b.i). The thiolate isomers of the parathion type, dialkyl aryl phosphorothionate insecticides, are often formed from the thionates during manufacturing and also in the field after the spraying of the insecticides. The phosphorothiolates are usually more toxic to mammals and less effective to insects than the original phosphorothionate insecticides (see Equation 91 in Section IV.B.3).

Acetoxon, acetofos
Diethyl S-ethoxycarbonylmethyl phosphorothiolate

$$\underset{(C_2H_5O)_2\,PSCH_2\,COC_2H_5}{\overset{O\qquad O}{\underset{\parallel\qquad\parallel}{}}}$$

Acetoxon is produced by the reaction of ethyl chloroacetate with ammonium diethyl phosphorothioate. It is a liquid; bp 120°C at 0.15 mmHg; and d_4^{20} 1.1840. It is soluble in water and most organic solvents. Acetoxon is a contact insecticide with low mammalian toxicity; the acute oral LD_{50} for various animals is 300 to 700 mg/kg. In warm-blooded animals, acetoxon is degraded by carboxyester hydrolysis.

The dimethyl phosphorothiolate homolog (methylacetofos) (bp 116°C at 0.35 mmHg; d_4^{20} 1.250) has a lower mammalian toxicity than acetoxon itself; the acute oral LD_{50} for rats is 1,000 mg/kg. It is more unstable than the ethyl homolog because of the high alkylation capability; consequently it is degraded during storage.

Cyanthoate, Tartan[®]
S-[N-(1-Cyano-1-methylethyl)carbamoylmethyl] diethyl phosphorothiolate

$$\underset{(C_2H_5O)_2P}{\overset{O}{\underset{\parallel}{}}}-SCH_2\underset{}{\overset{O}{\underset{\parallel}{C}}}-NH\underset{}{\overset{CN}{\underset{\mid}{C}}}(CH_3)_2$$

Cyanthoate is an insecticide and acaricide developed by Montecatini. It is a light yellow liquid, d_4^{20} 1.191. The solubility in water at 20°C is 7%. It is soluble in ether, ethanol, benzene, acetone, and chloroform. In spite of the existence of the carboxyamide group in the molecule as, for example, in dimethoate, the mammalian toxicity of cyanthoate is extremely high; the acute oral LD_{50} for rats is 3 to 4 mg/kg.

PTMD, Danifos[®]
S-(p-Chlorophenylthiomethyl) diethyl phosphorothiolate

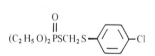

Danifos is the oxon analog of carbophenothion and has been developed as an acaricide by Kumiai Chemical Company, Japan. It is a pale yellowish liquid. It is useful for the control of spider mite, aphids, and scale on citrus. The acute oral LD_{50} for mice is 165 mg/kg.

DMCP, Fujithion[®]
S-(p-Chlorophenyl) dimethyl phosphorothiolate

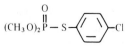

Fujithion was developed into a commercial insecticide in 1968 by Kumiai Chemical Company, Japan. The technical product is a yellowish liquid; bp 101 to 106°C at 0.006 mmHg; d_4^{20} 1.353; and n_D^{20} 1.5604. It is readily soluble in acetone, carbon tetrachloride, and toluene. The solubility in water at 21°C is 0.24%.

Fujithion is effective to control leaf-hoppers and plant-hoppers which transmit rice virus diseases. The acute oral LD_{50} for male rats is 100 mg/kg.

S-Aryl dialkyl phosphorothiolates are generally powerful anticholinesterase agents and relatively labile to alkaline hydrolysis. The I_{50} value of Fujithion for housefly head chlolinesterase is $8.6 \times 10^{-8} M$. The structure-activity relationship of S-aryl dialkyl phosphorothiolates has been reported.[198] The p-chloro derivative is more active than the p-nitro derivative as anticholinesterase agent and insecticide. However, the aryl substitution in S-aryl phosphorothiolates affects the insecticidal activity to a much smaller degree than in the series of the corresponding phosphates and phosphorothionates.[198] Thus, diethyl S-phenyl phosphorothiolate has almost the same insecticidal activity as the p-nitro derivative and about half the activity of the p-chloro derivative. For the remarkable effect of the substituent on the insecticidal activity in the series of diethyl phenyl phosphorothionates, see Table 56.

e. Phosphoramidates

Although the derivatives of phosphoramidic acid have been actively investigated since the early stages of the development of organophosphorus chemistry and organophosphorus pesticides, only a small number of compounds in this class were developed into practical insecticides, probably due to the relative difficulty in syntheses and their high mammalian toxicity. However, recent research has overcome some of the problems and disclosed a great possibility of phosphoramidates for use as pesticides, including insecticides, miticides, nematocides, anthelmintics, herbicides, and others. (Readers refer also to the corresponding Sections.)

Since the amide group imparts more or less systemic properties, several compounds of this class are useful as systemic insecticides (see V.A.2). With the exceptions of the derivatives of pyrophosphoric acid and phosphorofluoridic acid, the phosphoramidates derived from secondary amines are generally much less active as insecticides than those derived from primary amines or ammonia.

Amidothioate, Mitemate®
2-Chloro-4-methylthiophenyl methyl N-ethylphosphoramidothionate

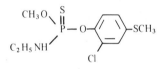

Amidothioate was introduced in 1965 by Nippon Kayaku Company, Japan, as a miticide. It is a light yellow oil with slight odor and dissolves readily in most organic solvents but hardly in water. Amidothioate is relatively persistent against alkali hydrolysis so that it can be used mixed with Bordeau mixture and has a relatively long residual activity (about 40 days) on plants. It is effective in the control of mites on citrus, apple, and pear, killing both eggs and active mites. The acute oral LD_{50} for mice is 33 mg/kg.

Although p-methylthiophenyl phosphoramidothionates having no second substituent group on the benzene ring display a high ovicidal activity, they also have a high mammalian toxicity. The introduction of a methyl group on the 3-position decreases the toxicity somewhat (to about 1/4) but not so effectively as in the phosphorothionate fenthion (1/50) (see Table 54). However, the introduction of a chlorine atom in the 2-position effectively causes a substantial decrease in toxicity.

The following alkyl phenyl phosphoramidates are promising new insecticides. Methyl 2,4,5-trichlorophenyl phosphoramidate (Dow ET-15) shows a high insecticidal activity, but has a low degree of mammalian toxicity; the oral LD_{50} for rats is 710 mg/kg.[914] The structure-activity relationship of the oxon analogs was studied in detail.[475,476] o-(Isopropoxycarbonyl) phenyl methyl phosphoramidothionate (isocarbophos, Optunal®) is effective in the control of the Colorado potato beetle and useful for cotton insect control.[373,915] It also exhibits systemic activity against aphids and leafrollers. The oral LD_{50} for rats is 50 to 100 mg/kg.

Among related alkyl phenyl phosphoramidates, phenamiphos is a nematocide (see Section V.B), Narlene® has animal systemic activity (see Section V.A.3), and DMPA is useful as a herbicide (see Section V.E).

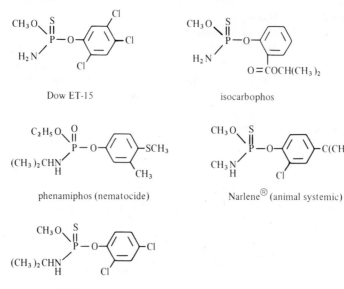

Dow ET-15

isocarbophos

phenamiphos (nematocide)

Narlene® (animal systemic)

DMPA (herbicide)

Recently, vinyl phosphoramidothionates and S-alkyl phosphoramidothiolates were developed into promising new insecticide classes, as shown by the following examples.

1-Isoproproxycarbonyl-1-propen-2-yl methyl N-ethylphosphoramidothionate

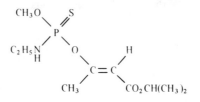

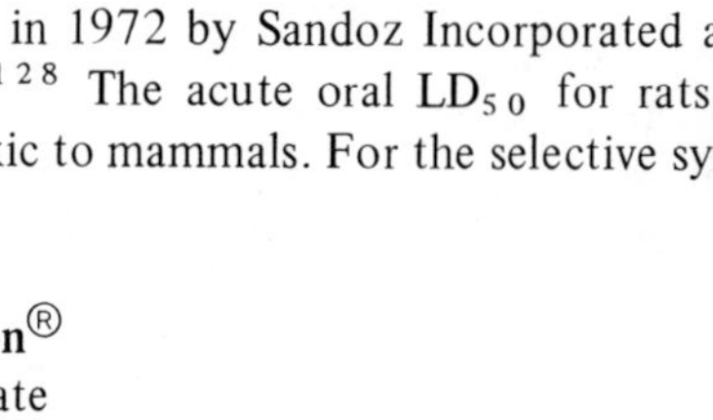

This compound was introduced in 1972 by Sandoz Incorporated as an experimental insecticide under the code number Compound 577.[128] The acute oral LD_{50} for rats of the *cis*-crotonate isomer is 122 mg/kg. The *trans*-isomer is more toxic to mammals. For the selective synthesis of the *cis*-isomer, see Section II.E.1.

Methamidophos, Monitor®, Tamaron®
O,S,-Dimethyl phosphoramidothiolate

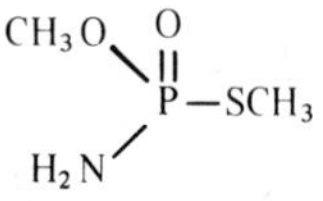

This simple but highly active insecticidal compound was first synthesized and applied for a patent in 1964 by Bayer AG and one year later, independently, by Chevron Research Company, which introduced it in 1969 as an experimental insecticide and acaricide.[916]

Methamidophos is synthesized according to the following sequence of reactions:

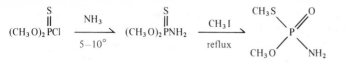

$$(CH_3O)_2PCl \xrightarrow[5-10°]{NH_3} (CH_3O)_2PNH_2 \xrightarrow[reflux]{CH_3I}$$

(16)

Methamidophos is a colorless crystalline solid, mp 44.5°C. It is readily soluble in water, alcohols, ketones, and aliphatic hydrocarbons, slightly soluble in ether, but practically insoluble in petroleum ether.

Methamidophos is a broad spectrum insecticide particularly effective to control caterpillars and aphids. It has acaricidal activity too. It has not only a good contact activity, but also a systemic action against both biting and sucking pests. The mammalian toxicity is very high; the acute oral LD_{50} for rats is about 30 mg/kg. In spite of the high insecticidal activity, methamidophos shows only a weak in vitro anticholinesterase activity; the I_{50} value for flyhead cholinesterase is $3.9 \times 10^{-5}M$.[121] By chemical or biochemical oxidation, methamidophos is converted into a potent inhibitor of cholinesterase.[917]

The structure-activity relationship of O,S-dialkyl phosphoramidothiolates was reported by Quistad et al.[121] The higher alkyl (ethyl, n-propyl) ester homologs also show high insecticidal activity. O-Ethyl S-methyl phosphoramidothiolate (Bay 65258) is more effective than the methyl homolog against some insects like tobacco budworms.[918] N-Alkylation causes a great decrease in the insecticidal activity, whereas N-acetylation does not cause any distinct change in the insecticidal activity, but reduces greatly the mammalian toxicity. The N-acetyl derivative of methamidophos is named acephate and is a very promising systemic insecticide (see Section V.A.2.a). The thionothiolate analog of methamidophos is also much less toxic to mammals than the thiolate; the oral LD_{50} for rats is 250 to 500 mg/kg. The thiono isomer of methamidophos is almost inactive as an insecticide.

f. Pyrophosphates

TEPP, tepp, Tetron®, Bladan®
Tetraethyl pyrophosphate

$$\begin{matrix} O & & O \\ \parallel & & \parallel \\ (C_2H_5O)_2P & -O- & P(OC_2H_5) \end{matrix}$$

Tetraethyl pyrophosphate is the first commercial organophosphorus insecticide. It was marketed in 1943 in Germany by Bayer AG under the name "Bladan." Since the first synthesis of TEPP by Moschnine, about 100 years had passed till the discovery of its insecticidal activity in 1938 by Schrader. He originally thought that hexaethyl tetrapyrophosphate was the active principle of Bladan. Many methods for the synthesis of TEPP are known.[13] The best method is probably the one developed in 1948 by Toy, who obtained pure TEPP by controlled hydrolysis of diethyl phosphorochloridate in the presence of base (see Section II.G).[147]

TEPP is a colorless, odorless, and hygroscopic liquid; bp 124°C at 1 mmHg; d_4^{20} 1.185; and n_D^{20} 1.4196. It is miscible with water and most organic solvents except petroleum oils, in which TEPP is practically insoluble. TEPP decomposes so rapidly in the presence of moisture that it vanishes within 48 hr after application, forming completely harmless compounds.

TEPP is a contact insecticide and acaricide particularly effective to control aphids and mites in active stages. The mammalian toxicity is extremely high; the oral LD_{50} for rats is 1.2 to 2.0 mg/kg. It is absorbed quickly into the skin and the vapor is also highly toxic. Because of this high toxicity, the use of TEPP is very restricted, and it is now only historically important as an insecticide.

The insecticidal activity of homologs decreases in the following order: ethyl > methyl > propyl.

Sulfotep, sulfotepp, Bladafum®
O,O,O,O-Tetraethyl dithiopyrophosphate

$$\begin{matrix} S & & S \\ \parallel & & \parallel \\ (C_2H_5O)_2P & -O- & P(OC_2H_5)_2 \end{matrix}$$

The insecticidal activity of sulfotep was discovered in 1944 by Schrader. It is synthesized by the reaction of sulfur on TEPP or by the action of water on diethyl phosphorochloridothionate in the presence of pyridine (see Equation 145 in Section II.G).[919] The latter method is widely used for the synthesis of all homologous dithiopyrophosphate esters except the methyl ester.

$$2\,(RO)_2\overset{\overset{\displaystyle S}{\|}}{P}Cl + H_2O + 2C_5H_5N \longrightarrow (RO)_2\overset{\overset{\displaystyle S}{\|}}{P}O\overset{\overset{\displaystyle S}{\|}}{P}(OR)_2 + 2C_5H_5N\cdot HCl \tag{17}$$

Sulfotep is a light yellow liquid; bp 136 to 139°C at 2 mmHg; d^{25} 1.196; and n_D^{25} 1.4758. The vapor pressure at 20°C is 1.7×10^{-4} mmHg. It is soluble in most organic solvents, but only slightly in water. Sulfotep is very stable against hydrolysis.

Sulfotep is an insecticide and acaricide with high contact and fumigant activity. It is also effective to mollusks. The mammalian toxicity is very high, even though it is lower than that of TEPP; the acute oral LD_{50} for rats is 5 mg/kg. The corresponding monothiono analog is more toxic than TEPP (LD_{50} rats = 0.5 mg/kg).

Propyl thiopyrophosphate, Aspon®, NPD®
O,O,O,O-Tetra-*n*-propyl dithiopyrophosphate

$$(C_3H_7O)_2\overset{\overset{\displaystyle S}{\|}}{P}-O-\overset{\overset{\displaystyle S}{\|}}{P}(OC_3H_7)_2$$

Propyl thiopyrophosphate was synthesized in 1950 by Schrader and introduced in 1963 by Stauffer Chemical Company as an insecticide. It is synthesized from dipropyl phosphorochloridothionate by the action of water in the presence of pyridine.[919] It is an amber liquid; bp 104°C at 0.01 mmHg; d_{25}^{25} 1.121; and n_D^{25} 1.4712. It is practically insoluble in water (0.16%) but miscible with most organic solvents except petroleum oils. Propyl thiopyrophosphate is relatively stable against hydrolytic action.

Propyl thiopyrophosphate is an insecticide with contact and fumigant actions, particularly effective in controlling aphids and chinch bugs. The mammalian toxicity is very low; the acute oral LD_{50} for rats is 900 to 1,700 mg/kg. The tetraisopropyl isomer has much higher mammalian toxicity.

g. Phosphonates and Phosphinates

Although only two phosphonate insecticides were commercialized in the 1950's, many insecticides of this class have been developed into commercial or semi-commercial use in this decade.

Phosphonate esters have some characteristic properties in comparison with corresponding phosphate esters. Phosphonates are more reactive (see Section III.A.2.e) and more biologically active than phosphate esters. The variation of the alkyl ester group has less effect on biological activities in alkylphosphonate esters than in phosphate esters; for example, *n*-butyl *p*-nitrophenyl ethylphosphonate has almost the same insecticidal activity as the methyl ester homolog. Therefore, some alkyl groups higher than ethyl are also involved in phosphonate insecticides. Moreover, phenylphosphonates as well as alkylphosphonates are active. Thus, a greater variety of structure for possible insecticides is expected in the phosphonate class than in the phosphate class.

The effect of the methyl ester group of decreasing the mammalian toxicity, which is common in dialkyl phenyl phosphorothionates, is not observed in alkyl phenyl alkyl- and phenyl-phosphonothionates (see Tables 31, 58, and 59); phosphonothionate esters having an *O*-methyl group are biologically more active toward both insects and mammals than those having an *O*-ethyl ester group.[614] This may be due to the fact that demethylation contributes little to the in vivo degradation of methyl phosphonothionate esters in contrast with its high contribution to the degradation of dimethyl phenyl phosphorothionate esters.[616]

Introduction of the second substituent group to the *p*-substituted phenyl ester ring of alkyl phenyl phenylphosphonothionates does not cause such a remarkable selective toxicity as is seen in dimethyl substituted phenyl phosphorothionates, but depresses the toxicity towards both insects and mammals

TABLE 58

Biological Activities of Some Alkyl Phenyl Phenylphosphonothionates

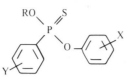

R	X	Y	Oral LD$_{50}$ mouse mg/kg	LC$_{50}$ Callosobruchus chinensis dipping method (dilution)
C$_2$H$_5$	4-NO$_2$	–	EPN	20	X 40,000
	3-NO$_2$	–		46	X 1,300
	2-NO$_2$	–		7	X 8,000
CH$_3$	4-NO$_2$	–		7	X 74,000
n-C$_3$H$_7$	4-NO$_2$	–		300	X 1,300
C$_2$H$_5$	4-NO$_2$-2-Cl	–		15	X 24,000
	4-NO$_2$-3-Cl	–		66	X 13,500
	4-NO$_2$-2-CH$_3$	–		30	X 17,000
	4-NO$_2$-3-CH$_3$	–		30	X 4,000
	4-CN	–	CYP	50	X 27,000
CH$_3$	4-CN	–		9	X 52,000
n-C$_3$H$_7$	4-CN	–		300	X 1,300
C$_2$H$_5$	4-CN-3-Cl	–		30	X 17,000
	4-CN-3-CH$_3$	–		21	X 1,500
	4-CN	3-Cl		67	X 4,400
	4-CN	3-CH$_3$		142	X 1,400

Table 58.[614] Introduction of any substituent on the phenyl group attached directly to phosphorus in the EPN-type phenylphosphonate esters causes a reduction in insecticidal activity. The structure-activity relationship of some EPN-type compounds and parathion related compounds is shown in Tables 58 and 59.[494,614,862]

Only one compound of phosphinate esters is practically used as an insecticide. Phosphinate esters generally have a higher susceptibility to alkaline hydrolysis than the corresponding phosphonate esters.[151] Many phosphinothionate esters have insecticidal activity, but are less potent, in general, than corresponding phosphorothionate or phosphonothionate esters (see Table 59). The mammalian toxicity of phosphinates is variable by the structure; the diethylphosphinothionate analog of parathion is more toxic than parathion, whereas the dimethyl analog has a much lower toxicity (oral LD$_{50}$ rats = 100 mg/kg). It is amazing to note that the dimethylphosphinate analog of azinphosmethyl does not produce any symptoms of poisoning to rats orally administered a dose of 1000 mg/kg, in contrast with the high toxicity of azinphosmethyl (LD$_{50}$ rats = 15 mg/kg).[151]

Only a few reviews on phosphonate insecticides have been published. Schrader's monograph includes many possible insecticides of the phosphonate and phosphinate classes.[13] Fukuto has reviewed the structure-activity relationship of phosphonate esters.[151] The metabolism of phosphonate insecticides was reviewed recently by Menn.[920]

Trichlorfon, trichlorphon, metrifonate, chlorophos, Dipterex®, Neguvon®, Tugon®
Dimethyl 1-hydroxy-2,2,2-trichloroethylphosphonate

$$(CH_3O)_2 \overset{\overset{\displaystyle O}{\|}}{P} - \underset{\underset{\displaystyle OH}{|}}{CH} - CCl_3$$

TABLE 59

Biological Activities of Phosphonate and Phosphinate Analogs of Parathion and Fenitrothion

Structure	Name	LD_{50} rat oral mg/kg	LD_{50} mouse oral mg/kg	LD_{50} housefly topical μg/g
$(C_2H_5O)_2\overset{S}{P}O$—⟨ ⟩—$NO_2$	Parathion	6.8		0.93
C_2H_5O\ $\overset{S}{P}O$—⟨ ⟩—NO_2 / C_2H_5		2.5		0.93
$(C_2H_5)_2\overset{S}{P}O$—⟨ ⟩—$NO_2$		5		
$(CH_3O)_2\overset{S}{P}O$—⟨ ⟩—NO_2	Parathion-methyl	14	23	1.2
CH_3O\ $\overset{S}{P}O$—⟨ ⟩—NO_2 / CH_3	Methyl paraphonothion	1	3.7	0.73
$(CH_3)_2\overset{S}{P}O$—⟨ ⟩—NO_2		100	51	6.4
$(CH_3O)_2\overset{S}{P}O$—⟨ ⟩(CH_3)NO_2	Fenitrothion	250	1250	3.1
CH_3O\ $\overset{S}{P}O$—⟨ ⟩(CH_3)NO_2 / CH_3	Sumiphonothion		14	1.9
$(CH_3)_2\overset{S}{P}O$—⟨ ⟩(CH_3)NO_2			370	6.3

Trichlorfon was first prepared in 1952 by Lorenz and then independently by Barthel in 1954 by the reaction of dimethyl phosphite with chloral (see Equation 183 in Chapter II).[174] Another method for the synthesis is shown in Equation 184 in Chapter II. Trichlorfon is a white crystalline powder; mp 83 to 84°C; and bp 100°C at 0.1 mmHg. It is soluble in water (15.4 g in 100 ml), ether, chloroform, and benzene, but insoluble in petroleum oils. Two molecules of trichlorfon are associated together.[68] Trichlorfon is relatively stable in acid, but is readily converted by alkali to dichlorvos (see Equation 19 in Chapter II) and hydrolytic products such as dimethyl hydrogen phosphate and dichloroacetaldehyde. By acid hydrolysis, demethylation takes place.

Trichlorfon has a high insecticidal activity, particularly against *Diptera*. It is useful for the control of both sucking and chewing insects on field crops, vegetables, and seed crops. Trichlorfon is also useful for the control of insects of public health significance and animal ectoparasites because of its low toxicity. The acute oral LD_{50} for male rats is 630 mg/kg.

Trichlorfon itself is a poor inhibitor of acetylcholinesterase, but is readily converted into the strong inhibitor dichlorvos at physiological pH conditions (see Sections IV.A.3.b.iv and IV. B.1.f).[482,569] The transformation proceeds nonenzymatically, and at room temperature it takes only 1 and 7.5 hours for 50% conversion at pH 8 and 7, respectively. Thus, in the sense of the mode of action, trichlorfon is regarded as a phosphate ester, although no evidence has been obtained for the in vivo conversion of trichlorfon into dichlorvos. Demethylation is an important degradation route of trichlorfon in mammals.[921-923] In a dog, however, 63% of the trichlorfon administered intravenously was excreted in urine in the form of the glucuronide of trichloroethanol, indicating phosphorus-carbon bond cleavage (see Section IV.B.2.b.iii; Equation 74).[481] The main metabolic pathways of trichlorfon are shown below.

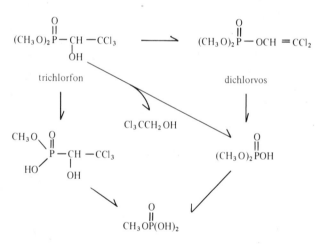

(18)

Butonate, tribuphon
Dimethyl 1-*n*-butyryloxy-2,2,2-trichloroethylphosphonate

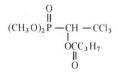

Butonate was discovered by Arthur and Casida in 1958 in the course of a study on the effect of acylation of the hydroxyl group of trichlorfon on selective toxicity. Their success gave a clue to make selective insecticides by introducing a biochemically removable protective group to a highly active compound (see Section IV.C.2). For the synthesis, see Equations 178 and 188 in Section II.H.3. Butonate is a liquid with bp 112 to 114°C at 0.03 mmHg: d_4^{20} 1.3998; and n_D^{20} 1.4740. It is moderately soluble in water and highly soluble in many organic solvents.

Butonate is useful for the control of cockroaches, ants, flies, and ectoparasites of domestic animals because of its low mammalian toxicity; the oral LD_{50} for rats is about 1100 mg/kg. It is converted enzymatically into trichlorfon by the action of esterases in insects and plants.[924] Degradation to dimethyl phosphoric acid and to the desmethyl butonate predominates in mammals. The high selective toxicity of butonate is attributed to the difference in metabolic pathways between insect pests and mammals.

EPN
Ethyl *p*-nitrophenyl phenylphosphonothionate

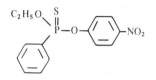

EPN was the first commercialized phosphonate insecticide and was introduced in 1949 by Du Pont. For the synthesis, see Equation 154 in Section II.H.1. EPN is a light yellow crystalline solid; mp. 36°C; d^{25} 1.27; and n_D^{30} 1.5978. It is soluble in many organic solvents but practically insoluble in water. EPN is relatively stable under neutral and acidic conditions, but is hydrolyzed in alkali solutions into phenylphosphonothioic acid, ethanol, and nitrophenol.

EPN is an insecticide and acaricide useful to control many phytophagus mites and insect species, including European corn borer, orange tortrix, rice stem borer, and boll weevil. It has rather high mammalian toxicity, being hazardous by skin contact and inhalation; the acute oral LD_{50} for male and female rats is 40 and 12 mg/kg, respectively.

EPN potentiates the toxicity of the insecticide malathion, as described in Section IV.D.2. EPN also shows delayed neurotoxicity (see Section IV. E).

Cyanofenphos, CYP, Surecide[R]
p-Cyanophenyl ethyl phenylphosphonothionate

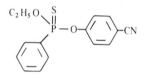

Cyanofenphos was introduced in 1962 and commercialized in 1966 by Sumitomo Chemical Company, Japan.[879] The active ingredient is synthesized by the reaction of *p*-cyanophenol with ethyl phenylphosphonochloridothionate. It is a white crystalline powder; mp 83°C; and moderately soluble in ketones and aromatic hydrocarbons.

Cyanofenphos is an insecticide effective to control rice stem borer, cotton boll worm, cabbage worm, and aphids. The acute oral LD_{50} for rats is about 80 mg/kg.

The methyl ester homolog is more toxic to both mammals and insects.[614] The introduction of a second substituent to the cyanophenyl ring is not effective in decreasing mammalian toxicity.

EPBP, S-Seven[R]
2,4-Dichlorophenyl ethyl phenylphosphonothionate

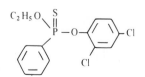

This compound was almost inactive against Azuki-bean weevils (*Collosobruchus chinensis*) employed during the first screening by Sumitomo Chemical Company.[614] However, high activity against soil inhabiting insects such as onion maggot was found by Nissan Chemical Company, Japan, who developed it as a soil insecticide in 1964.[925] It is a liquid, n_D^{31} 1.5970.

The mammalian toxicity is lower in comparison with other EPN type insecticides, alkyl (substituted phenyl) phenylphosphonothionates; the acute oral LD_{50} for mice is 274 mg/kg.

The methyl ester homolog, like other EPN type compounds, is more toxic (LD_{50} mice = 75 mg/kg) than the ethyl homolog (see Tables 31 and 58). In the series of ethyl chlorinated phenyl phenylphosphonothionates, the insecticidal activity increases as the number of chlorine atoms introduced is increased.[614]

Leptophos, Phosvel®, VCS-506
4-Bromo-2,5-dichlorophenyl methyl phenylphosphonothionate

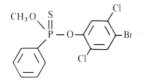

Leptophos was introduced in 1965 as an experimental insecticide by Velsicol Chemical Corporation. It is a white solid; mp 70.2°C; and soluble in aromatic hydrocarbons and ketones, but practically insoluble in water (2.4 ppm). It is stable to acid, but is hydrolyzed slowly under strongly alkaline conditions.

Leptophos is particularly useful to control lepidopteran, such as tobacco budworms, cotton leafworms, and rice stem borers. It has some fungicidal effect, too. Actually, a closely related compound, methyl 2,4,5-trichlorophenyl phenylphosphonate (H 6034), is a fungicide.[863] Leptophos is a relatively stable insecticide and persists in plants for a long time. For example, leptophos remained in tomato plants without degradation for a period of 28 days after treatment.[926] In mammals, however, leptophos is rapidly metabolized and excreted, principally in the urine, as degradation products: 4-bromo-2,5-dichlorophenol, O-methyl phenylphosphonothioate, methyl phenylphosphonate, and phenylphosphonic acid.[926a]

The oral LD_{50} for rats is 90 mg/kg. Neither the mutagenic nor teratogenic potential was found in experimental animals fed with leptophos.

Colep®
p-Nitrophenyl phenyl methylphosphonothionate

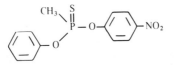

Colep was developed in 1962 by Monsanto Company as an insecticide, but it is not marketed. The metabolic degradation in both animals and plants occurs through the cleavage of the phenyl and *p*-nitrophenyl ester linkages. The liberated phenols are conjugated in plants to the β- or α-glucosides, and in rats to glucuronides.[927]

Trichloronate, Agritox®, Phytosol®, Agrisil®
Ethyl 2,4,5-trichlorophenyl ethylphosphonothionate

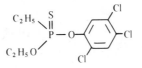

Trichloronate was developed in 1960 by Bayer AG as a soil insecticide. It is an amber liquid; bp 108°C at 0.01 mmHg; d_4^{20} 1.365; n_D^{22} 1.561; and soluble in most organic solvents, but practically insoluble in water.

Trichloronate is a nonsystemic insecticide with a long residual activity, being particularly effective to control soil pests like root maggots and wireworms. The acute oral LD_{50} for rats is 50 mg/kg.

Some selected alkylphosphonothionates, which have been introduced as experimental insecticides, are listed as follows:

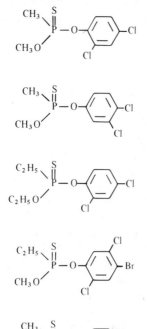

BAY 30911
systemic and contact insecticide
with nematocidal activity
LD_{50} rats = 140 mg/kg

BAY 80833
soil insecticide

Stauffer N-3,054
insecticide
LD_{50} rats = 75 mg/kg

CELA K-41
insecticide and acaricide,
7 times more active than the trichloro
analog; LD_{40} rats = 80 mg/kg

Stauffer B-10341
effective to resistant strain
houseflies

Fonofos, Dyfonate®
Ethyl S-phenyl ethylphosphonothiolothionate

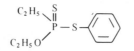

Fonofos was introduced in 1967 by Stauffer Chemical as a soil insecticide. It is synthesized by the reaction of ethyl ethylphosphonochloridate with thiophenol (Equation 160 in Section II.E.2).[161] Fonofos is a light yellow liquid with a mercaptan-like odor; bp 130°C at 0.1 mmHg; n_D^{20} 1.5883; and d_{25}^{25} 1.16. It is practically insoluble in water (13 ppm at 25°C), but is miscible with ketones, kerosene, and xylene. Fonofos has a high degree of hydrolytic stability, persisting in soil to control soil insects like corn rootworms, wireworms, cutworms, and maggots.

The oral LD_{50} for rats is 16.5 mg/kg. The metabolism of fonofos in the rat and the potato plant has been thoroughly investigated (see Equation 90 in Section IV.B).[381,660] The main route of degradation is the cleavage of the P-S bonding.

Both of the alkyl groups of fonofos can be changed to a certain extent without a great decrease in insecticidal activity as shown in Table 60. For example, the methyl ester homolog (Stauffer N-3794) and the methyl methylphosphonothiolothionate (Stauffer N-3727) have much lower mammalian toxicity than

TABLE 60

Structure and Biological Activity of Alkyl S-Phenyl Alkylphosphonothiolothionates and Related Compounds[161],[928]

Compound		LD_{50}	
		Rat (mg/kg)	Housefly ($\mu g/25$ ♀)
C_2H_5, C_2H_5O — P(=S) — S — C$_6$H$_5$	Fonofos	16	4.3
C_2H_5, C_2H_5O — P(=O) — S — C$_6$H$_5$		2.7	4.8
CH_3, CH_3O — P(=S) — S — C$_6$H$_5$	N-3727	141	7.5
C_2H_5, CH_3O — P(=S) — S — C$_6$H$_5$	N-3794	90	5.8
$n\text{-}C_3H_7$, C_2H_5O — P(=S) — S — C$_6$H$_5$		55	8.8
C_2H_5, C_2H_5O — P(=S) — S — C$_6$H$_4$ — Cl	N-2596	6	5.0
C_2H_5, C_2H_5O — P(=S) — S — C$_6$H$_4$ — CH$_3$	BAY 38156	123	4.3
$ClCH_2$, C_2H_5O — P(=S) — S — C$_6$H$_4$ — CH$_3$		336	5.4
$ClCH_2$, $i\text{-}C_3H_7O$ — P(=S) — S — C$_6$H$_4$ — CH$_3$	B-10119	204	5.7
$ClCH_2$, $i\text{-}C_3H_7O$ — P(=S) — S — C$_6$H$_4$ — Cl	B-10046	26	5.5
$(C_2H_5O)_2$ P(=S) — S — C$_6$H$_5$		233	56
$(C_2H_5O)_2$ P(=S) — S — C$_6$H$_4$ — CH$_3$		316	330

fonofos has, but high insecticidal activity. Even higher alkyl derivatives such as the n-propylphosphono-thiolothionate homolog have a similar insecticidal activity. Introduction of a substituent on the S-phenyl group causes a remarkable change in mammalian toxicity. A chlorine atom on the p-position increases the mammalian toxicity about three times (N-2596), whereas a methyl group causes an eight-fold decrease in the toxicity without any depression in the insecticidal activity (BAY 38,156). The corresponding derivatives of phosphoric acid are much less toxic towards both insects and mammals than the phosphonate derivatives. For example, diethyl S-phenyl phosphorothiolothionate has an oral LD_{50} for rats of 233 mg/kg and is about 10 times less toxic to houseflies than fonofos.

Alkyl S-aryl chloromethylphosphonothiolothionates appear to be more favorable than ethylphosphono-thiolothionate esters in mammalian toxicity and insecticidal activity (see Table 60).[928] Introduction of a methyl group or a chlorine atom in the para position of the S-phenyl moiety enhances this favorable biological activity. The variation of the O-alkyl group reaches the optimum activity with iso-propyl for both p-methyl and p-chloro derivatives.[928] For example, i-propyl S-(p-chlorophenyl) chloromethyl-phosphonothiolothionate (B-10046) is 30-fold as active as parathion against the salt-marsh caterpillar.

Isobutyl S-(phthalimidomethyl) ethylphosphonothiolothionate

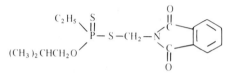

This compound was introduced in 1967 by Stauffer Chemical Company as an experimental foliar insecticide-acaricide under the code number N-4543. It is synthesized by the reaction of N-bromomethyl phthalimide with sodium O-isobutyl ethylphosphonodithioate (see Equation 161 in Section II.E.2).[160] N-4543 is a white crystalline powder; mp 58 to 60°C; bp 127°C at 0.001 mmHg; and sparingly soluble in kerosene, but soluble in acetone and xylene. The solubility in water at room temperature is less than 15 ppm.

N-4543 is a nonsystemic insecticide and acaricide promising for the control of pests on deciduous and citrus fruit crops and on some field and vegetable crops. The acute oral LD_{50} for male rats is 75 mg/kg.

Alkyl S-(phthalimidomethyl) alkylphosphonothiolothionates surpass the closely related phosphoro-thiolothionate insecticide phosmet in insecticidal activity. The variation of the O-alkyl group little affects the insecticidal activity but greatly the mammalian toxicity.[160] The lowest toxicity in this series was achieved by the isobutyl ester.

The metabolic fate of N-4543 is similar to that of phosmet. When rats are administered N-4543, the phthaloyl portion formed by hydrolytic cleavage is excreted in urine as phthalamic acid and phthalic acid. The phosphorus moiety is excreted mainly as isobutyl ethylphosphonic acid.[920]

Agvitor
2,4,5-Trichlorophenyl diethylphosphinothionate

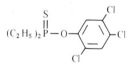

Agvitor is the phosphinate analog of the phosphorothionate animal system insecticide ronnel (see V.A.3) and the phosphonothionate insecticide trichloronate. It was filed for a patent in 1958 by Bayer AG.[929] It is a liquid and practically insoluble in water but highly soluble in organic solvents. Agvitor is effective to control carrot and onion flies by application to soils. The oral LD_{50} for rats is 100 mg/kg. The toxicity is somewhat less than the phosphonate analog trichloronate.

The phosphinate analogs of fenthion (XVII) and azinphosmethyl (XVIII) seem to be interesting compounds as insecticides.

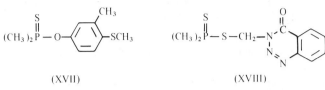

(XVII)

LD_{50} rat oral 175 mg/kg

(XVIII)

LD_{50} rat oral 1000 mg/kg

2. Plant Systemic Insecticides

Systemic insecticides are absorbed by the plant and translocated and stored for a limited time in transport tissues of the plant in quantities sufficient to make the plant, or at least the site of translocation, insecticidal.[930] Many organophosphorus insecticides can penetrate more or less into plant tissue but are not translocated and stored in insecticidal quantities; these may be called penetrating insecticides and include, for example, parathion, malathion, diazinon, and azinphosmethyl.

Systemic insecticides can be applied to treat seeds by soaking or coating to protect the young seedling, foliage or bark by spray, trunks by implanting, and the soil for root uptake. Systemic insecticides have some advantages compared with contact insecticides: they are highly selective ecologically by having little or no effect on the pest predators; they are very effective in protecting new plant growth formed subsequent to application and in killing insects living inside the plant, particularly trees, which cannot be controlled by spray of ordinary contact insecticides; and they are more persistent because the absorbed insecticides are not exposed to weathering. Systemic insecticides should have appropriate stability for a reasonable time, 3 to 5 weeks, in the plant to exert residual action, but should be decomposed into nontoxic compounds before harvest if they were applied to food crops.

Some systemic insecticides are not absorbed through leaves or stem, but through roots, with the transpiration stream operating from the roots. As the root system does not contain a cuticle, it absorbs foreign compounds more readily than the leaves and the stem. Translocation from roots occurs through the route of the plant nutrients in the xylem tissue, and lateral diffusion is limited. The stomata on the leaf and lenticels on the stem serve for entry of pesticides. Cuticle covering the outer surface of leaf and stem contains a high proportion of waxes and resists the entry, but lipid soluble insecticides may enter directly through the cuticle. Translocation from leaves occurs along with a slow downward assimilate stream in the phloem, and a faster upward transpiration stream in the xylem. Entry of systemic insecticides into the seed may proceed through the micropyle and the hilum.

C. E. Crisp proposed to classify systemic insecticides into two categories: symplastic and apoplastic.[931] The symplastic systemic insecticides pass through the plasmalemma into the living symplast and move along with the assimilate stream in the phloem tissue. The apoplastic systemic insecticides are transported almost exclusively by the transpiration stream within the apoplast continuum, which consists of nonliving cell wall and constitutes with the xylem tissue a continuous permeable system through which water and solutes may move freely. Since the volume of phloem is smaller than that of xylem tissue, the dilution effect is less in symplastic systemic insecticides than in apoplastic insecticides. Most organophosphorus systemic insecticides are apoplastic.

In order to enable the insecticide to move in the plant sap, sufficient water solubility is a function which the systemic insecticide should have, at least after biotransformation in plants. For penetration into the plant by foliar application, a proper lipophilic-hydrophilic balance is critical; too lipophilic a molecule may remain in the cuticular waxes and too hydrophilic a molecule may not penetrate into the cytoplasm.

For example, in the series of O,S-dimethyl N-n-alkylphosphoramidates, i.e., N-alkyl derivatives of methamidophos (Monitor), a parabolic relationship was found between systemic movement and the logarithmic octanol-water partition coefficient (log P) or relative hydrophobic constant π (for the definition of π, see Section IV.A.3.a).[931a] This means that there is an optimum lipophilic-hydrophilic balance for the systemic movement. The optimum π was computed as 1.19 in this series. Actually, the N-n-propyl derivative which has π of 1.31, nearest the optimum value, displayed the highest systemic movement in the cotton leaf petiole in this series. However, this is not the case in the series of the branched *N*-alkyl derivatives, in which steric effects reduce the systemic property.

Some systemic insecticides exert their insecticidal action in the form in which they entered the plant as, for example, mevinphos and some others in the biotransformed form as schradan and disulfoton (endometatoxic systemic insecticides). Many systemic insecticides have high mammalian toxicity and are effective only against sucking insects and mites, but not against chewing insects. However, some recently developed systemic insecticides are designed to overcome these disadvantages. Organophosphorus systemic insecticides have one of the following functional groups, with a few exceptions: phosphoramido, carboxylamido, amino, sulfide or its oxidized forms, and nitrogen containing hetero ring.

a. Phosphoramides

The systemic insecticides in the early time of insecticide development were phosphorodiamidic anhydrides or fluorides, as shown by the general formula XIX.

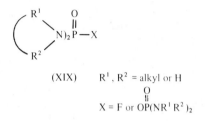

$$(XIX) \qquad R^1, R^2 = \text{alkyl or H}$$
$$X = F \text{ or } OP(NR^1R^2)_2$$

Schradan, dimefox, and mipafox belong to this class. They are soluble in water and polar organic solvents. They may be used for the control of sucking plant pests like aphids and mites, but are very toxic to mammals.

Recently, *N*-acylated phosphoramidate esters, shown by the general formula XX, were actively investigated with success in developing some new systemic insecticides with low mammalian toxicity, exemplified by acephate, avenin, and demuphos. Another new group of related systemic insecticides are the *N*-phosphoryl cyclic imidocarbonate derivatives (XXI) like phosfolan.

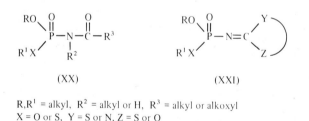

$$R,R^1 = \text{alkyl}, \quad R^2 = \text{alkyl or H}, \quad R^3 = \text{alkyl or alkoxyl}$$
$$X = O \text{ or } S, \quad Y = S \text{ or } N, \quad Z = S \text{ or } O$$

Schradan, OMPA, Systam®

N,N,N′,N′-Tetramethylphosphorodiamidic anhydride, octamethyl pyrophosphoramide

$$[(CH_3)_2N]_2P\overset{\text{O}}{\overset{\|}{-}}O-\overset{\text{O}}{\overset{\|}{P}}[N(CH_3)_2]_2$$

In 1941, Schrader discovered the systemic insecticidal activity of octamethyl pyrophosphoramide (OMPA), which was later named schradan after the discoverer. For the synthesis, see Equation 146 in Section II.G. Schradan is a colorless viscous liquid; bp 118 to 122°C at 0.3 mmHg; mp 14 to 20°C; d^{25} 1.1343; n_D^{25} 1.4612; and vp 1×10^{-3} mmHg at 25°C. It is miscible with water and most organic solvents. Schradan is hydrolyzed under acid conditions, but is stable in water and alkali. The crude product is a dark brown liquid and contains polyphosphoric acid derivatives such as triphosphoramide (25 to 50%) (XXII) and XXIII.

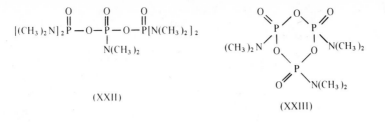

(XXII)   (XXIII)

The acute oral LD_{50} to male and female rats is 13.5 and 35.5 mg/kg, respectively. Schradan itself has no anticholinesterase activity in vitro, but is biotransformed in vivo into a strong inhibitor of cholinesterases. For the metabolism, see Equation 43 in Section IV.B.1.d. Schradan is used for the control of aphids and mites on citrus, apple, and hop.

Dimefox, Hanane®, Terra-systam®
N,N,N′,N′-Tetramethylphosphorodiamidic fluoride

$$[(CH_3)_2N]_2 \overset{\displaystyle O}{\overset{\|}{P}} F$$

Dimefox was introduced in 1949 by Fisons Pest Control Ltd. as a systemic insecticide. It is a colorless liquid; bp 80°C at 10 mmHg; d_4^{20} 1.1151; and n_D^{20} 1.4267. It is highly soluble in water and polar organic solvents. The partition coefficient between chloroform and water is about 15. Dimefox is stable in aqueous alkaline solutions, but is hydrolyzed in acid.

As dimefox is extremely toxic to mammals (LD_{50} rat = 1 to 5 mg/kg), it may be formulated in gelatine ampoules for safe application in soils. Dimefox is a poor inhibitor of cholinesterases but is metabolically activated.[473] For the synthesis, see Section II.F.

Mipafox, Isopestox®
N,N′-Diisopropylphosphorodiamidic fluoride

$$[(CH_3)_2C\overset{H}{N}]_2 \overset{\displaystyle O}{\overset{\|}{P}} F$$

Mipafox was introduced in 1950 by Fisons Pest Control Ltd. It is a white crystalline substance; mp 61 to 62°C; bp 125°C at 2 mmHg; and vp 0.0025 mmHg at 25°C. The solubility in water is about 8%. It is soluble in most organic solvents, except paraffinic hydrocarbons. Mipafox is not used in practice now because of its delayed neurotoxicity (see Section IV.E.1).

Avenin
Dimethyl *N*-(isopropoxycarbonyl)phosphoramidate

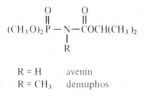

R = H   avenin
R = CH₃   demuphos

Russian scientists found the systemic insecticidal activity of phosphoryl urethanes in 1966.[709] These compounds have an interesting structure in which phosphoramidate moiety and carbamate moiety hold a nitrogen atom in common. For synthesis, see Section II.E.3. Avenin is an undistillable oil which is slightly soluble in water and more soluble in most organic solvents.

TABLE 61

Insecticidal Activity of Some Phosphoryl Urethanes

| | | LD_{50} (μg/insect) | | |
| | | Culex pipens | | |
Compound	Musca domestica	Male	Female	Bothynoderes punctiventris
$(CH_3O)_2PNHCO_2CH(CH_3)_2$ (with O double bond on P)	8.05	0.137	2.55	0.125
$(CH_3O)_2PNCO_2CH(CH_3)_2$ (with O double bond on P, CH_3 below)	2.21	0.12	0.51	0.069
CH_3O/CH_3S—PNHCO_2CH(CH_3)_2 (with O double bond on P)	1.8	0.007	0.22	>150

Avenin is almost nontoxic to mammals; the oral LD_{50} values for various experimental animals are more than 5000 mg/kg. It is highly selective to the sugar beet weevil, which can be controlled for 10 to 12 days by treatment of sugar beet seeds before planting. Bean aphids are also highly sensitive to this insecticide.

The N-methyl homolog of avenin is named demuphos and was introduced in 1968 and is more effective and persistent than the parent compound.[932] It can control sugar beet weevils for 20 to 25 days. The insecticidal activities of some phosphoryl urethanes are shown in Table 61.[710]

Acephate, Orthene®, Ortran®
O,S-Dimethyl N-acetylphosphoramidothiolate

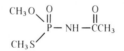

Acephate is the N-acetyl derivative of the insecticide methamidophos (Monitor®) (see p. 262) and was introduced by Chevron Chemical Company in 1971. N-Acetylation is accomplished in an ordinary manner by acetyl anhydride in the presence of a Lewis acid such as $ZnCl_2$, $FeCl_3$, or BF_3-etherate.

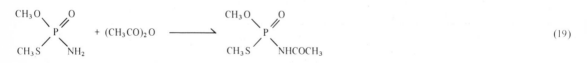

(19)

Acephate is a white crystalline powder, mp 91 to 92°C. It is highly soluble in water to about 65%.

Although methamidophos itself is highly toxic to mammals (LD_{50} mice = 27 mg/kg), the acetylation causes a dramatic decrease in the mammalian toxicity; the oral LD_{50} of acephate for mice and rats is 361 and 945 mg/kg, respectively. Acephate is a systemic insecticide with a moderate persistence and is effective not only to sucking pests, but also to chewing insects. Acephate is a poorer anticholinesterase than methamidophos, which itself is not a very strong inhibitor. Both compounds must be activated metabolically in plant or animal.

C. E. Crisp designed a systemic insecticide by modifying methamidophos. It is the N-malonyl ethyl ester derivative (XXIV).[931] According to his hypothesis, the lipophilic ester may be a good cuticular penetrant when applied to the foliage and will be hydrolyzed by esterases in the plant, yielding a carboxyl group which will bestow symplastic properties on the molecule; the acid may be transported to other sites with the assimilate stream, and the transported acid will be decomposed there oxidatively to produce methamidophos, which may exert the insecticidal activity.

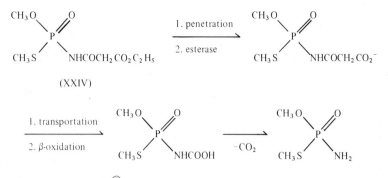

(20)

Phosfolan, Cyolane®
2-(Diethoxyphosphinylimino)-1,3-dithiolane

$$(C_2H_5O)_2 \overset{O}{\underset{\|}{P}} - N = C \overset{S - CH_2}{\underset{S - CH_2}{<}}$$

American Cyanamid Company found, in 1963, several dithiolane iminophosphates with systemic insecticidal activity. Phosfolan is synthesized by the reaction of diethyl phosphorochloridate and 2-imino-1,3-dithiolane.

$$(C_2H_5O)_2 \overset{O}{\underset{\|}{P}} Cl + HN = C \overset{S}{\underset{S}{<}} \xrightarrow[-HCl]{} (C_2H_5O)_2 \overset{O}{\underset{\|}{P}} - N = C \overset{S}{\underset{S}{<}}$$

(21)

Phosfolan is soluble in water and many organic solvents except aliphatic hydrocarbons.

Phosfolan is absorbed through roots and leaves and is particularly effective to protect cotton from the attacks of leaf feeding larvae, such as cotton leafworms, as well as sucking insects and mites. It is also useful for the control of soil insects, being persistent for a relatively long time in soil. The mammalian toxicity is very high; the acute oral LD_{50} for rats is 9 mg/kg.

It is interesting to note that phosfolan is less active as an anticholinesterase in vitro than its thiono analog.[32]

The 4-methyldithiolane derivative (mephosfolan, Cytrolane®) is also effective as an insecticide with systemic activity. It shows a significant activity against some borer insects. The mammalian toxicity is almost the same as that of phosfolan.

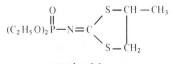

mephosfolan

The oxazolidine derivative, 2-(methoxymethylthiophosphinylimino)-3-ethyl-5-methyl-1,3-oxazoline (Stauffer R-16,661), is very effective as a systemic insecticide against both sucking and chewing pests on cotton. However, it is too toxic to mammals (LD_{50} mice = 0.1 to 0.2 mg/kg) to be utilized in practice. It is converted metabolically into a ketone which is more toxic to mammals but less effective to insects (see Equation 51 in Section IV.B.1.e).[565]

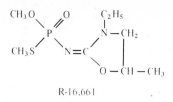

R-16,661

b. Carboxyamides and Related Compounds

One of the representative compounds of this class is dimethoate, which is the first systemic insecticide with low mammalian toxicity. Since the introduction of this compound by American Cyanamid Company and Montecatini in 1956, many similar S-acetamide derivatives of phosphorodithioates (XXV) have been developed as systemic insecticides and acaricides. They are generally synthesized by the reaction of an alkali salt of O,O-dialkyl phosphorodithioic acid and a corresponding haloacetamide (see Section II.C.1).

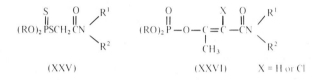

(XXV)   (XXVI)   X = H or Cl

Furthermore, the crotonamide derivatives of phosphate esters (XXVI) were found to have systemic activity. They are produced by the Perkow reaction from trialkyl phosphites and appropriate α-chloroacetoacetamides (see Section II.A.4.a). The methyl crotonate ester analog, mevinphos, is also active as a systemic insecticide. The *cis* isomers of these crotonamides and esters are the major components and essentially the active principles. These enol phosphate esters are active as anticholinesterase agents without the necessity of metabolic activation.

Another systemic insecticide of low mammalian toxicity is Amiphos®, S-(2-acetamidoethyl) dimethyl phosphorothiolothionate. This structure is reminiscent of another systemic insecticide, amiton, which is a salt of a tertiary amine and is very toxic to mammals.

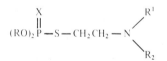

Amiphos: $R = CH_3$, $R^1 = H$, $R^2 = COCH_3$, $X = S$
amiton: $R = C_2H_5$, $R^1 = R^2 = CH_3$, $X = O$

Dimethoate, Rogor®, Cygon®
Dimethyl S-(N-methylcarbamoylmethyl) phosphorothiolothionate

$$\underset{(CH_3O)_2\overset{\displaystyle S}{\overset{\|}{P}}-SCH_2\overset{\displaystyle O}{\overset{\|}{C}}NHCH_3}{}$$

Dimethoate is a colorless crystalline substance; mp 51 to 52°C; bp 107°C at 0.05 mmHg; and vp 8.5 × 10^{-6} mmHg at 20°C. It is soluble in polar organic solvents. The solubility in water is about 3 to 4%. In addition to the above-mentioned general method for synthesis, aminolysis of the carboxy ester or anhydride can be utilized after the P-S-C bond formation (see Equations 55 and 56 in Section II.C.1).

Technical products of dimethoate may contain several by-products: O,O,S-trimethyl phosphoro-thiolothionate and dimethoate acid methyl ester (dimethyl S-(methoxycarbonylmethyl) phosphorothiolo-thionate) are the major impurities. A trace amount of the oxygen analog of dimethoate also may be present. The impurities increase the toxicity of dimethoate (see Section IV.D.2.a; Table 40). Thus, the oral

LD_{50} value of pure dimethoate for rats is more than 600 mg/kg, whereas that of the technical preparations is between 150 and 300 mg/kg.

Dimethoate is relatively unstable and decomposes on storage, particularly at elevated temperature and in the presence of impurities such as bases and dimethyl phosphorodithioic acid (see Equation 73 in Section III.C.2). Polar solvents increase the rate of decomposition. Alkylation reaction is the main degradation mechanism. As iron accelerates the decomposition, iron containers should be avoided for storage. When an alcohol such as methyl cellosolve is present in the formulation, transesterification as well as oxidation and thiono-thiolo rearrangement may occur during storage to yield more toxic products (see IV.D.1).[735,736]

When sprayed on plants, dimethoate may be oxidized to form the oxygen analog both enzymatically and nonenzymatically. O-Demethylation occurs preferably in plants including rice,[621] whereas amide cleavage proceeds in cotton to a great extent.[933]

In animals, the hydrolysis by amidase is important for the selective species toxicity of dimethoate. The cleavage of the P-O, P-S, and S-C linkages also occurs to a considerable degree (see Section IV.B.2.c.ii).

Dimethoate is useful for the control of sucking insects and mites on ornamentals, vegetables, cotton, and fruit crops.

Omethoate, Folimat®
Dimethyl S-(N-methylcarbamoylmethyl) phosphorothiolate

$$(CH_3O)_2 \overset{\overset{O}{\|}}{P} SCH_2 \overset{\overset{O}{\|}}{C} NHCH_3$$

Omethoate, the oxygen analog of dimethoate, was introduced in 1965 by Bayer AG as a systemic insecticide and acaricide. It is often called dimethoxon. Omethoate is an undistillable oil; d^{20} 1.3943; and n_D^{20} 1.4987. It is very soluble in water and alcohols, but insoluble in petroleum ether.

Omethoate may be used against sucking pests on fruit crops. The oral LD_{50} for rats is 50 mg/kg.

Ethoate-methyl, Fitios B/77
Dimethyl S-(N-ethylcarbamoylmethyl) phosphorothiolothionate

$$(CH_3O)_2 \overset{\overset{S}{\|}}{P} - SCH_2 \overset{\overset{O}{\|}}{C} NHC_2H_5$$

This is the N-ethyl homolog of dimethoate; mp 67 to 68°C; d_4^{20} 1.1670; and soluble in polar solvents. The solubility in water is 8.5 g/l at 25°C. Ethoate-methyl is somewhat more stable than dimethoate and is longer acting. The oral LD_{50} for mice is 350 mg/kg. It is manufactured by Bombrini Parodi-Delfino, Italy.

Prothoate, Fostion®, Fac 20®
Diethyl S-(N-isopropylcarbamoylmethyl) phosphorothiolothionate

$$(C_2H_5O)_2 \overset{\overset{S}{\|}}{P} SCH_2 \overset{\overset{O}{\|}}{C} NHCH(CH_3)_2$$

Prothoate was introduced in 1956 by Soc. Montecatini, Italy, as an experimental insecticide. The pure prothoate is a colorless crystalline substance; mp 28.5°C; d^{32} 1.151; and n^{32} 1.5128. It is practically insoluble in water, but is soluble in most organic solvents. It may be applied to control olive fly and cherry maggot. It has a strong ovicidal activity to mites. Prothoate is very toxic to mammals; the oral LD_{50} for rats is about 10 mg/kg. The O-methyl homolog (mp 77 to 78°C) is much less toxic to mammals (LD_{50} rats = 250 mg/kg).

Morphothion, Ekatin M®, Ekatin F®, Morphotox®

Dimethyl S-(morpholinocarbonylmethyl) phosphorothiolothionate

$$(CH_3O)_2 \overset{\overset{\displaystyle S}{\|}}{P}SCH_2\overset{\overset{\displaystyle O}{\|}}{C}-N\big\langle\ \ \big\rangle O$$

Morphothion was introduced in 1956 by Sandoz Ltd., Switzerland, and is a colorless crystalline substance, mp 63 to 64°C. It is poorly soluble in water, highly soluble in polar organic solvents. The oral LD_{50} for rats is 190 mg/kg. Morphothion is effective as a systemic and contact insecticide.

Amidithion, Thiocron®, Medithionat

Dimethyl S-(N-methoxyethylcarbamoylmethyl) phosphorothiolothionate

$$(CH_3O)_2 \overset{\overset{\displaystyle S}{\|}}{P}SCH_2\overset{\overset{\displaystyle O}{\|}}{C}NHC_2H_4OCH_3$$

Amidithion was introduced in 1963 by Ciba Ltd. It is a crystalline substance; mp 46°C; vp 8×10^{-6} mmHg at 20°C; moderately soluble in water (2%); soluble in polar organic solvents. It is hydrolyzed by alkali and is incompatible with alkaline pesticides. Amidithion is useful for the control of sucking pests on cherries and vegetables. It has low mammalian toxicity and phytotoxicity; the oral LD_{50} for rats is 600 to 660 mg/kg. Amidithion is rapidly decomposed in soils, in which the half-life is 2 to 3 days.

Formothion, Antio®, Aflix®

Dimethyl S-(N-formyl-N-methylcarbamoylemethyl) phosphorothiolothionate

$$(CH_3O)_2 \overset{\overset{\displaystyle S}{\|}}{P}SCH_2\overset{\overset{\displaystyle O}{\|}}{C}N\overset{\diagup CH_3}{\diagdown CHO}$$

This was introduced in 1963 by Sandoz Ltd. Pure formothion is a yellowish liquid which freezes at 25 to 26°C; d^{26} 1.361. It cannot be distilled without decomposition. It is poorly soluble in water, but highly soluble in organic solvents. Formothion is more stable than dimethoate in storage, but breaks down readily to dimethoate in alcohol or water.

Formothion is of relatively low toxicity to mammals (oral LD_{50} rats = 330 to 530 mg/kg) and is effective to both sucking and chewing pests by systemic and contact actions.

The main metabolite of formothion is dimethoate acid.

$$(CH_3O)_2 \overset{\overset{\displaystyle S}{\|}}{P}SCH_2\overset{\overset{\displaystyle O}{\|}}{C}N\overset{\diagup CH_3}{\diagdown CHO} \longrightarrow (CH_3O)_2 \overset{\overset{\displaystyle S}{\|}}{P}SCH_2COOH \tag{22}$$

Mecarbam, Pestan®, Murfotox®

Diethyl S-(N-ethoxycarbonyl-N-methylcarbamoylmethyl) phosphorothiolothionate

$$(C_2H_5O)_2 \overset{\overset{\displaystyle S}{\|}}{P}SCH_2\overset{\overset{\displaystyle O}{\|}}{C}-\underset{\underset{\displaystyle CH_3}{|}}{N}-\overset{\overset{\displaystyle O}{\|}}{C}OC_2H_5$$

Mecarbam has a carbamate moiety holding the nitrogen atom in common with the dimethoate-type carboxyamide. Mecarbam was introduced in 1961 by Murphy Chemical Ltd. It is a liquid; bp 144°C at 0.02

mmHg; mp 9°C; d^{20} 1.223; n_D^{20} 1.5138; and volatility at 40°C 6 × 10^{-6} mg/m^3. Mecarbam is highly soluble in many organic solvents, but slightly soluble in water (0.1%) and alkanes (2 to 4%).

Mecarbam is rather a contact insecticide and acaricide with ovicidal activity, though it has some systemic activity. It persists in soils for several weeks. Mecarbam is useful for the control of sucking and chewing plant pests and root maggots of vegetables.

The absence of the *N*-methyl group is more favorable for the systemic insecticidal activity, increasing greatly the root uptake.[163] The *N*-desmethyl homolog (Bay 19596) is effective against the oriental fruit fly and the melon fly.

$$\underset{(C_2H_5O)_2}{}\overset{S}{\underset{\|}{P}}SCH_2\overset{O}{\underset{\|}{C}}NHC\overset{O}{\underset{\|}{O}}C_2H_5$$

Bay 19596

In contrast, in the corresponding phosphonate series which is more effective than the phosphate series as systemic insecticides, the presence of *N*-methyl group appears to be more favorable to the systemic insecticidal activity. Thus, mecarphos or mecarphon (*S*-(*N*-methoxycarbonyl-*N*-methylcarbamoylmethyl) methyl methylphosphonothiolothionate) was recently introduced as a new experimental systemic insecticide by Murphy Chemical Ltd. The oral LD$_{50}$ for rats of mecarphon is 57 mg/kg.[163]

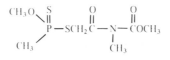

mecarphos

DAEP, Amiphos®
S-(2-Acetamidoethyl) dimethyl phosphorothiolothionate

$$(CH_3O)_2\overset{S}{\underset{\|}{P}}SCH_2CH_2NH\overset{O}{\underset{\|}{C}}CH_3$$

This compound was found in 1962 and commercialized in 1966 by Nippon Soda Company, Japan, under the trade name Amiphos. The active ingredient of Amiphos is a colorless crystalline substance, mp 22 to 23°C. It is soluble in most organic solvents, but is scarcely soluble in water. Amiphos has systemic properties as an insecticide and acaricide, and low mammalian toxicity; the oral LD$_{50}$ for rats is 400 mg/kg. It is useful to control sucking pests on fruits, vegetables, and ornamentals.

Amiton, Tetram®, Citram®
Diethyl *S*-(2-diethylaminoethyl) phosphorothiolate hydrogen oxalate

$$(C_2H_5O)_2\overset{O}{\underset{\|}{P}}SCH_2CH_2\overset{+}{\underset{H}{N}}(C_2H_5)_2 \cdot HOCOCOO^-$$

The free base of amiton is a colorless liquid; bp 80°C at 0.01 mmHg; and highly soluble in water and most organic solvents. The oxalate salt melts at 100 to 101°C. Amiton is a long acting systemic insecticide and acaricide. It is very toxic to mammals and the LD$_{50}$ values of the free base for various animals are 0.5 to 7 mg/kg. Owing to the ionization at physiological pH, amiton has little activity against invertebrates except for aphids, scale insects, and mites (see Section IV.C.4).

Amiton is manufactured by Imperial Chemical Industries. It is produced by the thiono-thiolo rearrangement from the thiono isomer (see Section II.D.4) or by the reaction of sodium diethyl phosphite with 2-diethylaminoethyl thiocyanate.

$$(C_2H_5O)_2PONa \; + \; NCSCH_2CH_2N(C_2H_5)_2 \; \longrightarrow \; (C_2H_5O)_2\overset{\overset{\displaystyle O}{\|}}{P}SCH_2CH_2N(C_2H_5)_2 \; + \; NaCN \tag{23}$$

Phosphamidon, Dimecron®
1-Chloro-1-N,N-diethylcarbamoyl-1-propen-2-yl dimethyl phosphate

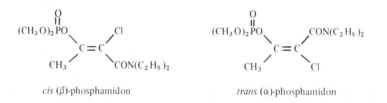

cis (β)-phosphamidon   trans (α)-phosphamidon

Phosphamidon was first introduced by Ciba Ltd., Switzerland, in 1956. It is synthesized bv the Perkow reaction from trimethyl phosphite and α,α-dichloro-N,N-diethylacetoacetamide (Equation 24), which is obtained by chlorination of diethylacetoacetamide with sulfuryl chloride. The cis-trans proportion remains almost constant regardless of change in reaction conditions or exposure of the reaction mixture to ultraviolet light. The technical product is a mixture of cis and trans isomers in the proportion 73:27. The cis isomer is more biologically active than the trans isomer.

$$(CH_3O)_3P \; + \; CH_3\overset{\overset{\displaystyle O}{\|}}{C}CCl_2\overset{\overset{\displaystyle O}{\|}}{C}N(C_2H_5)_2 \; \xrightarrow[-CH_3Cl]{} \; (CH_3O)_2\overset{\overset{\displaystyle O}{\|}}{P}O\underset{\underset{\displaystyle CH_3}{|}}{C}{=}CClCON(C_2H_5)_2 \tag{24}$$

In the process of the synthesis of dichlorodiethylacetoamide, small quantities of monochloro and α,α,γ-trichloro derivatives are also formed. Theretore, tneir Perkow reaction products, i.e., dechlorophosphamidon and γ-chlorophosphamidon, are found at about 1% each in the technical preparations of phosphamidon. Although γ-chlorophosphamidon has a higher anticholinesterase activity than phosphamidon, its toxicity is less than the latter because of its higher biodegradability.[934]

$$(CH_3O)_2\overset{\overset{\displaystyle O}{\|}}{P}{-}O{-}\underset{\underset{\displaystyle CH_3}{|}}{C}{=}CHCON(C_2H_5)_2 \qquad\qquad (CH_3O)_2\overset{\overset{\displaystyle O}{\|}}{P}{-}O{-}\underset{\underset{\displaystyle CH_2Cl}{|}}{C}{=}CClCON(C_2H_5)_2$$

dechlorophosphamidon   γ-chlorophosphamidon

Phosphamidon is a liquid; bp 150°C at 1 mmHg; d_4^{25} 1.2131; n_D^{25} 1.4721; and vp 2.5 × 10⁻⁵ mmHg at 20°C. It is miscible with water and most organic solvents, but is slightly soluble in alkanes.

Phosphamidon is effective against sucking plant pests, particularly on cotton, with both systemic and contact actions. The oral LD_{50} for rats is 15 to 27 mg/kg.

The major metabolic breakdown process of phosphamidon is P-O-vinyl hydrolysis, resulting in the formation of dimethyl phosphate and α-chlorodiethylacetoacetamide, which can be dehalogenated and further degraded. Demethylation occurs also as a minor degradation pathway. Another metabolic pathway is oxidative N-deethylation which yields toxic N-desethylphosphamidon and phosphamidon amide. The N-hydroxyethyl intermediates are also converted into carbohydrate conjugates. (See also Equations 46 and 57 in Section IV.B.) The main metabolic pathways are represented as follows:[644]

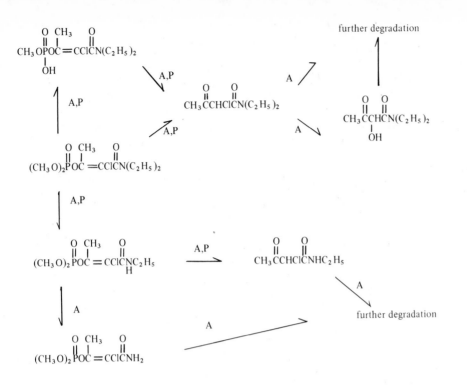

$$A = \text{animals}, \quad P = \text{plants} \tag{25}$$

A comprehensive review on phosphamidon has been published as a single volume of *Residue Reviews*.[935]

Dicrotophos, Bidrin®, Carbicron®, Ektafos®
3-(Dimethoxyphosphinyloxy)-*N,N*-dimethyl-*cis*-crotonamide

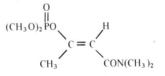

This compound was introduced in 1963 by Ciba AG (Carbicron) and then in 1965 by Shell Development Company (Bidrin). Dicrotophos is synthesized by the Perkow reaction from trimethyl phosphite and *N,N*-dimethyl-α-chloroacetoacetamide according to the following reaction sequence. The main product of the Perkow reaction is the *cis*-crotonamide.

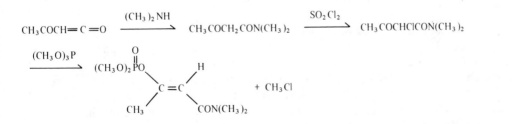

Methyl chloride liberated in the course of the Perkow reaction can react with trimethyl phosphite to yield dimethyl methylphosphonate as a by-product by the Michaelis-Arbuzov reaction (see Section II.H.3).

$$(CH_3O)_3P + CH_3Cl \longrightarrow (CH_3O)_2\overset{\overset{\displaystyle O}{\|}}{P}CH_3 + CH_3Cl \qquad (27)$$

Dicrotophos is a yellow to brown liquid; bp 90 to 95°C at 10^{-3} mmHg; n_D^{23} 1.4680; and d 1.216. It is miscible with water and most organic solvents except kerosenes. It is rather stable to heat, but is decomposed by adsorbents with relative ease. The half-life at 37°C is 50 days at pH 9 and 100 days at pH 1.

Dicrotophos is a moderately persistent systemic insecticide and is effective for the control of aphids, scale insects, thrips, bugs, leaf hoppers, and mites on cotton, seed crops, and ornamental trees. The oral LD_{50} for rats is 22 mg/kg.

In biological systems, the N-methyl group of dicrotophos may be oxidatively removed via the methylol intermediate giving the insecticide monocrotophos and finally the nonsubstituted amide (see Section IV.B.1.d). All these metabolites are active as insecticides and anticholinesterase agents.

The N-methoxy-N-methylcrotonamide analog (C-2307) is an experimental insecticide introduced by Ciba AG and is about 10 times as toxic to both mammals and insects as dicrotophos. The acute oral LD_{50} for rats is 2 mg/kg.

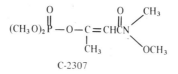

C-2307

Monocrotophos, Azodrin®, Nuvacron®
3-(Dimethoxyphosphinyloxy)-N-methyl-cis-crotonamide

$$(CH_3O)_2\overset{\overset{\displaystyle O}{\|}}{P}-O-\underset{\underset{\displaystyle CH_3}{|}}{C}=\overset{\overset{\displaystyle H}{|}}{C}-CONHCH_3$$

Monocrotophos was introduced in 1965 by Ciba. It is produced by the Perkow reaction from trimethyl phosphite and N-methyl-2-chloroacetoacetamide. The pure compound is a crystalline substance, mp 53 to 55°C. The technical product melts at 25 to 30°C. It is soluble in water and polar organic solvents, but only slightly soluble in alkanes. It is stable to heat but rather readily decomposed by adsorbents.

Monocrotophos is effective against a variety of foliage insects, particularly against bollworms on cotton, and has a good systemic activity. It may also be used for soil insect control. The acute oral LD_{50} for rats is 21 mg/kg.

Mevinphos, Phosdrin®
Dimethyl 1-methoxycarbonyl-1-propen-2-yl phosphate

$$(CH_3O)_2\overset{\overset{\displaystyle O}{\|}}{P}-O-\underset{\underset{\displaystyle CH_3}{|}}{C}=CH\overset{\overset{\displaystyle O}{\|}}{C}OCH_3$$

Mevinphos was introduced in 1953 by Shell Development Company. It is a yellow liquid; bp 106 to 107.5°C at 1 mmHg; d_4^{20} 1.25; and n_D^{20} 1.4494. It is highly soluble in water, alcohols, and benzene, but is only slightly soluble in petroleum ether.

Mevinphos is produced by the Perkow reaction from trimethyl phosphite and methyl α-chloroaceto-acetate. The product consists of the cis (60%) and trans (40%) isomers. The cis isomer is much more insecticidal (about 100 times more active) than the trans. The technical product may contain a small amount of the 1-chloro derivative of mevinphos.

In spite of the presence of a carboxy ester group in the molecule, mevinphos is very toxic to mammals; the oral LD_{50} for rats is 4 to 7 mg/kg. It is effective to control both sucking and chewing insects and mites by systemic and contact actions. It is a short acting insecticide degrading 90% within 4 days in plants. For the metabolism and toxicity, see Sections IV.B.2.b.ii and IV.B.2.c.i.

c. Sulfides and Related Compounds

Organophosphorus compounds containing a sulfide or the oxidized functional group in the molecule is one of the most important classes of systemic insecticides. Since the systemic insecticide demeton was introduced in 1951 by Bayer AG, many similar phosphorus compounds of this class have been developed as systemic insecticides.

Many of them are 2-alkylthioethyl esters of phosphorothioic acid or dithioic acid. Phosphorothioate esters having this moiety are synthesized by the reaction of a dialkyl phosphorochloridothionate with a 2-alkylthioethanol. The product readily undergoes the thiono-thiolo rearrangement to form the thiolo isomer.

$$
\underset{\substack{\|\\(RO)_2PCl}}{\overset{S}{}} + HOCH_2CH_2SR' \longrightarrow \underset{\substack{\|\\(RO)_2POCH_2CH_2SR'}}{\overset{S}{}} \longrightarrow \underset{\substack{\|\\(RO)_2PSCH_2CH_2SR'}}{\overset{O}{}} \tag{28}
$$

The thiolo isomer is more biologically active than the thiono isomer. The pure thiolo isomer may be produced by alkylation of dialkyl phosphorothioic acid with alkyl-2-chloroethyl sulfide.

$$
\underset{\substack{\|\\(RO)_2PSNa}}{\overset{O}{}} + ClCH_2CH_2SR' \longrightarrow \underset{\substack{\|\\(RO)_2PSCH_2CH_2SR'}}{\overset{O}{}} \tag{29}
$$

Phosphorothiolothionates are similarly synthesized by the alkylation of O,O-dialkyl phosphorodithioate. In some compounds, a methylene bridge connects the ester-S and the sulfide-S atoms. They are synthesized by the reaction of an O,O-dialkyl phosphorodithioate, formaldehyde, and an appropriate mercaptan (see Section II.C.2).

$$
\underset{\substack{\|\\(RO)_2PSH}}{\overset{S}{}} + CH_2O + R'SH \longrightarrow \underset{\substack{\|\\(RO)_2PSCH_2SR'}}{\overset{S}{}} \tag{30}
$$

Except for phosphorothiolate esters, these compounds are generally inactive as cholinesterase inhibitors, unless they are activated. In contrast with the above-mentioned phosphoramidate and carboxyamide systemic insecticides, the sulfide systemic insecticides are only slightly soluble in water. Some low water soluble compounds like phorate, disulfoton, and thiometon are suited for the protection of young seedlings by seed or soil treatment. In or on plants, the systemic sulfide insecticides are converted enzymatically or nonenzymatically into oxidation products (Equation 31). The sulfide function is transformed into sulfoxide and then sulfone. The thiophosphoryl group is oxidatively desulfurized to form a phosphoryl group. The sulfide oxidation is generally more rapid than the desulfuration. The oxidation increases the water solubility and anticholinesterase activity. The sulfide oxidation allows the insecticides to be translocated in lethal amounts in a relatively stable state before being hydrolyzed by plant enzymes. This function of the metabolite sulfoxide appears to be reasonable, analogous to the high penetrating and solvation properties of dimethylsulfoxide. Some related sulfoxides and sulfones are utilized as systemic insecticides.

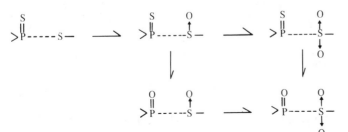

$$\tag{31}$$

Oxidation of organophosphorus sulfides

287

Demeton, Systox®, mercaptophos

O,O-Diethyl 2-ethylthioethyl phosphorothioate

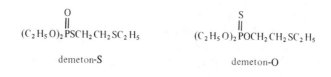

$$\begin{array}{ccc} & O & & & S \\ & \| & & & \| \\ (C_2H_5O)_2PSCH_2CH_2SC_2H_5 & & & (C_2H_5O)_2POCH_2CH_2SC_2H_5 \\ \\ \text{demeton-S} & & & \text{demeton-O} \end{array}$$

Demeton was developed as an insecticide by Bayer AG in 1951. The technical grade product is a mixture of thiono (70%) and thiolo (30%) isomers. The thiono and thiolo isomers are named demeton-O and demeton-S, respectively. Isosystox® is the trade name of demeton-S. Demeton-O is a colorless oil; bp 106°C at 0.4 mmHg; vp 2.48×10^{-4} mmHg; d_4^{20} 1.1193; the solubility in water is 60 ppm; and it is highly soluble in most organic solvents including aliphatic hydrocarbons. The oral LD_{50} for rats is 30 mg/kg. Demeton-S is a colorless liquid; bp 100°C at 0.25 mmHg; vp 2.6×10^{-4} mmHg at 20°; d_4^{20} 1.1325; water solubility 2000 ppm at 20°C; and highly soluble in most organic solvents. The thiolo isomer is much more toxic than its counterpart; the oral LD_{50} for rats is 1.5 mg/kg.

Demeton is synthesized by the reaction of diethyl phosphorochloridothionate with 2-ethylthioethanol. This procedure is always accompanied by the thiono-thiolo rearrangement yielding the thiolo isomer. The pure thiolo isomer is prepared by the alkylation of diethyl phosphorothioate (see Section II.D.2).

Demeton is effective for 4 to 6 weeks against sucking plant pests when applied as a foliage spray or soil soak. The metabolic routes of demeton are the same in plants, insects, and mammals and are presented with the anticholinesterase activity (I_{50} for fly-head cholinesterase) of the intermediates in the following Equation.[204] In cotton plants the major metabolite is the thiolo isomer sulfoxide, which is then metabolized further to the sulfone at a much slower rate.

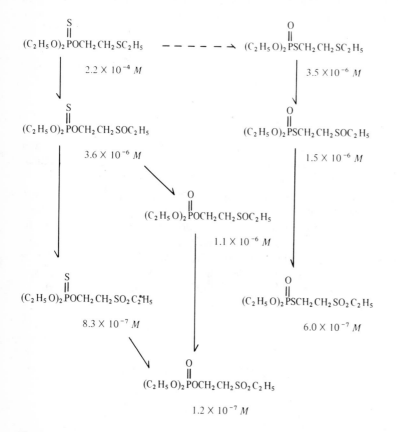

(32)

Metabolic routes of demeton and anticholinesterase activity of metabolites (I_{50} fly-head ChE, M)

Demeton-methyl, Methyldemeton, Metasystox®
O,O-Dimethyl 2-ethylthioethyl phosphorothioate

$$\underset{\text{demeton-O-methyl}}{(CH_3O)_2 \overset{\overset{\text{S}}{\|}}{P}OCH_2CH_2SC_2H_5} \qquad \underset{\text{demeton-S-methyl}}{(CH_3O)_2 \overset{\overset{\text{O}}{\|}}{P}SCH_2CH_2SC_2H_5}$$

This is the methyl homolog of demeton and was introduced in 1954 by Bayer AG as a systemic insecticide with less mammalian toxicity in comparison with the ethyl homolog demeton. Demeton-methyl is manufactured by the reaction of 2-ethylthioethanol with dimethyl phosphorochloridothionate. The product is a 70:30 mixture of the thiono isomer (demeton-O-methyl) and the thiolo isomer (demeton-S-methyl). Demeton-O-methyl is a liquid; bp 78°C and 0.2 mmHg; n_D^{20} 1.5063; d_4^{20} 1.1904; water solubility 330 ppm; and highly soluble in most organic solvents. The physical properties of demeton-S-methyl are bp 92°C at 0.2 mmHg; n_D^{20} 1.5065; d_4^{20} 1.207; water solubility 3300 ppm; and highly soluble in most organic solvents. Pure demeton-S-methyl is synthesized by the reaction of 2-chlorodiethyl sulfide with potassium dimethyl phosphorothioate or with trimethyl phosphorothionate (Equation 111 in Chapter II). It has been commercialized under the trade name Metasystox(i) and Isometasystox.

The oral LD_{50} values of the thiono and thiolo isomers for rats are 180 and 40 mg/kg, respectively.

The technical preparations of demeton-methyl may contain trimethyl phosphorothioate and unreacted chlorodiethyl sulfide. The thiono-thiolo rearrangement occurs more rapidly than with demeton; it takes 91 days with demeton-O for 10% rearrangement to the thiolo isomer, whereas it takes only 8 days with demeton-O-methyl. The rearrangement reaction proceeds rapidly in plants, too, and the produced thiolate undergoes further sulfide oxidation. Intermolecular transmethylation takes place readily in the presence of traces of water (see Equation 64 in Section III.C.1).[257] The sulfonium product (XXVII) is highly toxic to mammals. It is about 1100 times as toxic as demeton-S-methyl; the intravenous LD_{50} for rats is 0.06 mg/kg. The anticholinesterase activity of the sulfonium XXVII is 1600 times higher than the sulfide demeton-*S*-methyl

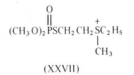

$$(CH_3O)_2 \overset{\overset{\text{O}}{\|}}{P}SCH_2CH_2\overset{\overset{+}{\underset{\underset{CH_3}{|}}{}}}{S}C_2H_5$$

(XXVII)

Demephion, methyldemeton-methyl, Tinox®, Cymetox®
O,O-Dimethyl 2-methylthioethyl phosphorothioate

$$\underset{}{(CH_3O)_2 \overset{\overset{\text{O}}{\|}}{P}SCH_2CH_2SCH_3} \qquad \underset{}{(CH_3O)_2 \overset{\overset{\text{S}}{\|}}{P}OCH_2CH_2SCH_3}$$

Demephion is the S-methyl homolog of demeton-methyl and is produced by Farbenfabriken Wolfen, East Germany. It is a mixture of the thiono (30 to 40%) and the thiolo (60 to 70%) isomers. The thiono isomer boils with decomposition at 107°C at 0.1 mmHg; n_D^{25} 1.488; and d_4^{20} 1.198. The thiolo isomer boils with decomposition at 65°C at 0.1 mmHg; n_D^{25} 1.508; and d_4^{20} 1.218. The water solubility at room temperature of the thiono and thiolo isomers is 300 and 3000 ppm, respectively. The mixture is miscible with many organic solvents except aliphatic hydrocarbons. The oral LD_{50} for rats is more than 50 mg/kg for the thiono isomer and 40 mg/kg for the thiolo isomer.

Thiometon, Ekatin®, Dithiometasystox®
Dimethyl S-(2-ethylthioethyl) phosphorothiolothionate

$$\underset{\displaystyle (CH_3O)_2 \overset{\displaystyle \overset{S}{\|}}{P}SCH_2CH_2SC_2H_5}{}$$

Thiometon is manufactured by Sandoz Ltd., Switzerland. It is synthesized by the alkylation of O,O-dimethyl phosphorodithioate with 2-ethylthioethyl chloride or tosylate (see Section II.C.1).

Thiometon is a colorless oil; bp 104°C at 0.3 mmHg; vp 3×10^{-4} mmHg at 20°C; d_4^{20} 1.208; water solubility 200 ppm; and highly soluble in most organic solvents except alkanes.

Thiometon is a systemic insecticide and acaricide with contact activity. The mammalian toxicity is relatively low; the oral LD_{50} for rats is 70 to 120 mg/kg. It is effective against aphids, sawflies, thrips, and mites.

Thiometon is metabolized oxidatively in plants, forming demeton-S-methyl sulfoxide and sulfone, which appear to be the active principles.

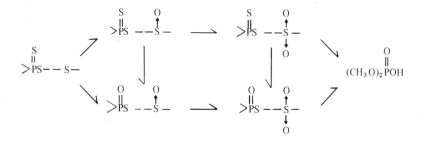

$$(33)$$

Disulfoton, thiodemeton, Disyston®, Dithiosystox®
Diethyl S-(2-ethylthioethyl) phosphorothiolothionate

$$\underset{\displaystyle (C_2H_5O)_2 \overset{\displaystyle \overset{S}{\|}}{P}SCH_2CH_2SC_2H_5}{}$$

Disulfoton was introduced in 1956 by Bayer AG. It is a colorless liquid; bp 113°C at 0.4 mmHg; d_4^{20} 1.144; and n_D^{20} 1.5348. The water solubility is 25 ppm. It is soluble in most organic solvents.

Disulfoton is much more toxic to mammals than the O-methyl homolog thiometon; the oral LD_{50} for male and female rats is 12.5 and 2.5 mg/kg, respectively. Disulfoton reacts with alkylating agents such as alkyl halides and dimethyl sulfate to yield the corresponding S-alkyl sulfonium compound which has a strong anticholinesterase activity and high mammalian toxicity, but has no insecticidal activity because of its low penetrative ability.

Tetrathion
Ethyl S-(2-ethylthioethyl) methyl phosphorothiolothionate

$$\underset{\displaystyle C_2H_5O}{\overset{\displaystyle CH_3O}{}} \!\!\! \underset{\displaystyle}{\overset{\displaystyle \overset{S}{\|}}{P}} \!\!-\! SCH_2CH_2SC_2H_5$$

Tetrathion was developed in Czechoslovakia. It has chemical and insecticidal properties similar to thiometon and disulfoton. It may be applied in soils. The LD_{50} values for various animals are between 1.2 and 22.3 mg/kg.

The starting material, O-ethyl O-methyl hydrogen phosphorodithioate, is produced by the reaction of phosphorus pentasulfide with a mixture of methanol and ethanol (Equation 34). Dimethyl and diethyl hydrogen phosphorodithioates are also formed in this process; consequently, the technical tetrathion contains both thiometon and disulfoton as impurities.

$$P_2S_5 + 2CH_3OH + 2 C_2H_5OH \longrightarrow 2 \underset{C_2H_5O}{\overset{CH_3O}{>}}P\underset{SH}{\overset{S}{<}} + SH_2 \tag{34}$$

Isothioate, Hosdon®
Dimethyl S-(2-isopropylthioethyl)phosphorothiolothionate

$$(CH_3O)_2\overset{S}{\overset{||}{P}}SCH_2CH_2SCH(CH_3)_2$$

Isothioate, the S-isopropyl homolog of thiometon, was developed into commercialization in 1972 by Nihon Noyaku Company, Japan, as a systemic insecticide and acaricide with low mammalian toxicity. It is a liquid; bp 53 to 56°C at 0.01 mmHg; vp 2×10^{-4} mmHg at 20°C; d_4^{21} 1.19; and n_D^{21} 1.5350. The water solubility is 97 ppm at 25°C. It is miscible with most organic solvents.

The oral LD_{50} for male rats is 200 mg/kg. Isothioate is effective against sucking plant pests by soil treatment or foliage spray. It is less persistent than disulfoton.

Phorate, Thimet®, timet
Diethyl S-(ethylthiomethyl) phosphorothiolothionate

$$(C_2H_5O)_2\overset{S}{\overset{||}{P}}SCH_2SC_2H_5$$

Phorate was introduced in 1954 by American Cyanamid Company. It is a liquid of bp 100°C at 0.4 mmHg; n_D^{25} 1.5349. Phorate is produced by the reaction of O,O-diethyl hydrogen phosphorodithioate with formaldehyde, followed by the reaction with ethylmercaptan (see Section II.C.2). It is also synthesized by the reaction of sodium O,O-diethyl phosphorodithioate with chloromethylethyl sulfide.

Phorate has a high mammalian toxicity; the oral LD_{50} for rats is 2 to 4 mg/kg. It is effective against sucking plant pests as a systemic insecticide and also has good contact and vapor actions. Phorate is relatively unstable to hydrolysis; the half-life at pH 8 and 70°C is only 2 hr. However, it protects plants for a relatively long time because of the greater persistency of the sulfoxide metabolite in plants and in soils. For the metabolism, see Section IV.B.1.c.

The S-isopropyl homolog, diethyl S-(isopropylthiomethyl) phosphorothiolothionate, was introduced in 1954 by American Cyanamid Company as an experimental systemic insecticide with a contact activity under the code number Am. Cy. 12008. The oral LD_{50} for mice is 4 to 16 mg/kg.

$$(C_2H_5O)_2\overset{S}{\overset{||}{P}}SCH_2SCH(CH_3)_2$$

Am. Cy. 12008

Aphidan, IPSP, PSP-204
Diisopropyl S-(ethylsulfinylmethyl) phosphorothiolothionate

Aphidan was introduced in 1964 by Hokko Chemical Company, Japan, as a systemic insecticide. It is synthesized by the oxidation of the corresponding sulfide with hydrogen peroxide.

$$(C_3H_7O)_2 \overset{\overset{\displaystyle S}{\|}}{P}SCH_2SC_2H_5 \xrightarrow{\;H_2O_2\;} (C_3H_7O)_2 \overset{\overset{\displaystyle S}{\|}}{P}SCH_2SOC_2H_5 \tag{35}$$

Aphidan is a liquid of d_4^{20} 1.696. The water solubility at 15°C is 1,500 ppm. Aphidan is closely related to the sulfoxide metabolite of phorate, but has the isopropyl ester group, which probably contributes to the low mammalian toxicity of aphidan in comparison with phorate; the oral LD_{50} for rats is 84.5 mg/kg.

Aphidan applied in soils is absorbed through the root to protect the plant from the attack of aphids. It can protect potatoes for 60 days from the aphid vectors of virus diseases by treating the seed potatoes, although the α-sulfinyl ester is rather unstable in contrast with β-sulfinyl esters such as oxydisulfoton (see Section III.A.2.a).

Oxydisulfoton, Disyston S®, Disyston-sulfoxide®
Diethyl S-(2-ethylsulfinylethyl) phosphorothiolothionate

$$(C_2H_5O)_2 \overset{\overset{\displaystyle S}{\|}}{P}SCH_2CH_2 \overset{\overset{\displaystyle O}{\uparrow}}{S}C_2H_5$$

Oxydisulfoton, the sulfoxide derivative of disulfoton, was introduced in 1965 by Bayer AG. It is a light brown undistillable liquid; d^{20} 1.209; n^{20} 1.5402; water solubility 10 ppm; and soluble in organic solvents except ligroin. It is useful for seed treatment to protect seedlings from vectors of virus diseases, but is highly toxic to mammals (LD_{50} for rats = 3.5 mg/kg). Oxydisulfoton is one of the most persistent organophosphorus insecticides (see Section IV.B.3).

Oxydemeton-methyl, Metasystox R®, Isometasystox-sulfoxide, demeton-S-methyl-sulfoxide
Dimethyl S-(2-ethylsulfinylethyl) phosphorothiolate

$$(CH_3O)_2 \overset{\overset{\displaystyle O}{\|}}{P}SCH_2CH_2 \overset{\overset{\displaystyle O}{\uparrow}}{S}C_2H_5$$

Oxydemeton-methyl was introduced in 1960 by Bayer AG as a systemic insecticide. It is synthesized by the alkylation of potassium dimethyl phosphorothioate with 2-bromodiethyl sulfoxide or by the oxidation of demeton-S-methyl with bromine in water or other suitable oxidation agents (see Equation 85 in Section II.D.2). Oxydemeton-methyl is a clear amber liquid; bp 106°C at 0.01 mmHg; d_4^{20} 1.28; miscible with water; and soluble in most organic solvents except petroleum ether. Being relatively stable to alkaline hydrolysis, it is compatible with alkaline pesticides.

Oxydemeton-methyl is effective against sucking plant insects on cotton, sugar beet, and vegetables. The oral LD_{50} for rats is 70 mg/kg.

Metasystox S®, Estox®
Dimethyl S-(2-ethylsulfinylisopropyl) phosphorothiolate

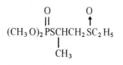

The active ingredient of Metasystox S, which was introduced by Bayer AG in 1955, is a yellowish oily liquid; practically odorless; bp 115°C at 0.02 mmHg; vp 3.5 × 10^{-6} mmHg at 20°C; and d_4^{20} 1.257. It is highly soluble in water and most organic solvents except aliphatic hydrocarbons. It is synthesized by oxidation of the corresponding sulfide with hydrogen peroxide. The oxidation product is much more stable than the sulfide against hydrolysis (see Section III.A.2.a). The insecticidal property resembles that of oxydemeton-methyl. The oral LD_{50} for rats is 105 mg/kg.

Dioxydemeton-S-methyl, Demeton-S-methyl-sulphone, Metaisosystox-sulfon
Dimethyl S-(2-ethylsulfonylethyl) phosphorothiolate

$$(CH_3O)_2 \overset{\overset{\displaystyle O}{\|}}{P}SCH_2CH_2SO_2C_2H_5$$

This is a systemic insecticide introduced by Bayer AG in 1965. It is produced by the potassium permanganate oxidation of demeton-S-methyl into the sulfone. The sulfone is a white to yellow crystalline substance; mp 60°C; and bp 120°C at 0.03 mmHg. It is readily soluble in alcohols, but is poorly soluble in aromatic hydrocarbons. Under alkaline conditions, the sulfone is more unstable than the corresponding sulfoxide because of the alkali catalyzed elimination reaction (see Section III.A.2.a). The oral LD_{50} for rats is 37.5 mg/kg.

Intration
S-(2-Acetylthioethyl) dimethyl phosphorothiolothionate

$$(CH_3O)_2 \overset{\overset{\displaystyle S}{\|}}{P}SCH_2CH_2\overset{\overset{\displaystyle O}{\|}}{S}CCH_3$$

This was introduced in 1967 as an experimental systemic insecticide for aphid control on hop in Czechoslovakia.[936]

Vamidothion, Kilval®
Dimethyl S-[2-(1-methylcarbamoylethylthio)ethyl] phosphorothiolate

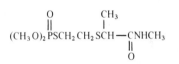

Vamidothion is a systemic insecticide introduced in 1962 by Rhone-Poulenc, France. The structure appears to be a combination of demeton-S-methyl and dimethoate. Vamidothion is synthesized according to the following route:

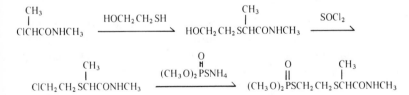

(36)

Vamidothion is a white crystalline substance; mp 46 to 48°C; and highly soluble in water and polar organic solvents. It resembles demeton-methyl in the insecticidal activity, but persists longer. The oral LD_{50} for rats is 64 to 100 mg/kg.

In contrast with other sulfide type insecticides, the biological oxidation of vamidothion does not appear to proceed beyond the sulfoxide stage.[210] The oral toxicity of the sulfoxide is about half that of the parent compound.

$$\underset{\text{(CH}_3\text{O})_2\text{PSCH}_2\text{CH}_2\overset{\overset{\displaystyle\text{CH}_3}{|}}{\text{SCHCONHCH}_3}}{\overset{\overset{\displaystyle\text{O}}{\|}}{}} \longrightarrow \underset{\text{(CH}_3\text{O})_2\text{PSCH}_2\text{CH}_2\overset{\overset{\displaystyle\text{CH}_3}{|}}{\underset{\underset{\displaystyle\text{O}}{\downarrow}}{\text{SCHCONHCH}_3}}}{\overset{\overset{\displaystyle\text{O}}{\|}}{}} \tag{37}$$

An experimental insecticide with the similar structure, isopropyl S-[(2-ethylcarbamoyloxyethyl) thiomethyl] methyl phosphorothiolothionate (R-6790), is promising for the control of boll weevils and two-spotted mites.[937]

R-6790

d. Phosphorus Esters Containing Hetero Aromatic Ring

Pyrazothion®
Diethyl 3-methylpyrazol-5-yl phosphorothionate

$$\underset{\text{(C}_2\text{H}_5\text{O})_2}{\text{P}}\overset{\overset{\displaystyle\text{X}}{\|}}{}-\text{O}-\!\!\!\!\overset{\text{CH}_3}{\underset{\underset{\text{N}}{}}{\overbrace{}}}\!\!\!\text{NH}$$

X = S   Pyrazothion®
X = O   Pyrazoxon®

Pyrazothion was introduced in 1952 by Geigy Chemical Corporation as a systemic insecticide. It is a dark yellow liquid; miscible with most organic solvents; slightly soluble in water; and decomposes on heating. Pyrazothion has good activity against aphids and spider mites. The acute oral LD_{50} for rats is 36 mg/kg.

The oxygen analog Pyrazoxon® shows a high systemic activity but is very toxic to mammals; the oral LD_{50} for mice is 4 mg/kg.

Menazon, azidithion, Saphizon®, Sayfos®
S-(4,6-Diamino-1,3,5-triazin-2-ylmethyl) dimethyl phosphorothiolothionate

$$\underset{\text{(CH}_3\text{O})_2\text{PSCH}_2}{\overset{\overset{\displaystyle\text{S}}{\|}}{}}-\!\!\!\!\overset{\text{N}=\!\!=\!\!\text{NH}_2}{\underset{\text{N}-\!\!\!\text{NH}_2}{\overbrace{}\text{N}}}$$

Menazon is a systemic insecticide with low mammalian toxicity introduced in 1959 by Plant Protection Ltd. It is synthesized by the reaction of 2-chloromethyl-4,6-diamino-1,3,5-triazine with sodium O,O-dimethyl phosphorodithioate (Equation 38) or by the reaction of biguanide with O,O-dimethyl S-ethoxycarbonylmethyl phosphorodithioate (see Section II.C.1).

$$\underset{\text{(CH}_3\text{O})_2\text{PSNa}}{\overset{\overset{\displaystyle\text{S}}{\|}}{}} + \text{ClCH}_2\!\!-\!\!\!\overset{\text{N}-\!\!\!\text{NH}_2}{\underset{\text{N}=\!\!\!\text{NH}_2}{\overbrace{}\text{N}}} \longrightarrow \underset{\text{(CH}_3\text{O})_2\text{PSCH}_2}{\overset{\overset{\displaystyle\text{S}}{\|}}{}}\!\!-\!\!\!\overset{\text{N}-\!\!\!\text{NH}_2}{\underset{\text{N}=\!\!\!\text{NH}_2}{\overbrace{}\text{N}}} + \text{NaCl} \tag{38}$$

Menazon is a colorless crystalline substance with a mercaptan-like odor; mp 160°C (decomp.); water solubility 1000 ppm; sparingly soluble in most organic solvents; and moderately soluble in ethyl-cellosolve and tetrahydrofurfuryl alcohol. It is a weak base and gives picrate and hydrochloride salts.

Menazon is a selective insecticide effective particularly against aphids by foliar spray, soil treatment, seed dressing, and root dip. The oral LD_{50} for rats is 900 to 1,950 mg/kg. For the metabolism of menazon, see Section IV.B.2.b.iii. Both menazon and the oxon are poor cholinesterase inhibitors in vitro. An unknown mechanism may be involved for the in vivo activation of menazon. Although in plants detoxication by hydrolysis occurs and no metabolite that could account for the insecticidal activity is found, the in vitro anticholinesterase activity of menazon is increased after administration to aphids.[937a]

Endothion, Exothion®, Endocide®, Phosphopyron®
Dimethyl S-(5-methoxy -4-oxo-4H-pyran-2-ylmethyl) phosphorothiolate

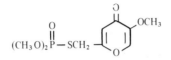

Endothion was introduced in 1957 by Rhone-Poulenc as a systemic insecticide and acaricide, but is no longer available. It is a white crystalline substance; mp 90 to 91°C; and highly soluble in water (150 g/100 ml), but slightly soluble in hydrocarbons.

Endothion is synthesized by the reaction of 5-methoxy-2-chloromethylpyrone with sodium O,O-diethyl phosphorothioate.

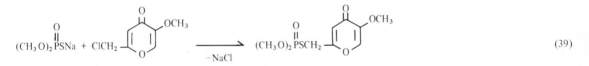

(39)

The oral LD_{50} for rats is 30 to 50 mg/kg. The primary metabolic pathway in plants is demethylation.[210]

endothion   ⟶

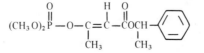

(40)

3. Animal Systemic Insecticides and Veterinary Pesticides
Some organophosphorus pesticides, can be used for the control of endo- or ectoparasites of domestic animals. The systemics, administered orally or dermally to the host, move through the body tissues to kill the parasites. Veterinary pesticides however, are not always systemics. Parasites include not only insects, but also mites, ticks, and helminths, i.e., intestinal nematodes.

Several reviews on pesticides for use on animals have been published.[938-941]

a. Phosphates

Crotoxyphos, Ciodrin®
Dimethyl cis-1-methyl-2-(1-phenylethoxycarbonyl)vinyl phosphate

Crotoxyphos was introduced in 1962 by Shell Development Company for the control of ectoparasites on livestock.[942] Crotoxyphos is synthesized by the reaction of trimethyl phosphite with α-methylbenzyl α-chloroacetoacetate. The technical product is a straw-colored liquid; bp 135°C at 0.03 mmHg; and n_D^{25} 1.5005. It is miscible with acetone, ethanol, and xylene, and slightly soluble in aliphatic hydrocarbons and water (0.1%). The half-life in aqueous solutions at pH 9 and 1 at 38°C is 35 and 87 hr, respectively.

Crotoxyphos has no systemic properties, but is effective for the control of flies, lice, mites, and ticks on cattle and pigs by spray. It may be applied to dairy cattle because it is not excreted into milk. The acute oral LD_{50} to rats is 125 mg/kg.

The p-chlorobenzyl crotonate ester analog has similar insecticidal activity; the oral LD_{50} for rats is 140 mg/kg. The methyl crotonate ester analog (mevinphos) and some crotonamide analogs like dicrotophos are active as plant systemic insecticides and were described in the preceding Section, V.A.2.b.

Chlorfenvinphos and some related vinyl phosphate esters may also be used as veterinary pesticides (see V.A.1.a).

Coroxon
3-Chloro-4-methylcoumarin-7-yl diethyl phosphate

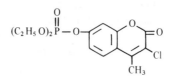

Coroxon was introduced in 1961 by Cooper Technical Bureau as a synergist of the anthelmintic phenothiazine for ruminants.[943] It is synthesized by oxidation of the corresponding thiono analog that is the animal systemic insecticide coumaphos. Coroxon is a white crystalline solid, mp 72°C. The solubility in water at 25°C is 230 ppm. The mammalian toxicity is very high; the oral LD_{50} for rats is 12 mg/kg.

Haloxon
Bis(2-chloroethyl) 3-chloro-4-methylcoumarin-7-yl phosphate

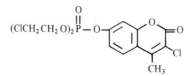

Haloxon was introduced in 1962 by Cooper Technical Bureau as an anthelmintic agent.[63] It is synthesized by the reaction of 3-chloro-4-methyl-7-hydroxycoumarin with bischloroethyl phosphite in the presence of carbon tetrachloride and triethylamine (see Equation 10 in Section II.A.1).[63] It is a white crystalline solid, mp 92°C, soluble in benzene and chloroform but practically insoluble in water (15 ppm).

Haloxon displays high anthelmintic properties in a variety of animals, but has no useful insecticidal activity. The acute toxicity is very low; the oral LD_{50} for rats is 900 mg/kg. It is below one tenth of the toxicity of the corresponding ethyl ester. This is partly due to the high rate of spontaneous reactivation of mammalian acetylcholinesterase inhibited with haloxon (see Section IV.A.2.b.i and Table 17). In spite of the low acute toxicity, haloxon shows a delayed neurotoxicity (see Section IV.E).

Maretin®, Rametin®, naphthalophos
N-(Diethoxyphosphinyloxy) -naphthalimide

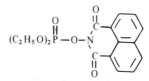

This is the first pesticide in the class of phosphate esters of hydroxylamine derivatives and was introduced by Bayer AG as an anthelmintic.[61] For the synthesis, see Equation 5 in Section II.A. It is a solid with a melting point of 196°C. It is practically insoluble in water and kerosene. Maretin has insecticidal activity, too. The acute oral LD_{50} for rats is 75 mg/kg.

The thiono analog (BAY 22408) is an insecticide with lower mammalian toxicity; the oral LD_{50} for rats is 500 mg/kg. It is highly effective to control Mediterranean fruit flies, but is less effective than malathion against Oriental fruit flies and melon flies.

b. Phosphorothionates

Coumaphos, Co-Ral®, Asuntol®, Muscatox®
3-Chloro-4-methylcoumarin-7-yl diethyl phosphorothionate

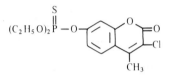

Coumaphos was synthesized first by Schrader in 1951 by condensation of diethyl phosphoro-chloridothionate with 3-chloro-4-methyl-7-hydroxycoumarin.[13] It is a colorless crystalline power; mp 95°C; insoluble in water (1.5 ppm); and highly soluble in esters, ketones, and aromatic hydrocarbons.

Coumaphos is useful to control ectoparasites on cattle, goats, sheep, and poultry by feed or spray. It is particularly effective to flies and mosquito larvae. Fly larvae in fecal material of poultry are killed by feeding coumaphos. Coumaphos is active against gastro-intestinal nematodes and also has synergistic activity with the anthelmintic agent phenothiazine.[943] As it is relatively stable to acid and alkali, it may be used as a dip.

Coumaphos is of relatively low toxicity to mammals except mice; the oral LD_{50} for rats is 90 to 110 mg/kg, whereas that for mice is 55 mg/kg. Although the degradation takes place rapidly in the liver of the cow and rat, the activation is rather significant in mice.[944] The principal metabolite excreted in urine is diethyl phosphorothioic acid. Deethylation products are also found in lesser amounts.[638]

Of the related compounds, the unchlorinated analog (Potasan®) is a crop insecticide with a higher mammalian toxicity (see Section V.A.1.b).

Coumithoate, Dition®
Diethyl 3,4-tetramethylene-7-coumarinyl phosphorothionate

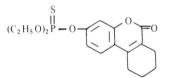

Coumithoate was introduced in 1956 by Montecatini as an insecticide and miticide usable for livestock pest control and for public health. It is a solid, melting at 88 to 89°C, and is highly soluble in most organic solvents but insoluble in water. The acute oral LD_{50} for rats is 150 mg/kg.

Ronnel, fenchlorphos, Nankor®, Trolene®, Korlan®
Dimethyl 2,4,5-trichlorophenyl phosphorothionate

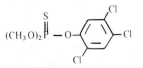

Ronnel was introduced in 1954 by the Dow Chemical Company as an experimental insecticide and was the first animal systemic insecticide.[945] Two synthetic routes are available: 1) the reaction of dimethyl phosphorochloridothionate with 2,4,5-trichlorophenol in the presence of potassium carbonate or with the sodium salt of the phenol (Equation 41); and 2) the reaction of the trichlorophenyl phosphorodichloridothionate with methanol in an aqueous caustic alkali solution (Equation 42). The phosphorodichloridothionate is produced by the addition of sulfur to the trichlorophenyl phosphorodichloridite obtained by the reaction of trichlorophenol with phosphorus trichloride.

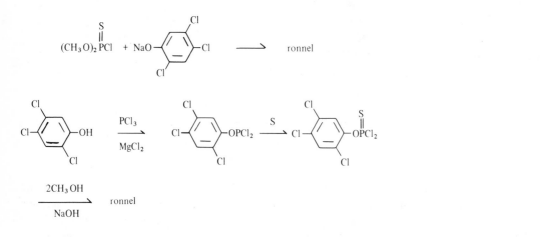

$$(41)$$

$$(42)$$

Ronnel is a white crystalline power; mp 40 to 42°C; bp 97°C at 0.01 mmHg; d_4^{25} 1.4850; highly soluble in most organic solvents; water solubility 44 ppm. It is hydrolyzed by strong alkali to trichlorophenol and dimethyl phosphorothioic acid, whereas demethylation takes place in a weak alkaline solution.[533]

Ronnel is effective to control cattle grub, lice, horn fly, screwworm, and ticks on livestock by oral administration. It has also a contact insecticidal activity against flies and cockroaches. The mammalian toxicity is very low; the acute oral LD_{50} for rats is 1250 to 1750 mg/kg. Ronnel is rapidly degraded and excreted in the rat and cow. The degradation proceeds through the cleavage of both the P-O-phenyl bond and the P-O-methyl bond.[533]

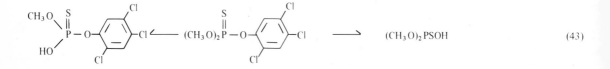

$$(43)$$

The great importance of demethylation in mammals for the selective toxicity of methyl phosphorothionate insecticides (see IV.B.2.b) was first found with ronnel by Plapp and Casida in 1958.[533]

The related unsymmetrical homolog, ethyl methyl 2,4,5-trichlorophenyl phosphorothionate (trichlormetaphos-3), was utilized in the USSR for the control of warble flies on cattle.[34] It is also used for plant protection against sucking insects and mites and for public health against midge larvae.

Trichlormetafos-3 is somewhat more toxic to mammals than ronnel; the acute oral LD_{50} for various experimental animals is 330 to 800 mg/kg. It is a liquid; bp 127°C at 0.15 mmHg; d_4^{20} 1.4345; n_D^{20} 1.5520.

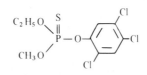

trichlormetafos-3

Famphur, Famophos®
Dimethyl p-(dimethylsulfamoyl)phenyl phosphorothionate

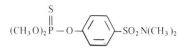

Famphur was introduced by American Cyanamid Company as a systemic insecticide for cattle grubs. It is crystalline powder; mp 55°C; very soluble in chlorinated hydrocarbons, slightly soluble in polar solvents, and insoluble in aliphatic hydrocarbons. The water solubility is about 100 ppm.

The acute oral LD_{50} for rats is 35 mg/kg and for sheep 400 mg/kg. It may be applied to cattle intramuscularly, by pour-on, or by feeding.

The principal degradation routes of famphur in mammals are P-O-phenyl bond cleavage and O-demethylation. Oxidative desulfuration and N-demethylation, which yield toxic products, take place to a small extent.[552,946]

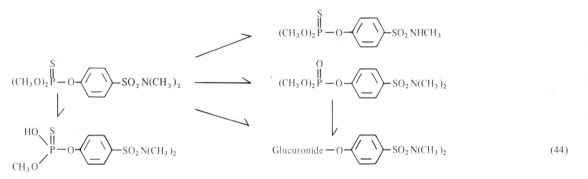

(44)

The O-ethyl homolog of famphur (Kaya-ace®) is useful as a nematocide (see Section V.B). The N-unsubstituted homolog of famphur is a systemic insecticide cythioate (Proban®). It appears to be useful for the control of animal ectoparasites such as ticks.[947] The acute oral LD_{50} of cythioate for rats is 160 mg/kg. It is a solid, melting at 43 to 44°C.

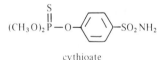

cythioate

Fenthion and related compounds

The insecticide fenthion (see Section V.A.1.b) appears to be promising for cattle grub control by feeding.[948] Its O-ethyl homolog, Lucijet® (diethyl 3-methyl-4-methylthiophenyl phosphorothionate), is effective against sheep blowfly larvae. The toxicity to mammals is considerably higher than that of the methyl homolog fenthion; the acute oral LD_{50} for rats is 25 to 100 mg/kg.

Introduction of an additional methyl group at another *meta* position of fenthion results in further decrease in mammalian toxicity. Dimethyl 3,5-dimethyl-4-methylthiophenyl phosphorothionate (Bay 37342) is one of the most promising animal systemic insecticides for the control of sheep botfly (*Oestrus ovis*).[949] It was 100% effective at 50 mg/kg dose. The acute oral LD_{50} to rats is 1,000 mg/kg.

The insecticide diazinon (see Section V.A.1.b) may also be used as a veterinary drug for the control of ticks on cattle and of blowflies and mites on sheep.

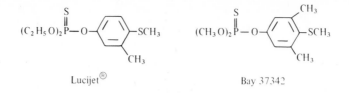

Lucijet® Bay 37342

Recently, a quinoline derivative has been added to the list of veterinary pesticides. Quinothion, diethyl 2-methylquinolin-4-yl phosphorothionate, which was developed by Cooper Technical Bureau, is effective against the sheep parasite *Psoroples communis ovis scab* by spraying or dipping.[949a] It is formulated with an amine stabilizer.

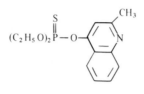

quinothion

c. Phosphorothiolothionates

R-3,828
S-(*p*-Chloro-α-phenylbenzyl) diethyl phosphorothiolothionate

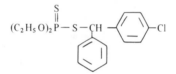

This is a new animal systemic insecticide introduced by Stauffer Chemical Company under the code number R-3,828. It has a very low mammalian toxicity; the acute oral LD_{50} for rats is 10,000 mg/kg. The principal detoxication in mammals is due to the cleavage of the P-S-C linkage to form *p*-chloro-α-phenylbenzyl alcohol, which is converted into the glucuronide and excreted.[950] In addition, deethylation and oxidative desulfuration also occur.

The insecticide phosmet (see Section V.A.1.c) is active as an animal systemic to control cattle grub, too.[951] Malathion is effective without systemic properties against horn flies, lice, and ticks on cattle.[940]

d. Phosphoramidates

Crufomate, Ruelene®
2-Chloro-4-*tert*-butylphenyl methyl *N*-methylphosphoramidate

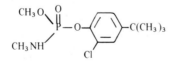

Crufomate was introduced in 1959 by Dow Chemical Company as a promising anthelmintic for cattle and sheep.[952] For the synthesis, see Equation 119 in Section II.E.1. The pure product is a white odorless crystalline powder; mp 60°C; bp 117 to 118°C at 0.01 mmHg. The technical preparations are yellow viscous oils. It is soluble in polar organic solvents. The solubility in water is 0.5%.

Crufomate is useful for the control of intestinal worms and warble flies by feeding the animals 37 to 150 mg/kg doses. It can be also used by spray to control ectoparasites of cattle. It does not accumulate in body fat like the thiono analog Dowco 109. However, it cannot be used for lactating dairy cattle. Crufomate has only low toxicity to plant pests. The acute oral LD_{50} for rats is 770 to 1,000 mg/kg. Crufomate inhibits cholinesterase in vitro with a high specificity to the insect enzyme; housefly-head cholinesterase is about 800 times as sensitive to crufomate as human erythrocyte acetylcholinesterase.[32] Crufomate is degraded in animals by cleavage of both ester and amide bonds (see Equation 73 in Section IV.B.2.b).[626,953]

Narlene®, Dowco 109®
2-Chloro-4-*tert*-butylphenyl methyl *N*-methylphosphoramidothionate

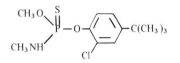

This is the thiono analog of crufomate described in the preceding paragraphs. It was introduced in 1958 by Dow Chemical Company as a systemic insecticide for livestock. Although it has only a poor contact insecticidal activity, it is effective to control cattle grubs and botflies on livestock and also as an anthelmintic. In spite of its low mammalian toxicity (the acute oral LD_{50} for rats is 1,000 mg/kg), it has the defect of accumulating in body fat, differing from the oxon analog crufomate in this respect.

e. Phosphonates
The insecticide trichlorfon (see V.A.1.g) is useful for the control of cattle grubs, warble flies, and intestinal worms, too. 4-Bromo-2,5-dichlorophenyl propyl methylphosphonate (CELA K-159) is a new animal systemic insecticide of promise for the control of cattle grubs; more than 84% of the grubs were eliminated by a dose of 5 mg/mg.[954]

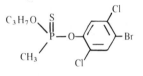

CELA K-159

Aromatic heterocycles may also be introduced into the phosphonate ester class. For example, oxinothiophos (ethyl quinolin-8-yl phenylphosphonothionate; Bacdip®) was recently developed by Bayer AG as one of promising new acaricides.[2] As it is relatively stable to alkaline media, it can be used in a dip to control ticks on livestock. The acute oral LD_{50} for rats is 150 mg/kg.

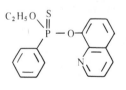

oxinothiophos

B. NEMATOCIDES

Most crop plants are attacked by phytopathogenic nematodes and often suffer a serious damage in yield. In order to prevent the damage, rotation has been recommended because of the host specificity of the nematode. Chemical control of nematodes, however, makes it feasible to plant the same crop successively for several seasons. Many nematocides are soil fumigants which volatilize and penetrate throughout the soil.

They include carbon disulfide, chlorpicrin, dichloropropene-dichloropropane mixture (D-D), ethylene bromide, and methyl bromide. Although they have the advantage of having powerful initial activity, they are poor in plant tolerance so that a safety interval of several weeks after field treatment is necessary before planting crops.

Organophosphate nematocides, which have been recently developed, differ from the fumigants in their basic characteristics: they are spread in the soil through the water system; they display high plant tolerance so that they can be used on established crops; they disperse and persist for a relatively long time in the soil; and they are anticholinesterase agents. Some of them are water soluble and possess systemic properties, being effective against nematodes that have entered roots. Most organophosphorus nematocides are also effective in the control of soil-borne insects.

Dichlofenthion, VC-13 Nemacide®

2,4-Dichlorophenyl diethyl phosphorothionate

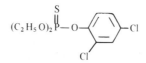

Dichlofenthion was introduced in 1956 by the Virginia-Carolina Chemical Corporation as the first organophosphorus nematocide. The pure material is a colorless liquid; bp 108°C at 0.01 mmHg; d_4^{20} 1.313; n_D^{25} 1.5318. It is highly soluble in most organic solvents. The solubility in water is 245 ppm. Dichlofenthion is relatively stable against hydrolysis and heat in comparison with corresponding nitrophenyl esters; it persists for a long time in soils, being effective for 1 to 2 years to control nematodes and soil-borne insects. It has no systemic activity, so it is not effective against nematodes which have penetrated into roots. The toxicity to mammals is relatively low; the acute oral LD_{50} for rats is 270 mg/kg; dogs fed 0.75 mg/kg/day for 90 days suffer no toxicological effect.

Fensulfothion, Terracur P®, Dasanit®

Diethyl p-methylsulfinylphenyl phosphorothionate

$$(C_2H_5O)_2\overset{\overset{\displaystyle S}{\|}}{P}-O-\underset{}{\bigcirc}-SOCH_3$$

Fensulfothion was developed by Schrader in 1957 and commercialized in 1965. It may be synthesized by the condensation of diethyl phosphorochloridothionate and p-methylsulfinylphenol or by the oxidation of the corresponding sulfide with hydrogen peroxide (see Section III.D; Equation 102). It is a liquid; bp 138 to 141°C at 0.01 mmHg; d 1.202; n_D^{24} 1.540; soluble in most organic solvents except aliphatic hydrocarbons. Water solubility is 1600 ppm. The oxidation of the sulfoxide group to the sulfone and the thiono-thiolo rearrangement take place rather readily with fensulfothion.

Fensulfothion is effective as a nematocide and soil insecticide with a long residual action. It displays both systemic and contact actions. The mammalian toxicity is very high; it is also absorbed through skin. The acute oral LD_{50} for rats is 4 to 10 mg/kg.

DSP, Kaya-ace®

Diethyl p-dimethylsulfamoylphenyl phosphorothionate

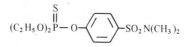

This was introduced in 1963 and commercialized in 1966 by Nippon Kayaku Company as a nematocide under the trade name Kaya-ace®. The pure compound is a colorless crystalline solid; mp 68 to 69°C; readily soluble in alcohols and benzene; almost insoluble in water.

When applied to soil, it persists for a relatively long time to control a variety of nematodes. It has insecticidal activity, too. The acute oral LD_{50} for mice is 65.4 mg/kg.

Many similar sulfamoylphenyl phosphorothionates have been synthesized and evaluated for insecticides.[955] Some of them are useful as systemic insecticides for animals, as exemplified by famphur (see V.A.3).

Thionazin, Nemafos®, Zinophos®, Cynem®
Diethyl 2-pyrazinyl phosphorothionate

$$(C_2H_5O)_2\overset{\overset{\displaystyle S}{\|}}{P}-O-\text{[pyrazinyl]}$$

Thionazin was introduced by American Cyanamid Company in 1966 for the control of soil insects and nematodes. It is an almost colorless liquid; n_D^{25} 1.5148; solubility in water 1.14 g/l; highly soluble in most organic solvents. Thionazin persists about 4 weeks in soils to protect plants from soil-borne pests and is absorbed through roots, exhibiting systemic nematocidal activity. Thionazin is hydrolyzed in soils to form pyrazinol, which is decomposed further into smaller fragments.[503a] Thionazin has a high mammalian toxicity; the acute oral LD_{50} for rats is 12 mg/kg.

Prophos, Mocap®
S,S-Dipropyl ethyl phosphorodithiolate

$$C_2H_5O\overset{\overset{\displaystyle O}{\|}}{P}(SC_3H_7)_2$$

Prophos is a nematocide-soil insecticide developed by the Mobil Chemical Company. It is a liquid; bp 86 to 91°C at 0.2 mmHg; soluble in most organic solvents; insoluble in water.

Prophos is effective to control nematodes as well as soil insects like corn rootworms. It acts only with contact. The acute oral LD_{50} for rats is 61.5 mg/kg.

Prophos is degraded in vivo by P-S bond cleavage and deethylation. The methylation of the liberated mercaptan occurs rapidly and is followed by oxidation to produce methyl propyl sulfide, sulfoxide, and sulfone.[956]

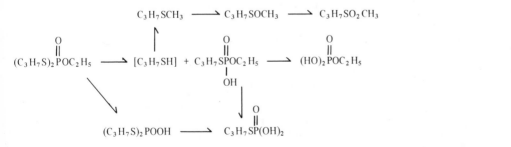

$$\tag{45}$$

Of related phosphorodithiolate compounds, S,S-diphenyl derivative (edifenphos) is useful as a fungicide (see Section V.D); S,S-dipropyl methyl phosphorothionodithiolate (VC 3-668) is effective as a larvicide of manure from chicks by oral administration. Some alkylphosphonodithiolate esters, exemplified by the following chemical structure, are active as nematocides.

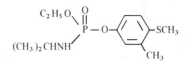

Phenamiphos, Nemacur®
Ethyl 3-methyl-4-methylthiophenyl *N*-isopropylphosphoramidate

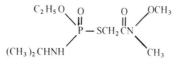

This is a new nematocide developed by Bayer AG.[957] It is a white crystalline substance; mp 49°C. The solubility in water is 700 ppm at 20°C. It is highly soluble in most organic solvents except aliphatic hydrocarbons.

Phenamiphos is effective against almost all important nematode species below 5 ppm in soil for a long time (3 to 4 months). It has systemic properties; it is absorbed by the roots and the leaves and translocated in the whole plant. Thus, it displays not only protective but also curative activity. The acute oral LD_{50}s for male and female rats are 15.3 and 19.4 mg/kg, respectively.

Phenamiphos can be regarded as the isopropylphosphoramidate analog of the insecticide phenthion. A similar modification of the insecticide dimethoate results in an effective nematocide, ethyl *S*-(*N*-methoxy-*N*-methylcarbamoylmethyl) *N*-isopropylphosphoramidothiolate (FCS 13).

FCS 13

Nellite®, Dowco 169
Phenyl *N,N′*-dimethylphosphorodiamidate

This nematocide was introduced by Dow Chemical Company under the code number Dowco 169. It is a solid; mp 105 to 106°C; bp 162°C at 0.5 mmHg. It is soluble in water (11.6 g/100 g at 25°C).

Nellite has a systemic activity and is specifically effective for root-knot nematode larvae. The acute oral LD_{50} is 250 mg/kg. It is rapidly metabolized in the cucumber plant into the conjugates of phenol and methylamine.[958]

C. INSECT CHEMOSTERILANTS

The successful eradication of the screwworm, *Cochliomyia hominivorax*, a serious livestock pest, from the island of Curacao and then the southeastern United States by the release of male insects sterilized by gamma radiation[959,960] has been shown. The demonstration that some chemicals could also induce sexual sterility in insects[961] suggested the potential usefulness of such chemicals (chemosterilants) for practical insect control and eradication. Chemosterilants may have an advantage in inducing sterility more economically in a high proportion of a natural population, without the necessity of rearing and releasing large numbers of insects. Because of competition between sterile males and normal males for mating with fertile females, chemosterilants have an advantage over insecticides. For example, when 50% of the population are killed by an insecticide (50% males and 50% females), the reproduction will decrease by 50%. When, however, the same percentage of the insect population is sterilized (50% males and 50% females), half of the remaining fertile females can be expected to mate with sterile males, provided sterilization does not affect the mating ability; consequently, the total population decrease will be 75%. As male insects are generally polygamous, sterilization of males is more effective than that of females. In many insect species females are monogamous.

Insect chemosterilants may be classified into three groups: 1) antimetabolites, 2) alkylating agents, and 3) miscellaneous agents. Aminopterin, methotrexate, 5-fluorouracil, and 5-fluoroorotic acid are examples of antimetabolite chemosterilants. These compounds are mostly female sterilants and are not so effective, in general, as aziridine derivatives which are alkylating agents.

The term "alkylating agents" does not necessarily mean chemicals which introduce an alkyl group into a molecule in the organic chemistry sense, but rather chemicals which introduce a substituted alkyl group such as aminoalkyl, hydroxyalkyl, thioalkyl, or even more complex groups under physiological conditions. They are preferably called biological alkylating agents.[413] Representative chemosterilants of this class are tepa, thiotepa, metepa, and apholate. They are all aziridine derivatives of phosphoric acid.

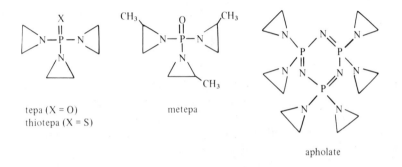

tepa (X = O)
thiotepa (X = S)

metepa

apholate

Aziridine derivatives having chemosterilant activity are not limited to phosphoric acid derivatives; some derivatives of carbonate, carboxylate, sulfinate, sulfate, sulfide, disulfide, thiazine, and quinone are also known to be effective sterilants.[962] Most of the effective aziridine derivatives have (or will have after biotransformation) electron withdrawing functional groups attached to the aziridine nitrogen, such as P=O, C=O, C=N, P=S, S=O, or SO$_2$. They have at least two aziridine groups in the molecule; monofunctional derivatives are generally less effective as sterilants. In general, male insects are more susceptible than females to these compounds. This does not mean that they are strictly specific to males; all the known "male chemosterilants" are effective to females too.[963]

Many aziridine derivatives are water soluble and highly reactive compounds. Aziridine or ethyleneimine is a highly strained three-membered cyclic amine. Protonation on the nitrogen atom makes it more reactive toward ring opening and alkylation of a nucleophile.

$$R-N \xrightarrow{H^+} R-\overset{+}{\underset{H}{N}} \xleftarrow{A^-} R-NH-CH_2CH_2-A \tag{45}$$

Ethyleneimine itself is not active enough as a chemosterilant, though it is mutagenic and oncolytic.[964] Ethyleneimides, i.e., N-acyl derivatives of aziridine such as tepa and other related sterilants, are more reactive than uncharged N-alkyl aziridines because electron withdrawal by the acyl group makes the methylene groups more reactive with nucleophiles.[261]

$$\longrightarrow >P-NHCH_2CH_2-A + OH^- \tag{46}$$

305

Aziridine chemosterilants are rather stable in alkaline solutions but are very unstable under acid conditions; for example, the half-life of tepa at pH 9 is more than 50 hr, while that at pH 5.5 is about 30 min.[965] NMR study on the decomposition of tepa in heavy water indicated that two hydrolytic routes are possible: the cleavage of P-N bonds to liberate aziridine and the opening of aziridine rings.

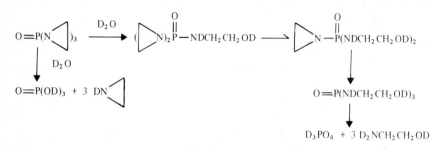

$$D_3PO_4 \; + \; 3 \; D_2NCH_2CH_2OD \tag{47}$$

Aziridines are readily polymerized by the initial action of alkylating agents or acids.

$$R-N\triangleleft \xrightarrow{R'X} R-\overset{R'}{\underset{+}{N}}\triangleleft \xrightarrow{R-N\triangleleft} R-\overset{R'}{N}\diagdown\diagup\overset{R}{\underset{+}{N}}\triangleleft \longrightarrow \text{polymer} \tag{48}$$

In addition to aziridine compounds, several groups of alkylating agents are known. They involve alkyl halides, sulfate esters, sulfonate esters, phosphate esters, chloromethyl ethers, ammonium and sulfonium compounds, 2-chloroethyl sulfides, 2-chloroethylamines, epoxides, β-lactones, diazoalkanes, activated ethylenic compounds, and halogenomethyl ketones and esters.[413] However, only two groups of these, 2-chloroethylamines (nitrogen mustards) and alkane sulfonates, have received considerable attention as candidate chemosterilants.[962] Some sulfonates appear to be promising, as exemplified by busulfan, which was more effective than apholate against the boll weevil.[966] 2-Chloroethylamines are regarded to be essentially the same as aziridine derivatives in the manner in which they react (Equation 49), but are less active as chemosterilants in comparison with aziridines.

$$CH_3SO_3(CH_2)_4OSO_2CH_3$$

busulfan

$$R_2NCH_2CH_2Cl \longrightarrow R_2\overset{+}{N}\underset{CH_2}{\overset{CH_2}{\diagup\diagdown}} \xrightarrow{A^-} R_2NCH_2CH_2A \tag{49}$$

Alkylating agents may react with a thiol group, amino group, or nitrogen atom in heterocyclic bases in biologically important molecules.[413] The hydroxyl group, in sugars for example, can react with alkylating agents, but the reaction is not significant under physiological conditions in the presence of excess water molecules. Most of the molecular changes induced by alkylating chemosterilants involve alteration of nucleic acid or protein metabolism.

Administration of thiotepa caused a decrease of the thiol group content in blood serum and in tumor tissue of rats. The SH enzyme dehydrogenase in Ehrlich ascite cells was inhibited by thiotepa. The prolonged metaphase of dividing cells in tissue culture induced by thiotepa is probably due to the sensitivity of thiol groups in mitotic apparatus.[964] On the other hand, alkaline phosphatase activity, which is specific for DNA metabolism, in gonads of Southern corn rootworm was reduced by apholate. However, Turner insisted that induction of sterility was more closely related to an alteration of ovarian protein metabolism than to the specific enzyme inhibition.[964]

The DNA content and lactate dehydrogenase (LDH) activity, which increased during incubation of normal housefly eggs, did not increase in nonviable eggs deposited by females that had fed on a diet containing aziridine chemosterilants. The behavior of LDH is very characteristic; other dehydrogenases such as glucose-6-phosphate dehydrogenase and malic acid dehydrogenase appeared normal as judged by enzyme activity measurement and acrylamide gel electrophoresis pattern.[967] In chemosterilized eggs, only two LDH isozymes were detected, whereas in normal eggs during embryogenesis, the pattern changed from two to five LDH isozymes. Painter and his co-workers suggested that LDH in chemosterilized eggs may be bound in some way, because the activity was increased by purification procedure using DEAE-cellulose column chromatography.[967]

DNA is not synthesized in any of the chemosterilized eggs as determined by methylated albumin-kieselguhr column chromatography.[968] The inhibition of DNA biosynthesis was accompanied by an increase of acid soluble DNA precursor content, possibly owing to either the inhibition of DNA polymerase or the stimulation of DNA degradation.[969] Although it has been known that the primary reaction of alkylating agents with nucleic acids in vitro occurs particularly on the *N*-7 of guanine moieties and bifunctional agents yield di(guanin-7-yl) derivatives in addition to 7-alkylguanines,[413] no such guanine derivatives were found in the RNA, the purines of which had been labeled by glycine-C^{14}, isolated from eggs of the houseflies treated with apholate or thiotepa.[970] Whether such reaction occurs in the DNA or nucleic acid precursors of chemosterilized eggs is not known.

Formate-C^{14} uptake by housefly testes (probably into purine nucleotides) was inhibited by thiotepa (see Figure 18).[971] Besides the inhibition of *de novo* synthesis of purine ribonucleotides, thiotepa also appears to inhibit DNA synthesis by inhibition of the conversion of purine ribonucleotides to deoxycompounds for incorporation into DNA.[972] On the other hand, in stable flies treated topically with apholate, thymidine-^{3}H incorporation into nuclei of nurse cells and follicular cells of ovaries was inhibited.[973]

There was no change in the RNA content of eggs obtained from apholate sterilized female houseflies, but the adenylic acid content was significantly lower than in the RNA of normal eggs. The RNA of eggs from thiotepa sterilized flies was a little lower in guanilic acid content and contained an unidentified component not present in normal fly egg RNA.[970] The maxima of the UV absorption spectra of the ribosomes (271 nm) and ribosomal RNA (257 nm) from chemosterilized fly eggs differ from those of ribosomes (267 nm) and ribosomal RNA (263 nm) from normal eggs.[974] In a cell free system, the ribosomes from chemosterilized eggs were 1.6 times less active in inducing protein synthesis than the normal egg ribosomes.[975] Alteration of both RNA and DNA in eggs as a result of aziridinyl chemosterilization of the adult houseflies was further confirmed by a DNA-RNA hybridization technique.[976] The RNA from normal or chemosterilized eggs hybridized with the DNA from the same type

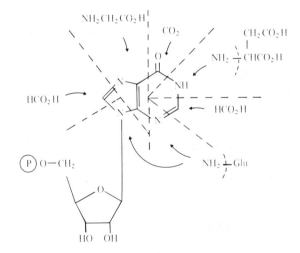

FIGURE 18.   Biosynthesis of purine nucleotides.

of eggs to almost the same degree. However, a lesser degree of hybridization was observed between RNA from one type of eggs and DNA from the other. These facts indicate the RNA and DNA from chemosterilized eggs were altered so that they were complementary to each other, but were only partially complementary to the RNA and DNA from normal eggs.

The alkylating agents are not only very effective for sterilization of insects, but also have anticancer activity. The anticancer agents of this category are exemplified by cyclophosphamide, phosphazine, and some related compounds.

cyclophosphamide

phosphazine

alkyl N-[bis(1-aziridinyl)phosphinyl] carbamates

However, the alkylating agents also have deleterious effects on a variety of organisms. The compounds exhibit little acute toxic action, but are notable for their delayed effect, accompanied by a selective action against some proliferating tissues. In addition to sexual sterilization, the alkylating agents may produce such unusual biological effects as mutagenesis, teratogenesis, and carcinogenesis.[977] This, then, is the limitation for practical applications of the alkylating chemosterilants for insect control, though they still might be useful to reared populations and in combination with attractants, baits, or lures for the treatment of natural populations.

Recently, "nonalkylating sterilants" have been developed.[963] This was initiated by the discovery of the sterilizing activity of the nonaziridine containing phosphoramide hempa (hexamethylphosphoric triamide) in 1964.[978] Many phosphoramides have been examined for sterilizing activity and some of them were found to be active as chemosterilants.[979,980] Even in non-nitrogen containing phosphorus esters, some compounds showed sterilization activity. For example, treatment of the adult females of spider mites, *Tetranichus urticae,* with trimethyl and triethyl phosphates caused a reduction both in egg production and hatching percentage.[981] However, this can be due to alkylation. Some kinds of nonphosphorus compounds having nonalkylating sterilant activity have been found. They include melamines, dithiobiurets, thiourea, dithiazolium salts, organotin compounds, organoboron compounds, alkaloids, steroids, and antibiotics.[963,964]

Hempa is enzymatically demethylated in the organisms via a highly reactive methylol intermediate (XXVIII) which appears to be the active principle for sterilization.[963,982,983] The intermediate was synthesized and confirmed to be active as a sterilant.[984] The N-methylol hempa XXVIII is rapidly converted into the methylene compound (XXIX) liberating formaldehyde by heat, and is changed to the demethyl derivative (XX) in aqueous solutions. Any decomposed products, i.e., the demethyl compound, the methylene compound, and formaldehyde, have no or only poor sterilizing activity. Only the methylol intermediate is as active as the parent compound hempa.

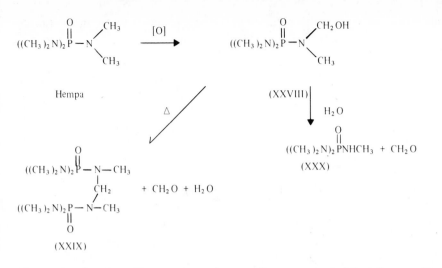

Hempa

(XXVIII)

Δ

H_2O

$((CH_3)_2N)_2\overset{\overset{\displaystyle O}{\|}}{P}-N-CH_3$
$\overset{|}{CH_2}$ $+$ CH_2O $+$ H_2O
$((CH_3)_2N)_2P-\overset{|}{N}-CH_3$
$\overset{\|}{O}$

(XXIX)

$((CH_3)_2N)_2\overset{\overset{\displaystyle O}{\|}}{P}NHCH_3$ + CH_2O

(XXX)

(50)

The mode of action of hempa is not known. Turner suggested that the coordination of essential metals by hempa might be related to the sterilant activity,[964] while Borkovec considered that the active N-methylol metabolite is a kind of alkylating agent.[963]

Several excellent reviews and books on chemosterilants have been published.[21,962–964,966,972,986–990]

Tepa, APO, Aphoxide
Tris-(1-aziridinyl)phosphine oxide, trimethylenephosphoramide

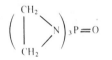

Tepa is a colorless, odorless, and hygroscopic crystalline solid; mp 45°; bp 90 to 91°(0.3 mmHg); extremely soluble in water and very soluble in acetone, alcohol, and ether. Tepa is unstable in aqueous solutions but is stable as long as 12 months in anhydrous polyethyleneglycol. Fifty percent effective dose for sterilization of houseflies by injection (ED_{50}) is 0.1 μg/fly. It is a highly selective sterilant, as shown by the ratio LD_{50}/ED_{90} for houseflies of 416.[964] However, the mammalian toxicity is relatively high; the oral LD_{50} for rats is 37 mg/kg.[977]

One aziridine group of tepa can be replaced by an appropriate group without a great change in insect sterilizing activity. The activity of P-substituted bis(1-aziridinyl)phosphine oxide (XXXI) decreases in the following order: X = RNH > RS > RO > R.[991] Within each homologous series, activity decreases with increasing size of the alkyl group. The methylamino derivative is as active as tepa but is less selective as a sterilant; its selectivity ratio for housefly LD_{50}/ED_{90} is 88. In the sulfide series, a similar relationship was observed between the activity and the structure of the substituent.

(XXXI)

For houseflies, high sterilizing activity is more likely to be associated with more polar compounds,[966] whereas for the mosquito pupae, *Anopheles albimanus,* the less polar compounds are the more effective sterilants.[992] Thus, sulfides are generally more effective than the corresponding oxides against the mosquito pupae. Substitution on the carbon of the aziridine ring generally decreases the sterilizing activity.[966]

Thiotepa
Tris(1-aziridinyl)phosphine sulfide

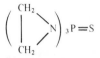

Thiotepa is a white crystalline solid with moderate hygroscopicity; mp 51.5°; soluble in water (19%), benzene, and acetone. It is suitable for sterilization of *Culex pipiens quinquefasciatus*; tepa and metepa are ineffective as pupal sterilants. Thiotepa is biotransformed rapidly into tepa in mammals and insects.

Metepa, Methaphoxide, MAPO
Tris(2-methyl-1-aziridinyl)phosphine oxide

Metepa is a straw-colored liquid; bp 118 to 125° (1 mmHg); d_{25}^{25} 1.079; n_D^{25} 1.4798. It is miscible with water and most organic solvents. Metepa is stable in alkaline solutions but extremely unstable in acid solutions; the half-life in N-NaOH and in N-HCl is more than 100 hr and less than 1 min, respectively. The ED_{50} for houseflies by injection is 1.31 μg/fly and the oral LD_{50} for rats is 136 mg/kg.

Apholate, APN, Omaflora[®]
2,2,4,4,6,6,-Hexakis (1-aziridinyl) -2,2,4,4,6,6-hexahydro -1,3,5,2,4,6-triazatriphosphorine

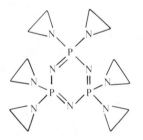

Apholate is a white odorless crystalline solid, mp 155°. As apholate polymerizes at high temperature, it is difficult to establish a precise melting point. It is soluble in water (20%) and chloroform (20%), but only slightly soluble in methanol and acetone. Some samples may contain as much as 5% water insoluble impurity which has the same sterilizing capacity as apholate. The ED_{50} for male adult houseflies by injection is 0.404 μg/fly.

Hempa, HMPA, HMP
Hexamethylphosphoric triamide

$$[(CH_3)_2 N]_3 P{=}O$$

Hempa is a colorless liquid; mp 7.2°; bp 68 to 70° at 1 mm Hg; n_D^{25} 1.4582; d_4^{20} 1.0253. It is soluble in water and in both polar and nonpolar solvents. In contrast to aziridine compounds, hempa is chemically stable; it resists alkaline and dilute acid hydrolysis. By strong acids, however, P-N bond cleavage occurs rapidly. The ED_{50} value for male houseflies by injection is 5.42 μg/fly.[978] In comparison to tepa, hempa is about 50 times less effective as a chemosterilant. Hempa has, however, a greater advantage in mammalian toxicity; its minimum lethal dose for rats by oral administration is 2640 mg/kg.[978]

The thiono analog of hempa, thiohempa, has almost the same activity. Replacement of one or more methyl groups in hempa or thiohempa with higher alkyl or aryl groups or with hydrogen led to a decrease in activity.[21,979] However, moderate activity towards some species of insects was observed in some derivatives in which one dimethylamino group of hempa was displaced by a nitrogen containing heterocycle.[980]

The physiological and cytological properties of these nonaziridinyl chemosterilants resemble the effects of alkylating agents.[21] Like the alkylating agent tepa, hempa induced a high frequency of recessive lethal mutations in the parasitic wasp, *Bracon hebetor*.[985]

In addition to the above mentioned compounds, a number of phosphorus compounds have been reported to be effective as insect chemosterilants.[21] Selected compounds are listed below.

$(\triangleright N)_3 P{=}N{-}SO_2C_6H_5$	*N*-[Tris(1-aziridinyl)phospha-nylidene] benzenesulfonamide[993]
$(\overset{CH_3}{\triangleright}N)_3 P{=}S$	Tris(2-methyl-1-aziridinyl) phosphine sulfide, methiotepa[994]
$(\triangleright N)_2 \overset{O}{\underset{\|}{P}}{-}NHCH_3$	*P,P*-Bis(1-aziridinyl)-*N*-methyl-phosphinic amide[995]
$(\triangleright N)_2 \overset{S}{\underset{\|}{P}}NHCH_2CH_2CH_2OCH_3$	*P,P*-Bis(1-aziridinyl)-*N*-(3-methoxy-propyl)phosphinothioic amide[995]
$(\triangleright N)_2 \overset{S}{\underset{\|}{P}}{-}N\overset{\frown}{\underset{\smile}{}}O$	Bis(1-aziridinyl)morpholino-phosphine sulfide, morzid[964]
$(\triangleright N)_2 \overset{O}{\underset{\|}{P}}N\underset{CH_3}{CH_2}CH_2\underset{CH_3}{N}\overset{O}{\underset{\|}{P}}(N\triangleleft)_2$	*N,N'*-Ethylenebis[*P,P*-bis(1-aziridinyl)-*N*-methylphosphinic amide] aphamid[961]
$(\triangleright N)_2 \overset{S}{\underset{\|}{P}}NHCH_2CH_2NHP(\overset{S}{\underset{\|}{}}N\triangleleft)_2$	*N,N'*-Ethylenebis[*P,P*-bis(1-aziridinyl)phosphinothioic amide][996]
$(\triangleright N)_2 \overset{O}{\underset{\|}{P}}NHCONH{-}\bigcirc{-}Cl$	1-[Bis(1-aziridinyl)phosphinyl]-3-(3,4-dichlorophenyl)urea[997]
$(\triangleright N)_2 \overset{O}{\underset{\|}{P}}{-}\underset{H}{N}{-}$ pyrimidine${-}N\overset{\frown}{\underset{\smile}{}}O$	*P,P*-Bis(1-aziridinyl)-*N*-(6-morpholinopyrimidine-4-yl)-phosphinic amide[998]
$(\triangleright N)_2 \overset{S}{\underset{\|}{P}}CH_3$	Bis(1-aziridinyl)methylphosphine sulfide[999]

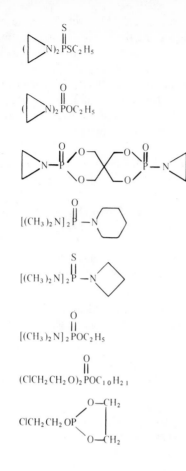

S-Ethyl bis(1-aziridinyl)-
phosphinodithioate[1000]

O-Ethyl bis(1-aziridinyl)-
phosphinate[996]

3,9-Bis(1-aziridinyl)-2,4,8,10-
tetraoxa-3,9-diphosphaspiro[5.5]
undecane-3,9-dioxide[1001]

N,N,N',N'-Tetramethyl-P-piperidino-
phosphonic diamide[979]

N,N,N',N'-Tetramethyl-P-azetidino-
phosphonothioic diamide[980]

Ethyl N,N,N',N'-tetramethyl-
phosphorodiamidate[1002]

Bis(2-chloroethyl) decyl
phosphate[1003]

2-(2-Chloroethoxy)-1,3,2-
dioxaphospholane[1004]

D. FUNGICIDES

Plants may be attacked by diseases caused by the infection of microorganisms or viruses. Fungal diseases are particularly serious. Inorganic copper compounds, sulfur, and metallic polysulfides were exclusively used as fungicides before 1940. Modern organic fungicides, including organomercury compounds, organotin compounds, organoarsenic compounds, dithiocarbamates, and trichloromethylthioamides, have been developed in the last three decades. It is only very recently that organophosphorus compounds have attracted much attention for their fungicidal activity because the biological activity of most organophosphorus compounds was believed to be due to their anticholinesterase activity by phosphorylation reaction, and the inhibition of esterases appears to cause no critical damage in fungi.

Environmental pollution problems by organomercury fungicides applied for the control of rice blast disease, which is one of the most serious diseases of rice, particularly in Japan (1952 to 1967), stimulated scientists to develop new biodegradable fungicides which are effective against the pathogen *Piricularia oryzae*. One great success was the finding of the organophosphorus fungicide Kitazin (*S*-benzyl diethyl phosphorothiolate) in 1963.[1005] Following this initial work, many phosphoro- and phosphono-thiolate esters were evaluated for fungicidal activity, and some of them were developed into commercial fungicides.

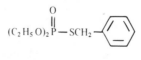

Kitazin

Another important group of organophosphorus fungicides is the derivatives of phosphoramidic acid, exemplified by triamiphos and phosbutyl.

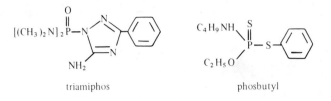

triamiphos   phosbutyl

Certain phosphites and phosphonium salts are also active as fungicides.

It is believed that the fungicidal mechanism of the *S*-benzyl phosphorothiolate Kitazin is due to the block of the biosynthesis of fungal cell wall chitin. However, the mechanisms of the other organophosphorus fungicides are not well established yet. Only a few reviews on organophosphorus fungicides have been published.[132,302,1006,1007]

1. Phosphorothiolates, Phosphonothiolates and Related Esters

Several phosphorothiolates and phosphonothiolates have been commercialized as fungicides. These phosphorothiolate type fungicides are particularly effective to control *Piricularia oryzae*. Although the *S*-benzyl ester moiety is common to some fungicides, for example Kitazin P, Conen, and Inezin T, it is not necessarily required for fungicidal activity: the number of carbon atoms between S and benzene ring can be altered from zero to three without significant change in the activity.[101,1008] Thus, phenyl phosphorothiolate esters such as Cerezin and edinfenphos are also very effective fungicides. Moreover, the benzyl group can be replaced by an alkyl group containing five to eight carbon atoms. The thiolate ester form, however, appears to be important for the anti-rice blast activity. The activity of phosphorothionate, phosphorothiolothionate, and phosphate esters corresponding to the phosphorothiolate ester Kitazin is extremely low. A similar structure-activity relationship was also observed in a series of saligenin cyclic phosphorus esters: only *S*-alkyl cyclic phosphorothiolate (XXXII) showed considerable fungicidal activity.[902] However, the phosphate ester of dichlorophenol, phosdiphen, was recently found to be effective as a fungicide for the control of *P. oryzae*. A few phosphonothionate esters like the compounds XXXIII and XXXIV also have fungicidal activity against this fungus.[132]

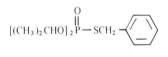

Kitazin P

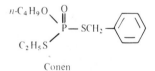

Conen

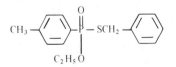

Inezin T

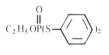

edifenphos

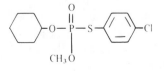

Cerezin

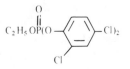

phosdiphen

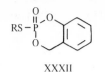

XXXII

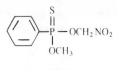

XXXIII

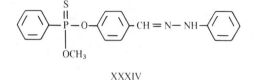

XXXIV

Some simpler trialkyl esters of phosphorothioic acids are effective against certain fungi in soils. Trimethyl phosphorothionate and trimethyl phosphorotetrathioate appear to be useful for the control of *Pythium* sp.[132],[1007] The fungicidal activity of trialkyl phosphorotetrathioates decreases with increasing length of the alkyl chain.

$$(CH_3O)_3P=S \qquad\qquad (CH_3S)_3P=S$$

In general, organophosphorus compounds are highly species selective in fungicidal activity. Thus, trimethyl phosphorothionate is almost ineffective against *Rhizoctonia*, *Fusarium*, and *Verticillium*.[1007] On the other hand, an analogous compound, diethyl *S*-methyl phosphorothiolothionate, is a good fungicide against *Rhizoctonia solani*.[132] The oral LD_{50} of this compound for rats is 156 mg/kg. It is interesting to note that the structurally related insecticide phorate shows a marked fungicidal effect at high temperatures (27 to 33°C) against *R. solani*, whereas at lower temperatures against *F. moniliforme*.[1009]

phorate

The structural characteristics of these fungicidal organophosphorus esters differ distinctively from those of insecticidal ones as seen in Section V.A. This suggests, apparently, that the mode of fungicidal action may not be related to esterase inhibition. In this context, it is interesting to note that the fungicidal thiol ester analog of salithion (XXXII; R = CH$_3$) has not only phosphorylating activity but also high alkylating activity towards SH compounds and inhibits SH enzymes (see Sections III.C.1 and IV.E.3).[1010] It was also suggested that the bactericidal activity of the insecticide chlorthion may be due to its inhibitory activity to the SH enzyme triosephosphate dehydrogenase.[1011]

On the other hand, the fungicide Kitazin inhibits the incorporation of ^{14}C-glucosamine into the cell wall chitin, that is the β-(1 → 4)-linked polymer of *N*-acetyl-D-glucosamine, of *Piricularia oryzae*.[838],[1012] The mycelia of the fungus treated with Kitazin P accumulated UDP-*N*-acetyl glucosamine.[1012] Thus, the fungicidal action of Kitazin may be due to the inhibition of one or more biochemical processes beyond the formation of the intermediate, i.e., permeation of an intermediate through cytoplasmic membrane, formation of *N*-acetyl glucosamine-β-1,4-linkage and further polymerization, or formation of cell wall chitin layer. Maeda et al. suggested that the site of the action is related more probably to permeation than polymerization.[1012] The biosynthetic pathway of cell wall chitin and the possible primary site of action of the fungicide Kitazin may be schematically illustrated as follows:

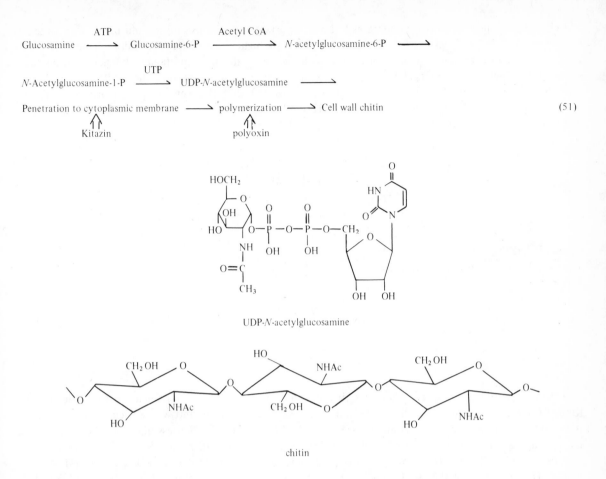

$$\text{Glucosamine} \xrightarrow{\text{ATP}} \text{Glucosamine-6-P} \xrightarrow{\text{Acetyl CoA}} N\text{-acetylglucosamine-6-P} \longrightarrow$$

$$N\text{-Acetylglucosamine-1-P} \xrightarrow{\text{UTP}} \text{UDP-}N\text{-acetylglucosamine} \longrightarrow$$

$$\text{Penetration to cytoplasmic membrane} \longrightarrow \text{polymerization} \longrightarrow \text{Cell wall chitin} \qquad (51)$$

$$\underset{\text{Kitazin}}{\Uparrow} \qquad \underset{\text{polyoxin}}{\Uparrow}$$

UDP-*N*-acetylglucosamine

chitin

The antibiotic polyoxins, which are produced by *Streptomyces cacaoi* and are now in practical use as agricultural fungicides in Japan, also inhibit the biosynthesis of cell wall chitin of some phytopathogenic fungi.[1013,1014] For example, polyoxin D, which is the most active component of polyoxins, inhibits the activity of chitin synthetase (chitin-UDP acetylglucosaminyltransferase) from *Piricularia oryzae* more than 80% at $2 \times 10^{-7}M$. Polyoxins are almost nontoxic to mammals; the mouse does not suffer any adverse effect by the oral administration of 15 g polyoxins/kg.[1015] On the other hand, the new bactericidal antibiotic phosphonomycin, (-)(1R,2S)1,2-epoxypropylphosphonic acid, which is produced by strains of *Streptomyces,* blocks the synthesis of bacterial cell walls by inhibiting irreversibly pyruvate-UDP-*N*-acetylglucosamine transferase of both Gram-positive and Gram-negative bacteria.[1016] The bacterial cell walls are not made of chitin but of more complicated mucopeptide, in which the polysaccharide part consists of acetylglucosamine and acetylmuramic acid, the derivative of glucosamine.[1017] The mammalian toxicity of phosphonomycin is very low; the intraperitoneal LD_{50} for mice is 4,000 mg/kg.

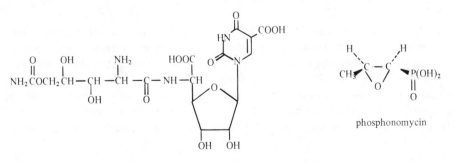

polyoxin D

phosphonomycin

Cell walls are not present in vertebrates but are important for microorganisms and plants to maintain life. The cell walls of fungi consist of chitin but those of higher plants are of cellulose. Therefore, the inhibitors of chitin biosynthesis are considered to be some of the most preferable fungicides from the standpoint of selective toxicity. Moreover, as chitin occurs also in arthropods, the inhibition of chitin synthesis might be lethal for insects too. It was suggested that the new nonphosphorus insecticide 1-(4-chlorophenyl)-3-(2,6-difluorobenzoyl)urea (TH-6040) exerts its insecticidal action by blocking the chitin formation of insects.[1018]

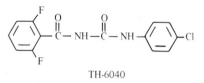

TH-6040

Kitazin P®, IBP

S-Benzyl diisopropyl phosphorothiolate

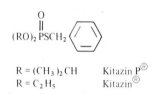

$$R = (CH_3)_2CH \quad \text{Kitazin P}^®$$
$$R = C_2H_5 \quad \text{Kitazin}^®$$

S-Benzyl diethyl phosphorothiolate was first introduced in 1965 by Ihara Chemical Company (now Kumiai Chemical Company) as a fungicide under the trade name Kitazin, but was replaced in 1967 by the isopropyl homolog Kitazin P for commercialization. Pure Kitazin P is a colorless liquid; bp 126°C at 0.04 mmHg; mp 22.5 to 23.8°C; n_D^{20} 1.5106. It is slightly soluble in water (0.1% at 18°C) but readily soluble in most organic solvents. The acute oral LD_{50} for mice is 660 mg/kg; dermal LD_{50} for mice is 4,000 mg/kg. The properties of the ethyl homolog, Kitazin, are as follows: bp 120 to 130°C at 0.1 to 0.5 mmHg; n_D^{20} 1.5240; oral LD_{50} for mice 237.7 mg/kg.

The Kitazins are manufactured by the following sequence of reactions:

$$PCl_3 + 3ROH \xrightarrow[-RCl, 2HCl]{} (RO)_2POH \xrightarrow{S. NH_3} (RO)_2 \overset{O}{\underset{\|}{P}}SNH_4 \xrightarrow[-NH_4Cl]{ClCH_2C_6H_5} (RO)_2 \overset{O}{\underset{\|}{P}}SCH_2C_6H_5 \quad (52)$$

In the homologous series of dialkyl S-benzyl phosphorothiolates, the maximum fungicidal activity is obtained when the number of carbon atoms in the alkyl group is three or four. The dimethyl homolog has poor activity. In the analogous series, the phosphorothiolate esters are much more effective than corresponding phosphorothionate, phosphorothiolothionate, and phosphate esters. Introduction of substituents such as chlorine atom or nitro group on the benzene ring has little effect on the increase of fungicidal activity.

Kitazin and Kitazin P inhibit more strongly the mycelial growth and the spore formation of *Piricularia oryzae* than the spore germination. Thus, they are effective curatively rather than prophylactically. Kitazin P has systemic properties; it is absorbed by the roots of rice plants and transferred over the entire foliage to protect the plant from rice blast disease. The fungicidal action of Kitazins is believed to be due to the inhibition of chitin formation. The corresponding phosphate analogs similarly inhibit the chitin formation and mycelial growth of *P. oryzae* in culture media, but are not effective to control the disease. The poor in vivo activity of the phosphates was attributed to the rapid degradation of the phosphate esters in plants.[1012]

The Kitazins also have insecticidal activity, but the activity is too weak to be utilized for insecticides. However, Kitazins have a considerable synergistic activity with the insecticide fenitrothion.

Kitazin is rather stable in neutral aqueous solutions (the half-life at $27°C$ is 80 days), but is decomposed rapidly under UV light by the cleavage of the S-C and P-S linkages.

Kitazin P is metabolically degraded, mainly by cleavage of the S-C bonding, in rats, cockroaches, rice plants, *Piricularia oryzae*, and soils, yielding O,O-diisopropyl hydrogen phosphorothioate.[302,567,1019] Diisopropyl hydrogen phosphate was also found as one of the major degradation products in rice plants, but was only a minor metabolite in rats, insects, and fungi. O-Dealkylation occurs in fungi but is a minor pathway of metabolism. The most interesting metabolite in *P. oryzae* is the *meta* hydroxy derivative of Kitazin P. Neither *ortho* nor *para* hydroxylated metabolite was found. The fungicidal activity of the *m*-hydroxy Kitazin P is less than the parent compound. In rice plants the isomerized product, diisopropyl benzyl phosphorothionate, was found.[1019] It is probably produced photochemically (see Equation 118 in Section III.E). The metabolic pathways of Kitazin P are shown below.

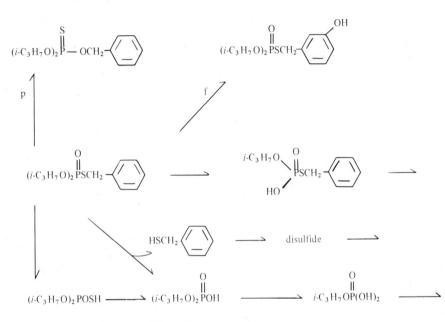

Metabolic pathway of Kitazin P   (53)
   f: This transformation is observed only in fungus.
   p: This transformation is observed only in plants (probably due to light).

Several clones of *P. oryzae* which are resistant to the Kitazins were obtained from colonies grown on media containing the fungicide.[1020] They are resistant to other organophosphorus fungicides too, but are susceptible to the antibiotic kasugamycin and to the organochlorine fungicide pentachlorophenol. Some phosphoramidates, however, are effective to control selectively these resistant clones and have a synergistic effect with the Kitazins and other phosphorothiolate type fungicides.[1021] The most active one among tested phosphoramidates was ethyl 2,4,5-trichlorophenyl *N*-methyl-*N*-phenylphosphoramidate (XXXV). In contrast, the phosphoramidate herbicide amiprophos (ethyl 2-nitro-4-methyl *N*-isopropylphosphoramido-thionate) acts antagonistically with the fungicide to the resistant clones.

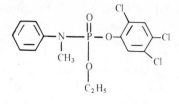

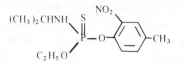

(XXXV)

amiprophos

Conen®, BEBP
S-Benzyl *n*-butyl *S*-ethyl phosphorodithiolate

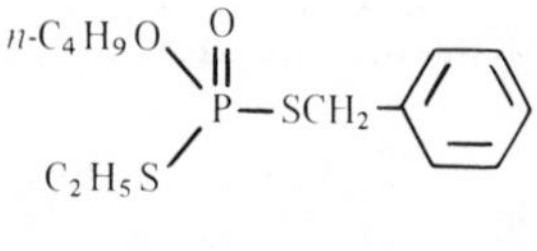

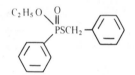

This was developed by Sumitomo Chemical Company as a fungicide to control rice blast disease. It is synthesized by the reaction of ethyl bromide with sodium *S*-benzyl *O*-butyl phosphorodithioate. The oral LD_{50} for mice is 118 mg/kg.

Inezin®, ESBP
S-Benzyl ethyl phenylphosphonothiolate

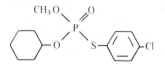

Inezin was introduced in 1967 by Nissan Chemical Company, Japan, as a fungicide to control rice blast and rice sheath blight. It is synthesized according to the following reaction sequence:

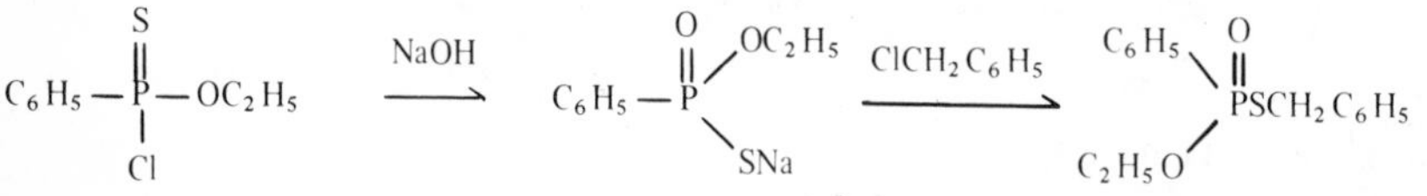

$$C_6H_5-\overset{\overset{S}{\|}}{\underset{\underset{Cl}{|}}{P}}-OC_2H_5 \xrightarrow{NaOH} C_6H_5-\overset{\overset{O}{\|}}{\underset{\underset{SNa}{}}{P}}\overset{OC_2H_5}{} \xrightarrow{ClCH_2C_6H_5} \overset{C_6H_5}{\underset{C_2H_5O}{}}\overset{\overset{O}{\|}}{P}SCH_2C_6H_5 \qquad (54)$$

It is a liquid and is soluble in most organic solvents. The oral LD_{50} for mice is 720 mg/kg.

The degradation of Inezin proceeds mainly through the cleavage of both the P-S and S-C linkages by *P. oryzae*, UV light, and in neutral and acid solutions.[302] In rice plants, S-C bond cleavage and deethylation occur preferentially. As in the case of Kitazin, hydroxylation occurs on the *meta* position of the benzyl group by fungal action.[568]

Cerezin®
S-p-Chlorophenyl cyclohexyl methyl phosphorothiolate

The active ingredient of Cerezin was discovered by Schrader in 1961 and developed into a fungicide in 1966 by Nihon Tokushu Noyaku Seizo Company, Japan.[1006] It is synthesized by the reaction of cyclohexyl methyl phosphate with chlorophenylsulfenyl chloride (see Equation 96 in Section II.D). Cerezin has a curative effect on rice blast disease. It also has insecticidal activity against two hopper species, *Nephotettix cincticeps* and *Delphacodes striatella,* which transmit virus disease to rice plants. The acute oral LD_{50} for rats is 160 mg/kg.

Edifenphos, Hinosan®, EDDP
S,S-Diphenyl ethyl phosphorodithiolate

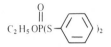

Edifenphos was first examined as an insecticide by Bayer AG, but its high activity against rice blast disease was found by Nihon Tokushu Noyaku Seizo Company. It is synthesized by the reaction of sodium thiophenolate and ethyl phosphorodichloridate or by the reaction of phenylsulfenyl chloride with diethyl *S*-phenyl phosphorothioite as shown in Section II.D (Equations 78 and 95).

Edifenphos has the advantage of a high prophylactic effect against *P. oryzae*. The *n*-propyl and isopropyl homologs have almost the same fungicidal activity as the ethyl ester edifenphos, but the methyl and butyl homologs are much less active than the latter. The introduction of a chlorine atom into the benzene ring causes a remarkable decrease in fungicidal activity.

Edifenphos is rather unstable in aqueous solutions (half-life at pH 7 and 27°C is 60 hr), but is much more stable than Kitazin in dry film under UV light or on rice plants.[1022] The main degradation pathway is the P-S bond cleavage to yield ethyl *S*-phenyl hydrogen phosphorothiolate and then ethyl dihydrogen phosphate. This pathway is common to metabolism in various organisms, chemical hydrolysis, and photochemical reactions.[302] Dealkylation occurs to some extent particularly in rice plants. *S*-Phenyl dihydrogen phosphorothiolate was often found as the dominant degradation product by photoreaction and in rice plants.[1023] *p*-Hydroxylation of the benzene ring takes place metabolically in rats, cockroaches, and *P. oryzae*.[1024] In rice plants, however, the formation of the *p*-hydroxy metabolite was not detected. *S,S,S*-Triphenyl phosphorotrithiolate and diethyl *S*-phenyl phosphorothiolate were detected in rice plants sprayed with Hinosan.[1023] These products appear not to be the transformation products of ethyl *S,S*-diphenyl phosphorodithiolate but impurities in the technical product.[1025] The metabolic pathways of edifenphos are as follows:

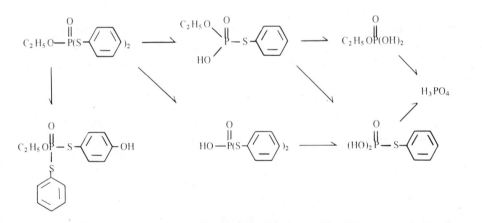

(55)

In addition to the above-mentioned compounds, many kinds of neutral esters of phosphoric or phosphonic acids and their thio analogs have been reported or patented as effective fungicides. A heteroaromatic ester of phosphorothioic acid, diethyl 6-carbethoxy-5-methylpyrazolo-[1,5-*a*]-pyrimidin-2-yl phosphorothionate (pyrazophos), has been recently introduced as a systemic fungicide against mildew by Farbwerke Hoechst AG under the name HOE 2873 or Afugan®.[1026] The oral LD_{50} for rats is 140 mg/kg.

The structures of some other selected compounds are shown below. They involve polyhaloalkyl esters (XXXVI),[1027] halovinyl esters (XXXVII),[1007] haloallyl esters (XXXVIII),[1028] halophenyl phenylphosphonate esters (XXXIX),[863] bicyclic esters (XL),[1029] oxime esters (XLI),[1007] phosphinyl haloalkyl disulfides (XLII),[1030] alkylaminosulfinylphosphonates (XLIII) (for synthesis, see Equation 106 in Chapter II),[116] and antimony (XLIV)[1007] or tin containing phosphorus compounds (XLV).[1031]

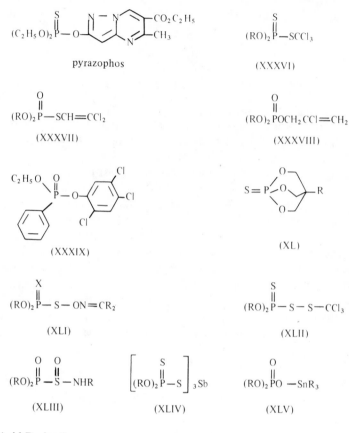

pyrazophos

(XXXVI)

(XXXVII)

(XXXVIII)

(XXXIX)

(XL)

(XLI)

(XLII)

(XLIII)

(XLIV)

(XLV)

2. Phosphoramidic Acid Derivatives

Introduction of an amide group in place of an alkyl ester group often affords organophosphorus esters with fungicidal activity. For instance, although the insecticide dimethoate (dimethyl S-(N-methyl-carbamoylmethyl) phosphorothiolothionate) has no fungicidal activity, its dialkylphosphoramidate analogs, such as compound XLVI, show some fungicidal as well as acaricidal activity.[1032]

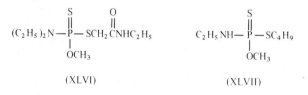

(XLVI)

(XLVII)

The S-alkylcarbamoylmethyl group is not required for the fungicidal effect, and some S-phenyl phosphoramidothiolothionates are more active than XLVI. Phosbutyl (ethyl S-phenyl N-butylphosphoramidothiolothionate) is a practically useful fungicide in this series.[1007] It is effective selectively against mycelial cells of *Fusarium* but is much less effective against fungal spores. On the other hand, S-alkyl analogs, exemplified by S-butyl methyl N-ethylphosphoramidothiolothionate (XLVII), are more active against spores of *Botrytis cinerea*.[1033]

An amido analog of Kitazin, S-benzyl ethyl N-isopropylphosphoramidothiolate, was patented as a fungicide.[1034]

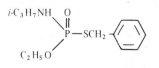

Some amido esters of phosphonic acid such as XLVIII are also fungicidal.[1007] For their activity, a polychlorinated phenyl ester group appears to be required. The phenol liberated by ester hydrolysis may contribute, at least in part, to the fungicidal activity of the phosphonamidate. Pentachlorophenol liberated from XLVIII is a well-known fungicide.

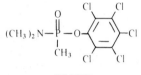

(XLVIII)

With phosphorodiamidate esters, at least one aryl group, preferably substituted with chlorine atom or nitro group, is required for high fungicidal activity. The following compounds are examples in this which have the prominent activity series.[132,1007,1035]

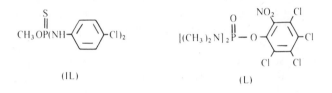

(IL)

(L)

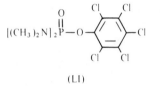

(LI)

The mode of action of the dianilide (IL) may differ from that of polychlorinated phenol derivatives L and LI. The compound L has an acaricidal activity, too. The derivatives of polychloro- or nitro-phenols, which are known as uncouplers of oxidative phosphorylation, show in general a variety of biocidal activities with low species selectivity.

Some phosphoramides or phosphonamides in which the phosphorus atom is attached directly to the nitrogen atom of a heterocycle such as phthalimide, imidazole, or triazole represent a unique class of fungicides and show an interesting structure-fungicidal activity relationship. Some examples of the fungicidal N-phosphoryl heterocycles will be described in detail later.

Phosbutyl
Ethyl S-phenyl N-butylphosphoramidothiolothionate

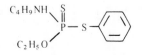

Phosbutyl is synthesized from ethyl S-phenyl phosphorochloridothiolothionate and butylamine in the presence of sodium hydroxide.[34] It is a liquid; bp 150 to 151°C at 0.3 mmHg; d_4^{20} 1.1445; highly soluble in most organic solvents but poorly soluble in water. The mammalian toxicity is relatively low; the oral LD_{50} for rats is 300 mg/kg. Phosbutyl is not active against spore germination, but is highly active against mycelial cells. Being absorbed rapidly by the plant, it thus shows a good curative activity for many plants infected with pathogenic fungi.

For the fungicidal activity, the S-phenyl group appears very important, but the introduction of any group in the ring decreases the activity. Replacement of the O-ethyl group or the N-butyl group by another alkyl group results in a decrease in the activity against mycelial cells, though the activity against spores is sometimes increased.

Dowco 199

Diethyl phthalimidophosphonothionate, N-(Diethoxyphosphinothioyl) phthalimide

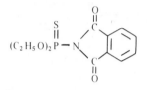

Dowco 199 is a fungicide introduced by the Dow Chemical Co.[1036] It is a solid, melting at 83 to 84°C, and is highly soluble in aromatic hydrocarbons and ethyl acetate but only slightly soluble in aliphatic hydrocarbons. For the synthesis, see Section II.E, Equations 123 and 132. The oral LD_{50} for rats is 5,660 mg/kg. The low mammalian toxicity is attributed to the weak anticholinesterase activity. Dowco 199 is effective to control diseases of fruit crops and ornamental plants, particularly powdery mildew. It is curative as well as protective in its action.

The isopropyl homolog has similar fungicidal activity but is about three times more toxic to mammals. The methyl homolog and the methylamide analog are much less active in fungicidal action. The aromatic ring is necessary for the fungicidal activity, but any substitution on the ring causes a remarkable decrease in the activity. The ethoxy groups are replaceable by ethylthio or ethyl groups without great depression in the fungicidal activity. Thus, the corresponding phosphorotrithioate (LII)[132] and phosphonodithioate (LIII)[1037] analogs are also effective fungicides.

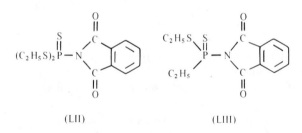

(LII)   (LIII)

It is interesting to note that the fungicidal activity of Dowco 199 is lost by replacing the thiophosphoryl sulfur atom with an oxygen atom. Furthermore, if the phthalimide-N is not directly attached to the phosphorus but through an S-CH₂ bridge or an oxygen atom, the phosphorus compounds are not fungicidal but insecticidal; compare the structures of the fungicide Dowco 199 and the insecticide phosmet (Imidan).

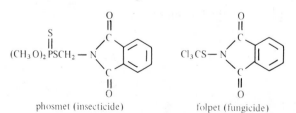

phosmet (insecticide)   folpet (fungicide)

On the other hand, some nonphosphorus containing *N*-substituted phthalimides like the *N*-trichloro-methylthio derivative (folpet) are active as fungicides. Tolkmith inferred from this that the fungitoxiphoric group in phthalimidophosphonothionates is not the phosphorus containing moiety but the unsaturated carboximide ring, and that the fungicidal activity may not be brought about by the phosphorylative inhibition of enzymes but probably by acylation accompanied with the opening of the carboximide ring.[1036]

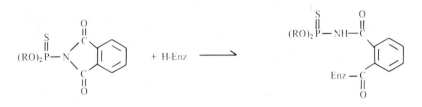

$$(56)$$

1-(*N*,*N*-Diethylaminophenylphosphinothioyl)-imidazole

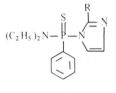

(LIV)

Some imidazolylphosphinamidothionates were found to be effective in the control of *Erysiphe cichoracearum* on cucumbers and *Phytophthora infestans* on potato plants, for which the above-mentioned phthalimidophosphonothionates are not active.[1036,1038] The compound LIV (R = H; mp 43°C) has a sufficiently low mammalian toxicity; the acute oral LD_{50} for rats is about 1,000 mg/kg. It is synthesized from phenylphosphonothioic dichloride, diethylamine, and imidazole as the following reaction sequence:

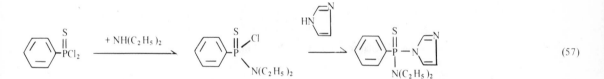

$$(57)$$

The structure LIV involves an asymmetric phosphorus atom, and the individual optical isomers of the compounds having a methyl group in place of R were isolated. The *l*-isomer was about twice as toxic to mammals as the *d*-isomer, though there was no distinct difference in the fungicidal activity between the isomers.

Of several phosphorus free imidazole derivatives tested for fungicidal activity, the *N*-triphenylmethyl derivative was almost the same as LIV in both fungicidal activity and mammalian toxicity. Tolkmith concluded, therefore, that the fungicidal activity of imidazolylphosphinamidothionates does not arise from initial phosphorylation of an enzyme essential to the normal functioning of the fungus involved.

Triamiphos, Wepsyn®, Wepsin®
P-(5-Amino-3-phenyl-1,2,4-triazol-1-yl)-*N*,*N*,*N'*,*N'*-tetramethyl phosphonic diamide, 5-amino-1-[bis-(dimethylamino)phosphinyl]-3-phenyl-1,2,4-triazole

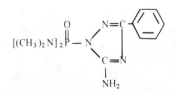

The fungicidal and insecticidal properties of triamiphos were first reported by Van den Bos and others in 1960.[1039] This may be the first reported organophosphorus fungicide. For the synthesis, see Equation 124 in Section II.E. The pure compound is a white crystalline substance; mp 167 to 168°C. The solubility in water is 250 ppm at 20°C; it is moderately soluble in most organic solvents except petroleum ether. Triamiphos is highly toxic to mammals; the acute oral LD_{50} for rats is 20 mg/kg. It is effective to control powdery mildew on apples and roses. It is also useful as an acaricide with systemic properties.

The 5-anilino-3-alkyl analogs are also active as fungicides and insecticides.[132]

Besides the above-mentioned compounds, many fungicidal or bactericidal phosphorus compounds which contain the P-N bond(s) in the molecule appear in patent literature. Some structures of interesting compounds are shown below:

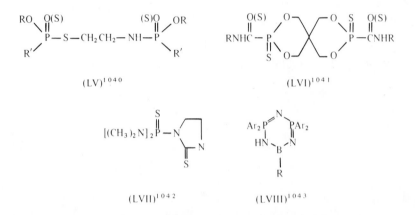

(LV)[1040]

(LVI)[1041]

(LVII)[1042]

(LVIII)[1043]

3. Miscellaneous Fungicidal Phosphorus Compounds

Certain phosphites, phosphate salts, phosphonium salts, and phosphine alkylenes are known to have fungicidal activity. Structures of some selected compounds are listed in Table 62.

TABLE 62

Some Miscellaneous Fungicidal Phosphorus Compounds

Structure	Remark	Ref.
CH_3 — (◯ — O)$_3$P	Effective for wood preserving	1044
Cl_n — (◯ — S)$_2$POR	Effective against *Piricularia oryzae*	1045
(◯ — CH_2O)$_2$POH	Effective against *Piricularia oryzae*	132
Cl_3CCH—P—C(CH$_3$)$_2$ with O, OH, HO, OH	Effective against *Stachybotrys alternans*	1046
$\begin{bmatrix} C_2H_5O \\ \\ C_6H_5 \end{bmatrix}$ P (S, S$^-$) $\begin{bmatrix} H_2N(C_2H_5)_2 \end{bmatrix}^+$	Effective against *Xanthomonus oryzae*	1047

TABLE 62 (continued)

Some Miscellaneous Fungicidal Phosphorus Compounds

Structure	Remark	Ref.
$[(C_6H_5)_3\overset{+}{P}C_{10}H_{21}] \ [(C_6H_5)_3SnBrCl]^-$	Fungicide decafentin oral LD_{50} rat 450 to 515 mg/kg	867
$[(C_6H_5)_3P=\underset{\underset{P(C_6H_5)_2}{\mid}}{C}-\overset{+}{P}(C_6H_5)_3] \ Cl^-$		1048
$(C_6H_5)_3P=\underset{\underset{C\equiv N}{\mid}}{C}-\overset{\overset{S}{\parallel}}{C}-SR$		1049

E. HERBICIDES AND PLANT GROWTH REGULATORS

The use of herbicides in agriculture has been increasing remarkably in this decade all over the world. The production of organic herbicides in the United States increased five times from 1959 to 1969.[1050] Many kinds of organic herbicides are now used.[1051] They include chlorophenoxy acids, chlorobenzoic acids, chloroaliphatic acids, amides, carbamates, ureas, triazines, bipyridylium salts, biphenylethers, phenols, and so on. Several organophosphorus compounds were very recently added to the list of commercial herbicides. Some organophosphorus compounds are useful as plant growth regulators. Herbicidal compounds or plant growth regulators have been found among derivatives of phosphorothioic acids, phosphoramidic acids, phosphonic acids, phosphonium salts, and phosphites.

The herbicidal mechanism of these organophosphorus compounds is not well known. Recently, an enzyme similar to animal acetylcholinesterase has been discovered in plants and partially purified.[1052] The enzyme is inhibited by organophosphates such as paraoxon. Although acetylcholine was suggested to participate in phytochrome mediated processes in plants,[1053] it is not certain whether or not the inhibition of the plant esterase is essential for the herbicidal action of organophosphorus herbicides. For the mode of action of the phosphonate herbicide glyphosate, the block of aromatic amino acid biosynthesis was suggested.[861] Only a few reviews have been published specifically dealing with organophosphorus herbicides.[1054]

1. Derivatives of Phosphorothioic Acids

Some aryl phosphorothionates (LIX to LXII) display herbicidal activity.[1055-1058] It appears that no electron withdrawing group on the aromatic ring is required for the herbicidal activity, in contrast to the requirement for insecticidal activity. The compound LXII is regarded as the modification of the herbicide 6-*sec*-butyl-2,4-dinitrophenol (dinoseb): the phosphorus moiety may act as a carrier of the active ingredient which will be liberated by hydrolysis. The compounds LIX and LX are reminiscent of the plant growth regulator naphthyl acetic acid and the urea type herbicides such as monuron (3-(*p*-chlorophenyl)-1,1-dimethylurea), respectively.

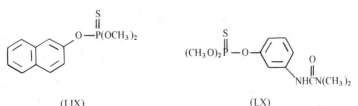

(LIX)   (LX)

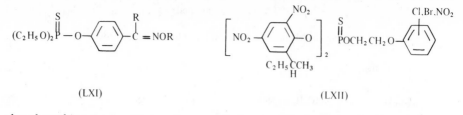

(LXI)   (LXII)

Certain phosphorothionates containing an aromatic heterocycle in the molecule, like LXIII, are also patented as herbicides.[1059] An interesting type of herbicidal compounds, phosphoryl carbimidyl disulfides (LXIV), has been patented.[1060]

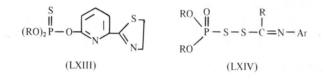

(LXIII)   (LXIV)

Among phosphorodithioate esters, N-[2-(diisopropoxyphosphinothioylthio)ethyl] benzenesulfonamide (bensulide) has been utilized in practice as a herbicide. Another type of herbicidal phosphorodithioates involves the analogs of dimethoate like LXV and LXVI.[1060a,1061] S,S,S-Tributyl phosphorotrithiolate (DEF) is a useful defoliant for cotton.

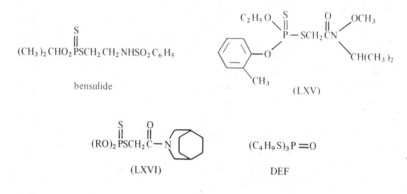

(CH₃)₂CHO₂PSCH₂CH₂NHSO₂C₆H₅

bensulide

(LXV)

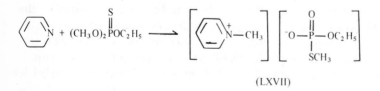

(RO)₂PSCH₂C—N

(LXVI)   (C₄H₉S)₃P=O

DEF

Besides these fully esterified compounds, certain tertiary ammonium salts of partial phosphoric esters (LXVII) have herbicidal or plant growth regulating activity.[1062] Such salts may be obtained by the reaction of a methyl phosphorothionate with an amine; alkylation and the thiono-thiolo rearrangement take place (see Section III.C.3).

$$\text{(pyridine)} + (CH_3O)_2\overset{\text{S}}{\underset{}{P}}OC_2H_5 \longrightarrow \left[\text{(pyridine)}\overset{+}{N}-CH_3\right]\left[{}^-O-\overset{O}{\underset{SCH_3}{P}}-OC_2H_5\right] \tag{58}$$

(LXVII)

Bensulide, Betasan®, Prefar®

N-[2-(Diisopropoxyphosphinothioylthio) ethyl] benzenesulfonamide

$$\left(\overset{CH_3}{\underset{CH_3}{\diagup}}CHO\right)_2\overset{\text{S}}{\underset{}{P}}SCH_2CH_2NHSO_2\text{(phenyl)}$$

Bensulide was introduced in 1964 by Stauffer Chemical Company as a selective herbicide. The synthesis is performed according to the general method for phosphorothiolothionates (see Section II.C.1) as follows:

$$(C_3H_7O)_2 \overset{\overset{S}{\|}}{P}SNa \; + \; ClCH_2CH_2NHSO_2C_6H_5 \quad \xrightarrow[-NaCl]{} \quad (C_3H_7O)_2 \overset{\overset{S}{\|}}{P}SCH_2CH_2NHSO_2C_6H_5 \tag{59}$$

It is a colored liquid or crystalline solid; mp $34.5°C$; d^{25} 1.224; n_D^{30} 1.5438. Bensulide is essentially insoluble in water (25 ppm) and miscible in methylisobutylketone and xylene. Formulations containing certain bases such as abiethylamine (0.1 to 1%) are greatly stabilized.[1063]

The mammalian toxicity is low; the acute oral LD_{50} for rats is 770 mg/kg. Bensulide is effective for the preemergence control of weeds in turf, principally crabgrass (*Digitaria* spp.), with a long action (4 to 12 months). Bensulide reduces proteolytic activity in squash seedlings to 35% of normal at $5 \times 10^{-5} M$ concentration.[1064]

DEF®, butifos
S,S,S-Tributyl phosphorotrithiolate

$$(C_4H_9S)_3P{=}O$$

Tributyl phosphorotrithiolate was first evaluated in 1954 by Ethyl Corporation as a defoliant. It may be synthesized by the reaction of butyl mercaptan with phosphorus oxychloride in the presence of a base or by oxidation of tributyl phosphorotrithioite, which is also a defoliant named merphos (see Section II.D.5; Equation 113). It is a colorless to pale yellow liquid; bp $150°C$ at 0.3 mmHg; n_D^{25} 1.532; insoluble in water; soluble in most organic solvents.

DEF is an effective defoliant for use on cotton. The oral LD_{50} for rats is 325 mg/kg. Inhibiting carboxyesterase, DEF acts as a synergist of malathion (see IV.D). It was also suggested that DEF inhibits glutathion dependent deethylation of diazinon by housefly enzyme.[604] DEF has a delayed neurotoxicity (see Section IV.E.1).

2. Derivatives of Phosphoramidic Acid

Some alkyl aryl *N*-alkylphosphoramidothionates display high herbicidal activity. Since DMPA (2,4-dichlorophenyl methyl *N*-isopropylphosphoramidothionate; Zytron®) was introduced in 1958 as the first organophosphorus herbicide, several compounds of this class have been developed into promising experimental herbicides. It is interesting to compare the structures of herbicidal and insecticidal phosphoramidates (see Table 63). The variation of the *N*-alkyl group influences greatly the biological activity. Phosphoramidates having a bulky alkyl group as *iso*-propyl, *sec*-butyl, and cyclohexyl on the amido-nitrogen atom show high herbicidal activity,[1065] whereas *N*-ethyl, -methyl, or -nonsubstituted phosphoramidates are suitable as insecticides. Dialkylphosphoramidates are generally less effective than *N*-monoalkyl derivatives in herbicidal activity. The variation of the *O*-alkyl group little affects the activity. The ring substituents appear to be important for the biological activities. Aryl phosphoramidates, the *ortho* position of which is substituted with chlorine atom or the nitro or cyano group, show high herbicidal activity. *p*-Nitrophenyl derivatives have much less herbicidal activity. The second substituent plays a role in increasing the activity. It is interesting to note that most aryl phosphoramidate insecticides also have a substituent on the *o*-position.

Herbicidal or plant growth regulating activity is also found in some phosphorodiamides,[1066] phosphoric triamides,[1067] and cyclotriphosphazenes (LXVIII).[1068] An herbicide patent was taken out for phosphoryl derivatives made from *s*-triazines (LXIX), the latter being a well-known class of herbicides.[1069] Isocyanate phosphate derivatives (LXX) and (LXXI) are patented as herbicides.[146,1070]

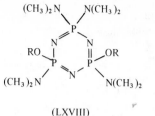

(LXVIII)

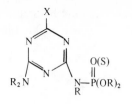

(LXIX)

X = halogen, CN, RO, RS, etc.

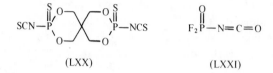

(LXX)   (LXXI)

DMPA, Zytron®

2,4-Dichlorophenyl methyl *N*-isopropylphosphoramidothionate

DMPA was the first organophosphorus herbicide, introduced in 1959 by Dow Chemical Company.[1071] For the structure, see Table 63. DMPA is a white crystalline substance; mp 51°C; slightly soluble in water (5 ppm at 25°C); soluble in most organic solvents. For the synthesis, see Section II.E.1; Equation 120. It is relatively stable in acid and alkaline media and at temperatures up to 70°C. For the acid catalyzed hydrolysis of the oxon analog, a detailed discussion may be found in Section III.A.2.b.

DMPA is a preemergence selective contact herbicide for control of crabgrass in turf. It has a considerable insecticidal activity at the rates used for crabgrass control; ant control is obtained in turf treated for crabgrass control. Although it has a relatively low acute mammalian toxicity (the oral LD_{50} for rats is 270 mg/kg), a sign of delayed neurotoxicity was observed (see Section IV.E.1).[743]

TABLE 63

Structures of Some Phosphoramidate Herbicides and Insecticides

Herbicides   Insecticides

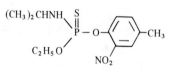

DMPA

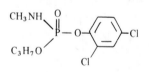

Dowco 175

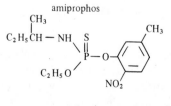

amiprophos

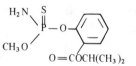

isocarbophos

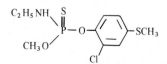

cremart   amidothioate

Cremart, S-2846

Ethyl 3-methyl-6-nitrophenyl *N-sec*-butylphosphoramidothionate

This is a new herbicide developed by Sumitomo Chemical Company. For the structure, see Table 63. Cremart is a brownish liquid; n_D^{20} 1.5340. The water solubility is 10 to 30 mg/l at 20°C. It is readily soluble in most organic solvents. The mammalian toxicity is low; the oral LD_{50} for male and female rats is 790 and 630 mg/kg, respectively.

By preemergence treatment, cremart is effective for the control of a variety of annual weeds. Cereal plants, beans, cotton, carrot, and parsley are tolerant to this treatment. It has been suggested that cremart disturbs the mitosis of plant cells.[1072]

Amiprophos

Ethyl 2-nitro-4-methyl *N*-isopropylphosphoramidothionate

Amiprophos was introduced in 1971 by Nihon Tokushu Noyaku Company as an experimental herbicide.[1065] The acute oral LD_{50} for rats is 720 mg/kg.

3. Phosphonates and Phosphinates

Recently, two interesting phosphonic acids have been introduced as plant growth regulators. They are 2-chloroethylphosphonic acid (ethephon; Ethrel®) and ethyl hydrogen propylphosphonate (NIA 10637).

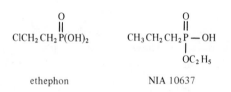

$$\begin{array}{cc} & O \\ & \parallel \\ ClCH_2CH_2P(OH)_2 & CH_3CH_2CH_2P-OH \\ & | \\ & OC_2H_5 \end{array}$$

ethephon   NIA 10637

Ethephon generates ethylene in plants to induce fruit ripening, abscission, flowering, and some other physiological effects.[1051] Ethyl hydrogen propylphosphonate is a different type of plant growth regulator:[1037] it retards the growth of plants by reducing the internode elongation. Propylphosphonic acid yields a similar effect.[1073]

More recently, another phosphonic acid, *N*-(phosphonomethyl)glycine (glyphosate), has been introduced as a herbicide.

$$\begin{array}{cc} O & O \\ \parallel & \parallel \\ (HO)_2P-CH_2NHCH_2COH \end{array}$$

glyphosate

In addition to these free acids, many herbicidal neutral esters and ester amides of phosphonic acid have been reported or patented. Structures of some selected compounds are shown below. They include the derivatives of alkylthioalkenyl- (and corresponding sulfoxide or sulfone) (LXXII) (LXXIII),[1074-1076] α-haloalkyl- (LXXIV) (LXXV),[1077,1078] α-hydroxyalkyl- (LXXVI),[1079] acyl- (LXXVII),[1080] and carbamoyl- or thiocarbamoyl- (LXXVIII to LXXX)[1081-1083] phosphonic acids. In a series of aryl *N*-alkylchloroalkylphosphonamidates (LXXV), *p*-chloro derivatives are more effective than *o*-derivatives in contrast to the phosphoramidothionate series (see the preceding Section). The compound LXXVII is a modification of the herbicide 2,4-dichlorophenoxyacetic acid.

$$\begin{array}{cc} O & Cl \quad O \\ \parallel & | \quad \parallel \\ RSCH=CH-P(SR)_2 & CH_3SCH=CHC=CH-P(OCH_3)_2 \end{array}$$

(LXXII)   (LXXIII)

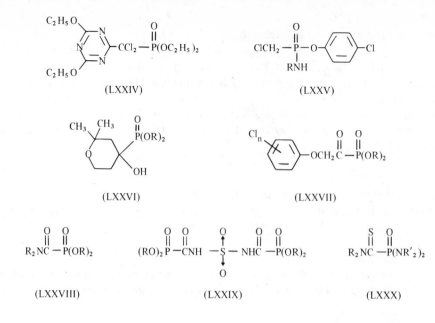

(LXXIV)

(LXXV)

(LXXVI)

(LXXVII)

$$R_2NC\overset{O}{\overset{\|}{}}-P(OR)_2\overset{O}{\overset{\|}{}}$$

(LXXVIII)

$$(RO)_2\overset{O}{\overset{\|}{P}}-\overset{O}{\overset{\|}{C}}NH-\overset{O}{\overset{\uparrow}{\underset{\downarrow}{S}}}-NH\overset{O}{\overset{\|}{C}}-\overset{O}{\overset{\|}{P}}(OR)_2$$

(LXXIX)

$$R_2N\overset{S}{\overset{\|}{C}}-P(NR'_2)_2\overset{O}{\overset{\|}{}}$$

(LXXX)

Herbicidal activity is also found in certain phosphinates, as exemplified by ethyl bis(2-ethylhexyl)phosphinate (**LXXXI**).[867] Moreover, a few phosphine oxides such as tricyclopentylphosphine oxide (**LXXXII**) have been patented as herbicides.[1084]

(LXXXI)

(LXXXII)

Ethephon, Ethrel®
2-Chloroethylphosphonic acid

2-Chloroethylphosphonic acid was first synthesized by the following reaction sequence.[1085] It is manufactured by Amchem Products Incorporated.

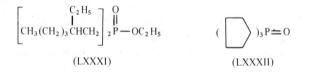

$$PCl_3 + 3\,O\!\!\triangleleft \longrightarrow (ClCH_2CH_2O)_3P \longrightarrow ClCH_2CH_2\overset{O}{\overset{\|}{P}}(OCH_2CH_2Cl)_2 \overset{H^+}{\longrightarrow} ClCH_2CH_2\overset{O}{\overset{\|}{P}}(OH)_2 \qquad (60)$$

Ethephon is a white hygroscopic crystalline solid; mp 74 to 75°C; very soluble in water, alcohol, and propylene glycol but slightly soluble in aromatic solvents. It is stable under pH 3.5, but is decomposed rapidly to release ethylene in aqueous solutions at pH higher than 4 to 5 (see also III.A.2.e). Ethylene produces numerous physiological effects on plant tissues, for example, flowering, vegetative growth and dormancy, abscission, maturity and ripening, disease and freeze resistance, and latex production. As the pH of cytoplasm of plant cells is generally greater than 4, ethephon can be utilized to regulate various phases of plant metabolism, growth, and development.[1086]

$$ClCH_2CH_2-\overset{O}{\underset{OH}{\overset{\|}{P}}}-O^- + OH^- \longrightarrow Cl^- + CH_2{=}CH_2 + {}^-OP(OH)_2\overset{O}{\overset{\|}{}} \qquad (61)$$

It was observed that ethephon treated plants have twice the peroxidase activity of the controls.[1087] The acute oral LD_{50} of ethephon for rats is 4,200 mg/kg.

Some neutral esters of 2-chloroethylphosphonic acid, as illustrated by compound LXXXIII, are also useful for the same purpose as ethephon.[1088]

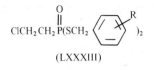

(LXXXIII)

Glyphosate

N-(Phosphonomethyl)glycine

Glyphosate was introduced by Monsanto Company as a herbicide in 1971. It is synthesized from chloromethylphosphonic acid and glycine by the action of sodium hydroxide (Equation 62) or by the oxidation of *N*-phosphinomethylglycine (Equation 63).

$$(HO)_2 \overset{O}{\underset{\|}{P}} CH_2 Cl + NH_2 CH_2 COOH \xrightarrow{NaOH} (HO)_2 \overset{O}{\underset{\|}{P}} CH_2 NHCH_2 COOH \qquad (62)$$

$$(HO)_2 PCH_2 NHCH_2 COOH \xrightarrow{[O]} (HO)_2 \overset{O}{\underset{\|}{P}} CH_2 NHCH_2 COOH \qquad (63)$$

Glyphosate melts at 230°C with decomposition and is soluble at 1% in water at 25°C. The acute oral LD_{50} for rats is 4,320 mg/kg. Glyphosate is a postemergence herbicide for broad spectrum control of both annual and perennial weeds. It was suggested that *N*-phosphonomethylglycine interferes with the biosynthesis of phenylalanine in plants, probably by inhibiting or repressing chorismate mutase and/or prephenate dehydratase.[861]

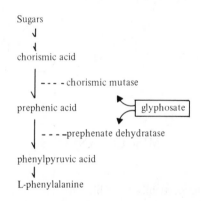

4. Phosphites and Phosphonium Salts

A mixture of tris- and bis-β-(2,4-dichlorophenoxy)ethyl phosphites, which are produced by the reaction of β-dichlorophenoxyethanol with phosphorus trichloride, is useful for preemergence weed control. It is named 2,4-DEP (Falone®, Falodin®) and was introduced by Uniroyal Incorporated. The acute oral LD_{50} for rats is about 850 mg/kg. 2,4-DEP is slowly hydrolyzed in soils to release 2,4-dichlorphenoxyethanol, which may be converted into the herbicide 2,4-dichlorphenoxyacetic acid (2,4-D).

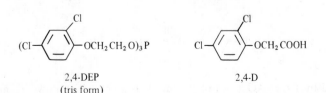

2,4-DEP
(tris form)

2,4-D

Tributyl phosphorotrithioite (merphos, Folex®) is a very effective cotton defoliant. Though the acute toxicity to mammals is very low (LD$_{50}$ for rats: 1,270 mg/kg), it has a delayed neurotoxicity (see Section IV.E.1). The phosphite may be converted biologically or photochemically into the corresponding phosphoryl compound. The phosphoryl analog of merphos is also an effective defoliant, DEF (see Section V.E.1).

$$(C_4H_9S)_3P$$

merphos

Some phosphonium salts show certain plant growth regulating activities. 2,4-Dichlorobenzyltributyl-phosphonium chloride (Phosfon®) (mp 114 to 120°C; oral LD$_{50}$ for rats 178 mg/kg) is effective as a height retardant for chrysanthemums and other horticultural plants. Butyltriethylphosphonium chloride is a defoliant. On the other hand, the yield of snap beans (*Phaseolus vulgaris*) is increased by a foliar spray of 5-chloro-2-thenyltributylphosphonium chloride (CTBP).[1089]

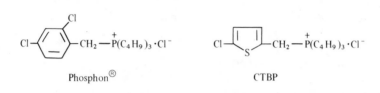

Phosphon®   CTBP

Certain quaternary ammonium compounds, exemplified by CCC and AMO-1618, have similar ability to retard plant growth. There is evidence that these growth retardants block the biosynthesis of gibberellins.[1090] They also inhibit a plant esterase, which has a high affinity for acetylcholine, at concentrations comparable to those which retard the plant growth.[1091] Acetylcholine appears to participate in phytochrome mediated processes in plants.[1053]

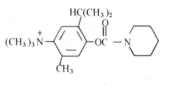

CCC   AMO-1618

F. MISCELLANEOUS

1. Rodenticides

Although many organophosphorus pesticides have high toxicity to rodents, only a few compounds are useful as rodenticides. This may be due to the fact that the onset of the cholinergic effects by most organophosphates is so rapid that the ingestion of lethal doses is prevented.[1092] Gophacide® is one of the rare examples of organophosphorus compounds useful as a rodenticide.

Gophacide®

N-(Bis-*p*-chlorophenoxyphosphinothioyl)acetamidine, di-(*p*-chlorophenyl) *N*-acetimidophosphora-midothionate

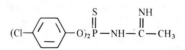

This was introduced by Bayer AG as a rodenticide. It is a crystalline powder, mp 104 to 106°C, and is

soluble in chlorinated hydrocarbons and slightly in alcohol, benzene, and ether. The acute oral LD_{50} for male rats is 7.5 mg/kg. Female rats are twice as susceptible as the male.

Gophacide is not active as an anticholinesterase in vitro, but is slowly activated in vivo;[1092] the slow onset of symptoms may be responsible for the acceptability of this organophosphorus compound to rodents as a constituent of baits, because it makes it possible for the rodent to consume a lethal dose.

2. Insecticide Synergists

Some propynol derivatives are effective as inhibitors of mixed function oxidases and as synergists of pyrethroids, as mentioned in Section IV.D.2.b (see also Figure 17). Several phosphonate esters of propynol, for example, propyl 2-propynyl phenylphosphonate (NIA 16388), were introduced as synergists by Niagara Chemical Division FMC.

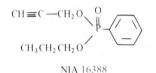

NIA 16388

NIA 16388 is a yellow liquid. It is immiscible with water, but is completely miscible with most organic solvents except aliphatic hydrocarbons. The acute oral LD_{50} to rats is 362 mg/kg. It shows juvenile hormone activity like other synergists, for example, piperonyl butoxide.[1093,1094]

Some phosphinate esters of another unsaturated alcohol involving a triple bond (LXXXIV) are also patented as synergists of pyrethroids.[1095]

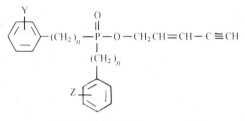

(LXXXIV) Y,Z = halogen or methyl

Diethyl phenyl phosphorothionate (SV_1) is another type of an insecticide synergist. It synergizes organophosphate and carbamate insecticides (see Section IV.D.2.b; Table 43).

3. Insect Repellents

Some phosphoramidates are reported to have repellent activity against insects. Butyl cyclohexenyl N,N,-diethylphosphoramidate and the isobutyl isomer (LXXXV) are most promising as repellents for fleas, mosquitoes, and houseflies.[1096] They surpass in effectiveness such known repellents as diethyl toluamide.

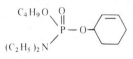

(LXXXV)

n-butyl isomer: bp 173 to 175° (9 mmHg); n_D^{20} 1.4620; d_4^{20} 1.0170
iso-butyl isomer: bp 165 to 170° (8 mmHg); n_D^{20} 1.4544; d_4^{20} 1.0300

REFERENCES

1. **Fisher, E. B. and Van Wazer, J. R.,** Uses of organic phosphorus compounds, in *Phosphorus and Its Compounds,* Vol. II, Van Wazer, J. R., Ed., Interscience, New York, 1961, 1897.
2. **Fest, C. and Schmidt, K.-J.,** Insektizide Phosphorsäureester, in *Chemie der Pflanzenschutz- und Schädlingsbekämpfungsmittel,* Vol. 1, Wegler, R., Ed., Springer-Verlag, Berlin, 1970, 246.
3. **Kosolapoff, G. M.,** *Organophosphorus Compounds,* John Wiley, New York, 1950.
4. **Michaelis, C. A. A.,** Über die organischen Verbindungen des Phosphorus mit Stickstoff, *Liebigs Ann. Chem.,* 326, 129, 1903.
5. **Arbuzov, A.,** Isomerization process of some phosphorous compounds, *J. Russ. Phys. Chem. Soc.,* 42, 395, 1910; *C. A.,* 5, 1397, 1911.
6. **Harden, H.,** *Alcholic Fermentation,* Langmans, Green & Co., London, 1932, 42.
7. **Todd, A.,** Some aspects of phosphate chemistry, *Proc. Natl. Acad. Sci.,* 45, 1389, 1959.
8. **Cramer, F.,** Darstellung von Estern, Amiden und Anhydriden der Phosphorsäure, *Angew, Chem.,* 72, 236, 1960.
9. **Lange, W. and Krueger, B.,** Über Ester der Monofluorphosphorsäure, *Chem. Ber.,* 65, 1598, 1932.
10. **Holmstedt, B.,** Structure-activity relationships of the organophosphorus anticholinesterase agents, in *Cholinesterases and Anticholinesterase Agents,* Koelle, G. B., Ed., Springer-Verlag, Berlin, 1963, 428.
11. **Saunders, B. C.,** *Some Aspects of The Chemistry and Toxic Action of Organic Compounds Containing Phosphorus and Fluorine,* Cambridge University Press, London, 1957.
12. **Schrader, G.,** *Die Entwicklung neuer Insektizide auf Grundlage organischer Fluor- und Phosphor-Verbindungen,* Monographie Nr. 62 zu Angewandte Chemie und Chemie- Ingenieur-Technik, Verlag Chemie, Weinheim, 1952.
13. **Schrader, G.,** *Die Entwicklung neuer insektizider Phosphorsäure-Ester,* Verlag Chemie, Weinheim, 1963, 281.
14. **Schrader, G.,** Zur Kenntnis neuer, wenig toxischer Insektizide auf der Basis von Phosphorsäureestern, *Angew. Chem.,* 73, 331, 1961.
15. **Nishizawa, Y.,** New low toxic organophosphorus insecticide, *Bull. Agric. Chem. Soc. Jap.,* 24, 744, 1960.
16. **Perkow, W.,** Umsetzungen mit Alkylphosphiten. I. Mitteil: Umlagerungen bei der Reaktion mit Chloral und Bromal, *Chem. Ber.,* 87, 755, 1954.
17. **Adrian, E. D., Feldberg, W., and Kilby, B. A.,** The cholinesterase inhibitory action of fluorophosphonates, *Br. J. Pharmacol.,* 2, 56, 1947.
18. **Balls, A. K. and Jansen, E. F.,** Stoichiometric inhibition of chymotrypsin, *Adv. Enzymol.,* 13, 321, 1952.
19. **Gage, J. C.,** A cholinesterase inhibition derived from *O,O*-diethyl *O-p*-nitrophenyl thiophosphate in vivo, *Biochem. J.,* 54, 426, 1953.
20. **Kado, M. and Yoshinaga, E.,** Fungicidal action of organophosphorus compounds, *Residue Rev.,* 25, 133, 1969.
21. **Borkovec, A. B.,** Insect Chemosterilants, *Adv. Pest Control Res.,* Vol. 7, 1966.
22. **Arnold, H., Bourseaux, F., and Brock, N.,** Über Beziehungen zwischen chemischer Konstitution und cancerotoxischer Wirkung in der Reihe der Phosphamidester des Bis-(β-chloräthyl)amins, *Arzneim.-Forsch.,* 11, 143, 1961.
23. **Sasse, K.,** *Organische Phosphorverbindungen,* Teil 1 (1963), Teil 2 (1964), in Houben-Weyl, *Methoden der organischen Chemie,* Band XII, Georg Thieme, Stuttgart.
24. **Van Wazer, J. R., Ed.,** *Phosphorus and its Compounds,* Vol. I (1958), Vol. II (1961), Interscience, New York.
25. **Hudson, R. F.,** *Structure and Mechanism in Organo-Phosphorus Chemistry,* Academic Press, London, 1965.
26. **Kirby, A. J. and Warren, S. G.,** *The Organic Chemistry of Phosphorus,* Elsevier, Amsterdam, 1967.
27. **Geffter, Ye. L.,** *Organophosphorus Monomers and Polymers,* Pergamon, Oxford, 1962.
28. **Mukaiyama, T.,** *Yukirin-Kagobutsu no Kagaku (Chemistry of Organophosphorus Compounds),* Hirokawa, Tokyo, 1968.
29. **Yukigoseikagaku Kyokai, Ed.,** *Yukirin-Kagobutsu (Organophosphorus Compounds),* Gihodo, Tokyo, 1971.
30. **Metcalf, R. L.,** *Organic Insecticides – Their Chemistry and Mode of Action,* Interscience, New York, 1955.
31. **O'Brien, R. D.,** *Toxic Phosphorus Esters, Chemistry, Metabolism, and Biological Effects,* Academic Press, New York, 1960.
32. **O'Brien, R. D.,** *Insecticides Action and Metabolism,* Academic Press, New York, 1967.
33. **Heath, D. F.,** *Organophosphorus Poisons Anticholinesterases and Related Compounds,* Pergamon, Oxford, 1961.
34. **Melnikov, N. N.,** *Chemistry of pesticides, Residue Rev.,* 36, 303, 1971.
35. **Kohrana, H. G.,** *Some Recent Developments in the Chemistry of Phosphate Esters of Biological Interest,* John Wiley, New York, 1961.
36. **Koelle, G. B., Ed.,** *Cholinesterases and Anticholinesterase Agents,* in *Handbuch der Experimentellen Pharmakologie,* Vol. XV, Springer-Verlag, Berlin, 1963.
37. **Anon.,** Nomenclature of compounds containing one phosphorus atom, *J. Chem. Soc.,* 5122, 1952.
38. **Kabachnik, M. I. and Golubeva, E. I.,** Prisoedinenie seri k dialkilfosfitam, *Doklad, Akad. Nauk SSSR,* 105, 1528, 1955.
39. **Horiguchi, M. and Kandatsu, M.,** Ciliatine: a new aminophosphonic acid contained in rumen *Ciliate protozoa, Bull. Agric. Chem. Soc. Jap.,* 24, 565, 1960.
40. **Kittredge, J. S., Roberts, E., and Simonsen, D. G.,** The occurrence of free 2-aminoethylphosphonic acid in the sea anemone, *Anthopleura elegantissima, Biochemistry,* 1, 624, 1962.

41. Kabachnick, M. I., Mastrukova, T. A., Shipov, A. E., and Melentyeva, T. A., The application of the Hammett equation to the theory of autometic equilibrium thione-thiol equilibrium, acidity, and structure of phosphorus thio-acids, *Tetrahedron,* 9, 10, 1960.

42. Dittmer, D. C., Ramsay, O. B., and Spalding, R. E., Reactivity of thiophosphates. II. Hydrolysis of S-n-butylphosphorothioate and S-(2-aminoethyl) phosphorothioate, *J. Org. Chem.,* 28, 1273, 1963.

43. Ramierez, F. and Desai, N. B., Crystalline 1:1 adducts from the reaction of tertiary phosphite esters with ortho-quinones and with alpha-diketones. New routes to quinol-monophosphates and to ketol-monophosphates, *J. Am. Chem. Soc.,* 82, 2652, 1960.

44. Hamilton, W. C., LaPlaca, S. J., and Ramieretz, F., The structure of pentaoxyphosphorane by X-ray analysis, *J. Am. Chem. Soc.,* 87, 127, 1965.

45. Hudson, R. F. and Green, M., Die Stereochemie von Substitutionsreaktionen am Phosphor, *Angew. Chem.,* 75, 47, 1963.

46. Meisenheimer, J. and Lichtenstadt, L., Über optischaktive Verbindungen des Phosphors, *Chem. Ber.,* 44, 356, 1911.

47. Hilgetag, G. and Lehmann, G., Beiträg zur Chemie der Thiophosphate. VII. Optisch aktive Thiophosphate, *J. Prakt. Chem.,* 8, 224, 1959.

48. Ooms, A. J. J. and Boter, H. L., Stereospecificity of hydrolytic enzymes in their reaction with optically active organophosphorus compounds. I. The reaction of cholinesterases and paraoxonase with S-alkyl p-nitrophenyl methylphosphonothiolates, *Biochem. Pharmacol.,* 14, 1839, 1965.

49. Aaron, H. S., Michel, H. O., Witten, B., and Miller, J. I., The stereochemistry of asymmetric phosphorus compounds. II. Stereospecificity in the irreversible inactivation of cholinesterases by the enantiomorphs of an organophosphorus inhibitor, *J. Am. Chem. Soc.,* 80, 456, 1958.

50. Fukuto, T. R. and Metcalf, R. L., Insecticidal activity of the enantiomorphs of O-ethyl S-2-(ethylthio)ethyl ethylphosphonothiolate, *J. Econ. Entomol.,* 52, 739, 1959.

51. Seiber, J. N. and Tolkmith, H., Optically active organophosphoramides O-2,4-dichlorophenyl O-methyl isopropyl-phosphoramidothioate and some related compounds, *Tetrahedron,* 25, 381, 1969.

52. Boter, H. L. and Van Dijk, C., The inhibition of acetyl cholinesterase and butylcholinesterase by enantiomeric forms of sarin, *Biochem. Pharmacol.,* 18, 2403, 1969.

53. Horner, L., Winkler, H., Rapp, A., Mentrup, A., Hoffmann, H., and Beck, P., Phosphorus organic compounds. XXX. Optically active tertiary phosphines from optically active quaternary phosphonium salts, *Tetrahedron Lett.,* 161, 1961; *C. A.,* 55, 25819, 1961.

54. Brewster, R. Q. and McEwen, W. E., *Organic Chemistry,* Prentice-Hall, Englewood Cliffs, 1961, 375.

55. Hassan, A. and Dauterman, W. C., Studies on the optically active isomers of O,O-diethyl malathion and O,O-diethyl malaoxon, *Biochem. Pharmacol.,* 17, 1431, 1968.

56. McEwen, W. E., Stereochemistry of reactions of organophosphorus compounds, in *Topics in Phosphorus Chemistry,* Vol. 2, Grayson, M. and Griffith, E. J., Eds., Interscience, New York, 1966, 1.

57. Gallagher, M. J. and Jenkins, I. D., Stereochemical aspects of phosphorus chemistry, in *Topics in Stereochemistry,* Vol. 3, Eliel, E. L. and Allinger, N. L., Eds., Interscience, New York, 1968, 1.

58. Ugi, I. and Ramirez, F., Stereochemistry of five-coordinate phosphorus, *Chem. Br.,* 8, 198, 1972.

59. Kozolapoff, G. M. and Maier, L., Eds., *Organic Phosphorus Compounds,* Vol. 1 to 6, Interscience, New York, 1972–3.

60. Casida, J. E., Isomeric substituted-vinyl phosphates as systemic insecticides, *Science,* 122, 597, 1955.

61. Lorenz, W. and Wegler, R., Phosphoric and thiophosphoric esters, German Patent, 962,608; *C. A.,* 51, 15588, 1957.

62. Steinberg, G. M., Reactions of dialkyl phosphites. Synthesis of dialkyl chlorophosphates, tetraalkyl pyrophosphates, and mixed orthophosphate esters, *J. Org. Chem.,* 15, 637, 1950.

63. Brown, N. C., Hollinshead, D. T., Kingsbury, P. A., and Malone, J. C., A new class of compounds showing anthelmintic properties, *Nature,* 194, 379, 1962.

64. Tsai, C. Y. and Cheu, K. L., Synthesis of dialkyl hydrogen phosphites, *K'o Hsueh T'ung Pao,* 1964, 1003; *C. A.,* 62, 11676, 1965.

65. Arbuzov, B. A. and Vinogradova, V. S., Viscosity and structure of dialkyl phosphites, *Izv. Akad. Nauk SSSR Otd. Khim. Nauk,* 1952, 507; *C. A.,* 47, 4834, 1953.

66. Sakamoto, H., Nakagawa, M., and Nishizawa, Y., Oxidation of O-alkyl O-(subst. phenyl) phenylphosphonothioates, *Agric. Biol. Chem.* (Tokyo), 26, 128, 1962.

67. Berkelhammer, G., Dauterman, W. C., and O'Brien, R. D., Conversion of phosphorothionates to their oxygen analogs with dinitrogentetroxide, *J. Agric. Food Chem.,* 11, 307, 1963.

68. Lorenz, W., Henglein, A., and Schrader, G., The new insecticide O,O-dimethyl 2,2,2-trichloro-1-hydroxyethylphosphonate, *J. Am. Chem. Soc.,* 77, 2554, 1955.

69. Timmler, H. and Kurz, J., Zur Bildung von Phosphorsäure-estern und α-Hydroxyphosphonsäureestern bei der Umsetzung von Dialkylphosphiten mit aromatischen Ketonen, *Chem. Ber.,* 104, 3740, 1971.

70. Chopard, P. A., Clark, V. M., Hudson, R. F., and Kirby, A. J., The mechanism of the reaction between trialkyl phosphites and α-halogenated ketones, *Tetrahedron,* 21, 1961, 1965.

71. Stiles, A. R., Reilly, C. H., Polland, G. R., Tieman, C. H., Ward, L. F., Jr., Phillips, D. D., Soloway, S. B., and Whetstone, R. R., Preparation, physical properties, and configuration of the isomers in Phosdrin® insecticide, *J. Org. Chem.,* 26, 3960, 1961.

72. **Cram, D. J. and Flhafez, F. A. A.,** Studies in stereochemistry. X. The rule of "steric control of asymmetric induction" in the syntheses of acyclic systems, *J. Am. Chem. Soc.,* 74, 5828, 1952.

73. **Anliker, R. and Beriger, E.,** Chemical and physical properties of phosphamidon, *Residue Rev.,* 37, 1, 1971.

74. **Trippett, S.,** A new synthesis of acetylenes. II. The reaction of triphenylphosphine with α-halogenocarbonyl compounds, *J. Chem. Soc.,* p. 2337, 1962.

75. **Ramirez, F. and Dershowitz, S.,** Reaction of trialkyl phosphites and of triaryl phosphites with chloranil. A new synthesis of hydroquinone monoalkyl ethers, *J. Am. Chem. Soc.,* 81, 587, 1959.

76. **Nishizawa, Y.,** Studies on organophosphorus compounds. I. The reaction of *O,O*-dimethyl phosphonate with *p*-benzoquinone, *Bull. Chem. Soc. Jap..,* 34, 688, 1961.

77. **Thompson, Q. E.,** Ozone oxidation of nucleophilic substances. I. Tertiary phosphite esters, *J. Am. Chem. Soc.,* 83, 845, 1961.

78. **Mukaiyama, T., Mitsunobu, O., and Obata, T.,** Oxidation of phosphites and phosphines via quaternary phosphonium salts, *J. Org. Chem.,* 30, 101, 1965.

79. **Nishizawa, Y.,** Studies on organophosphorus insecticides. VIII. Reaction of phosphorothioates having CH_3-O-P linkage with pyridine, *Agric. Biol. Chem.* (Tokyo), 25, 820, 1961.

80. **Nishizawa, Y., Nakagawa, M., Suzuki, Y., Sakamoto, H., and Mizutani, T.,** Studies on the organophosphorus insecticides. VI. Development of new low toxic phosphorus insecticides, *Agric. Biol. Chem.* (Tokyo), 25, 597, 1961.

81. **Sanpei, N., Tomita, K., Yanai, T., and Oka, H.,** Method of preparation of new phosphate or phosphonate esters, Japanese Patent 43-16137, 1968.

82. **Lorenz, W. and Hammann, I.,** Insecticidal phosphoric, phosphonic, and thiophosphoric or -phosphonic acid esters, British Patent 1,171,836; *C. A.,* 74, 63485, 1971.

83. **Eto, M., Kinoshita, Y., Kato, T., and Oshima, Y.,** Saligenin cyclic alkyl phosphates and phosphorothionates with insecticidal activity, *Agric. Biol. Chem.* (Tokyo), 27, 789, 1963.

84. **Kobayashi, K., Eto, M., Hirai, S., and Oshima, Y.,** Studies on saligenin cyclic phosphorus esters. XI. An improved method for the preparation of 2-substituted 4H-1,3,2-benzodioxaphosphorin-2-sulfides, *J. Agric. Chem. Soc. Jap.,* 40, 315, 1966.

85. **Fukuto, T. R. and Metcalf, R. L.,** Isomerization of β-ethylmercaptoethyl diethyl thionophosphate (Systox), *J. Am. Chem. Soc.,* 76, 5103, 1954.

86. **Nuretdinov, I. A., Neklesova, I. D., Kudrina, M. A., Iradiova, I. S., and Buina, N. A.,** Synthesis and properties of diethylaryl selenophosphates and thiophosphates, *Izv. Akad. Nauk SSSR Ser. Khim.,* 1971, 1266; *C. A.,* 75, 76308, 1971.

87. **Gottlieb, H. B.,** Sulfur addition with the aid of thiophosphoryl chloride and the catalysis of triaryl thiophosphate formation, *J. Am. Chem. Soc.,* 54, 748, 1932.

88. **Fletcher, J. H., Hamilton, J. C., Hechenbleikner, I., Hoegberg, E. I., Sertl, B. J., and Cassaday, J. T.,** The synthesis of parathion and some closely related compounds, *J. Am. Chem. Soc.,* 72, 2461, 1950.

89. **Spencer, E. Y.,** *Guide of the Chemicals Used in Crop Protection,* 5th ed., Canada Department of Agriculture, 1968.

90. **Overberger, C. G., Michelotti, F. W., and Carbateas, P. M.,** Preparation of triazines by the reaction of biguanide and esters, *J. Am. Chem. Soc.,* 79, 941, 1957.

91. **Lorenz, W.,** Esters of thiophosphoric acids, German Patent 927, 270; *C. A..,* 52, 2908, 1958.

92. **Rüfenacht, K.,** Zur Chemie von GS 13005, einem neuen insektiziden Phosphorsaureester, *Helv. Chim. Acta,* 51, 518, 1968.

93. **Clark, E. L., Johnson, G. A., and Mattson, E. L.,** Evaluation of *O,O*-dialkyl *S*-alkylthiomethyl phosphorodithioates, *J. Agric. Food Chem.,* 3, 834, 1955.

94. **Oswald, A. A.,** Insecticidal and fungicidal α-aminoethyl thiophosphates, German Patent 1,953,370; *C. A.,* 73, 77388, 1970.

95. **Bacon, W. E. and LeSuer, W. M.,** Chemistry of the aliphatic esters of phosphorothioic acids. III. Alkoxide cleavage of *O,O,S*-trialkyl phosphorodithioates, *J. Am. Chem. Soc.,* 76, 670, 1954.

96. **Drively, W. R., Haubein, A. H., Lohr, A. D., and Moseley, P. B.,** Two organophosphorus derivatives of *p*-dioxane with insecticidal and acaricidal activity, *J. Am. Chem. Soc.,* 81, 139, 1959.

97. **Oswald, A. A. and Lesser, J. H.,** Pesticidal S-vinyl dithiophosphates and *S,S'*-vinylene bis(thiophosphates), Ger. Offen., 2,032,494; *C. A.,* 74, 124812, 1971.

98. **Schrader, G., Hoffman, H., Homyer, B., and Hammann, I.,** Insecticidal and acaricidal unsymmetrical *O,O*-dialkyl S-phenyl phosphorodithioates, Ger. Offen 1,936,401; *C. A.,* 74, 86807, 1971.

99. **Almasi, L., Hantz, A., and Baicu, T.,** Notiz über *O,O*-dialkyl-*S*-aziridinyl-(1)-dithiophosphate, *Chem. Ber.,* 104, 3982, 1971.

100. **Schrader, G., Mannes, K., and Scheinpflug, H.,** Preparation of phosphorodithiol diesters, Japanese Patent 43-5747, 1968.

101. **Kawada, H., Okuda, I., Yoshinaga, E., and Kado, M.,** IBP (Kitazin P®), *Noyaku Seisan Gijitsu (Pesticide and Technique),* 22, 70, 1970.

102. **Kabachnik, M. I., Ioffe, S. T., and Mastryukova, T. A.,** Theory of tautomeric equilibrium in solutions. Tautomerism of dialkyl thiophosphates, *Zh. Obshch. Khim.,* 25, 684, 1955; *C. A.,* 50, 3850, 1956.

103. **Pearson, R. G.,** Hard and soft acids and bases, *J. Am. Chem. Soc.,* 85, 3533, 1963.

104. **Miller, B.,** Ease of displacement of thiol and oxide anions from methyl groups: carbon basisities of anions of oxygen and sulphur, *Proc. Chem. Soc.,* 1962, 303.

105. **Kano, S., Kasako, S., and Kamimura, H.,** *O,O*-Dialkyl *S*-(2-acetamidoethyl) phosphorothioates, Japanese Patent 68-3,088; *C. A.,* 69, 86355, 1968.

106. **Morrison, D. C.,** The reaction of sulfenyl chlorides with trialkyl phosphites, *J. Am. Chem. Soc.,* 77, 181, 1955.

107. **Hoffman, F. E., Moore, T. R., and Kogan, B.,** The reaction between triethyl phosphite and alkyl arylsulfonyl chlorides, *J. Am. Chem. Soc.,* 78, 6413, 1956.

108. **Hoffman, H. and Scheinpflug, H.,** Fungicidal *O*-alkyl *O*-cyclohexyl *S*-(4-halophenyl) thiophosphates, German Patent 1,816,566; *C. A.,* 73, 65436, 1970.

109. **Lippman, A. E.,** *O,S*-Dialkyl phosphorothioates, *J. Org. Chem.,* 30, 3217, 1965.

110. **Sheppard, W. A.,** Reaction of thiocyanates with trialkyl phosphites, *J. Org. Chem.,* 26, 1460, 1961.

111. **Murdock, L. L. and Hopkins, T. L.,** Synthesis of *O,O*-dialkyl *S*-aryl phosphorothiolates, *J. Org. Chem.,* 33, 907, 1968.

112. **Michalski, J., Modro, T., and Wieczorkowski, J.,** Organophosphorus compounds of sulphur and selenium. XV. Reactions of organic thiosulphonates with trialkyl phosphites and dialkyl phosphites, *J. Chem. Soc.,* p. 1665, 1960.

113. **Jacobson, H. I., Harvey, R. G., and Jensen, E. V.,** The reaction of triethyl phosphite with a dialkyl disulfide, *J. Am. Chem. Soc.,* 77, 6064, 1955.

114. **Michalski, J. and Wasiak, J.,** Organophosphorus compounds of sulphur and selenium. XXII. Reactions of organic disulfides with dialkyl phosphites and thiophosphites, *J. Chem. Soc.,* p. 5056, 1962.

115. **Pilgram, K., Phillips, D. D., and Korte, F.,** Reaction of cyclic phosphoramidites with disulfides. I. A novel synthesis of phosphoramidothioates, *J. Org. Chem.,* 29, 1844, 1964.

116. **Birum, G. H. and Heininger, S. A.,** Parasiticidal reaction products of sulfinylamines with dialkyl phosphites, U.S. Patent 2,893,910; *C. A.,* 54, 1434, 1960.

117. **Fukuto, T. R. and Stafford, E. M.,** The isomerization of *O,O*-diethyl *O*-2-diethylaminoethyl phosphorothionate, *J. Am. Chem. Soc.,* 79, 6083, 1957.

118. **Burn, A. J. and Cadogan, J. I. G.,** The reactivity of organophosphorus compounds. IX. The reaction of thionates with alkyl iodides, *J. Chem. Soc.,* p. 5532, 1961.

119. **Eto, M., Sasaki, M., Iio, M., Eto, M., and Ohkawa, H.,** Synthesis of 2-methylthio-4H-1,3,2-benzodioxaphosphorin-2-oxide by thiono-thiol conversion and its use as phosphorylating agent, *Tetrahedron Lett.,* 45, 4263, 1971.

120. **Sasaki, M., Ohkawa, H., and Eto, M.,** The conversion of saligenin cyclic methyl phosphorothionate into the thiolate analogs and a new rearrangement reaction, *J. Fac. Agric. Kyushu Univ.,* 17, 173, 1973.

121. **Quistad, G. B., Fukuto, T. R., and Metcalf, R. L.,** Insecticidal anticholinesterase, and hydrolytic properties of phosphoramidothiolates, *J. Agric. Food Chem.,* 18, 189, 1970.

122. **Miller, B. and O'Leary, T. P.,** Preferential alkylation of nitrogen rather than sulfur phosphoramidothioate anions, *J. Org. Chem.,* 27, 3382, 1962.

123. **Greenbaum, S. B.,** Cyclic sulfone insecticides and arthropodicides, U.S. Patent 3,288,671; *C. A.,* 66, 45785, 1967.

124. **Greenbaum, S. B.,** Cyclic sulfone organophosphate pesticides, U.S. Patent 3,341,407; *C. A.,* 69, 96452, 1968.

125. **Greenbaum, S. B.,** Unsaturated carbamoyl- or acyloxy-containing phosphorothioates and phosphorodithioates pesticides, U.S. Patent 3,431,325; *C. A.,* 70, 105957, 1969.

126. **Mueller, W. H., Rubin, R. M., and Buller, P. E.,** The addition of *O,O'*-dimethylphosphorylsulfenyl chloride to unsaturated hydrocarbons, *J. Org. Chem.,* 31, 3537, 1966.

127. **Teichmann, H. and Hilgetag, G.,** Nucleophilic reactivity of the thiophosphoryl group, *Angew. Chem. Int. Ed.,* 6, 1013, 1967.

128. **Leber, J. P.,** New class of vinyl thionophosphate insecticides, in *Pesticide Chemistry Proceedings 2nd International IUPAC Congress,* Vol. 1, Tahori, A. S., Ed., Gordon & Breach, London, 1972, 381.

129. **Arnold, H. and Bourseaux, F.,** Synthese und Abbau cytostatisch wirksamer cyclischer *N*-Phosphamidester des Bis-(β-chloräthyl)amins, *Angew. Chem.,* 70, 539, 1958.

130. **Tolkmith, H. and Senkbeil, H. O.,** Fungicidal phthalimidophosphonothionates, *Science,* 155, 85, 1967.

131. Dow Chem. Co., Preparation of dicarboximidophosphorus compounds, Dutch Patent 6,514,464; *C. A.,* 66, 38046, 1967.

132. **Schlör, H.,** Chemie der Fungizide, in *Chemie der Pflanzenschutz- und Schädlings-bekämpfungsmittel,* Band 2, Wegler R., Ed., Springer-Verlag, Berlin, 1970, 45.

133. **Atherton, F. R., Openshaw, H. T., and Todd, A. R.,** Studies on phosphorylation. II. The reaction of dialkyl phosphites with polyhalogen compounds in presence of bases. A new method for the phosphorylation of amines, *J. Chem. Soc.,* p. 660, 1945.

134. **Kabachnik, M. I. and Gilyarov, V. A.,** Imides of phosphorus acids. V. Reaction of trialkyl phosphites with hydrazoic acid, *Izv. Akad. Nauk SSSR Otd. Khim. Nauk,* 1961, 816; *C. A.,* 55, 27014, 1961.

135. **Blackburn, G. M. and Cohen, J. S.,** Chemical oxidative phosphorylation, in *Topics in Phosphorus Chemistry,* Vol. 6, Grayson, M. and Griffith, E. J., Eds., Interscience, New York, 1969, 187.

136. **Shokol, V. A. and Derkach, G. I.,** Preparation of dimethyl esters in N-methylalkylurethanephosphoric acid, USSR Patent 199,878; *C. A.,* 68, 77715, 1968.

137. **Stopkan, V. V.,** Systemic insecticide demuphos, *Khim. Prom. Ukr.,* 1968, 18; *C. A.,* 69, 85698, 1968.

138. **Kabachnik, M. I. and Gilyarov, V. A.,** Imides of phosphorus acids. 6. Trialkyl-*N*-acylimidophosphates, *Izv. Akad. Nauk SSSR Otd. Khim. Nauk,* 1961, 819.

139. **Kabachnik, M. I., Gilyarov, V. A., and Popov, E. M.,** On imides of phosphorus acids. 7. Amido-imido tautomerism of pentavalent phosphoric acid amides, *Izv. Akad. Nauk SSSR Otd. Khim. Nauk,* 1961, 1022.

140. **Stopkan, V. V., Lobov, V. P., Shokol, V. A., Mikhailyuchenko, N. K., and Derkach, G. I.,** Derivatives of phosphazophosphonyls as new insecticides, *Fiziol. Aktiv. Veshchestva, Akad. Nauk, Ukr. SSR, Respub. Mezhvedom. Sb.,* 1966, 86; *C. A.,* 67, 31879, 1967.

141. **Fahmy, M. A. H., Fukuto, T. R., Myers, R. O., and March, R. B.,** The selective toxicity of new *N*-phosphorothioylcarbamate esters, *J. Agric. Food Chem.,* 18, 793, 1970.

142. **Derkach, G. I. and Slyusanko, E. I.,** Mixed diesters of urethanphosphate, *Zh. Obshch. Khim.,* 35, 2220, 1965; *C. A.,* 64, 11082, 1966.

143. **Crystal, M. M.,** Sexual sterilization of screw-worm flies by orally administered 1-(bis(1-aziridinyl)phosphinyl)-3-(3, 4-dichlorophenyl)urea: effects of feeding times and concentrations of vehicle, *J. Econ. Entomol.,* 61, 140, 1968.

144. **Raetz, R. F. W. and Gruber, M. J.,** Sexual sterilization of insects with *N*-substituted ureidodi-l-aziridinylphosphine oxides, U.S. Patent 3,314,848; *C. A.,* 67, 54030, 1967.

145. **Addor, R. W. and Aliman, D. E.,** Pesticides, S. African Patent 68 01, 720; *C.A.,* 70, 86572, 1969.

146. **Raetz, R. F. W. and Bliss, A. D.,** 3,9-Dichloro-2,4,8,10-tetraoxa-3,9-diphosphaspiro(5.5)undecane-3,9-disulfide, U.S. Patent 3,325,566; *C. A.,* 67, 90391, 1967.

146a.**Derkatch, G. I.,** Isocyanates des acides du phosphore, in *Chimie Organique du Phosphore,* Centre de la Recherche Scientifique, Paris, 1970, 313.

147. **Toy, A. D. F.,** The preparation of tetraethyl pyrophosphate and other tetraalkyl pyrophosphates, *J. Am. Chem. Soc.,* 70, 3882, 1948.

148. **Schrader, G., Lorenz, W., and Mühlmann, R.,** Monothio-pyrophosphosäure-tetraalkylester — ihre Herstellung und ihre Eigenschaften, *Angew. Chem.,* 70, 690, 1958.

149. **Miller, B.,** Phosphorothioates; II. The effect of the nucleophilicity of the attacking anion upon rates of displacement on *O,O*-diphenyl phosphorochloridothioate, *J. Am. Chem. Soc.,* 84, 403, 1962.

150. **Almasi, L. and Paskucz, L.,** Synthese der asymmetrischen Dithiopyrophosphorsäure-*O,O,O,O*-tetraalkylester, *Angew. Chem.,* 79, 859, 1967.

151. **Fukuto, T. R.,** The chemistry of organic insecticides, *Ann. Rev. Entomol.,* 6, 313, 1961.

152. **Okazaki, R.,** Tables for synthesis of organophosphorus compounds, in *Yuki Rin Kagobutsu (Organophosphorus Compounds),* Gihodo, Tokyo, 1971, 288.

153. **Buchener, B. and Lockhart, L. B., Jr.,** Phenyldichlorophosphine, *Org. Synth.,* 31, 88, 1951.

154. **Shindo, N., Wada, S., Ota, K., Suzuki, F., and Ohta, Y.,** *O*-Ethyl-*O*-(p-nitrophenyl)benzenethionophosphate, U.S. Patent 3,327,026; *C. A.,* 67, 64531, 1967.

155. **Kinnear, A. M. and Perren, E. A.,** Formation of organophosphorus compounds by the reaction of alkyl chlorides with phosphorus trichloride in the presence of aluminum chloride, *J. Chem. Soc.,* p. 3437, 1952.

156. **Kabachnik, M. I. and Godovikov, N. N.,** Synthesis of phosphonothioic chlorides, *DokP. Akad. Nauk SSSR,* 110, 217, 1956.

157. **Komov, I. P., Ivin, S. Z., and Karavanov, K. V.,** Reaction of sulfur and inorganic sulfides with complex compounds of alkyltetrachlorophosphines and aluminum chloride, *Zh. Obshch. Khim.,* 28, 2960, 1958; *C. A.,* 53, 9035, 1959.

158. **Komkov, I. P., Karavanov, K. V., and Ivin, S. Z.,** New methods of preparation of alkyldichlorophosphines and dialkylchlorophosphines, *Zh. Obshch. Khim.,* 28, 2963, 1958; *C. A.,* 53, 9035, 1959.

159. **Eto, M., Kishimoto, K., Matsumura, K., Ohshita, N., and Oshima, Y.,** Studies on saligenin cyclic phosphorus esters with insecticidal activity. IX. Derivatives of phosphonic and phosphonothionic acids, *Agric. Biol. Chem.* (Tokyo), 30, 181, 1966.

160. **Szabo, K. and Menn, J. J.,** Synthesis and biological properties of insecticidal *N*-(mercaptomethyl)phthalimide-*S*-(*O*-alkyl)alkylphosphonodithioates and thiolates, *J. Agric. Food Chem.,* 17, 863, 1969.

161. **Menn, J. J. and Szabo, K.,** The synthesis and biological properties of new *O*-alkyl *S*-aryl alkylphosphonodithioates, *J. Econ. Entomol.,* 58, 734, 1965.

162. **Fluck, E. and Binder, H.,** Über eine einfache Darstellung der Dithiophosphonsäureester-halogenide und des neuartigen 1-Thia-3-aza-2,4-diphosphacyclobutan-*P,P'*-disulfid-Systems, *Angew. Chem.,* 79, 243, 1967.

163. **Pianka, M.,** The effect of substitution in certain chemical systems on their systemic insecticidal activity in plants, in *Pesticide Chemistry, Proceedings 2nd International IUPAC Congress,* Vol. 1, Tahori, A. S., Ed., Gordon & Breach, London, 1972, 265.

164. **Mitsunobu, O.,** Synthesis of organo-phosphorus compounds, *J. Synth. Org. Chem. Jap.,* 28, 206, 1970.

165. **Maier, L.,** Process for exchanging halogen atoms for hydrocarbon radicals in phosphorus halides, U.S. Patent 3,321,557; *C. A.,* 67, 73684, 1967.

166. **Kosolapoff, G. M.,** The synthesis of phosphonic and phosphinic acids, *Org. React.,* 6, 273, 1951.

167. **Harvey, R. G. and Sombre, E. R.,** The Michaelis-Arbusov and related reactions, in *Topics in Phosphorus Chemistry,* Vol. 1, Interscience, New York, 1964, 57.

168. **Michaelis, A. and Becker, T.,** Ueber die Constitution der phosphorigen Säure, *Chem. Ber.,* 30, 1003, 1897.

169. **Pelchowitz, Z.,** Organic phosphorus compounds. I. The reaction of dialkyl methylphosphonates and methylphosphonothioates with inorganic acid chlorides, *J. Chem. Soc.,* p. 238, 1961.

170. **Coe, D. G., Perry, B. J., and Brown, R. K.,** The structure of dialkylthiopyrophosphonates and related compounds, *J. Chem. Soc.,* p. 3604, 1957.

171. **Sittig, M.,** *Agricultural Chemicals Manufacture,* Noyes Data Corp., Park Ridge, p. 39, 1971.

172. **Zaripov, R. K., Azebaev, I. N., and Shamgunov, G. Sh.,** Trichloroacetylphosphonic acid esters and some of their derivatives, *Tr. Khim. Met. Inst., Akad. Nauk Kaz. SSR.,* 8, 48, 1969; *C. A.,* 72, 31,930, 1970.

173. **Popoff, I. C.,** Dialkylcarbamoylphosphonic and dialkylthiocarbamoylphosphonic diamides as acaricides, U.S. Patent 3,294,628; 3,321,516; *C. A.,* 66, 65632, 1967; 67, 73678, 1967.

174. **Barthel, W. F., Giang, P. A., and Hall, S. A.,** Dialkyl α-hydroxyphosphonates derived from chloral, *J. Am. Chem. Soc.,* 76, 4186, 1954.

175. **Fricke, G. and Georgi, W.,** Über die Inhaltstoffe eines Reaktionsproduktes aus Chloral, Phosphortrichlorid und Methanol, *J. Pr. Chem.,* 20, 4 Reihe, 250, 1963.

176. **Botts, M. F. and Regel, E. K.,** Nematocidal alkylthioalkylphosphonodithioic acid esters, U.S. Patent 3,463,840; *C. A.,* 71, 91643, 1969.

177. **Arthur, B. W. and Casida, J. E.,** Biological activity of several *O,O*-dialkyl alpha-acyloxyethyl phosphonates, *J. Agric. Food Chem.,* 6, 360, 1958.

178. **Abramov, V. S., Barabanov, V. I., and Long, L. I.,** Reactions of phosphinic acids with aldehydes and ketones. Esters of (α-(diethoxyphosphonyl)-α-hydroxyethyl)-butylphosphinic acid and of (α-carbalkoxy-α-hydroxyethyl)phosphonic acid, *Zh. Obshch. Khim.,* 37, 714, 1967; *C. A.,* 67, 73652, 1967.

179. **Ulrich, H.,** Fungicidal arylsulfonyl carbamoyl phosphonates, U.S. Patent 3,413,382; *C.A.,* 70, 58015, 1969.

180. **Guenther, E. and Loettge, W.,** Herbicidal dialkyl (1-aminocyclohexyl)phosphonates, German Patent 2,022,228; *C. A.,* 74, 100219, 1971.

181. **Maynard, J. A. and Swan, J. M.,** 2-Halogenoalkylphosphonic acids: a new class of phosphorylating agent, *Proc. Chem. Soc.,* 1963, 61.

182. **Beeby, M. H. and Mann, F. G.,** The preparation of 1-substituted 1:2:3:4-tetrahydrophosphinolones and 2-substituted 1:2:3:4-tetrahydroisophosphinolines, *J. Chem. Soc.,* p. 411, 1951.

183. **Ivin, S. Z. and Karavanov, K. V.,** Reaction of sulfur and inorganic sulfide with complex compounds of dialkyltrichlorophosphines and aluminum chloride, *Zh. Obshch. Khim.,* 28, 2958, 1958; *C. A.,* 53, 9035, 1959.

184. **Maier, L.,** Organische Phosphorverbindungen, II. Darstellung von unsymmetrischen Phosphinsäuren und unsymmetrischen Thiophosphinsäurehalogeniden, *Chem. Ber.,* 94, 3051, 1961.

185. **Maier, L.,** Organische Phosphorverbindungen, III. Darstellung von unsymmetrischen Phosphinen und unsymmetrischen Phosphinsäurechloriden, *Chem. Ber.,* 94, 3056, 1961.

186. **Parshall, G. W.,** Tetramethylbiphosphine disulfide, *Org. Synth.,* 45, 102, 1965.

187. **Nowtony, K.,** Phosphinic acid esters, French Patent 1,445,042; *C. A.,* 66, 55587, 1967.

188. **Abramov, V. S. and Barabanov, V. I.,** Reactions of dialkyl phosphonates with aldehydes and ketones. XXXI. Alkyl esters of α-hydroxy β, β, β-trichloroethyl (butyl) phosphinic acid, *Khim. Org. Soedin. Fosfora, Akad. Nauk SSSR, Otd. Obshch. Tekh. Khim.,* 1967, 135; *C. A.,* 69, 67469, 1968.

189. **Beermann, C. and Reuter, M.,** Alkyl (2-carbamoylethyl) methylphosphinates, German Patent 1,946,574; *C. A.,* 74, 125849, 1971.

190. **Melton, J. M.,** Phosphinothioate esters, U.S. Patent 3,376,365; *C. A.,* 69, 52289, 1968.

191. **Ruzicka, J. H., Thomson, J., and Wheals, B. B.,** The gas chromatographic determination of organophosphorus pesticides. II. A comparative study of hydrolysis rates, *J. Chromatogr.,* 31, 37, 1967.

192. **Mühlmann, R. and Schrader, G.,** Hydrolyse der insektiziden Phosphorsäureester, *Z. Naturforsch.,* 12b, 196, 1957.

193. **Büchler, W.,** Analytical methods for phosphamidon, *Residue Rev.,* 37, 15, 1971.

194. **Bruice, T. C. and Benkovic, S. J.,** *Bioorganic Mechanisms,* Vol. II, Benjamin, New York, 1966, 16.

194a.**Gillespie, P., Ramirez, F., Ugi, I., and Marquarding, D.,** Displacement reactions of phosphorus(V) compounds and their pentacoordinate intermediates, *Angew. Chem. Int. Ed.,* 12, 91, 1973.

195. **Barnard, P. W. C., Bunton, C. A., Llewellyn, D. R., Vernon, C. A., and Welch, V. A.,** The reactions of organic phosphates. V. The hydrolysis of triphenyl and trimethyl phosphates, *J. Chem. Soc.,* p. 2670, 1961.

196. **Bunton, C. A., Mhala, M. M., Oldham, K. G., and Vernon, C. A.,** The reactions of organic phosphates. III. The hydrolysis of dimethyl phosphate, *J. Chem. Soc.,* p. 3293, 1960.

197. **Fukuto, T. R. and Metcalf, R. L.,** Structure and insecticidal activity of some diethyl substituted phenyl phosphates, *J. Agric. Food Chem.,* 4, 930, 1956.

198. **Murdock, L. L. and Hopkins, T. L.,** Insecticidal, anticholinesterase, and hydrolytic properties of *O,O*-dialkyl *S*-aryl phosphorothiolates in relation to structure, *J. Agric. Food Chem.,* 16, 954, 1968.

199. **Pullman, B. and Valdemoro, C.,** Electronic structure and activity of organophosphorus inhibitors of esterases, *Biochim. Biophys. Acta,* 43, 548, 1960.

200. **Brady, U. E., Jr. and Arthur, B. W.,** Biological and chemical properties of dimethoate and related derivatives, *J. Econ. Entomol.,* 56, 477, 1963.

201. **Plapp, F. W. and Casida, J. E.,** Hydrolysis of alkylphosphate bond in certain dialkyl aryl phosphorothioate insecticides by rats, cockroaches, and alkali, *J. Econ. Entomol.,* 51, 800, 1958.

202. **Hudson, R. F. and Keay, L.,** The hydrolysis of diisopropyl methylphosphonodithiolate, *J. Chem. Soc.,* p. 3269, 1956.

203. **Benkovic, S. J.,** Hydrolytic reactions of inorganic esters, in *Comprehensive Chemical Kinetics,* Vol. 10, Bannford, C. H. and Tipper, C. F. M., Eds., Elsevier, Amsterdam, 1972, 1.

204. **Fukuto, T. R., Metcalf, R. L., March, R. B., and Maxon, M. G.,** Chemical behavior of Systox isomers in biological systems, *J. Econ. Entomol.,* 48, 347, 1955.

205. **Bowman, J. S. and Casida, J. E.,** Further studies on the metabolism of Thimet by plants, insects, and mammals, *J. Econ. Entomol.,* 51, 838, 1958.

206. **MacDougall, D., Archer, T. E., and Winterlin, W. L.,** Systox, in *Analytical Methods for Pesticides, Plant Growth Regulators, and Food Additives,* Zweig, G., Ed., Vol. 2, Academic Press, New York, 1964, 451.

207. **Norris, M. V., Vail, W. A., and Averall, P. R.,** Colorimetric estimation of malathion residues, *J. Agric. Food Chem.,* 2, 570, 1954.

208. **Konrad, J. G., Chesters, G., and Armstrong, D. E.,** Soil degradation of malathion, a phosphorodithioate insecticide, *Soil Sci. Soc. Amer. Proc.,* 33, 259, 1969.

209. **Giang, P. A. and Schechter, H. S.,** Colorimetric method for estimation of Guthion residues in cotton seeds and cottonseed oil, *J. Agric. Food Chem.,* 6, 845, 1958.

210. **Metivier, J.,** Chemical structure and biological activity relationship. Mode of action and selectivity of insecticides and acaricides, in *Pesticide Chemistry Proceedings Second International IUPAC Congress,* Vol. 1, Tahori, A. S., Ed., Gordon & Breach, London, 1972, 325.

211. **Traylor, P. S. and Westheimer, F. H.,** Mechanism in the hydrolysis of phosphorodiamidic chlorides, *J. Am. Chem. Soc.,* 87, 553, 1965.

212. **Gerrard, A. F. and Hamer, N. K.,** Evidence for planar intermediate in alkaline solvolysis of methyl *N*-cyclohexylphosphoramidic chloride, *J. Chem. Soc.(B),* 1968, 539.

213. **Gerrard, A. F. and Hamer, N. K.,** Mechanism of the rapid alkaline hydrolysis of a phosphoramidothioate ester, *J. Chem. Soc.(B),* 1967, 1122.

214. **Fukuto, T. R.,** Relationships between the structure of organophosphorus compounds and their activity as acetylcholinesterase inhibitors, *Bull. W. H. O.,* 44, 31, 1971.

215. **Sanborn, J. R. and Fukuto, T. R.,** Insecticidal, anticholinesterase, and hydrolytic properties of *S*-aryl phosphoramidothioates, *J. Agric. Food Chem.,* 20, 926, 1972.

216. **Metcalf, R. L.,** *Organic Insecticides — Their Chemistry and Mode of Action,* Interscience, New York, 1955, 263.

217. **Garrison, A. W. and Boozer, C. E.,** The acid-catalyzed hydrolysis of a series of phosphoramidates, *J. Am. Chem. Soc.,* 90, 3486, 1968.

218. **Lichtenthaler, F. W. and Cramer, F.,** Zur Chemie der "energiereichen Phosphate". XVII. Über die Reaktivität von Enolphosphaten, *Chem. Ber.,* 95, 1971, 1962.

219. **Gatterdam, P. E., Casida, J. E., and Stoutamire, D. W.,** Relation of structure to stability, antiesterase activity and toxicity with substituted-vinyl phosphate insecticides, *J. Econ. Entomol.,* 52, 270, 1959.

220. **Fukuto, T. R., Horning, E. O., Metcalf, R. L., and Winton, M. Y.,** The configuration of the α and β isomers of methyl 3-(dimethoxyphosphinyloxy) crotonate (Phosdrin®), *J. Org. Chem.,* 26, 4620, 1961.

221. **Spencer, E. Y.,** Biochemistry and structure of organophosphorus pesticides, in *Toxicology, Biodegradation and Efficacy of Livestock Pesticides,* Kahn, M. A. and Haufe, W. O., Eds., Swets & Zeitlinger, Amsterdam, 1972, 23.

222. **Clark, V. M., Hutchinson, D. W., Kirby, A. J., and Warren, S. G.,** The design of phosphorylating agents, *Angew. Chem. Int. Ed.,* 3, 678, 1964.

223. **Gomaa, H. M., Suffet, I. H., and Faust, S. D.,** Kinetics of hydrolysis of diazinon and diazoxon, *Residue Rev.,* 29, 171, 1969.

224. **Schmidt, K. J.,** Structure and activity of some phosphates and phosphonates in the series of azanaphthols, in *Pesticide Chemistry,* Proceedings Second International IUPAC Congress, Vol. 1, Tahori, A. S., Ed., Gordon & Breach, London, 1972, 365.

225. **Kumamoto, J., Cox, J. R., and Westheimer, F. H.,** Barium ethylene phosphate, *J. Am. Chem. Soc.,* 78, 4858, 1956.

226. **Haake, P. and Westheimer, F. H.,** Hydrolysis and exchange in esters of phosphoric acid, *J. Am. Chem. Soc.,* 83, 1102, 1961.

227. **Covitz, F. and Westheimer, F. H.,** The hydrolysis of methyl ethylene phosphate: steric hindrance in general base catalysis, *J. Am. Chem. Soc.,* 85, 1773, 1963.

228. **Eberhard, A. and Westheimer, F. H.,** Hydrolysis of phostonates, *J. Am. Chem. Soc.,* 87, 253, 1965.

229. **Boyd, D. B.,** Mechanism of hydrolysis of cyclic phosphate esters, *J. Am. Chem. Soc.,* 91, 1200, 1969.

230. **Blackburn, G. M., Cohen, J. S., and Todd, L.,** Cyclic phosphate and phosphite triesters – A P^{31} NMR study, *Tetrahedron Lett.,* p. 2873, 1964.

231. **Dennis, E. A. and Westheimer, F. H.,** The geometry of the transition state in the hydrolysis of phosphate esters, *J. Am. Chem. Soc.,* 88, 3432, 1966.

232. **Fukuto, T. R. and Metcalf, R. L.,** Reactivity of some 2-*p*-nitrophenoxy-1,3,2-dioxaphospholane 2-oxides and -dioxaphosphorinane 2-oxides, *J. Med. Chem.,* 8, 759, 1965.

233. **Eto, M., Kinoshita, Y., Kato, T., and Oshima, Y.,** Saligenin cyclic methyl phosphate and its thiono analogue: new insecticides related to the active metabolite of tri-*o*-cresyl phosphate, *Nature,* 200, 171, 1963.

234. **Eto, M., Hanada, K., Namazu, Y., and Oshima, Y.,** The correlation between antiesterase activities and chemical structure of saligenin cyclic phosphates, *Agric. Biol. Chem.* (Tokyo), 27, 723, 1963.

235. Oyama, H., Fakuhara, N., and Eto, M., Unpublished data.

236. Kaiser. E. T., Lee, T. W. S., and Boer, F. P., Structure and enzymatic reactivity of an aromatic five-membered cyclic phosphate diester. Biological implication, *J. Am. Chem. Soc.*, 93, 2351, 1971.

237. Fukuto, T. R. and Metcalf, R. L., The effect of structure on the reactivity of alkylphosphonate esters, *J. Am. Chem. Soc.*, 81, 372, 1959.

238. Frank, D. S. and Usher, D. A., A mechanism involving pseudorotation for the hydrolysis of dimethylphosphoacetoin, *J. Am. Chem. Soc.*, 89, 6360, 1967.

239. Conant, J. B. and Cook, A. A., A new type of addition reaction, *J. Am. Chem. Soc.*, 42, 830, 1920.

240. Pratt, H. K. and Goeschl, J. D., Physiological roles of ethylene in plants, *Annu. Rev. Plant Physiol.*, 20, 541, 1969.

241. Augustinsson, K.-B. and Heimbürger, G., Enzymatic hydrolysis of organophosphorus compounds. VI. Effect of metallic ions on the phosphorylphosphatases of human and swine kidney, *Acta Chem. Scand.*, 9, 383, 1955.

242. Wagner-Jauregg, T., Hackley, B. E., Jr., Lies, T. A., Owens, O. O., and Proper, R., Model reactions of phosphorus-containing enzyme inactivators. IV. The catalytic activity of certain metal salts and chelates in the hydrolysis of diisopropyl fluorophosphate, *J. Am. Chem. Soc.*, 77, 922, 1955.

243. Ketelaar, J. A. A., Gersmann, H. R., and Beck, M. M., Metal-catalyzed hydrolysis of thiophosphoric esters, *Nature*, 177, 392, 1956.

244. Mortland, M. M. and Raman, K. V., Catalytic hydrolysis of some organic phosphate pesticides by copper (II), *J. Agric. Food Chem.*, 15, 163, 1967.

245. Green, A. L., Sainsbury, G. L., Saville, B., and Stansfield, M., The reactivity of some active nucleophilic reagents with organophosphorus anticholinesterases, *J. Chem. Soc.*, p. 1583, 1958.

246. Green, A. L. and Saville, B., The reaction of oximes with isopropyl methylphosphonofluoridate (Sarin), *J. Chem. Soc.*, p. 3887, 1956.

247. Hilgetag, G. and Teichmann, H., The alkylating properties of alkyl thiophosphates, *Angew. Chem. Int. Ed.*, 4, 914, 1965.

248. Steeger, O., Koetz, G., and Seeger, P., Pesticide triesters of β-alkylthioethyl dialkyl thiophosphoric acids, German Patent 1,242,602; *C. A.*, 67, 63765, 1967.

249. Nagasawa, K., Organic phosphates. VI. Alcoholysis of catechol cyclic phosphates, *Chem. Pharm. Bull.* (Tokyo), 7, 397, 1959.

250. Lynen, F., Eggerer, H., Henning, U., and Kessel, I., Farnesyl-pyrophosphat und 3-Methyl-3-butenyl-1-pyrophoshat, die biologischen Vorstufen des Squalens, *Angew. Chem.*, 70, 738, 1958.

251. Eto, M. and Ohkawa, H., Alkylation reaction of organophosphorus pesticides: its chemical and biochemical significances, in *Biochemical Toxicology of Insecticides*, O'Brien, R. D. and Yamamoto, I., Eds., Academic Press, New York, 1970, 93.

252. Baddiley, J., Clark, V. M., Michalski, J. J., and Todd, A. R., Studies on phosphorylation. V. The reaction of tertiary bases with esters of phosphorus, phosphoric, and pyrophosphoric acids. A new method of selective debenzylation, *J. Chem. Soc.*, p. 815, 1949.

253. Clark, V. M. and Todd, A. R., Studies on phosphorylation. VII. The action of salts on neutral benzyl esters of the oxy-acids of phosphorus. A new method of selective debenzylation, *J. Chem. Soc.*, p. 2030, 1950.

254. Eto, M. and Oshima, Y., The reaction of cyclic phosphorus esters with some oximes, *Agric. Biol. Chem.* (Tokyo), 26, 834, 1962.

255. Ohkawa, H. and Eto, M., Alkylation of mercaptans and inhibition of "SH enzymes" by saligenin cyclic phosphate and phosphorothiolate esters, *Agric. Biol. Chem.* (Tokyo), 33, 443, 1969.

256. Ohkawa, H., Maruo, S., and Eto, M., Reaction of partially hydrolyzed products of saligenin cyclic phosphorus esters toward SH-compounds, *J. Fac. Agric. Kyushu Univ.*, 17, 13, 1972.

257. Heath, D. F. and Vandekar, M., Some spontaneous reactions of O,O-dimethyl S-ethylthioethyl phosphorothiolate and related compounds in water and on storage, and their effects on the toxicological properties of the compounds, *Biochem. J.*, 67, 187, 1957.

258. Porter, P. E., Vapona insecticide (DDVP), in *Analytical Methods for Pesticides, Plant Growth Regulators, and Food Additives*, Vol. II, Zweig, G., Ed., Academic Press, New York, 1964, 561.

259. Crosby, D. G., The nonmetabolic decomposition of pesticides, *Ann. N.Y. Acad. Sci.*, 160, 82, 1969.

260. Tammelin, L. E., Isomerization of β-dimethylaminoethyl-diethyl-thionophosphate, *Acta Chem. Scand.*, 11, 1738, 1957.

261. Stock, J. A., The design of tumor-inhibitory alkylating drugs, in *Drug Design*, Vol. II, Ariëns, E. J., Ed., Academic Press, New York, 1971, 531.

262. Eto, M., Tan, L. C., Oshima, Y., and Takehara, H., The isomerization of alkyl phosphorothionates induced by carboxylic acid amides, *Agric. Biol. Chem.* (Tokyo), 32, 656, 1968.

263. Dauterman, W. C., Casida, J. E., Knaak, J. B., and Kowalczyk, T., Bovine metabolism of organophosphorus insecticides. Metabolism and residues associated with oral administration of dimethoate to rats and three lactating cows, *J. Agric. Food Chem.*, 7, 188, 1959.

264. Hilgetag, G. and Teichmann, H., Beiträg zur Chemie der Thiophosphate. III. Zur Kenntnis des Natrium-O,S-dimethyl-thiophosphats, *J. Prakt. Chem.*, 8, 97, 1959.

265. Hilgetag, G. and Teichmann, H., Beiträg zur Chemie der Thiophosphate. IV. O,O-Dimethylthiophosphate, *J. Prakt. Chem.*, 8, 104, 1959.

266. **Dunn, C. L.,** Determination of 2,3-*p*-dioxanedithiol *S,S*-bis(*O,O*-diethyl phosphorodithioate), *J. Agric. Food Chem.*, 6, 203, 1958.

267. **Mel'nikov, N. N., Khaskin, B. A., and Tuturina, N. N.,** Organic insectofungicides. CVIII. Reactions of esters of thio- and dithiophosphoric acids with tertiary fatty aromatic and aromatic phosphines, *Khim. Org. Soedin. Fosfora Akad. Nauk SSR. Otd. Obshch. Tekh. Khim.*, 1967, 277; *C. A.*, 69, 43989, 1968.

268. **Mel'nikov, N. N., Khaskin, B. A., and Petruchenko, N. B.,** Organic insectofungicides. CIX. Reaction of esters of thio- and dithiophosphoric acids with hexaalkyltriaminophosphines, *Khim. Org. Soedin. Fosfora Akad. Nauk SSSR. Otd. Obshch. Tekh. Khim.*, 1967, 283; *C.A.*, 69, 43990, 1968.

269. **Metcalf, R. L. and March, R. B.,** The isomerization of organic thionophosphate insecticides, *J. Econ. Entomol.*, 46, 288, 1953.

270. **Cadogan, J. I. G. and Thomas, L. C.,** The reactivity of organophosphorus compounds. III. The decomposition of 2-diethylaminoethyl diethyl phosphate and of *S*-2-diethylaminoethyl diethyl phosphorothioate ("amiton"), *J. Chem. Soc.*, p. 2248, 1960.

271. **Cadogan, J. I. G.,** The reactivity of organophosphorus compounds. XI. High temperature decomposition of *S*-2-diethylaminoethyl diethyl phosphorothioate ("amiton"), *J. Chem. Soc.*, p. 18, 1962.

272. **Menzer, R. E., Iqbal, Z. M., and Boyd, G. R.,** Metabolism of *O*-ethyl *S,S*-dipropyl phosphorodithioate (Mocap) in bean and corn plants, *J. Agric. Food Chem.*, 19, 351, 1971.

273. **Pilgram, K., Phillips, D. D., and Korte, F.,** Reaction of cyclic phosphoramidites with disulfides. II. A novel synthesis of ethylene bis(sulfides) and bis(dithiocarbamates), *J. Org. Chem.*, 29, 1848, 1964.

274. **Benckhuijsen, C.,** Acid-catalysed conversion of triethyleneimine thiophosphoramide (thio-TEPA) to an SH compound, *Biochem. Pharmacol.*, 17, 55, 1968.

275. **Mendoza, C. E., Wales, P. J., Grant, D. L., and McCully, K. A.,** Effect of bromine and ultraviolet light on eight pesticides detected with liver esterases of five species, *J. Agric. Food Chem.*, 17, 1196, 1969.

276. **McBain, J. B., Yamamoto, I., and Casida, J. E.,** Oxygenated intermediate in peracid and microsomal oxidations of the organophosphonothionate insecticide Dyfonate®, *Life Sci. II*, 10, 1311, 1971.

277. **Wustner, D. A., Desmarchelier, J., and Fukuto, T. R.,** Structure for the oxygenated product of peracid oxidation of Dyfonate® insecticide (*O*-ethyl *S*-phenyl ethylphosphonodithioate), *Life Sci. II*, 11, 583, 1972.

278. **Herriott, A. W.,** Peroxy acid oxidation of phosphinothioates, a reversal of stereochemistry, *J. Am. Chem. Soc.*, 93, 3504, 1971.

279. **Patchett, G. G. and Batchelder, G. H.,** Determination of Trithion crop residues by cholinesterase inhibition measurement, *J. Agric. Food Chem.*, 8, 54, 1960.

280. **Stambach, K., Delley, R., Suter, R., and Szekely, G.,** Über analytische Bestimmunger von Phenkapton, *Z. Anal. Chem.*, 196, 332, 1963.

281. **Aichenegg, P. C. and Gillen, L. E.,** Sulfenyl- and sulfinyl-phosphonic dichloride insecticides, U.S. Patent 3,454,679; *C. A.*, 71, 90816, 1969.

282. **Wieland, T. and Lambert, R.,** Synthese und Eigenschaften des Barium-*S*-*n*-butylthiophosphats, *Chem. Ber.*, 89, 2476, 1956.

283. **Cook, A. F., Holman, M. J., and Nussbaum, A. L.,** Nucleoside S-alkyl phosphorothioates. II. Preparation and chemical and enzymatic properties, *J. Am. Chem. Soc.*, 91, 1522, 1969.

284. **Eto, M., Iio, M., Kobayashi, Y., Omura, H., and Eto, M.,** Syntheses of ribonucleoside 5'-*S*-methyl phosphoro-thiolates and ribonucleoside 3':5'-cyclic phosphates from nucleosides applying a new phosphorylating agent, MTBO, *Agric. Biol. Chem.* (Tokyo), in press, 1974.

285. **Tsuyuki, H., Stahmann, M. A., and Casida, J. E.,** Preparation, purification, isomerization and biological properties of octamethylpyrophosphoramide *N*-oxide, *J. Agric. Food Chem.*, 3, 922, 1955.

286. **Spencer, E. Y. and O'Brien, R. D.,** Enhancement of anticholinesterase activity in octamethylpyrophosphoamide by chlorine, *J. Agric. Food Chem.*, 1, 716, 1953.

287. **Spencer, E. Y., Todd, A. R., and Webb, R. F.,** Studies on phosphorylation. XVII. The hydrolysis of methyl 3-(*O,O*-dimethyl phosphonyl)but-2-enoate, *J. Chem. Soc.*, p. 2968, 1958.

288. **Zweig, G.,** *Analytical Methods for Pesticides, Plant Growth Regulators, and Food Additives*, Vol. II, Academic Press, New York, 1964.

289. **Friestad, H. O.,** Automated colorimetric determination of residues of parathion and similar compounds in plant extracts, in *Pesticide Chemistry, Proceedings Second International IUPAC Congress*, IV, Tahori, A. S., Ed., Gordon and Breach Science, London, 1971, 299.

290. **Allen, P. T. and Beckman, H.,** Polarography for the determination of organic feed medicaments, *Residue Rev.*, 5, 91, 1963.

291. **Crosby, D. G.,** Experimental approaches to pesticide photodecomposition, *Residue Rev.*, 25, 1, 1969.

292. **Frawley, J. P., Cook, J. W., Blake, J. R., and Fitzhugh, O. G.,** Effect of light on chemical and biological properties of parathion, *J. Agric. Food Chem.*, 6, 28, 1958.

293. **Ackermann, H.,** Dünnschichtchromatographisch-enzymatischer Nachweis phosphororganischer Insektizide, Aktivierung schwacher Esterasehemmer, *J. Chromatogr.*, 36, 309, 1968.

294. **Okada, K. and Uchida, T.,** The study on the decomposition compounds of EPN by the irradiation of ultraviolet ray, *J. Agric. Chem. Soc. Jap.*, 36, 245, 1962.

295. **Dauterman, W. C.,** Biological and nonbiological modifications of organophosphorus compounds, *Bull. W. H. O.,* 44, 133, 1971.

296. **Mitchell, T. H., Ruzicka, J. H., Thomson, J., and Wheals, B. B.,** The chromatographic determination of organophosphorus pesticides. III. The effect of irradiation on the parent compounds, *J. Chromatogr.,* 32, 17, 1968.

297. **Rosen, J. D.,** The photochemistry of several pesticides, in *Environmental Toxicology of Pesticides,* Matsumura, F., Boush, G. M., and Misato, T., Eds., Academic Press, New York, 1972.

298. **Niessen, H., Teitz, H., and Frehse, H.,** On the occurrence of biologically active metabolites of the active ingredient S1752 after application of Lebycid, *Pflanzenschutz Nachr. Bayer,* 15, 129, 1962.

299. **Pardue, J. R., Hansen, E. A., Barron, R. P., and Chen, J. T.,** Diazinon residues on field-sprayed kale. Hydroxydiazinon — a new alteration product of diazinon, *J. Agric. Food Chem.,* 18, 405, 1970.

300. **Doi, Y., Haba, K., Imai, M., Hayakawa, S., and Saito, S.,** Microdetermination of parathions by thermo- and ultraviolet decomposition products, *Acta Med. Okayama,* 22, 281, 1968.

300a. **Joiner, R. L. and Baetcke, K. P.,** Parathion: persistence on cotton and identification of its photoalteration products, *J. Agric. Food Chem.,* 21, 391, 1973.

301. **Frehse, H.,** Terminal residues of organophosphorus insecticides in plants, in *Pesticide Terminal Residues,* Buttersworths, London, 1971, 9.

302. **Uesugi, Y., Tomizawa, C., and Murai, T.,** Degradation of organophosphorus fungicides, in *Environmental Toxicology of Pesticides,* Matsumura, F., Boush, G. M., and Misato, T., Eds., Academic Press, New York, 1972, 327.

303. **Beynon, K. I. and Wright, A. N.,** Breakdown of carbon-14 labeled chlorfenvinphos insecticide on crops, *J. Sci. Food Agric.,* 19, 146, 1968.

304. **Smith, G. N.,** Ultraviolet light decomposition studies with Dursban and 3,5,6-trichloro-2-pyridinol, *J. Econ. Entomol.,* 61, 793, 1968.

305. **Gamrath, H. R., Halton, R. E., and Weesner, W. E.,** Chemical and physical properties of alkyl aryl phosphates, *Ind. Eng. Chem.,* 46, 208, 1954.

306. **Bellet, E. M. and Fukuto, T. R.,** Thermal rearrangement of substituted acetophenon *O*-(diethylphosphoryl)oximes and synthesis and biological activity of series of related phosphoramidates, *J. Agric. Food Chem.,* 20, 931, 1972.

307. **Blinn, R. C.,** A total phosphorus technique for determining organophosphorus pesticide residues using Schöniger flask combustion, *J. Agric. Food Chem.,* 12, 337, 1964.

308. **Manual, A. J.,** Measurement of residues of phorate and its oxygen analog sulfone in plant tissues by total phosphorus determination, *J. Agric. Food Chem.,* 16, 57, 1968.

309. **Ott, D. E. and Gunther, F. A.,** Automated analysis of organophosphorus insecticides by wet digestion-oxidation and colorimetric determination of the derived orthophosphate, *J. Assoc. Off. Agric. Chem.,* 51, 697, 1968.

310. **Broderick, E. J., Taschenberg, E. F., Hicks, L. J.., Avens, A. W., and Bourke, J. B.,** Rapid method for surface residues of organophosphorus pesticides by total phosphorus, *J. Agric. Food Chem.,* 15, 454, 1967.

311. **Nakamura, D.,** Colorimetric determination of phosphorus, *J. Agric. Chem. Soc. Japan,* 24, 1, 1950.

312. **Anliker, R. and Menzer, R. E.,** Method for phosphamidon residue analysis, *J. Agric. Food Chem.,* 11, 291, 1963.

313. **Allen, R. J. L.,** The estimation of phosphorus, *Biochem. J.,* 34, 858, 1940.

314. **Feigl, F.,** *Spot Tests in Organic Analysis,* 6th ed., Elsevier, Amsterdam, 1960.

315. **Batchelder, G. H.,, Patchett, G. G., and Menn, J. J.,** Imidan®, in *Analytical Methods for Pesticides, Plant Growth Regulators and Food Additives,* Vol. V., Zweig, G., Ed., Academic Press, New York, 1967, 257.

316. **Horwitz, W., Ed.,** *Official Methods of Analysis of the Association of Official Agricultural Chemists,* 8th ed., Association of Official Agricultural Chemists, Washington, 1955, 115.

317. **Pack, D. E., Ospenson, J. N., and Kohn, G. K.,** Phosphamidon, in *Analytical Methods for Pesticides, Plant Growth Regulators and Food Additives,* Vol. II, Zweig, G., Ed., Academic Press, 1964, 375.

318. **Hanes, C. S. and Isherwood, F. A.,** Separation of the phosphoric esters on the filter paper chromatogram, *Nature,* 164, 1107, 1949.

319. **March, R. B., Metcalf, R. L., and Fukuto, T. R.,** Paper chromatography of the systemic insecticides, demeton and schradan, *J. Agric. Food Chem.,* 2, 732, 1954.

320. **Suzuki, K., Goto, S., and Kashiwa, T.,** Microanalysis of some organophosphorus pesticides by a spectrophotometric method using safranine reagent, *Bunseki Kagaku,* 17, 1279, 1968.

321. **Suzuki, K., Arimura, M., and Watanabe, F.,** Determination of several organophosphorus pesticides by colorimetry after separation by thin layer chromatography, *Noyaku Seisan Gijutsu,* 19, 10, 1968.

322. **Murano, A. and Nagase, M.,** Colorimetric determination of *S*-benzyl-*O*-*n*-butyl-*S*-ethyl phosphorodithiolate and some other organophosphorus pesticides, *Bunseki Kagaku,* 20, 665, 1971.

323. **Gehauf, B., Epstein, J., and Wilson, G. B.,** Reaction for colorimetric estimation of some phosphorus compounds, *Anal. Chem.,* 29, 278, 1957.

324. **Gehauf, B. and Goldenson, J.,** Detection and estimation of nerve gases by fluorescence reaction, *Anal. Chem.,* 29, 276, 1957.

325. **Goldenson, J.,** Detection of nerve gases by chemiluminescence, *Anal. Chem.,* 29, 977, 1957.

326. **Kramer, D. N. and Gamson, R. M.,** Analysis of toxic phosphorus compounds, *Anal. Chem.,* 29(12), 21A, 1957.

327. **Watts, R. R.,** 4-(*p*-Nitrobenzyl)pyridine, a new chromogenic spray reagent for the organosphosphate pesticides, *J. Assoc. Off. Agric. Chem.,* 48, 1161, 1965.

328. **Getz, M. E. and Watts, R. R.,** Application of 4-(p-nitrobenzyl)pyridine as a rapid quantitative reagent for organophosphate pesticides, *J. Assoe. Off. Agric. Chem.,* 47, 1094, 1964.

329. **Epstein, R. W., Rosenthal, R. W., and Ess, R. J.,** Use of γ-(4-nitrobenzyl)pyridine as analytical reagent for ethylenimines and alkylating agents, *Anal. Chem.,* 27, 1435, 1955.

330. **Bäumler, J. and Rippstein, S.,** Dünnschichtchromatographischer Nachweis von Insektiziden, *Helv. Chim. Acta,* 44, 1162, 1961.

331. **Fujimoto, M. and Tsujino, Y.,** Colorimetric determination of organophosphorus insecticides with palladium chloride, *Sankyo Kenkyusho Nempo,* 18, 144, 1966; *C. A.,* 66, 114892, 1967.

332. **Menn, J. J., Erwin, W. R., and Gordon, H. T.,** Color reaction of 2,6-dibromo-*N*-chloro-*p*-quinoneimine with thiophosphate insecticides on paper chromatograms, *J. Agric. Food Chem.,* 5, 601, 1957.

333. **Watts, R. R.,** Chromogenic spray reagents for the organophosphate pesticides, *Residue Rev.,* 18, 105, 1967.

334. **Getz, M. E.,** Past, present, and future application of paper and thin-layer chromatography for determining pesticide residues, *Advances in Chemistry Series,* 108, 119, 1971.

335. **Lloyd, G. A. and Bell, G. J.,** Mobile laboratory methods for the determination of pesticides in air. I. Phosphorothiolothionates, *Analyst,* 91, 809, 1966.

336. **Schechter, M. S. and Hornstein, I.,** Chemical analysis of pesticide residues, *Adv. Pest Control Res.,* 1, 353, 1957.

337. **MacDougall, D.,** The potential of fluorescence for pesticide residue analysis, *Residue Rev.,* 5, 118, 1964.

338. **Averell, P. R. and Norris, M. V.,** Estimation of small amounts of *O,O*-diethyl *O-p*-nitrophenyl thiophosphate, *Anal. Chem.,* 20, 753, 1948.

339. **MacDougall, D.,** Guthion, in *Analytical Methods for Pesticides, Plant Growth Regulators and Food Additives,* Vol. II, Zweig, G., Ed., 1964, 231.

340. **Emerson, E.,** The condensation of aminoantipyrine. II. A new color test for phenolic compounds, *J. Org. Chem.,* 8, 417, 1943.

341. **MacDougall, D.,** The use of fluorometric measurements for determination of pesticide residues, *Residue Rev.,* 1, 24, 1962.

342. **Stammbach, K.,** Phenkapton, in *Analytical Methods for Pesticides, Plant Growth Regulators and Food Additives,* Vol. II, Zweig, G., Ed., 1964, 339.

343. **Lloyd, G. A. and Bell, G. J.,** Mobile laboratory methods for the determination of pesticides in air. III. Mevinphos, *Analyst,* 92, 578, 1967.

344. **Fujiwara, K.,** New reaction for the detection of chloroform, *Sitz. Nat. Ges. Rostock,* 6, 33, 1916; *C. A.,* 11, 3201, 1917.

345. **Giang, P. A., Barthel, W. F., and Hall, S. A.,** Colorimetric determination of *O,O*-dialkyl 1-hydroxyphosphonates derived from chloral, *J. Agric. Food Chem.,* 2, 1281, 1954.

346. **Ancher, T. E.,** Enzymatic methods, in *Analytical Methods for Pesticides, Plant Growth Regulators, and Food Additives,* Vol. I, Zweig, G., Ed., Academic Press, New York, 1963, 373.

347. **Gage, J. C.,** Residue determination by cholinesterase inhibition analysis, *Adv. Pest Control Res.,* 4, 183, 1961.

348. **Gunther, F. A., Ott, D. E., and Heath, F. E.,** The oxidation of parathion to paraoxon in aqueous media by silver oxide, *Bull. Environ. Contam. Toxicol.,* 3, 49, 1968.

349. **Ellman, G. L., Courtney, K. D., Andres, V., Jr., and Featherstone, R. M.,** A new and rapid colorimetric determination of acetylcholinesterase activity, *Biochem. Pharmacol.,* 7, 88, 1961.

350. **Booth, G. M. and Lee, A.-H.,** Distribution of cholinesterases in insects, *Bull. W. H. O.,* 44, 91, 1971.

351. **Asperen, K. Van,** A study of housefly esterases by means of a sensitive colorimetric method, *J. Ins. Physiol.,* 8, 401, 1962.

352. **Kramer, D. N. and Gamson, R. H.,** Colorimetric determination of acetylcholinesterase activity, *Anal. Chem.,* 30, 251, 1958.

353. **O'Brien, R. D.,** Binding sites of cholinesterases alkylation by an aziridinium derivative, *Biochem. J.,* 113, 713, 1969.

354. **O'Brien, R. D.,** The design of organophosphate and carbamate inhibitors of cholinesterases, in *Drug Design,* Vol. II, Ariens, E. J., Ed., Academic Press, New York, 1971, 161.

355. **Guilbault, G. C. and Kramer, D. N.,** Resorufin butyrate and indoxyl acetate as fluorogenic substrates for cholinesterase, *Anal. Chem.,* 37, 120, 1965.

356. **Sadar, M. H., Kuan, S. S., and Guilbault, G. G.,** Traces analysis of pesticides using cholinesterase from human serum, rat liver, electric eel, bean leaf beetle, and white fringe beetle, *Anal. Chem.,* 42, 1770, 1970.

357. **Schutzmann, R. L.,** Improved spray reagents for TLC fluorogenic detection of cholinesterase inhibitors, *J. Assoc. Off. Anal. Chem.,* 53, 1056, 1970.

358. **Mendoza, C. E., Wales, P. J., McLeod, H. A., and McKinley, W. P.,** Enzymatic determination of ten organophosphorus pesticides and carbaryl on thin-layer chromatograms: an evaluation of indoxyl, substituted indoxyl and 1-naphthyl acetates as substrates of esterases, *Analyst,* 93, 34, 1968.

359. **Coulson, D. M.,** Gas chromatography of pesticides, in *Advances in Pest Control Research,* Vol. V, Metcalf, R. L., Ed., Interscience, New York, 1962, 153.

360. **Westlake, W. E. and Gunther, F. A.,** Advances in gas chromatographic detectors illustrated from applications to pesticide residue evaluations, *Residue Rev.,* 18, 175, 1967.

361. **Sherma, J. and Zweig, G.,** Gas chromatography, in *Analytical Methods for Pesticides and Plant Growth Regulators,* Vol. II, Zweig, G., Ed., Academic Press, New York, 1972.

362. **Burchfield, H. P., Rhoades, J. W., and Wheeler, R. J.,** Simultaneous and selective detection of phosphorus, sulfur, and halogen in pesticides by microcoulometric gas chromatography, *J. Agric. Food Chem.,* 13, 511, 1965.

363. **Bache, C. A. and Lisk, D. J.,** Determination of organophosphorus insecticide residues using the emission spectrometric detector, *Anal. Chem.,* 37, 1477, 1965.

364. **Beroza, M. and Bowman, M. C.,** Instrumentation in determination of organophosphorus terminal residues, in *Pesticide Terminal Residues,* Butterworths, London, 1971, 79.

365. **Bowman, M. C. and Beroza, M.,** Gas-chromatographic detector for simultaneous sensing of phosphorus- and sulfur-containing compounds by flame photometry, *Anal. Chem.,* 40, 1448, 1968.

366. **St. John, L. E., Jr. and Lisk, D. J.,** Determination of hydrolytic metabolites of organophosphorus insecticides in cow urine using an improved thermionic detector, *J. Agric. Food Chem.,* 16, 48, 1968.

367. **Jaglan, P. S., March, R. B., Fukuto, T. R., and Gunther, F. A.,** Gas–liquid chromatographic determination of methyl parathion and metabolites, *J. Agric. Food Chem.,* 18, 809, 1970.

368. **Jaglan, P. S., March, R. B., and Gunther, R. A.,** Column esterification in the gas chromatography of the dealkyl metabolites of methyl parathion and methyl paraoxon, *Anal. Chem.,* 41, 1671, 1969.

369. **Kadoum, A. M.,** Extraction and cleanup methods to determine malathion and its hydrolytic products in stored grains by gas–liquid chromatography, *J. Agric. Food Chem.,* 17, 1178, 1969.

370. **St. John, L. E., Jr. and Lisk, D. J.,** Rapid, sensitive residue determination of organophosphorus insecticides by alkali thermionic gas chromatography of their methylated alkyl phosphate hydrolytic products, *J. Agric. Food Chem.,* 16, 408, 1968.

371. **Askew, J., Ruzicka, J. H. A., and Wheals, B. B.,** Organophosphorus pesticides – A gas chromatographic screening technique based on detection of methylated hydrolysis products, *J. Chromatogr.,* 41, 180, 1969.

372. **Shafik, M. T., Bradway, D., and Enos, H. F.,** A method for confirmation of organophosphorus compounds at the residue level, *Bull. Environ. Contam. Toxicol.,* 6, 55, 1971.

373. **Thornton, J. S. and Stanley, C. W.,** Determination of BAY 93820 residues in plant and animal tissues by alkali flame gas chromatography, *J. Agric. Food Chem.,* 19, 73, 1971.

374. **Beroza, M. and Bowman, M. C.,** Chromatographic determination of trace amounts of pesticide residues, *Proceedings University Missouri's 3rd Annual Conference on Trace Substances in Environmental Health,* p. 331, 1969.

375. **Biros, F. J.,** Recent applications of mass spectrometry and combined gas chromatography – mass spectrometry to pesticide residue analysis, *Residue Rev.,* 40, 1, 1971.

376. **Damico, J. H.,** The mass spectra of some organophosphorus pesticide compounds, *J. Assoc. Off. Anal. Chem.,* 49, 1027, 1966.

377. **Jorg, J., Houriet, R., and Spiteller, G.,** Massenspektren von Pflanzenschutzmitteln, *Monatsh. Chem.,* 97, 1064, 1966.

378. **Gillis, R. G. and Occoclowitz, J. L.,** The mass spectrometry of phosphorus compounds, in *Analytical Chemistry of Phosphorus Compounds,* Halmann, M., Ed., John Wiley, New York, 1972, 295.

379. **McLafferty, R. W.,** Mass spectrometric analysis broad applicability to chemical research, *Anal. Chem.,* 28, 306, 1956.

380. **Tatematsu, A., Yoshizumi, H., and Goto, T.,** Analysis of tetraethyl pyrophosphate by mass spectrometry, *Bunseki Kagaku,* 17, 774, 1968.

381. **McBain, J. B., Hoffman, L. J., and Menn, J. J.,** Dyfonate metabolism studies. II. Metabolic pathway of *O*-ethyl *S*-phenyl ethylphosphonodithioate in rats, *Pest. Biochem. Physiol.,* 1, 356, 1971.

382. **Cooks, R. G. and Gerrard, A. F.,** Electron impact induced rearrangements in compounds having the P=S bond, *J. Chem. Soc. (B),* p.1327, 1968.

383. **Biros, F. J.,** Application of combined gas–chromatography – mass spectrometry to pesticide residue identifications, in *Pesticides Identification at the Residue Level, Advances in Chemistry Series,* 104, 132, 1971.

383a.**Field, F. H.,** Chemical ionization mass spectrometry, *Acc. Chem. Res.,* 1, 42, 1968.

383b.**Holmstead, R. L. and Casida, J. E.,** Chemical ionization mass spectrometry of organophosphorus insecticides, *J. Assoc. Off. Anal. Chem.,* in press, 1974.

384. **Takagi, H., Ed.,** *Sites of Action of Drugs.* Nankodo, Tokyo, 1968, 68.

385. **Smallman, B. N. and Mansingh, A.,** The cholinergic system in insect development, *Annu. Rev. Entomol.,* 14, 387, 1969.

386. **Gahery, Y. and Boistel, J.,** Studies of some pharmacological substances which modify the electrical activity of the sixth abdominal ganglion of the cockroach, *Periplaneta americana,* in *The Physiology of the Insect Central Nervous Systems,* Treherene, J. E. and Beament, J. W. C., Eds., Academic Press, London, 1965, 73.

387. **Brown, B. E.,** Neuromuscular transmitter substance in insect visceral muscle, *Science,* 155, 595, 1967.

388. **Colhoun, E. H.,** The physiological significance of acetylcholine in insects and observations upon other pharmacologically active substances, in *Advances in Insect Physiology,* Vol. 1, Beament, J. W. L., Treherne, J. E., and Wigglesworth, V. B., Eds., Academic Press, London, 1963, 1.

389. **Kerkut, G. A., Shapira, A., and Walker, R. J.,** Effect of acetylcholine, glutamic acid, and GABA on the concentrations of the perfused cockroach leg, *Comp. Biochem. Physiol.,* 16, 37, 1965.

390. **Tashiro, S., Taniguchi, E., and Eto, M.,** L-Leucine: a neuroactive substance in insects, *Science,* 175, 448, 1972.
391. **Holmstedt, B.,** Distribution and determination of cholinesterases in mammals, *Bull. W. H. O.,* 44, 99, 1971.
392. **Meeter, E., Wolthuis, O. L., and Van Benthem, R. M. J.,** The anticholinesterase hypotherimia in the rat: its practical application in the study of the central effectiveness of oximes, *Bull. W. H. O.,* 44, 251, 1971.
393. **Burt, P. E., Gregory, G. E., and Molloy, F. M.,** A histochemical and electrophysiological study of the action of diazoxon on cholinesterase activity and nerve conduction in ganglia of the cockroach *Periplaneta americana* L., *Ann. Appl. Biol.,* 58, 341, 1966.
394. **Sakai, M. and Sato, Y.,** Metabolic conversion of the nereistoxin-related compounds into neristoxin as a factor of their insecticidal action, in *Pesticide Chemistry, Proceedings Second International IUPAC Congress,* Vol. 1, Tahori, A. S., Ed., Gordon & Breach, London, 1972, 455.
395. **Sakai, M.,** Nereistoxin and its derivatives; their ganglionic blocking and insecticidal activity, in *Biochemical Toxicology of Insecticides,* O'Brien, R. D. and Yamamoto, I., Eds., Academic Press, New York, 1970, 13.
396. **Yamamoto, I.,** Mode of action of pyrethroids, nicotinoids, and rotenoids, *Annu. Rev. Entomol.,* 15, 257, 1970.
397. **Fukami, J.,** Kongo no gaichu-bojozai no arikata, *Kagaku to Seibutsu,* 10, 506, 1972.
398. **Eldefrawi, M. E., Britten, A. G., and O'Brien, R. D.,** Action of organophosphates on binding of cholinergic ligands, *Pest. Biochem. Physiol.,* 1, 101, 1971.
399. **Wilson, I. B.,** Acetylcholinesterase, in *The Enzymes,* Vol. 4, Boyer, P. D., Lardy, H., and Myrbäck, K., Eds., Academic Press, 1960, 501.
400. **Augustinsson, K. B.,** Butyryl- and propionylcholinesterases and related types of eserine-sensitive esterases, in *The Enzymes,* Vol. 4, Boyer, P. D., Lardy, H., and Myrbäck, K., Eds., Academic Press, 1960, 521.
401. **Engelhard, N., Prachel, K., and Nenner, M.,** Acetylcholinesterase, *Angew. Chem. Int. Ed.,* 6, 615, 1967.
402. **O'Brien, R. D.,** The properties of acetylcholine receptor in vitro from torpedo electroplax, housefly head and rat brain, in *Biochemical Toxicology of Insecticides,* O'Brien, R. D. and Yamamoto, I., Eds., Academic Press, New York, 1970, 1.
403. **De Robertis, E.,** Molecular biology of synaptic receptors, *Science,* 171, 963, 1971.
404. **Bosmann, H. B.,** Acetylcholine receptor I. Identification and biochemical characteristics of a cholinergic receptor of guinea pig cerebral cortex, *J. Biol. Chem.,* 247, 130, 1972.
405. **Oosterbaan, R. A. and Jansz, H. S.,** Cholinesterases, esterases and lipases, in *Comprehensive Biochemistry,* Vol. 16, Florkin, M. and Stotz, E. H., Eds., Elsevier, Amsterdam, 1965, 1.
406. **Augustinsson, K. B.,** Comparative aspects of the purification and properties of cholinesterases, *Bull. W. H. O.,* 44, 81, 1971.
407. **Leuzinger, W., Goldberg, M. and Cauvin, E.,** Molecular properties of acetylcholinesterase, *J. Mol. Biol.,* 40, 217, 1969.
408. **Metcalf, R. L., March, R. B., and Maxon, M. G.,** Substrate preferences of insect cholinesterases, *Ann. Entomol. Soc. Am.,* 48, 222, 1955.
409. **Casida, J. E.,** Comparative enzymology of certain insect acetylesterases in relation to poisoning by organophosphorus insecticides, *Biochem. J.,* 60, 487, 1955.
410. **Dauterman, W. C., Talens, A., and Asperen, K. Van,** Partial purification and properties of flyhead cholinesterase, *J. Insect Physiol.,* 8, 1, 1962.
411. **Bergmann, F., Segal, R., Shimoni, A. and Wurzel, M.,** The pH-dependence of enzymic ester hydrolysis, *Biochem. J.,* 63, 684, 1956.
412. **Krupka, R. M.,** Chemical structure and function of the active center of acetylcholinesterase, *Biochemistry,* 5, 1988, 1966.
413. **Ross, W. C. J.,** *Biological Alkylating Agents,* Butterworths, London, 1962, 34.
414. **Weil, L., James, S., and Buchert, A. R.,** Photooxidation of crystalline chymotrypsin in the presence of methylene blue, *Arch. Biochem. Biophys.,* 46, 266, 1953.
415. **Murachi, T.,** A general reaction of diisopropyl phosphorofluoridate with proteins without direct effect on enzymic activities, *Biochim. Biophys. Acta,* 71, 239, 1963.
416. **Murachi, T., Inagami, T., and Yasui, M.,** Evidence for alkylphosphorylation of tyrosyl residues of stem bromelain by diisopropyl phosphorofluoridate, *Biochemistry,* 4, 2815, 1965.
417. **Cohen, J. A., Oosterbaan, R. A., Jansz, H. S., and Berends, F.,** The active site of esterases, *J. Cell Comp. Physiol.,* 54, (Suppl. 1), 231, 1959.
418. **Cunningham, L.,** The structure and mechanism of action of proteolytic enzymes, in *Comprehensive Biochemistry,* Vol. 16, Florkin, M. and Stotz, E. H., Eds., 1965, 85.
419. **Porter, G. R., Rydon, H. N., and Schofield, J. A.,** Nature of the reactive serine residue in enzyme inhibited by organophosphorus compounds, *Nature,* 182, 927, 1958.
420. **Cunningham, L. W.,** Proposed mechanism of action of hydrolytic enzymes, *Science,* 125, 1145, 1957.
421. **Koshland, D. E., Jr.,** Conformation changes at the active site during enzyme action, *Fed. Proc.,* 23, 719, 1964.
422. **Ingraham, L. L.,** *Biochemical Mechanisms,* John Wiley, New York, 1961, 37.
423. **Aldridge, W. N.,** The nature of the reaction of organophosphorus compounds and carbamates with esterases, *Bull. W. H. O.,* 44, 25, 1971.
424. **Aldridge, W. N. and Reiner, E.,** *Enzyme Inhibitors as Substrates,* North-Holland, Amsterdam, 1972.

425. **Aldridge, W. N.,** The differentiation of true and pseudo cholinesterases by organo-phosphorus compounds, *Biochem. J.,* 53, 62, 1953.

426. **Main, A. R.,** Affinity and phosphorylation constants for the inhibition of esterases by organophosphates, *Science,* 144, 992, 1964.

427. **Main, A. R. and Iversion, F.,** Measurement of the affinity and phosphorylation constants governing irreversible inhibition of cholinesterases by di-isopropyl phosphorofluoridate, *Biochem. J.,* 100, 525, 1966.

428. **Chiu, Y. C., Main, A. R., and Dauterman, W. C.,** Affinity and phosphorylation constants of a series of *O,O*-dialkyl malaoxons and paraoxons with acetylcholinesterase, *Biochem. Pharmacol.,* 18, 2171, 1969.

429. **Reiner, E. and Aldridge, W. N.,** Effect of pH on inhibition and spontaneous reactivation of acetylcholinesterase treated with esters of phosphorus acids and of carbamic acids, *Biochem. J.,* 105, 171, 1967.

430. **Hastings, F. L. and Dauterman, W. C.,** Phosphorylation and affinity constants for the inhibition of acetylcholinesterase by dimethoxon analogs, *Pest. Biochem. Physiol.,* 1, 248, 1971.

431. **Chiu, Y. C. and Dauterman, W. C.,** Effect of tetraethylammonium ions on the affinity and phosphorylation or carbamylation constants of malaoxon, tetram and temik with acetylcholinesterase, *Biochem. Pharmacol.,* 19, 1856, 1970.

432. **Chiu, Y. C. and Dauterman, W. C.,** The affinity and phosphorylation constants of the optical isomers of *O,O*-diethyl malaoxon and the geometric isomers of phosdrin with acetylcholinesterase, *Biochem. Pharmacol.,* 18, 359, 1969.

433. **Fukuto, T. R.,** Physico-organic chemical approach to the mode of action of organophosphorus insecticides, *Residue Rev.,* 25, 327, 1969.

434. **Reiner, E.,** Spontaneous reactivation of phosphorylated and carbamylated cholinesterases, *Bull. W. H. O.,* 44, 109, 1971.

435. **Wilson, I. B. and Froede, H. C.,** The design of reactivators for irreversibly blocked acetylcholinesterase, in *Drug Design,* Vol. II, Ariens, E. J., Ed., Academic Press, 1971, 213.

436. **Lee, R. M. and Hodsden, M. R.,** Cholinesterase activity in *Haemonchus contortus* and its inhibition by organophosphorus anthelmintics, *Biochem. Pharmacol.,* 12, 1241, 1963.

437. **Ahmad, S.,** Recovery of esterases in organophosphate-treated housefly (*Musca domestica*), *Comp. Biochem. Physiol.,* 33, 579, 1970.

438. **Mengel, D. C. and O'Brien, R. D.,** The spontaneous and induced recovery of fly-brain cholinesterase after inhibition by organophosphates, *Biochem. J.,* 75, 201, 1960.

439. **Heilbronn, E.,** Action of fluoride on cholinesterase II. In vitro reactivation of cholinesterases inhibited by organophosphorus compounds, *Biochem. Pharmacol.,* 14, 1363, 1965.

440. **Hagedorn, I., Stark, I., and Lorenz, H. P.,** Reactivation of phosphorylated acetylcholinesterase – Dependence upon activator acidity, *Angew. Chem. Int. Ed.,* 11, 307, 1972.

441. **Wilson, I. B., Ginsburg, S., and Quan, C.,** Molecular complementariness as basis for reactivation of alkyl phosphate-inhibited enzyme, *Arch. Biochim. Biophys.,* 77, 286, 1958.

442. **Poziomek, E. J., Kramer, D. N., Mosher, W. A., and Michel, H. O.,** Configurational analysis of 4-formyl-1-methyl-pyridinium iodide oximes and its relationship to a molecular complementarity theory on the reactivation of inhibited acetylcholinesterase, *J. Am. Chem. Soc.,* 83, 3916, 1961.

443. **Green, A. L. and Smith, H. J.,** The reactivation of cholinesterase inhibited with organophosphorus compounds, 1 and 2, *Biochem. J.,* 68, 28, 32, 1958.

444. **Namba, T.,** Cholinesterase inhibition by organophosphorus compounds and its clinical effects, *Bull. W. H. O.,* 44, 289, 1971.

445. **Mayer, O. and Michalek, H.,** Effect of DFP and Obidoxime on brain acetylcholine levels and on brain and peripheral cholinesterases, *Biochem. Pharmacol.,* 22, 3029, 1971.

446. **Schoene, K.,** Reaktivierung von *O,O*-Diäthylphosphoryl-Acetylcholinesterase Reaktivierungs-Rephosphorylierungs-Gleichgewicht, *Biochem. Pharmacol.,* 21, 163, 1972.

447. **Hackley, B. E., Jr., Steinberg, G. M., and Lamb, J. C.,** Formation of potent inhibitors of AChE by reaction of pyridinaldoximes with isopropyl methylphosphonofluoridate (GB), *Arch. Biochem. Biophys.,* 80, 211, 1959.

448. **Davies, D. A. and Green, A. L.,** The kinetics of reactivation, by oximes, of cholinesterase inhibited by organophosphorus compounds, *Biochem. J.,* 63, 529, 1956.

449. **Riley, G., Turnbull, J. H., and Wilson, W.,** Synthesis of some phosphorylated amino-hydroxy-acids and derived peptides related to the phosphoproteins, *J. Chem. Soc.,* p. 1373, 1957.

450. **Berends, F., Posthumus, C. H., Sluys, I. V. D., and Deierkauf, F. A.,** The chemical basis of the "aging process" of DFP-inhibited pseudocholinesterase, *Biochim. Biophys. Acta,* 34, 576, 1959.

451. **Harris, L. W., Fleisher, J. H., Clark, J., and Cliff, W. J.,** Dealkylation and loss of capacity for reactivation of cholinesterase inhibited by sarin, *Science,* 154, 404, 1966.

452. **Coult, D. B., Marsh, D. J., and Read, G.,** Dealkylation studies on inhibited acetylcholinesterase, *Biochem. J.,* 98, 869, 1966.

453. **Benshop, H. P. and Keijer, J. H.,** On the mechanism of aging of phosphonylated cholinesterases, *Biochim. Biophys. Acta,* 128, 586, 1966.

454. **Keijer, J. H. and Wolring, G. Z.,** Stereospecific aging of phosphonylated cholinesterases, *Biochim. Biophys. Acta,* 185, 465, 1969.

455. **Erlanger, B. F., Cohen, W., Vratsanos, S. M., Castleman, H., and Cooper, A. G.,** Postulated chemical basis for observed differences in the enzymatic behaviour of chymotrypsin and trypsin, *Nature,* 205, 868, 1965.

456. **Cadogan, J. J. G. and Maynard, J. A.,** The reaction of ethyl hydrogen methylphosphonate with *p*-nitrobenzonitrile oxide: its relevance to the possible reactivation of "aged" phosphonylated acetylcholinesterase, *Chem. Commun.,* p. 854, 1966.

457. **Aldridge, W. N. and Davison, A. N.,** The inhibition of erythrocyte cholinesterase by triesters of phosphoric acid, *Biochem. J.,* 51, 62, 1952.

458. **Fukuto, T. R.,** The chemistry and action of organic phosphorus insecticides, in *Advances in Pest Control Research,* Vol. 1, Metcalf, R. L., Ed., Interscience, New York, 1957, 147.

459. **Hammett, L.,** *Physical Organic Chemistry,* McGraw-Hill, New York, 1950.

460. **Fukui, K., Morokuma, K., Nagata, C., and Imamura, A.,** Electronic structure and biochemical activities in diethyl phenyl phosphates, *Bull. Chem. Soc. Jap.,* 34, 1224, 1961.

461. **Kier, L.,** *Molecular Orbital Theory in Drug Research,* Academic Press, New York, 1971.

462. **Schnaare, R. L.,** Electronic aspects of drug action, in *Drug Design,* Vol. I, Ariën, E. J., Ed., Academic Press, New York, 1971, 406.

463. **Taft, R.,** Separation of polar, steric and resonance effects in reactivity, in *Steric Effects in Organic Chemistry,* Newman, M. S., Ed., John Wiley, New York, 1956, 556.

464. **Hansch, C. and Fujita, T.,** ρ-σ-π Analysis, a method for the correlation of biological activity and chemical structure, *J. Am. Chem. Soc.,* 86, 1616, 1964.

465. **Hansch, C.,** Quantitative structure-activity relationships in drug design, in *Drug Design,* Vol. I, Ariën, E. J., Ed., Academic Press, New York, 1971, 271.

466. **Fujita, M.,** New approaches in structure-activity study of pesticides, *Kagaku no Ryoiki,* 22, 578, 1968.

467. **Darlington, W. A., Partos, R. D., and Ratts, K. W.,** Correlation of cholinesterase inhibition and toxicity in insects and mammals. I. Ethylphosphonates, *Toxicol. Appl. Pharmacol.,* 18, 542, 1971.

468. **Hansch, C. and Deutsch, E. W.,** The use of substituent constants in the study of structure-activity relationships in cholinesterase inhibitors, *Biochim. Biophys. Acta,* 126, 117, 1966.

469. **Hansch, C.,** The use of physicochemical parameters and regression analysis in pesticide design, in *Biochemical Toxicology of Insecticides,* O'Brien, R. D. and Yamamoto, I., Eds., Academic Press, New York, 1970, 33.

470. **Kutter, E. and Hansch, C.,** Steric parameters in drug design. Monoamine oxidase inhibitors and antihistamines, *J. Med. Chem.,* 91, 615, 1969.

471. **Bracha, P. and O'Brien, R. D.,** Trialkyl phosphate and phosphorothiolate anticholinesterases, *Biochemistry,* 7, 1545, 1555, 1968.

472. **Lovell, J. B.,** The relationship of anticholinesterase activity, penetration, and insect and mammalian toxicity of certain organophosphorus insecticides, *J. Econ. Entomol.,* 56, 310, 1963.

473. **Arthur, B. W. and Casida, J. E.,** Biological and chemical oxidation of tetramethyl phosphorodiamidic fluoride (dimefox), *J. Econ. Entomol.,* 51, 49, 1958.

474. **Eto, M., Kobayashi, K., Kato, T., Kojima, K., and Oshima, Y.,** Saligenin cyclic phosphoramidates and phosphoramidothionates as pesticides, *Agric. Biol. Chem.* (Tokyo), 29, 243, 1965.

475. **Fukuto, T. R., Winton, M. Y., and March, R. B.,** Structure and insecticidal activity of alkyl 2,4,5-trichlorophenyl N-alkylphosphoramidates, *J. Econ. Entomol.,* 56, 808, 1963.

476. **Neely, W. B. and Whitney, W. K.,** Statistical analysis of insecticidal activity in a series of phosphoramidates, *J. Agric. Food Chem.,* 16, 571, 1968.

477. **Cramer, F. and Gärtner, K. G.,** Zur Chemie der "energiereichen Phosphate," I. Darstellung von Enolphosphaten und Acylphosphaten, *Chem. Ber.,* 91, 704, 1958.

478. **Morello, A., Spencer, E. Y., and Vardanis, A.,** Biochemical mechanisms in the toxicity of the geometrical isomers of two vinyl organophosphates, *Biochem. Pharmacol.,* 16, 1703, 1967.

479. **Vinopal, J. H. and Fukuto, T. R.,** Selective toxicity of phoxim (phenylglyoxylonitrile oxime *O,O*-diethyl phosphorothioate), *Pest. Biochem. Physiol.,* 1, 44, 1971.

480. **Fukuto, T. R., Metcalf, R. L., Jones, R. L., and Myers, R. O.,** Structure, reactivity, and biological activity of *O*-(diethyl phosphoryl)oximes and *O*-(methylcarbamoyl)oximes of substituted acetophenones and α-substituted benzaldehydes, *J. Agric. Food Chem.,* 17, 923, 1969.

481. **Arthur, B. W. and Casida, J. E.,** Metabolism and selectivity of *O,O*-dimethyl 2,2,2-trichloro-1-hydroxyethyl phosphonate and its acetyl and vinyl derivatives, *J. Agric. Food Chem.,* 5, 186, 1957.

482. **Metcalf, R. L., Fukuto, T. R., and March, R. B.,** Toxic action of Dipterex and DDVP to the housefly, *J. Econ. Entomol.,* 52, 44, 1959.

483. **Melnikov, N. N.,** On the mode of action of organophosphorus insecticides, *Vth International Pesticide Congress, Pesticides Abstract News Summary,* p. 431, 1964.

484. **Edmundson, R. S. and Lambie, A. J.,** Cyclic organophosphorus compounds as possible pesticides. I. 1,3,2-Dioxaphospholans, *J. Chem. Soc. (C),* p. 1997, 1966.

485. **Eto, M. and Sakamoto, K.,** unpublished data.

486. **Tichý, V., Rattaj, V., Janok, J., and Valentinovà, I.,** Zmiesăné estery kyseliny fosforečnej a tiofosforečnej odvodené od pyrokatechínu, *Chem. Zvesti,* 11, 398, 1957.

487. **Eto, M.,** Specificity and mechanism in the action of saligenin cyclic phosphorus esters, *Residue Rev.,* 25, 187, 1969.

488. Becker, E. L., Fukuto, T. R., Canham, D. C., and Boger, E., The relationship of enzyme inhibitory activity to the structure of *n*-alkylphosphonate and phenylalkylphosphonate esters, *Biochemistry*, 2, 72, 1963.

489. Eto, M., Casida, J. E., and Eto, T., Hydroxylation and cyclization reactions involved in the metabolism of tri-*o*-cresyl phosphate, *Biochem. Pharmacol.*, 11, 337, 1962.

490. Metcalf, R. L. and March, R. B., Studies of the mode of action of parathion and its derivatives and their toxicity to insects, *J. Econ. Entomol.*, 42, 721, 1949.

491. Dauterman, W. C. and O'Brien, R. D., Cholinesterase variation as a factor in organophosphate selectivity in insects, *J. Agric. Food Chem.*, 12, 318, 1964.

492. Camp, H. B., Fukuto, T. R., and Metcalf, R. L., Selective toxicity of isopropyl parathion. Effect of structure on toxicity and anticholinesterase activity, *J. Agric. Food Chem.*, 17, 243, 1969.

493. Tammelin, L. E., Dialkoxy-phosphorylthiocholines, alkoxy-methylphosphorylthiocholines and analogous choline esters. Syntheses, pK_a of tertiary homologue and cholinesterase inhibition, *Acta Chem. Scand.*, 11, 1340, 1957.

494. Hollingworth, R. M., Fukuto, T. R., and Metcalf, R. L., Selectivity of sumithion compared with methyl parathion, *J. Agric. Food Chem.*, 15, 235, 1967.

495. Eto, M., Sakata, M., and Sasayama, T., Biological activities of *p*-ethylphenyl and *p*-acetylphenyl phosphates and their thiono analogs, *Agric. Biol. Chem.*, 36, 645, 1972.

495a. Michel, H. O., Kinetics of the reactions of cholinesterase, chymotrypsin and trypsin with organophosphorus inactivators, *Fed. Proc.*, 14, 255, 1955.

496. Boter, H. L. and Ooms, A. J. J., Stereospecificity of hydrolytic enzymes in their reaction with optically active organophosphorus compounds. II. The inhibition of aliesterase, acetylesterase, chymotrypsin, and trypsin by S-alkyl *p*-nitrophenyl methylphosphonothiolates, *Biochem. Pharmacol.*, 16, 1563, 1967.

497. Hathway, D. E., Brown, S. S., Chasseaud, L. F., and Hutson, D. H., *Foreign Compound Metabolism in Mammals*, Vol. 1, The Chemical Society, London, 1970.

498. Brooks, G. T., Pathway of enzymatic degradation of pesticides, in *Environmental Quality and Safety*, Vol. 1, Coulston, F. and Korte, F., Eds., 1972, 106.

499. Shuster, L., Metabolism of drugs and toxic substances, *Annu. Rev. Biochem.*, 23, 571, 1964.

500. Casida, J. E., Chemistry and metabolism of terminal residues of organophosphorus compounds and carbamates, in *Pesticide Chemistry, Proceedings 2nd International IUPAC Congress*, Vol. VI, Tahori, A. S., Ed., Gordon & Breach, London, 1972, 295.

501. Fukuto, T. R. and Metcalf, R. L., Metabolism of insecticides in plants and animals, *Ann. N.Y. Acad. Sci.*, 160, 97, 1969.

502. Miyamoto, J., Organophosphorus insecticides and environment, *Botyu-Kagaku*, 36, 135, 1971.

503. Esser, H. O., Terminal residues of organophosphorus insecticides in animals, in *Pesticide Terminal Residues*, Butterworths, London, 33, 1971.

503a. Menzie, C. M., *Metabolism of Pesticides*, Bureau of Sport, Fisheries and Wildlife Special Scientific Report Wildlife No. 127, 1969.

504. Menzer, R. E. and Dauterman, W. C., Metabolism of some organophosphorus insecticides, *J. Agric. Food Chem.*, 18, 1031, 1970.

505. Hollingworth, R. H., Comparative metabolism and selectivity of organophosphate and carbamate insecticides, *Bull. W. H. O.*, 44, 155, 1971.

506. Ernst, W., Der Stoffwechsel von Pesticiden in Säigetieren, *Residue Rev.*, 18, 131, 1967.

507. Casida, J. E. and Lykken, L., Metabolism of organic pesticide chemicals in higher plants, *Annu. Rev. Plant Physiol.*, 20, 607, 1969.

508. Rowlands, D. G., The metabolism of contact insecticides in stored grains, *Residue Rev.*, 17, 105, 1967.

509. Spencer, E. Y., The significance of plant metabolites of insecticides, *Residue Rev.*, 9, 153, 1965.

510. Hodgson, E., Ed., *Enzymatic Oxidations of Toxicants*, North Carolina State University, Raleigh, 1968.

511. Gillette, J. R., Conney, A. H., Cosmides, G. J., Estabrook, R. W., Fouts, J. R., and Mannering, G. J., Eds., *Microsomes and Drug Oxidations*, Academic Press, New York, 1969.

512. Mason, H. S., Mechanism of oxygen metabolism, *Adv. Enzymol.*, 19, 79, 1957.

513. Hayaishi, O., Enzymic hydroxylation, *Annu. Rev. Biochem.*, 38, 21, 1969.

514. Gillette, J. R., Biochemistry of drug oxidation and reduction by enzymes in hepatic endoplasmic reticulum, *Adv. Pharmacol.*, 4, 219, 1966.

515. Knowles, C. O. and Casida, J. E., Mode of action of organophosphate anthelmintics: cholinesterase inhibition in *Ascaris lumbricoides*, *J. Agric. Food Chem.*, 14, 566, 1966.

516. Casida, J. E., Mixed-function oxidase involvement in the biochemistry of insecticide synergists, *J. Agric. Food Chem.*, 18, 753, 1970.

517. Omura, T. and Sato, R., The carbon monoxide-binding pigment of liver microsomes, *J. Biol. Chem.*, 239, 2370, 2379, 1964.

518. Kamin, H. and Masters, B. S. S., Electron transport in microsomes, in *Enzymatic Oxidations of Toxicants*, Hodgson, E., Ed., North Carolina State University, Raleigh, 1968, 5.

519. Estabrook, R. W. and Cohen, B., Organization of the microsomal electron transport system, in *Microsomes and Drug Oxidations*, Gillette, J. R., Conney, A. H., Cosmides, G. J., Estabrook, R. W., Fouts, J. R., and Mannering, G. J., Eds., Academic Press, New York, 1969, 95.

520. **Sato, R., Hishibayashi, H., and Ito, A.,** Characterization of two hemoproteins of liver microsomes, in *Microsomes and Drug Oxidations,* Gillette, J. R., Conney, A. H., Cosmides, G. J., Estabrook, R. W., Fouts, J. R., and Mannering, G. J., Eds., Academic Press, New York, 1969, 111.

521. **Casida, J. E.,** Insect microsomes and insecticide chemical oxidations, in *Microsomes and Drug Oxidations,* Gillette, J. R., Conney, A. H., Cosmides, G. J., Estabrook, R. W., Fouts, J. R., and Mannering, G. J., Eds., Academic Press, New York, 1969, 517.

522. **Matthews, H. B. and Hodgson, E.,** Naturally occurring inhibitor(s) of microsomal inhibitors from the housefly, *J. Econ. Entomol.,* 59, 1286, 1964.

523. **Hook, G. E. R., Jordan, T. W., and Smith, J. N.,** Factors affecting insect microsomal oxidations, in *Enzymatic Oxidations of Toxicants,* Hodgson, E., Ed., North Carolina State University, Raleigh, 1968, 27.

524. **Tsukamoto, M. and Casida, J. E.,** Albumin enhancement of oxidative metabolism of methylcarbamate insecticide chemicals by the housefly microsome-NADPH$_2$ system, *J. Econ. Entomol.,* 60, 617, 1967.

525. **Schonbrod, R. D. and Terriere, L. C.,** Eye pigments as inhibitors of microsomal aldrin epoxidase in the housefly, *J. Econ. Entomol.,* 64, 44, 1971.

526. **Wilson, T. G. and Hodgson, E.,** Mechanism of microsomal mixed-function oxidase inhibitor from the housefly *Musca domestica* L, *Pest. Biochem. Physiol.,* 2, 64, 1972.

527. **Schonbrod, R. D. and Terriere, L. C.,** Inhibition of housefly microsomal epoxidase by the eye pigment, xanthommatin, *Pest. Biochem. Physiol.,* 1, 409, 1971.

528. **Knaak, J. B., Stahmann, M. A., and Casida, J. E.,** Peroxidase and ethylenediaminetetraacetic acid − ferrous iron − catalyzed oxidation and hydrolysis of parathion, *J. Agric. Food Chem.,* 10, 154, 1962.

529. **Nakatsugawa, T. and Dahm, P. A.,** Microsomal metabolism of parathion, *Biochem. Pharmacol.,* 16, 25, 1967.

530. **Nakatsugawa, T., Tolman, N. M., and Dahm, P. A.,** Degradation of parathion in the rat, *Biochem. Pharmacol.,* 18, 1103, 1969.

531. **Camp, H. B., Fukuto, T. R., and Metcalf, R. L.,** Selective toxicity of isopropyl parathion metabolism in the housefly, honey bee, and white mouse, *J. Agric. Food Chem.,* 17, 249, 1969.

532. **Morello, A., Vandanis, A., and Spencer, E. Y.,** Comparative metabolism of two vinyl phosphorothionate isomers (thiono phosdrin) by the mouse and the fly, *Biochem. Pharmacol.,* 17, 1795, 1968.

533. **Plapp, F. W. and Casida, J. E.,** Bovine metabolism of organophosphorus insecticides. Metabolic fate of *O,O*-dimethyl *O*-(2,4,5-trichlorophenyl) phosphorothioate in rats and a cow, *J. Agric. Food Chem.,* 6, 662, 1958.

534. **McBain, J. B., Yamamoto, I., and Casida, J. E.,** Mechanism of activation and deactivation of Dyfonate® (*O*-ethyl *S*-phenyl ethylphosphonodithioate) by rat liver microsomes, *Life Sci.,* II, 10, 947, 1971.

535. **Ptashne, K. A. and Neal, R. A.,** Reaction of parathion and malathion with peroxytrifluoroacetic acid, a model system for the mixed function oxidases, *Biochemistry,* 11, 3224, 1972.

536. **Leesch, J. G. and Fukuto, T. R.,** The metabolism of Abate in mosquito larvae and houseflies, *Pest. Biochem. Physiol.,* 2, 223, 1972.

537. **Bowman, J. S. and Casida, J. E.,** Metabolism of the systemic insecticide *O,O*-diethyl *S*-ethylthiomethyl phosphorodithioate (Thimet) in plants, *J. Agric. Food Chem.,* 5, 192, 1957.

538. **Waggoner, T. B.,** Metabolism of Nemacur, ethyl 4-(methylthio)-*m*-tolyl isopropylphosphoramidate and identification of two metabolites in plants, *J. Agric. Food Chem.,* 20, 157, 1971.

539. **Metcalf, R. L.,** The role of oxidative reactions in the mode of action of insecticides, in *Enzymatic Oxidations of Toxicants,* Hodgson, E., Ed., North Carolina State University, Raleigh, 1968, 151.

540. **Getzin, L. W. and Shanks, C. H., Jr.,** Persistence, degradation and bioactivity of phorate and its oxidative analog in soil, *J. Econ. Entomol.,* 63, 52, 1970.

541. **Takase, I. and Nakamura, H.,** The fate of disulfoton and its oxidized products in soils, *Ann. Meeting Agric. Chem. Soc. Jap.,* abstract of papers, p. 82, 1973.

542. **Lee, Y. C., Hayes, M. G. J., and McCormick, D. B.,** Microsomal oxidation of α-thiocarboxylic acids to sulfoxides, *Biochem. Pharmacol.,* 19, 2825, 1970.

543. **Casida, J. E., Allen, T. C., and Stahmann, M. A.,** Mammalian conversion of octamethylpyrophosphoramide to a toxic phosphoramide *N*-oxide, *J. Biol. Chem.,* 210, 607, 1954.

544. **Spencer, E. Y., O'Brien, R. D., and White, R. W.,** Permanganate oxidation products of schradan, *J. Agric. Food Chem.,* 5, 123, 1957.

545. **Dolnick, E. H., Lindahl, I. L., Terrill, C. E., and Reynolds, P. J.,** Cyclophosphamide as a chemical "defleecing" agent for sheep, *Nature,* 221, 467, 1969.

546. **Connors, T. A., Grover, P. L., and McLoughlin, A. M.,** Microsomal activation of cyclophosphamide in vivo, *Biochem. Pharmacol.,* 19, 1533, 1970.

547. **Tochino, Y., Iwata, T., Takamizawa, A., and Hamazima, Y.,** Studies on the metabolism and appearance of anti-tumor activity of cyclophosphamide, *Proc. Symp. Drug Metab. Action,* 3rd., p. 51, 1971.

548. **Bakke, J. E., Feil, V. J., Fjelstul, C. E., and Thacker, E. J.,** Metabolism of cyclophosphamide by sheep, *J. Agric. Food Chem.,* 20, 384, 1972.

549. **Menzer, R. E. and Casida, J. E.,** Nature of toxic metabolites formed in mammals, insects, and plants from 3-(dimethoxyphosphinyloxy)-*N,N*-dimethyl-*cis*-crotonamide and its *N*-methyl analog, *J. Agric. Food Chem.,* 13, 102, 1965.

550. **Bull, D. L. and Lindquist, D. A.,** Metabolism of 3-hydroxy-*N,N*-dimethylcrotonamide dimethyl phosphate by cotton plants, insects, and rats, *J. Agric. Food Chem.,* 12, 310, 1964.

551. **Lucier, G. W. and Menzer, R. E.,** Nature of oxidative metabolites of dimethoate formed in rats, liver microsomes, and bean plants, *J. Agric. Food Chem.,* 18, 698, 1970.

552. **O'Brien, R. D., Kimmel, E. C., and Sferra, P. R.,** Toxicity and metabolism of famphur in insects and mice, *J. Agric. Food Chem.,* 13, 366, 1965.

553. **Casida, J. E., Eto, M., and Baron, R. L.,** Biological activity of a tri-*o*-cresyl phosphate metabolite, *Nature,* 191, 1396, 1961.

554. **Eto, M., Oshima, Y., and Casida, J. E.,** Plasma albumin as a catalyst in cyclization of diaryl *o*-(α-hydroxy)tolyl phosphates, *Biochem. Pharmacol.,* 16, 295, 1967.

555. **Eto, M., Matsuo, S., and Oshima, Y.,** Metabolic formation of saligenin cyclic phosphates from *o*-tolyl phosphates in house flies, *Musca domestica, Agric. Biol. Chem.* (Tokyo), 27, 870, 1963.

556. **Eto, M.,** Relation of chemical structure to biological activity of saligenin cyclic phosphorus esters, in *Pesticide Chemistry, Proceedings 2nd International IUPAC Congress,* Vol. 1, Tahori, A. S., Ed., Gordon & Breach, London, 1972, 311.

557. **Eto, M. and Abe, M.,** Metabolic activation of alkylphenyl phosphates, *Biochem. Pharmacol.,* 20, 967, 1971.

558. **Eto, M., Abe, M., and Takahara, H.,** Metabolism of tri-*p*-ethylphenyl phosphate and neurotoxicity of the metabolites, *Agric. Biol. Chem.,* 35, 929, 1971.

559. **Hösl, H.,** Metabolismus und neurotoxische Wirksamkeit von Tri-(*p*-äthylphenyl)-phosphorsäureester, Inaugural-Dissertation, University of Würzburg, 1971.

560. **Shishido, T., Usui, K., and Fukami, J.,** Oxidative metabolism of diazinon by microsomes from rat liver and cockroach fat body, *Pest. Biochem. Physiol.,* 2, 27, 1972.

561. **Fukami, J. and Shishido, T.,** Selective toxicity of diazinon and other non-systemic insecticides, in *Pesticide Chemistry, Proceedings 2nd International IUPAC Congress,* Vol. 1, Tahori, A. S., Ed., Gordon & Breach, London, 1972, 29.

562. **Machin, A. F., Auick, M. P., Rogers, H., and Anderson, P. H.,** Conversion of diazinon to hydroxydiazinon in the guinea pig and sheep, *Bull. Environ. Contam. Toxicol.,* 6, 26, 1971.

563. **Miyazaki, H., Tojinbara, I., Watanabe, Y., Osaka, T., and Okui, S.,** Metabolism of diazinon in animals and plants, *Proc. Symp. Drug Metab. Action,* 1st., p. 135, 1969.

564. **Janes, N. F., Machin, A. F., Quick, M. P., Rogers, H., Mundy, D. E., and Cross, A. J.,** Toxic metabolites of diazinon in sheep, *J. Agric. Food Chem.,* 21, 121, 1973.

565. **Fukuto, T. R., Shrivastava, S. P., and Black, A. L.,** Metabolism of 2- methoxy(methylthio)phosphinylimino-3-ethyl-5-methyl-1,3-oxazolidine in the cotton plant and houseflies, *Pest. Biochem. Physiol.,* 2, 162, 1972.

566. **Eto, M., Hashimoto, Y., and Miyamoto, K.,** unpublished data.

567. **Tomizawa, C. and Uesugi, Y.,** Metabolism of *S*-benzyl *O,O*-diisopropyl phosphorothioate (Kitazin P) by mycelial cells of *Pyricularia oryzae, Agric. Biol. Chem.,* 36, 294, 1972.

568. **Uesugi, Y. and Tomizawa, C.,** Metabolism of *S*-benzyl *O*-ethyl phenylphosphonothioate (Inezin) by mycelial cells of *Pyricularia oryzae, Agric. Biol. Chem.,* 36, 313, 1971.

569. **Miyamoto, J.,** Non-enzymatic conversion of Dipterex into DDVP and their inhibitory action on enzymes, *Botyu-Kagaku,* 24, 130, 1959.

570. **Benjamini, E., Metcalf, R. L., and Fukuto, T. R.,** Contact and systemic insecticidal properties of *O,O*-diethyl *O-p*-methylsulfinylphenyl phosphorothionate and its analog, *J. Econ. Entomol.,* 52, 99, 1959.

571. **Lucier, G. W. and Menzer, R. E.,** Nature of neutral phosphorus ester metabolites of phosphamidon formed in rats and liver microsomes, *J. Agric. Food Chem.,* 19, 1249, 1971.

572. **Clemens, G. P. and Menzer, R. E.,** Oxidative metabolism of phosphamidon in rats and a goat, *J. Agric. Food Chem.,* 16, 312, 1968.

573. **Miller, R. W., Gordon, C. H., Morgan, N. O., Bowman, M. C., and Beroza, M.,** Coumaphos as a feed additive for the control of house fly larvae in cow manure, *J. Econ. Entomol.,* 63, 853, 1970.

574. **Neal, R. A.,** Studies on the metabolism of diethyl 4-nitrophenyl phosphorothionate (parathion) in vitro, *Biochem. J.,* 103, 183, 1967.

575. **Nakatsugawa, T., Tolman, N. M., and Dahm, P. A.,** Degradation and activation of parathion analogs by microsomal enzymes, *Biochem. Pharmacol.,* 17, 1517, 1968.

576. **Yang, R. S. H., Hodgson, E., and Dauterman, W. C.,** Metabolism in vitro of diazinon and diazoxon in rat liver, *J. Agric. Food Chem.,* 19, 10, 1971.

577. **Nakatsugawa, T., Tolman, N. M., and Dahm, P. A.,** Oxidative degradation of diazinon by rat liver microsomes, *Biochem. Pharmacol.,* 18, 685, 1969.

578. **Motoyama, N. and Dauterman, W. C.,** In vitro metabolism of azinphosmethyl in susceptible and resistant houseflies, *Pest. Biochem. Physiol.,* 2, 113, 1972.

579. **Motoyama, N. and Dauterman, W. C.,** The in vitro metabolism of azinphosmethyl by mouse liver, *Pest. Biochem. Physiol.,* 2, 170, 1972.

580. **Dahm, P. A.,** Some aspects of the metabolism of parathion and diazinon, in *Biochemical Toxicology of Insecticides,* O'Brien, R. D. and Yamamoto, I., Eds., Academic Press, New York, 1970, 51.

581. **Yang, R. S. H., Hodgson, E., and Dauterman, W. C.,** Metabolism in vitro of diazinon and diazoxon in susceptible and resistant houseflies, *J. Agric. Food Chem.,* 19, 14, 1971.

582. **Neal, R. A.,** Enzymic mechanism of metabolism of phosphorothionate insecticide, *Arch. Intern. Med.,* 128, 118, 1971.

583. **Hodgson, E. and Casida, J. E.,** Mammalian enzymes involved in the degradation of 2,2-dichlorovinyl dimethyl phosphate, *J. Agric. Food. Chem.,* 10, 208, 1962.

584. **Casida, J. E. and Augustinsson, K. B.,** Reaction of plasma albumin with 1-naphthyl N-methylcarbamate and certain other esters, *Biochim. Biophys. Acta,* 36, 411, 1959.

585. **Main, A. R.,** The differentiation of the A-type esterases in sheep serum, *Biochem. J.,* 75, 188, 1960.

586. **Mounter, L. A.,** Enzymic hydrolysis of organophosphorus compounds, in *The Enzyme,* Vol. 4, Boyer, P. D., Lardy, H., and Myrbäck, L., Eds., Academic Press, New York, 1960, 541.

587. **Skrinjarić-Spoljar, M. and Reiner, E.,** Hydrolysis of diethyl-*p*-nitrophenyl phosphate and ethyl-*p*-nitrophenyl ethylphosphonate by human sera, *Biochim. Biophys. Acta,* 165, 289, 1968.

588. **Kojima, K. and O'Brien, R. D.,** Paraoxon hydrolyzing enzymes in rat liver, *J. Agric. Food Chem.,* 16, 574, 1968.

589. **Shishido, T. and Fukami, J.,** Enzymatic hydrolysis of diazoxon by rat tissue homogenates, *Pest. Biochem. Physiol.,* 2, 39, 1972.

590. **Krueger, H. R. and Casida, J. E.,** Hydrolysis of certain organophosphate insecticides by house fly enzymes, *J. Econ. Entomol.,* 54, 239, 1961.

591. **Jarczyk, H. J.,** The influence of esterases in insects on the degradation of organophosphates of the RE 605 series, *Pflanzenschutz-Nachr.,* 19, 1, 1966.

592. **Miyata, T. and Matsumura, F.,** Partial purification of American cockroach enzymes degrading certain organophosphate insecticides, *Pest. Biochem. Physiol.,* 1, 267, 1971.

593. **Matsumura, F. and Hogendijk, C. J.,** The enzymatic degradation of parathion in organophosphate-susceptible and -resistant houseflies, *J. Agric. Food Chem.,* 12, 447, 1964.

594. **Nakatsugawa, T., Tolman, N. M., and Dahm, P. A.,** Metabolism of S^{35}-parathion in the housefly, *J. Econ. Entomol.,* 62, 408, 1969.

595. **Welling, W., Blaakmeer, P., Vink, G. J., and Voerman, S.,** In vitro hydrolysis of paraoxon by parathion resistant houseflies, *Pest. Biochem. Physiol.,* 1, 61, 1971.

596. **Oppenoorth, F. J. and Van Asperen, K.,** The detoxication enzymes causing organophosphate resistance in the housefly; properties, inhibition, and the action of inhibitors as synergists, *Entomol. Exp. Appl.,* 4, 311, 1961.

597. **Oppenoorth, F. J.,** Biochemical genetics of insecticide resistance, *Annu. Rev. Entomol.,* 10, 185, 1965.

598. **Oppenoorth, F. J.,** Resistance in insects: the role of metabolism and the possible use of synergists, *Bull. W. H. O.,* 44, 195, 1971.

599. **Oppenoorth, F. J.,** Degradation and activation of organophosphorus insecticides and resistance in insects, in *Toxicology, Biodegradation and Efficacy of Livestock Pesticides,* Khan, M. A. and Haufe, W. O., Eds., Swets & Zeitlinger, Amsterdam, 1972, 73.

600. **Donninger, C., Nobbs, B. T., and Wilson, K.,** An enzyme catalysing the hydrolysis of phosphoric acid diesters in rat liver, *Biochem. J.,* 122, 51p, 1971.

601. **Shishido, T., Usui, K., Sato, M., and Fukami, J.,** Enzymatic conjugation of diazinon with glutathione in rat and American cockroach, *Pest. Biochem. Physiol.,* 2, 51, 1972.

602. **Boyland, E. and Chasseaud, L. F.,** The role of glutathione and glutathione S-transferase in mercapturic acid biosynthesis, *Adv. Enzymol.,* 32, 173, 1969.

603. **Donninger, C.,** Species specificity of phosphate triester anticholinesterases, *Bull. W. H. O.,* 44, 265, 1971.

604. **Lewis, J. B. and Sawicki, R. M.,** Characterization of resistance mechanism to diazinon, parathion and diazoxon in the organophosphorus-resistant SKA strain of house flies (*Musca domestica* L.), *Pest. Biochem. Physiol.,* 1, 275, 1971.

605. **Fukami, J. and Shishido, T.,** Studies on the selective toxicities of organic phosphorus insecticides. III. The characters of the enzyme system in cleavage of methyl parathion to desmethyl parathion in the supernatant of several species of homogenates, *Botyu-Kagaku,* 28, 77, 1963.

606. **Fukami, J. and Shishido, T.,** Nature of soluble, glutathione-dependent enzyme system active in cleavage of methyl parathion to desmethyl parathion, *J. Econ. Entomol.,* 59, 1338, 1966.

607. **Fukunaga, K., Fukami, J., and Shishido, T.,** The in vitro metabolism of organophosphorus insecticides by tissue homogenates from mammal and insect, *Residue Rev.,* 25, 223, 1969.

608. **Morello, A., Vardanis, A., and Spencer, E. Y.,** Mechanism of detoxication of some organophosphorus compounds: the role of glutathione-dependent demethylation, *Can. J. Biochem.,* 46, 885, 1968.

609. **Stenersen, J.,** Demethylation of the insecticide bromophos by a glutathione-dependent liver enzyme and by alkaline buffers, *J. Econ. Entomol.,* 62, 1043, 1969.

610. **Hollingworth, R. M.,** The dealkylation of organophosphorus triesters by liver enzymes, in *Biochemical Toxicology of Insecticides,* O'Brien, R. D. and Yamamoto, I., Eds., Academic Press, New York, 1970, 75.

611. **Saito, T.,** Selective toxicity of systemic insecticides, *Residue Rev.,* 25, 175, 1969.

612. **Oppenoorth, F. J., Rupes, V., Elbashir, S., Houx, N. W., and Voerman, S.,** Glutathione-dependent degradation of parathion and its significance for resistance in the housefly, *Pest. Biochem. Physiol.,* 2, 262, 1972.

613. **Negherbon, W. O.,** *Handbook of Toxicology, Insecticides,* Vol. III, Saunders, Philadelphia, 1959.

614. **Sakamoto, H. and Nishizawa, Y.,** Preparation and biological properties of phenylphosphonothioates, *Agric. Biol. Chem.,* 26, 252, 1962.

615. **Stenersen, J.,** Thin-layer chromatography of diesters and some monoesters of phosphoric acid, *J. Chromatogr.*, 54, 77, 1971.
616. **Hollingworth, R. M., Metcalf, R. L., and Fukuto, T. R.,** The selectivity of Sumithion compared with methyl parathion. Metabolism in the white mouse, *J. Agric. Food Chem.*, 15, 242, 1967.
617. **Ueda, K. and Kawai, M.,** Therapy of organophosphate poisoning with oxime antidote and glutathione, *4th International Congress on Rural Medicine,* Usuda, abstract of papers, p. 10, 1969.
618. **Nolan, J. and O'Brien, R. D.,** Biochemistry of resistance to paraoxon in strains of houseflies, *J. Agric. Food Chem.*, 18, 802, 1970.
619. **Matsumura, F. and Ward, C. T.,** Degradation of insecticides by human and the rat liver, *Arch. Environ. Health*, 13, 257, 1966.
620. **Hirose, M., Miyata, T., and Saito, T.,** Residue, degradation and metabolism of ^{14}C-labeled Elsan® (*O,O*-dimethyl *S*-α-carboethoxybenzyl phosphorodithioate) in cabbages, Hime-apples and strawberries, *Botyu-Kagaku*, 36, 43, 1971.
621. **Morikawa, O. and Saito, T.,** Degradation of vamidothion and dimethoate in plants, insects, and mammals, *Botyu-Kagaku*, 31, 130, 1966.
622. **Hassan, H., Zayed, S. M. A., and Bahig, M. R. E.,** Metabolic fate of dimethoate in the rat, *Biochem. Pharmacol.*, 18, 2429, 1969.
623. **Gage, J. C.,** Metabolism of menazon [*O,O*-dimethyl *S*-(4,6-diamino-*s*-triazin-2-ylmethyl) phosphorodithioate] in the rat, *Food Cosmet. Toxicol.*, 5, 349, 1967.
624. **Arthur, B. W. and Casida, J. E.,** Biological activity and metabolism of Hercules AC-528 components in rats and cockroaches, *J. Econ. Entomol.*, 52, 20, 1959.
625. **McBain, J. B., Menn, J. J., and Casida, J. E.,** Metabolism of carbonyl-^{14}C-labeled Imidan[N-(mercaptomethyl)-phthalimide *S*-(*O,O*-dimethyl phosphorodithioate)] in rats and cockroaches, *J. Agric. Food Chem.*, 16, 813, 1968.
626. **Bauriedel, W. R. and Swank, M. G.,** Residue and metabolism of radioactive 4-*tert*-butyl-2-chlorophenyl methyl methylphosphoramidate administered as a single oral dose to sheep, *J. Agric. Food Chem.*, 10, 150, 1962.
627. **Miyamoto, J.,** Studies on the mode of action of dipterex. II. New glucuronides obtained from the urine of rabbit following administration of dipterex, *Agric. Biol. Chem.*, 25, 566, 1961.
628. **Main, A. R. and Braid, P. E.,** Hydrolysis of malathion by ali-esterases in vitro and in vivo, *Biochem. J.*, 84, 255, 1962.
629. **Chen, P. R., Tucker, W. P., and Dauterman, W. C.,** Structure of biologically produced malathion monoacid, *J. Agric. Food Chem.*, 17, 86, 1969.
630. **Welling, W. and Blaakmeer, P. T.,** Metabolism of malathion in a resistant and a susceptible strain of houseflies, in *Pesticide Chemistry, Proceedings 2nd International IUPAC Congress,* Vol. II. Tahori, A. S., Ed., Gordon & Breach, London, 1972, 61.
631. **Kojima, K.,** Studies on the Selective Toxicity and Detoxication of Organophosphorus Compounds with References to the Studies of Carboxyesterase as a Factor in the Selective Toxicity of Malathion, Special Report of Toa Noyaku Co., 1961.
632. **Matsumura, F. and Brown, A. W. A.,** Biochemistry of malathion resistance in *Culex tarsalis, J. Econ. Entomol.*, 54, 1176, 1961.
633. **Matsumura, F. and Hogendijk, C. J.,** The enzymatic degradation of malathion in organophosphate resistant and susceptible strains of *Musca domestica, Entomol. Exp. Appl.*, 7, 179, 1964.
634. **Kojima, K., Ishizuka, T., and Kitakata, S.,** Mechanism of resistance to malathion in the green rice leafhopper, *Nephotettix cincticeps, Botyu-Kagaku*, 28, 17, 1963.
635. **Takahashi, Y., Saito, T., Iyatomi, K., and Eto, M.,** Mechanism of synergistic action between malathion and K-1 (2-phenyl-4H-1,3,2-benzodioxaphosphorin-2-oxide) in organophosphate resistant citrus red mites, *Botyu-Kagaku*, 38, 13, 1973.
636. **Ozaki, K.,** The resistance to organophosphorus insecticides of the green rice leafhopper, *Nephotettix cincticeps* Uhler and the small brown planthopper, *Laodelphax striatellus* Fallén, *Rev. Plant Protec. Res.*, 2, 1, 1969.
637 **Ohkawa, H., Eto, M., Oshima, Y., Tanaka, F., and Umeda, K.,** Two types of carboxyesterase degrading malathion in resistant houseflies and their inhibition by synergists, *Botyu-Kagaku*, 33, 139, 1968.
638. **Krueger, H. R., Casida, J. E., and Niedermeier, R. P.,** Bovine metabolism of organophosphorus insecticides. Metabolism and residues associated with dermal application of Co-ral to rats, a goat, and a cow, *J. Agric. Food Chem.*, 7, 182, 1959.
639. **Uchida, T. and O'Brien, R. D.,** Dimethoate degradation by human liver and its significance for acute toxicity, *Toxicol. Appl. Pharmacol.*, 10, 89, 1967.
640. **Chen, P. R. S. and Dauterman, W. C.,** Studies on the toxicity of dimethoate analogs and their hydrolysis by sheep liver amidase, *Pest. Biochem. Physiol.*, 1, 340, 1971.
641. **Uchida, T., Zschintzsch, J., and O'Brien, R. D.,** Relation between synergism and metabolism of dimethoate in mammals and insects, *Toxicol. Appl. Pharmacol.*, 8, 259, 1966.
642. **Uchida, T., Dauterman, W. C., and O'Brien, R. D.,** The metabolism of dimethoate by vertebrate tissues, *J. Agric. Food Chem.*, 12, 48, 1964.
643. **Dauterman, W. C., Viado, G. B., Casida, J. E., and O'Brien, R. D.,** Persistence of dimethoate and metabolites following folian application to plants, *J. Agric. Food Chem.*, 8, 115, 1960.

644. **Geissbühler, H., Voss, G., and Anliker, R.**, The metabolism of phosphamidon in plants and animals, *Residue Rev.*, 37, 39, 1971.

645. **Ahmed, M. K., Casida, J. E., and Nichols, R. E.**, Bovine metabolism of organophosphorus insecticides: significance of rumen fluid with particular reference to parathion, *J. Agric. Food Chem.*, 6, 740, 1958.

646. **Matsumura, F.**, Metabolism of insecticides in microorganisms and insects, in *Environmental Quality and Safety*, Vol. I, Coulston, F. and Korte, F., Eds., Georg Thieme, Stuttgart, 1972, 96.

647. **Hitchcock, M. and Murphy, S. D.**, Enzymic reduction of *O,O*-diethyl *O*-(4-nitrophenyl) phosphorothioate, *O,O*-diethyl *O*-(4-nitrophenyl) phosphate and *O*-ethyl *O*-(4-nitrophenyl) benzenethiophosphonate by tissues from mammals, birds, and fishes, *Biochem. Pharmacol.*, 16, 1801, 1967.

648. **Lichtenstein, E. P. and Fuhremann, T. W.**, Activity of an NADPH-dependent nitroreductase in houseflies, *Science*, 172, 589, 1971.

649. **Hutson, D. H., Akintonwa, D. A. A., and Hathway, D. E.**, The metabolism of 2-chloro-1-(2′,4′-dichlorophenyl)vinyl phosphate (chlorfenvinphos) in the dog and rat, *Biochem. J.*, 102, 133, 1967.

650. **Beynon, K. J. and Wright, A. N.**, Breakdown of the insecticide, Gardona, on plants and in soils, *J. Sci. Food Agric.*, 20, 250, 1969.

651. **Gutenmann, W. H., St. John, L. E., Jr., and Lisk, D. J.**, Metabolic studies with Gardona insecticide in the dairy cow, *J. Agric. Food Chem.*, 19, 1259, 1971.

652. **Gatterdam, P. E., Wozniak, L. A., Bullock, M. W., Parks, G. L., and Boyd, J. E.**, Absorption, metabolism, and excretion of tritium-labeled famphur in the sheep and calf, *J. Agric. Food Chem.*, 15, 845, 1967.

653. **Pankaskie, J. E., Fountaine, F. C., and Dahm, P. A.**, The degradation and detoxication of parathion in dairy cows, *J. Econ. Entomol.*, 45, 51, 1952.

654. **Casida, J. E., McBride, L., and Niedermeier, R. P.**, Metabolism of 2,2-dichlorovinyl dimethyl phosphate in relation to residues in milk and mammalian tissues, *J. Agric. Food Chem.*, 10, 370, 1962.

655. **Wendel, L. E. and Bull, D. L.**, Systemic activity and metabolism of dimethyl *p*-(methylthio)phenyl phosphate in cotton, *J. Agric. Food Chem.*, 18, 420, 1970.

656. **Bull, D. L. and Stokes, R. A.**, Metabolism of dimethyl *p*-(methylthio)phenyl phosphate in animals and plants, *J. Agric. Food Chem.*, 18, 1134, 1970.

657. **Miyamoto, J., Wakimura, A., and Kadota, T.**, Biodegradation of Cyanox (*O,O*-dimethyl *O*-(4-cyanophenyl) phosphorothioate) in rats, in *Environmental Quality and Safety*, Vol. 1, Coulston, F. and Korte, F., Eds., Georg Thieme, Stuttgart, 1972, 235.

658. **Esser, H. O., Mücke, W., and Alt, K. O.**, Der Abbau des Insektizides GS 13005 in der Ratte. Strukturaufklärung der wichtigsten Metabolite, *Helv. Chim. Acta*, 51, 513, 1968.

659. **Dupuis, G., Muecke, W., and Esser, H. O.**, Metabolic behavior of the insecticidal phosphorus ester GS-13005, *J. Econ. Entomol.*, 64, 588, 1971.

660. **McBain, J. B., Hoffman, L. J., and Menn, J. J.**, Metabolic degradation of *O*-ethyl *S*-phenyl ethylphosphonodithioate (Dyfonate) in potato plants, *J. Agric. Food Chem.*, 18, 1139, 1970.

661. **McBain, J. B. and Menn, J. J.**, *S*-methylation, oxidation, hydroxylation and conjugation of thiophenol in the rat, *Biochem. Pharmacol.*, 18, 2282, 1969.

662. **Gunther, F. A.**, Insecticide residues in California citrus fruits and products, *Residue Rev.*, 28, 1, 1969.

663. **Faust, S. D. and Suffet, I. H.**, Recovery, separation and identification of organic pesticides from natural and potable waters, *Residue Rev.*, 15, 44, 1966.

664. **Menzer, R. E., Fontanilla, E. L., and Ditman, L. P.**, Degradation of disulfoton and phorate in soil influenced by environmental factors and soil type, *Bull. Environ. Contam. Toxicol.*, 5, 1, 1970.

665. **Edwards, C. A.**, Insecticide residues in soils, *Residue Rev.*, 13, 83, 1966.

666. **Bowman, B. T., Adams, R. S., Jr., and Fenton, S. W.**, Effect of water on malathion adsorption on five montmorillonite systems, *J. Agric. Food Chem.*, 18, 723, 1970.

667. **Harris, C. R.**, Influence of soil moisture on the toxicity of insecticides in a mineral soil to insects, *J. Econ. Entomol.*, 57, 946, 1964.

668. **Getzin, L. W. and Chapman, R. K.**, The fate of phorate in soils, *J. Econ. Entomol.*, 53, 47, 1960.

669. **Saltzman, S. and Yaron, B.**, Parathion adsorption from aqueous solutions as influenced by soil components, in *Pesticide Chemistry, Proceedings 2nd International IUPAC Congress*, Vol. VI, Tahori, A. S., Ed., Gordon & Breach, London, 1972, 87.

670. **Getzin, L. W. and Rosefield, I.**, Organophosphorus insecticide degradation by heat-labile substances in soil, *J. Agric. Food Chem.*, 16, 598, 1968.

671. **Getzin, L. W.**, Persistence of diazinon and zinophos in soil: effect of autoclaving, temperature, moisture, and acidity, *J. Econ. Entomol.*, 61, 1560, 1968.

672. **Bro-Rasmussen, F., Noddegaard, E., and Voldum-Clausen, K.**, Degradation of diazinon in soil, *J. Sci. Food Agric.*, 19, 278, 1968.

673. **Lichtenstein, E. P., Fuhreman, T. W., and Schulz, K. R.**, Effect of sterilizing agents on persistence of parathion and diazinon in soils and water, *J. Agric. Food Chem.*, 16, 870, 1968.

674. **Getzin, L. W.**, Metabolism of diazinon and zinophos in soils, *J. Econ. Entomol.*, 60, 505, 1967.

675. **Kearney, P. C. and Helling, C. S.**, Reactions of pesticides in soils, *Residue Rev.*, 25, 25, 1969.

676. **Sethunathan, N. and Yoshida, T.,** Fate of diazinon in submerged soil. Accumulation of hydrolysis product, *J. Agric. Food Chem.,* 17, 1192, 1969.

677. **Lichtenstein, E. P.,** Persistence and fate of pesticides in soils, water and crops: significance to humans, in *Pesticide Chemistry, Proceedings 2nd International IUPAC Congress,* Vol. VI, Tahori, A. S., Ed., Gordon & Breach, London, 1972, 1.

678. **Mick, D. L. and Dahm, P. A.,** Metabolism of parathion by two species of *Rhizobium, J. Econ. Entomol.,* 63, 1155, 1970.

679. **Miyamoto, J., Kitagawa, K., and Sato, Y.,** Metabolism of organophosphorus insecticides by *Bacillus subtilis,* with special emphasis on Sumithion, *Jap. J. Exp. Med.,* 36, 211, 1966.

680. **Garretson, A. L. and San Clemente, C. L.,** Inhibition of nitrifying chemolithotrophic bacteria by several insecticides, *J. Econ. Entomol.,* 61, 285, 1968.

681. **Sommer, K.,** Effect of various pesticides on nitrification and nitrogen transformation in soils, *Landwirtsch Forsch. Sonderh.,* 25, 22, 1970; *C. A.,* 74, 110863, 1971.

682. **Kobayashi, T. and Katsura, S.,** Soil application of insecticides. IV. Effects of systemic organophosphates on the nitrification of soil and on the growth and yield of potato, *Jap. J. Appl. Entomol. Zool.,* 12, 53, 1968.

683. **Kearney, P. C., Nash, R. G., and Isensee, A. R.,** Persistence of pesticide residues in soils, in *Chemical Fallout,* Miller, M. W. and Berg, G. G., Eds., Charles C Thomas, Springfield, 1969, 54.

684. **Harris, C. R. and Hitchon, J. L.,** Laboratory evaluation of candidate materials as potential soil insecticides, *J. Econ. Entomol.,* 63, 2, 1970.

685. **Harris, C. R.,** Laboratory studies on persistence of biological activity of some insecticides in soils, *J. Econ. Entomol.,* 62, 1437, 1969.

686. **Bro-Rasmussen, F., Noddegaard, E., and Voldum-Clausen, K.,** Comparison of the disappearance of eight organophosphorus insecticides from soil in laboratory and in outdoor experiments, *Pest. Sci.,* 1, 179, 1970.

687. **Oppenoorth, F. J.,** Two types of sesamex-suppressible resistance in the housefly, *Entomol. Exp. Appl.,* 10, 75, 1967.

688. **Plapp, F. W., Jr. and Casida, J. E.,** Genetic control of house fly NADPH-dependent oxidases: relation to insecticide chemical metabolism and resistance, *J. Econ. Entomol.,* 62, 1174, 1969.

689. **O'Brien, R. D.,** Selective toxicity of insecticides, in *Advances in Pest Control Research,* Vol. IV, Metcalf, R. L., Ed., Interscience, New York, 1961, 75.

690. **Brown, A. W. A.,** The spread of insecticide resistance in pest species, in *Advances in Pest Control Research,* Vol. II, Metcalf, R. L., Ed., Interscience, New York, 1958, 351.

691. **Brown, A. W. A.,** Mechanisms of resistance against insecticides, *Annu. Rev. Entomol.,* 5, 301, 1960.

692. **Georghiou, G. P.,** Genetic studies on insecticide resistance, in *Advances in Pest Control Research,* Vol. VI, Metcalf, R. L., Ed., Interscience, New York, 1965, 171.

693. **Tsukamoto, M.,** Biochemical genetics of insecticide resistance in the housefly, *Residue Rev.,* 25, 289, 1969.

694. **Wharton, R. H. and Roulston,** Resistance of ticks to chemicals, *Annu. Rev. Entomol.,* 15, 381, 1970.

695. **Plapp, F. W., Jr.,** On the molecular biology of insecticide resistance, in *Biochemical Toxicology of Insecticides,* O'Brien, R. D. and Yamamoto, I., Eds., Academic Press, New York, 1970, 179.

696. **Busvine, J. R. and Feroz, M.,** Biochemistry of resistance: introduction, with remarks on resistance of *Cimex lectularius,* in *Pesticide Chemistry, Proceedings 2nd International IUPAC Congress,* Vol. II, Tahori, A. S., Ed., Gordon & Breach, London, 1971, 1.

697. **Forgash, A. J., Cook, B. J., and Riley, R. C.,** Mechanism of resistance in diazinon-selected multi-resistant *Musca domestica, J. Econ. Entomol.,* 55, 544, 1962.

698. **Sawicki, R. S. and Lord, K. A.,** Some properties of a mechanism delaying penetration of insecticides into houseflies, *Pest. Sci.,* 1, 213, 1970.

699. **Krueger, H. R. and O'Brien, R. D.,** Relationship between metabolism and differential toxicity of malathion in insects and mice, *J. Econ. Entomol.,* 52, 1063, 1959.

700. **O'Brien, R. D., Thorn, G. D., and Fisher, R. W.,** New organophosphate insecticides developed on rational principles, *J. Econ. Entomol.,* 51, 714, 1958.

701. **Matsumura, F. and Voss, G.,** Mechanism of malathion and parathion resistance in the two-spotted spider mite, *Tetranychus urticae, J. Econ. Entomol.,* 57, 911, 1964.

702. **Dauterman, W. C. and Matsumura, F.,** Effect of malathion analogs upon resistant and susceptible *Culex tarsalis* mosquitoes, *Science,* 138, 694, 1962.

703. **Feroz, M.,** Toxicological and genetic studies of organophosphorus resistance in *Cimex lectularius, Bull. Entomol. Res.,* 59, 377, 1969.

704. **Uchida, T., Rahmati, H. S., and O'Brien, R. D.,** The penetration and metabolism of H^3-dimethoate in insects, *J. Econ. Entomol.,* 58, 831, 1965.

705. **Fraser, J., Clinch, P. G., and Reay, R. C.,** N-Acylation of N-methylcarbamate insecticides and its effect on biological activity, *J. Sci. Food Agric.,* 16, 615, 1965.

706. **Ueda, H., Ishikawa, H., Nakaomi, K., Kobayashi, S., and Takahashi, S.,** On the chemical structure and biological activity of low toxic organophosphoramidates, *Ann. Meeting Agric. Chem. Soc. Jap.,* abstract of papers, p. 77, 1973.

707. **Fukuto, T. R.,** Rational design and testing of new and improved anticholinesterase insecticide (discussion), *Bull. W. H. O.,* 44, 411, 1971.

708. **Stopkan, V. V., Kondratyuk, V. I., Shokol, V. A., and Derkach, G. I.,** Insecticidal and biological effect of N-alkyl-N-carbalkoxyphosphoramide diesters, *Fiziol. Aktiv. Veshchestava,* 1969 (2), 19; *C.A.,* 73, 13471, 1970.

709. **Stopkan, V. V., Derkach, G. I., and Slyusarenko, E. I.,** Insecticide activity of avenine analogs, *Fiziol. Aktiv. Veshchestva, Akad. Nauk Ukr. SSR, Repub. Mezhvedom. Sb.,* 1966, 92; *C.A.,* 66, 114881, 1967.

710. **Derkach, G. I., Odintsov, V. S., and Petrenko, V. S.,** Insecticidal and antiesterase action of some esters of alkylurethane phosphoric acids, *Dopov. Akad. Nauk Ukr. SSR, Ser. B,* 31, 261, 1969; *C.A.,* 71, 29543, 1969.

711. **Bull, D. L. and Whitter, C. J.,** Factors influencing organophosphorus insecticide resistance in tobacco budworms, *J. Agric. Food Chem.,* 20, 561, 1972.

712. **Motoyama, N., Rock, G. C., and Dauterman, W. C.,** Studies on the metabolism of azinphosmethyl resistance in the predaceous mite, *Neoseiulus (T.) fallacis, Pest. Biochem. Physiol.,* 1, 205, 1971.

713. **Sawicki, R. M. and Farnham, A. W.,** Examination of the isolated autosomes of the SKA strain of house-flies (*Musca domestica* L.) for resistance to several insecticides with and without pretreatment with sesamex and TBTP, *Bull. Entomol. Res.,* 59, 409, 1969.

714. **Oppenoorth, F. J. and Van Asperen, K.,** Allelic genes in the housefly produces modified enzymes that cause organophosphate resistance, *Science,* 132, 298, 1960.

715. **Needham, P. H. and Sawicki, R. M.,** Diagnosis of resistance to organophosphorus insecticides in *Myzus persicae, Nature,* 230, 125, 1971.

716. **Folsom, M. D., Hansen, L. G., Philpot, R. M., Yang, R. S. H., Dauterman, W. C., and Hodgson, E.,** Biochemical characteristics of microsomal preparations from diazinon-resistant and -susceptible houseflies, *Life Sci.,* (II), 9, 869, 1970.

717. **Ohkawa, H., Eto, M., and Oshima, Y.,** Metabolism and toxicity of Salithion, 2-methoxy-4H-1,3,2-benzodioxaphosphorin-2-sulfide, *Jap. J. Appl. Entomol. Zool.,* 14, 191, 1970.

718. **Tsumuki, H., Saito, T., Miyata, T., and Iyatomi, K.,** Acute and subacute toxicity of organophosphorus insecticides to mammals, in *Biochemical Toxicology of Insecticides,* O'Brien, R. D. and Yamamoto, I., Eds., Academic Press, 1970, 65.

719. **Saito, T.,** Distribution of P^{32}-labeled schradan in the American cockroach, *Botyu-Kagaku,* 25, 57, 1960.

720. **Whetstone, R. R., Phillips, D. D., Sun, Y. P., Ward, L. F., Jr., and Shellenberger, T. E.,** 2-Chloro-1-(2,4,5-trichlorophenyl)vinyl dimethyl phosphate, a new insecticide with low toxicity to mammals, *J. Agric. Food Chem.,* 14, 352, 1966.

721. **Miyamoto, J.,** Mechanism of low toxicity of Sumithion toward animals, *Residue Rev.,* 25, 251, 1969.

722. **Saito, T.,** Electron microscopy of the ganglionic sheath of insects, *Botyu-Kagaku,* 25, 71, 1960.

723. **Smissaert, H. R.,** Cholinesterase inhibition in spider mites susceptible and resistant to organophosphate, *Science,* 143, 129, 1964.

724. **Zahavi, M., Tahori, A. S., and Stolero, F.,** Sensitivity of acetylcholinesterase in spider mites to organophosphorus compounds, *Biochem. Pharmacol.,* 19, 219, 1970.

725. **Smissaert, H. R., Voerman, S., Oostenbrugge, L., and Rennoy, N.,** Acetylcholinesterases of organophosphate-susceptible and -resistant spider mites, *J. Agric. Food Chem.,* 18, 66, 1970.

726. **Nolan, J., Schnitzerling, H. J., and Schuntner, C. A.,** Multiple forms of acetylcholinesterase from resistant and susceptible strains of the cattle tick, *Boophilus microplus* (Can.), *Pest. Biochem. Physiol.,* 2, 85, 1972.

727. **Sun, Y.-P. and Johnson, E. R.,** Analysis of joint action of insecticides against house flies, *J. Econ. Entomol.,* 53, 887, 1960.

728. **Sun, Y.-P. and Johnson, E. R.,** Synergistic and antagonistic actions of insecticide-synergist combinations and their mode of action, *J. Agric. Food Chem.,* 8, 261, 1960.

729. **Metcalf, R. L.,** Mode of action of insecticide synergists, *Annu. Rev. Entomol.,* 12, 229, 1967.

730. **DuBois, K. P.,** Potentiation of the toxicity of organophosphorus compounds, in *Advances in Pest Control Research,* Vol. IV, Metcalf, R. L., Ed., Interscience, New York, 1961, 117.

731. **Murphy, S. D.,** Mechanisms of pesticide interactions in vertebrates, *Residue Rev.,* 25, 201, 1969.

732. **O'Brien, R. D.,** Synergism, antagonism, and other interactions, in *Insecticides Action and Metabolism,* Academic Press, New York, 1967, 209.

733. **Wilkinson, C. F.,** Effects of synergists on the metabolism and toxicity of anticholinesterases, *Bull. W. H. O.,* 44, 171, 1971.

734. **Takehara, H., Kotakemori, M., and Oishi, T.,** Effect of physical and chemical properties of mineral diluents on the degradation of organophosphorus pesticide dust formulation, *Nippon Nogei Kagaku Kaishi,* 41, 209, 1967.

735. **Casida, J. E. and Sanderson, D. M.,** Toxic hazard from formulating the insecticide dimethoate in methyl "cellosolve," *Nature,* 189, 507, 1961.

736. **Casida, J. E. and Sanderson, D. M.,** Reaction of certain phosphorothionate insecticides with alcohols and potentiation by breakdown products, *J. Agric. Food Chem.,* 11, 91, 1963.

737. **Durham, W. F.,** The interaction of pesticides with other factors, *Residue Rev.,* 18, 21, 1967.

738. **Olinger, L. D. and Kerr, S. H.,** Effects of dimethyl sulfoxide on the biological activity of selected miticides and insecticides, *J. Econ. Entomol.,* 62, 403, 1969.

739. **Solon, J. M. and Nair, J. H.,** Effect of a sublethal concentration of LAS (linear alkyl-benzenesulfonate detergent) on the acute toxicity of various phosphate pesticides to the fathead minnow (*Pimephales promelas*), *Bull. Environ. Contam. Toxicol.,* 5, 408, 1970.

740. **Frawley, J. P., Fuyat, H. N., Hagan, E. C., Blake, J. R., and Fitzhuch, O. G.,** Marked potentiation in mammalian toxicity from simultaneous administration of two anticholinesterase compounds, *J. Pharmacol. Exp. Ther.,* 121, 96, 1957.

741. **Cook, J. W., Blake, J. R., Yip, G., and Williams, M.,** Malathionase. I. Activity and inhibition, *J. Assoc. Off. Agric. Chem.,* 41, 399, 1958.

742. **Casida, J. E.,** Specificity of substituted phenyl phosphorus compounds for esterase inhibition in mice, *Biochem. Pharmacol.,* 5, 332, 1961.

743. **Casida, J. E., Baron, R. L., Eto, M., and Engel, J. L.,** Potentiation and neurotoxicity induced by certain organophosphates, *Biochem. Pharmacol.,* 12, 73, 1963.

744. **Seume, F. W. and O'Brien, R. D.,** Potentiation of the toxicity to insects and mice of phosphorothionates containing carboxyester and carboxyamide groups, *Toxicol. Appl. Pharmacol.,* 2, 495, 1960.

745. **Pellegrini, G. and Santi, R.,** Potentiation of toxicity of organophosphorus compounds containing carboxylic ester functions toward warm-blooded animals by some organophosphorus impurities, *J. Agric. Food Chem.,* 20, 944, 1972.

746. **Plapp, F. W., Jr., Bigley, W. S., Chapman, G. A., and Eddy, G. W.,** Synergism of malathion against resistant house flies and mosquitoes, *J. Econ. Entomol.,* 56, 643, 1963.

747. **Plapp, F. W., Jr. and Tong, H. H. C.,** Synergism of malathion and parathion against resistant insects: phosphorus esters with synergistic properties, *J. Econ. Entomol.,* 59, 11, 1966.

748. **Bell, J. D. and Busvine, J. R.,** Synergism of organophosphates in *Musca domestica* and *Chrysomya putoria, Entomol. Exp. Appl.,* 10, 263, 1967.

749. **Plapp, F. W., Jr. and Valega, T. M.,** Synergism of carbamate and organophosphate insecticides by noninsecticidal carbamates, *J. Econ. Entomol.,* 60, 1094, 1967.

750. **O'Brien, R. D.,** Mode of action of insecticides, *Annu. Rev. Entomol.,* 11, 369, 1966.

751. **Eto, M., Oshima, Y., Kitakata, S., Tanaka, F., and Kojima, K.,** Studies on saligenin cyclic phosphorus esters with insecticidal activity. X. Synergism of malathion against susceptible and resistant insects, *Botyu-Kagaku,* 31, 33, 1965.

752. **Ohkawa, H., Eto, M., and Oshima, Y.,** Comparative study of saligenin cyclic phosphorus esters and triphenyl phosphate in the specificity of esterase inhibition, *Botyu-Kagaku,* 33, 21, 1968.

753. **Dyte, C. E. and Rowlands, D. G.,** The metabolism and synergism of malathion in resistant and susceptible strains of *Tribolium castaneum, J. Stored Prod. Res.,* 4, 157, 1968.

754. **Takahashi, Y., Saito, T., Iyatomi, K., and Eto, M.,** Joint toxic action of various compounds with malathion and dimethoate to organophosphate resistant citrus red mite, *Botyu-Kagaku,* 37, 13, 1972.

755. **Matsunaka, S.,** Activation and inactivation of herbicides by higher plants, *Residue Rev.,* 25, 45, 1969.

756. **Cheng, F.-Y., Smith, L. W., and Stephenson, G. R.,** Insecticide inhibition of herbicide metabolism in leaf tissues, *J. Agric. Food Chem.,* 19, 1183, 1971.

757. **Rosenberg, P. and Coon, J. M.,** Potentiation between cholinesterase inhibitors, *Proc. Soc. Exp. Biol. Med.,* 97, 836, 1958.

758. **McCollister, D. D., Oyen, F., and Rowe, V. K.,** Toxicological studies of *O,O*-dimethyl *O*-(2,4,5-trichlorophenyl) phosphorothioate (Ronnel) in laboratory animals, *J. Agric. Food Chem.,* 7, 689, 1959.

759. **Braid, P. E. and Nix, M.,** Potentiation of toxicity of sumithion by phosphamidon in rat, *Can. J. Physiol. Pharmacol.,* 46, 133, 1968.

760. **Bull, D. L., Lindquist, D. A., and House, V. S.,** Synergism of organophosphorus systemic insecticides, *J. Econ. Entomol.,* 58, 1157, 1965.

761. **Bigley, W. S.,** Inhibition of cholinesterase and aliesterase in parathion and paraoxon poisoning in the house fly, *J. Econ. Entomol.,* 59, 60, 1966.

762. **Norment, B. R. and Chambers, H. W.,** Joint actions in organophosphorus poisoning in boll weevils, *J. Econ. Entomol.,* 63, 499, 1970.

763. **Sun, Y.-P., Johnson, E. R., and Ward, L. F., Jr.,** Evaluation of synergistic mixtures containing sesamex and organophosphorus or chlorinated insecticides tested against house flies, *J. Econ. Entomol.,* 60, 828, 1967.

764. **Sun, Y.-P. and Johnson, E. R.,** Relationship between structure of several Azodrin® insecticide homologues and their toxicities to house flies, tested by injection, infusion, topical application, and spray method with and without synergist, *J. Econ. Entomol.,* 62, 1130, 1969.

765. **Menzer, R. E.,** Effect of enzyme induction on the metabolism and selectivity of organophosphorus insecticides, in *Pesticide Chemistry, Proceedings 2nd International IUPAC Congress,* Vol. 1, Tahori, A. S., Ed., Gordon & Breach, London, 1971, 51.

766. **Bull, D. L. and Lindquist, D. A.,** Metabolism of 3-hydroxy-*N*-methyl-*cis*-crotonamide dimethyl phosphate (Azodrin) by insects and rats, *J. Agric. Food Chem.,* 14, 105, 1966.

767. **Bull, D. L., Lindquist, D. A., and Grabbe, R. R.,** Comparative fate of the geometric isomers of phosphamidon in plants and animals, *J. Econ. Entomol.,* 60, 332, 1967.

357

768. **Robbins, W. E., Hopkins, T. L., and Darrow, D. I.,** Synergistic action of piperonyl butoxide with Bayer 21/199 and its corresponding phosphate in mice, *J. Econ. Entomol.,* 52, 660, 1959.

769. **Chang, S. C. and Borkovec, A. B.,** Comparative metabolism of [14] C-labeled hempa by house flies susceptible or resistant to Isolan, *J. Econ. Entomol.,* 62, 1417, 1969.

770. **O'Brien, R. D.,** The effect of SKF 525A (2-diethylaminoethyl 2:2-diphenylvalerate hydrochloride) on organophosphate metabolism in insects and mammals, *Biochem. J.,* 79, 229, 1961.

771. **Dahm, P. A., Kopecky, B. E., and Walker, C. B.,** Activation of organophosphorus insecticides by rat liver microsomes, *Toxicol. Appl. Pharmacol.,* 4, 683, 1962.

772. **Nakatsugawa, T. and Dahm, P. A.,** Parathion activation enzymes in the fat body microsomes of the American cockroach, *J. Econ. Entomol.,* 58, 500, 1965.

773. **Hennesy, D. J.,** The potential of carbamate synergists as pest control agents, in *Biochemical Toxicology of Insecticides,* O'Brien, R. D. and Yamamoto, I., Eds., Academic Press, New York, 1970, 105.

774. **Cloney, R. D. and Scherr, V. M.,** Molecular orbital study of hydride-transferring ability of benzodiheterolines as a basis of synergistic activity, *J. Agric. Food Chem.,* 16, 791, 1968.

775. **Hansch, C.,** The use of homolytic, steric, and hydrophobic constants in a structure-activity study of 1,3-benzodioxole synergists, *J. Med. Chem.,* 11, 920, 1968.

776. **Wilkinson, C. F.,** Insecticide synergists and their mode of action, in *Pesticide Chemistry, Proceedings 2nd International IUPAC Congress,* Vol. II, Tahori, A. S., Ed., Gordon & Breach, London, 1971, 117.

777. **Casida, J. E., Engel, J. L., Essac, E. G., Kamienski, F. X., and Kuwatsuka, S.,** Methylene-C[14]-dioxyphenyl compounds: metabolism in relation to their synergistic action, *Science,* 153, 1130, 1966.

778. **Kamienski, F. X. and Casida, J. E.,** Importance of demethylation in the metabolism in vivo and in vitro of methylenedioxyphenyl synergists and related compounds in mammals, *Biochem. Pharmacol.,* 19, 91, 1970.

779. **Wilkinson, C. F. and Hicks, L. J.,** Microsomal metabolism of the 1,3-benzodioxole ring and its possible significance in synergistic action, *J. Agric. Food Chem.,* 17, 829, 1969.

780. **Kuwatsuka, S.,** Biochemical aspects of methylenedioxyphenyl compounds in relation to the synergistic action, in *Biochemical Toxicology of Insecticides,* O'Brien, R. D. and Yamamoto, I., Eds., Academic Press, New York, 1970, 131.

781. **Ohkawa, H., Ohkawa, R., Yamamoto, I., and Casida, J. E.,** Enzymatic mechanisms and toxicological significance of hydrogen cyanide liberation from various organothiocyanates and organonitriles in mice and houseflies, *Pest. Biochem. Physiol.,* 2, 95, 1972.

782. **Ball, W. L., Sinclair, J. W., Crevier, M., and Kay, K.,** Modification of parathion's toxicity for rats by pretreatment with chlorinated hydrocarbon insecticides, *Can. J. Biochem. Physiol.,* 32, 440, 1954.

783. **Welch, R. M. and Coon, J. M.,** Studies on the effect of chlorcyclizine and other drugs on the toxicity of several organophosphate anticholinesterases, *J. Pharmacol. Exp. Ther.,* 143, 192, 1964.

784. **DuBois, K. P. and Kinoshita, F. K.,** Influence of induction of hepatic microsomal enzymes by phenobarbital on toxicity of organic phosphate insecticides, *Proc. Soc. Exp. Biol. Med.,* 129, 699, 1968.

785. **Tahori, A. S., Ed.,** *Resistance, Synergism, Enzyme Induction, Pesticide Chemistry,* Proceedings 2nd International IUPAC Congress, Vol. II, Gordon & Breach, London, 1971.

786. **Main, A. R.,** The role of A-esterase in the acute toxicity of paraoxon, TEPP, and parathion, *Can. J. Biochem. Physiol.,* 34, 197, 1956.

787. **Triolo, A. J. and Coon, J. M.,** Toxicologic interactions of chlorinated hydrocarbon and organophosphate insecticides, *J. Agric. Food Chem.,* 14, 549, 1966.

788. **Brodeur, J.,** Studies on the mechanism on phenobarbital-induced protection against malathion and EPN, *Can. J. Physiol. Pharmacol.,* 45, 1061, 1967.

789. **Remmer, H.,** Enzyme induction phenomenon: effects in vertebrate livers, in *Resistance, Synergism, Enzyme Induction, Pesticide Chemistry,* Vol. II. Tahori, A. S., Ed., Gordon & Breach, London, 1971, 167.

790. **Chapman, S. K. and Leibman, K. C.,** Effects of chlordane, DDT, and 3-methylcholanthrene upon the metabolism and toxicity of diethyl 4-nitrophenyl phosphorothionate (parathion), *Toxicol. Appl. Pharmacol.,* 18, 977, 1971.

791. **Skrinjaric-Spolfar, M., Matthews, H. B., Engel, J. L., and Casida, J. E.,** Response of hepatic microsomal mixed-function oxidases to various types of insecticide chemical synergists administered to mice, *Biochem. Pharmacol.,* 20, 1607, 1971.

792. **Agosin, M.,** Microsomal mixed-function oxidases and insecticide resistance, in *Resistance, Synergism, Enzyme Induction, Pesticide Chemistry,* Vol. II, Tahori, A. S., Ed., Gordon & Breach, London, 1971, 167.

793. **Gillett, J. W.,** Induction in different species, in *Resistance, Synergism, Enzyme Induction, Pesticide Chemistry,* Vol. II, Tahori, A. S., Ed., Gordon & Breach, London, 1971, 197.

794. **Ahmad, N. and Brindley, W. A.,** Modification of parathion toxicity to wax moth larvae by chlorcyclizine, aminopyrine or phenobarbital, *Toxicol. Appl. Pharmacol.,* 15, 433, 1969.

795. **Smith, M. I., Elvove, E., and Frazier, W. H.,** Pharmacological action of certain phenol esters, with special reference to etiology of so-called ginger paralysis, *U.S. Public Health Rep.,* 45, 2509, 1930.

796. **Smith, H. V. and Spalding, J. M. K.,** Outbreak of paralysis in Morocco due to ortho-cresyl phosphate poisoning, *Lancet,* 2, 1019, 1959.

797. **Cavanagh, J. B.,** The toxic effects of tri-ortho-cresyl phosphate on the nervous system; experimental study in hens, *J. Neurol. Neurosurg. Psychiatry,* 17, 163, 1954.

798. **Baron, R. L., Bennett, D. R., and Casida, J. E.,** Neurotoxic syndrome produced in chickens by a cyclic phosphate metabolite of tri-*o*-cresyl phosphate – a clinical and pathological study, *Br. J. Pharmacol. Chemother.,* 18, 465, 1962.

799. **Aldridge, W. N., Barnes, J. M., and Johnson, M. K.,** Studies on delayed neurotoxicity produced by some organophosphorus compounds, *Ann. N.Y. Acad. Sci.,* 160, 314, 1969.

800. **Hine, C. H., Dunlap, M. K., Rice, E. G., Coursey, M. M., Gross, R. M., and Anderson, H. H.,** The neurotoxicity and anticholinesterase properties of some substituted phenyl phosphates, *J. Pharmacol. Exp. Ther.,* 116, 227, 1956.

801. **Henschler, D.,** Beziehungen zwischen chemischer Struktur und Lähmungswirking von Triarylphosphaten, *Naunyn-Schmiedeberg's Arch. Exp. Path. Pharmakol.,* 237, 459, 1959.

802. **Bondy, H. F., Field, E. J., Worden, A. N., and Hughes, J. P. W.,** A study on the acute toxicity of the tri-aryl phosphates used as plasticizers, *Br. J. Ind. Med.,* 17, 190, 1960.

803. **Aldridge, W. N. and Barnes, J. M.,** Neurotoxic and biochemical properties of some triaryl phosphates, *Biochem. Pharmacol.,* 6, 177, 1961.

804. **Taylor, J. D. and Butter, H. S.,** Evidence for the presence of 2-(*o*-cresyl)-4H-1:3:2-benzodioxaphosphoran-2-one in cat intestine following tri-*o*-cresyl phosphate administration, *Toxicol. Appl. Pharmacol.,* 11, 529, 1967.

805. **Taylor, J. D.,** A neurotoxic syndrome produced in cats by a cyclic phosphate metabolite of tri-*o*-cresyl phosphate, *Toxicol. Appl. Pharmacol.,* 11, 538, 1967.

806. **Aldridge, W. N. and Johnson, M. K.,** Side effects of organophosphorus compounds: delayed neurotoxicity, *Bull. W.H.O.,* 44, 259, 1971.

807. **Eto, M., Sakata, M., Sasayama, T., and Abe, M.,** Metabolic activation and insecticidal activity of *p*-ethylphenyl phosphates, in *Pesticide Chemistry, Proceedings 2nd International IUPAC Congress,* Vol. I., Tahori, A. S., Ed., Gordon & Breach, London, 1972, 403.

808. **Davies, D. R.,** Neurotoxicity of organophosphorus compounds, in *Cholinesterases and Anticholinesterase Agents,* Koelle, G. B., Ed., Springer-Verlag, Berlin, 1963, 860.

809. **Aldridge, W. W. and Barnes, J. M.,** Further observations on the neurotoxicity of organophosphorus compounds, *Biochem. Pharmacol.,* 15, 541, 1966.

810. **Gains, T. B.,** Acute toxicity of pesticides, *Toxicol. Appl. Pharmacol.,* 14, 515, 1969.

811. **Cavanagh, J. B.,** Toxic substances and the nervous system, *Br. Med. Bull.,* 25, 268, 1969.

812. **Earl, C. J. and Thompson, R. H. S.,** Cholinesterase levels in the nervous system in tri-ortho-cresyl phosphate poisoning, *Br. J. Pharmacol.,* 7, 685, 1952.

813. **Baron, R. L. and Casida, J. E.,** Enzymatic and antidotal studies on the neurotoxic effect of certain organophosphates, *Biochem. Pharmacol.,* 11, 1129, 1962.

814. **Johnson, M. K.,** A phosphorylation site in brain and the delayed neurotoxic effect of some organophosphorus compounds, *Biochem. J.,* 111, 487, 1969.

815. **Johnson, M. K.,** Delayed neurotoxic action of some organophosphorus compounds, *Br. Med. Bull.,* 25, 231, 1969.

816. **Johnson, M. K.,** Organophosphorus and other inhibitors of brain "neurotoxic esterase" and the development of delayed neurotoxicity in hens, *Biochem. J.,* 120, 523, 1970.

817. **Johnson, M. K.,** The delayed neurotoxic effect of some organophosphorus compounds identification of the phosphorylation site as an esterase, *Biochem. J.,* 114, 711, 1969.

818. **Porcellati, G.,** Demyelinating cholinesterase inhibitors: lipid and protein metabolism, *Handbook of Neurochemistry,* Vol. 6, Plenum, 1971, 457.

819. **Rogers, J. C., Chambers, H., and Casida, J. E.,** Nicotinic acid analogs: effect on response of chick embryos and hens to organophosphate toxicants, *Science,* 144, 539, 1964.

820. **Upshall, D. G., Roger, J. C., and Casida, J. E.,** Biochemical studies on the teratogenic action of Bidrin and other neuroactive agents in developing hen eggs, *Biochem. Pharmacol.,* 17, 1529, 1968.

821. **Roger, J. C., Upshall, D. G., and Casida, J. E.,** Structure-activity and metabolism studies on organophosphate teratogens and their alleviating agents in developing hen eggs with special emphasis on Bidrin, *Biochem. Pharmacol.,* 18, 373, 1969.

822. **Dunachie, J. F. and Fletcher, W. W.,** Effect of some insecticides on the hatching rate of hens' eggs, *Nature,* 212, 1062, 1966.

823. **Klotzsche, C.,** Teratologic and embryotoxic investigations with formothion and thiometon, *Pharm. Acta Helv.,* 45, 434, 1970; *C. A.,* 73, 65402, 1970.

824. **Kimbrough, R. E. and Gaines, T. B.,** Effect of organic phosphorus compounds and alkylating agents on the rat fetus, *Arch. Environ. Health,* 16, 805, 1968.

825. **Ryan, L. C., Endecott, B. R., Hanneman, G. D., and Smith, P. W.,** Effects of an organophosphorus pesticide on reproduction in the rat, *Aviation Med. Rep.,* AM 70-3, 1970.

826. **Styrzynska, B. and Krzeminska, A.,** Deformation of puparia of the housefly on treatment with organophosphorus insecticides, *Rocz. Panstw. Zakl. Hig.,* 18, 377, 1967; *C. A.,* 68, 11931, 1968.

827. **Litterst, F. L., Lichtenstein, E. P., and Kajiwara, K.,** Effects of insecticides on growth of HeLa cells, *J. Agric. Food Chem.,* 17, 1199, 1969.

828. **Gabbiks, J., Bantung-Jurilla, M., and Friedman, L.,** Responses of cell cultures to insecticides, *Proc. Soc. Exp. Biol. Med.,* 125, 1002; 1967.

829. **Sax, K. and Sax, H. J.,** Possible mutagenic hazards of some food additives, beverages, and insecticides, *Jap. J. Genet.,* 43, 89, 1968.

829a.**Satoh, T.,** A liver arylamidase extremely sensitive to organophosphorus compounds, *Life Sci.,* 13, 1181, 1973.

830. **Freedland, R. A. and McFarland, L. Z.,** Effect of various pesticides on purified glutamate dehydrogenase, *Life Sci.,* 4, 1735, 1965.

831. **Dedek, W. and Lohs, K.,** Zur alkierenden Wirkung von Trichlorphon in Warmblütern. II. Verteilung von ^{14}C in Organen und Leberproteinen bei Ratten nach Applikation von ^{14}C-Trichlorphon, *Z. Naturforsch. B,* 25, 1110, 1970.

832. **Kiermeier, F., Kern, R., and Wildbrett, G.,** Einfluss organischer Insecticide auf Enzyme. III. Über die Wirkung von Phosphor- und Thiophosphorsäureester auf Peroxydase in vitro, *Biochem. Z.,* 336, 421, 1962.

833. **Kiermeier, F., Wildbrett, G., and Lettenmayer, L.,** Einfluss organischer Insecticide auf Enzyme. V. Über die Wirkung auf Milch-Xanthindehydrase, *Z. Lebensm.-Unter.-Forsch.,* 133, 22, 1966.

834. **Syrowatka, T.,** Effect of organophosphorus insecticides on the oxidative phosphorylation and respiratory control of rat liver mitochondria, *Rocz. Panstw. Zakl. Hig.,* 20, 557, 1969; *C. A.,* 72, 120479, 1970.

835. **Gar, K. A., Kheiman, V. A., Guseva, N. A., Kaluzhina, T. N., and Selezneva, V. P.,** Oxidative phosphorylation and glucolysis in beetle tissues of beet pests poisoned by insecticides, *Khim. Sel'sk. Khoz.,* 5, 27, 1967; *C. A.,* 66, 114889, 1967.

836. **Jaaskelainen, A. J. and Alha, A.,** Histochemically observable alterations in enzyme pattern of rat myocardium, caused by parathion, *Acta Pharmacol. Toxicol.,* 27, 112, 1969.

837. **Disney, G. W. and Smith, J. T.,** Malathion intoxication and sulfation of mucopolysaccharides, *J. Agric. Food Chem.,* 18, 540, 1970.

838. **Kakiki, K., Maeda, T., Abe, H., and Misato, T.,** Mode of action of organophosphorus fungicide, Kitazin. I. Effect on respiration, protein synthesis, nucleic acid synthesis, cell wall synthesis, and leakage of intracellular substances from Mycelia of *Pyricularia oryzae, Nippon Nogei Kagaku Kaishi,* 43, 37, 1969.

838a.**Bellet, E. M. and Casida, J. E.,** Bicyclic phosphorus esters: high toxicity without cholinesterase inhibition, *Science,* 182, 1135, 1973.

839. **Ntiforo, C. and Stein, M.,** Labilization of lysosome as an aspect of the biochemical toxicology of anticholinesterase pesticides, *Biochem. J.,* 102, 44p, 1967.

840. **Clark, G. and Stavinoha, W. B.,** Permeability change in central nervous system tissue in chronic poisoning with disulfoton, *Life Sci.,* 10, 421, 1971.

841. **Williams, C. H. and Keys, J. E.,** β-Glucuronidase activity in the serum and liver of rats treated with parathion, *Toxicol. Appl. Pharmacol.,* 16, 533, 1970.

842. **Gerogiev, I. N.,** Changes in the enzyme-secretory function of the intestine under the influence of toxic and maximally permissible amounts of chlorophos introduced together with food, *Vop. Pitan.,* 26, 27, 1967; *C.A.,* 67, 42835, 1967.

843. **Murphy, S. D.,** Some relationships between effects of insecticides and other stress conditions, *Ann. N.Y. Acad. Sci.,* 160, 366, 1969.

844. **Brzezinski, J.,** Catechol amines in urine of rats poisoned with organophosphate insecticides, *Diss. Pharm. Pharmacol.,* 21, 381, 1969; *C.A.,* 71, 122765, 1969.

845. **Wysocka-Paruszewska, B.,** Urine level of 4-hydroxy-3-methoxymandelic acid in urine of rats poisoned with phosphorus organic insecticides, *Diss. Pharm. Pharmacol.,* 22, 485, 1970; *C.A.,* 74, 98844, 1971.

846. **Meiniel, R.,** Décharge en catécholamines des cellules adrenales de l'embryon de Poulet après intoxication aigue au Parathion, *C.R. Acad. Sci. Ser. D,* 272, 873, 1971; *C.A.,* 74, 110825, 1971.

847. **Sternburg, J.,** Autointoxication and some stress phenomena, *Annu. Rev. Entomol.,* 8, 19, 1963.

848. **Sternburg, J.,** Effect of insecticides on neurophysiological activity in insects, *J. Agric. Food Chem.,* 8, 257, 1960.

849. **Flattum, R. F. and Sternburg, J. G.,** Release of a synaptically active material by nicotine in the central nervous system of the American cockroach, *J. Econ. Entomol.,* 63, 67, 1970.

850. **Tashiro, S., Taniguchi, E., and Eto, M.,** Isolation of a neuroactive substance, L-leucine, from the blood of silkworm poisoned with DDT, *Agric. Biol. Chem.,* 36, 2465, 1972.

851. **Dahm, P. A.,** Toxic effects produced in insects by organophosphorus compounds, *Bull. W. H. O.,* 44, 215, 1971.

852. **Casida, J. E. and Maddrell, S. H. P.,** Diuretic hormone release on poisoning *Rhodnius* with insecticide chemicals, *Pest. Biochem. Physiol.,* 1, 71, 1971.

853. **Beye, F.,** Wirkungen von Insektiziden auf die Katalase-aktivität von Kresse-Keimlingen, *Z. Naturforsch.,* 15b, 470, 1960.

854. **Bobyreva, T. V.,** Effect of organophosphorus insecticides on cotton oxidation-reduction enzymes, *Dokl. Akad. Nauk Uzb. SSR.,* 23, 52, 1966; *C.A.,* 67, 10650, 1967.

855. **Chikov, V. I.,** Effect of organophosphorus insecticides on photosynthesis, *Funkts. Osob. Khloroplastov,* 123, 1969; *C.A.,* 74, 98752, 1971.

856. **Derby, S. B. and Ruber, E.,** Primary production: depression of oxygen evolution in algal cultures by organophosphorus insecticides, *Bull. Environ. Contam. Toxicol.,* 5, 553, 1970.

857. **Yokota, K.,** The influence of pesticides on the coloration of apples, *Chem. Reg. Plants,* 7, 33, 1972.

858. **Burg, S. P. and Burg, E. A.,** Ethylene action and the ripening of the fruits, *Science,* 148, 190, 1965.

859. **Iwahori, S.,** Role of ethylene in plant growth and development. Its possible practical use for horticultural crops, *Chem. Reg. Plants,* 4, 40, 1969.

860. **Edgerton, L. J. and Blanpied, G. D.,** Regulation of growth and fruit maturation with 2-chloroethanephosphonic acid, *Nature,* 219, 1064, 1968.

861. **Jaworski, E. G.,** Mode of action of *N*-phosphonomethylglycine: inhibition of aromatic amino acid biosynthesis, *J. Agric. Food Chem.,* 20, 1195, 1972.

862. **Schrader, G.,** The modification of biological activity by structural changes in certain organophosphorus compounds, *World Rev. Pest Control,* 4, 140, 1965.

863. **Hoffmann, H.,** Development and application of pesticidal phosphoric acid derivatives, in *Chemie Organique du Phosphore,* Centre National de la Recherche Scientifique, Paris, 1970, 357.

864. **Mattson, A. M., Spillane, J. T., and Pearce, G. W.,** Dimethyl 2,2-dichlorovinyl phosphate (DDVP), an organic phosphate compound highly toxic to insects, *J. Agric. Food Chem.,* 3, 319, 1955.

865. **Scherer, O. and Stachler, G.,** Insecticidal phosphates, German Patent 1,283,592; *C. A.,* 70, 56712, 1969.

866. **Corey, R. A., Dorman, S. C., Hall, W. E., and Glover, L. C.,** Translocation studies with two new phosphate insecticides, *J. Econ. Entomol.,* 46, 386, 1953.

867. **Frear, D. E. H.,** *Pesticide Index,* 4th ed., College Science, State College, Pennsylvania, 1969.

868. **Graham, O. H.,** The primary evaluation of three organic phosphorus compounds for possible use in the control of livestock insects, *J. Econ. Entomol.,* 54, 1046, 1961.

869. **Beynon, K. I. and Wright, A. N.,** The breakdown of C^{14}-chlorfenvinphos in soils and in crops grown in the soils, *J. Sci. Food Agric.,* 18, 143, 1967.

870. **Sherman, M. and Heraick, R. B.,** Fly control and chronic toxicity from feeding two chlorinated organophosphorus insecticides to laying hens, *J. Econ. Entomol.,* 66, 139, 1973.

871. **Miller, R. W. and Pickens, C. G.,** Feeding of coumaphos, ronnel, and Rabon to dairy cows: larvicidal activity against house flies and effect on insect fauna and biodegradation of fecal pats, *J. Econ. Entomol.,* 66, 1077, 1973.

872. **Akintowa, D. A. A. and Hutson, D. H.,** Metabolism of 2-chloro-1-(2,4,5-trichlorophenyl)vinyl dimethyl phosphate in dog and rat, *J. Agric. Food Chem.,* 15, 632, 1967.

873. **Drummond, R. O., Ernst, S. E., and Trevino, J. L.,** *Boophilus annulatus* and *B. microplus*: laboratory tests of insecticides, *J. Econ. Entomol.,* 66, 130, 1973.

874. **Newallis, P. E., Lombardo, P., Gilbert, E. E., Spencer, E. Y., and Morello, A.,** Preparation and biological activity of the isomers of dimethyl 1,3-dicarbomethoxy-1-propen-2-yl phosphate (Bomyl), *J. Agric. Food Chem.,* 15, 940, 1967.

875. **Metcalf, R. L.,** Selective toxicity of insecticides, *World Rev. Pest Control,* 3, 28, 1964.

876. **Kilgore, W. W., Marei, N., and Winterlin, W.,** Parathion in plant tissues: new considerations, in *Degradation of Synthetic Organic Molecules in the Biosphere,* National Academy of Sciences, Washington, D.C., 1972, 291.

877. **Davich, T. B. and Apple, J. W.,** Pea aphid control with contact and systemic insecticidal sprays, *J. Econ. Entomol.,* 44, 528, 1951.

878. **Bakanova, Z. M., Mandel'baum, Ya. A., Supin, G. S., Melnikov, N. N., and Abbakumova, N. V.,** Properties and methods for analyzing methylnitrophos, *Khim. Sel'sk Khoz.,* 8, 32, 1970; *C. A.,* 72, 120360, 1970.

879. **Nishizawa, Y., Kuramoto, S., Kadota, T., Miyamoto, J., Fujimoto, K., and Sakamoto, H.,** Chemical and biological properties of *O,O*-dimethyl *O*-(4-cyanophenyl) phosphorothioate and *O*-ethyl *O*-(4-cyanophenyl) phenylphosphono-thioate, *Agric. Biol. Chem.* (Tokyo), 26, 257, 1962.

880. **Stiasni, M., Rehbinder, D., and Deckers, W.,** Absorption, distribution, and metabolism of *O*-(4-bromo-2,5-dichloro-phenyl) *O,O*-dimethyl phosphorothioate (bromophos) in the rat, *J. Agric. Food Chem.,* 15, 474, 1967.

881. **Dedek, W. and Schwarz, H.,** Zum Verhalten des mindertoxischen Insektizids ^{32}P-Bromophos nach cutaner Applikation am Rind, *Z. Naturforsch.,* 24b, 744, 1969.

882. **Haddow, B. C. and Marks, T. G.,** Iodofenphos, a promising new insecticide, *Proc. Br. Insectic. Fungic. Conf.,* 5th, 2, 531, 1969.

883. **Niessen, H., Tietz, H., and Frehse, F.,** On the occurence of biologically active metabolites of the active ingredient S1752 after application of lebycid, *Pflanzenschutz-Nachr.,* 15, 125, 1962.

884. **Francis, J. I. and Barnes, J. M.,** Studies on the mammalian toxicity of fenthion, *Bull. W. H. O.,* 29, 205, 1963.

885. **Brady, U. E. and LaBrecque, G. C.,** Larvicides for the control of house flies in poultry houses, *J. Econ. Entomol.,* 59, 1521, 1966.

886. **Kenaga, E. E., Whitney, W. K., Hardy, J. L., and Doty, A. E.,** Laboratory tests with Dursban insecticide, *J. Econ. Entomol.,* 58, 1043, 1965.

887. **Smith, G. N., Watson, B. S., and Fischer, F. S.,** Investigations on Dursban insecticide. Metabolism of (^{36}Cl) *O,O*-diethyl *O*-3,5,6-trichloro-2-pyridyl phosphorothionate in rats, *J. Agric. Food Chem.,* 15, 132, 1967.

888. **Kenaga, E. E.,** Some physical, chemical, and insecticidal properties of some *O,O*-dialkyl *O*-(3,5,6-trichloro-2-pyridyl) phosphates and phosphorothioates, *Bull. W. H. O.,* 44, 225, 1971.

889. **Gysin, H.,** Über einige neue Insektizide, *Chimia,* 8, 221, 1954.

890. **Masuda, T. and Fukuda, H.,** Behavior on and in rice plants of diazinon applied onto the surface of paddy soil, *Botyu-Kagaku,* 35, 134, 1970.

891. **Snell, B. K. and Sharp, S. P.,** Insecticidal and fungicidal pyrimidine thiophosphate derivative, and its application, British Patent 1,205,000; *C. A.,* 74, 52482, 1971.

892. Udagawa, T., Saito, T., and Miyata, T., Metabolism of pyridafenthion, O,O-diethyl-O-(3-oxo-2-phenyl-2H-pyridazin-6-yl) phosphorothioate, in mouse and rat, *Botyu-Kagaku*, 38, 75, 1973.

893. Schmidt, K. J. and Hammann, I., Bayrusil, a new insecticidal and acaricidal organophosphate, *Pflanzenschutz-Nachr.*, 22, 314, 1969.

894. Vulic, M., Emmel, L., and Mildenberger, H., HOE 2960, a new insecticidal and acaricidal phosphoric acid ester, *VIIth Int. Congr. Plant Protection*, Paris, summaries of papers, p. 123, 1970.

895. Tomita, K., Nagano, M., Yanai, T., Oka, H., Murakami, T., and Sampei, N., Syntheses of 3-hydroxyisoxazoles, *Ann. Sankyo Res. Lab.*, 22, 215, 1970.

896. Sampei, N., Tomita, K., Tsuji, H., Yanai, T., Oka, H., and Yamamoto, T., Structure and activity of some 3-isoxazolyl phosphates, *Ann. Sankyo Res. Lab.*, 22, 221, 1970.

897. Takemoto, T., Yokobe, T., and Nakajima, T., Isolation of flycidal constituent from *Amanita strobiliformis*, *J. Pharm. Soc., Japan*, 84, 1186, 1232, 1964.

898. Drager, G., Studies on the metabolism of phoxim (Bay 77488), *Pflanzenschutz-Nachr.*, 24, 239, 1971.

899. Bailey, D. L., LaBrecque, G. L., and Whitfield, T. L., Laboratory evaluation of insecticides as contact sprays against adult house flies, *J. Econ. Entomol.*, 63, 275, 1970.

900. Sumitomo Chemical Co., Stabilization of saligenin cyclic phosphorus esters, British Patent 1,228,121; *C. A.*, 75, 34498, 1971.

901. Eto, M., Kobayashi, K., Sasamoto, T., Cheng, H. M., Aikawa, T., Kume, T., and Oshima, Y., Studies on saligenin cyclic phosphorus esters with insecticidal activity. XII. Insecticidal activity of ring-substituted derivatives, *Botyu-Kagaku*, 33, 73, 1968.

902. Kobayashi, K., Eto, M., Oshima, Y., Hirano, T., Hosoi, T., and Wakamori, S., Synthesis and biological activities as insecticides and fungicides of saligenin cyclic phosphorothiolates, *Botyu-Kagaku*, 34, 165, 1969.

903. Oshima, Y., Salithion®, *Noyaku-Seisangijitsu (Pesticide and Technique)*, 27, 1, 1972.

904. Thurston, R., Effect of insecticides on the green peach aphid, *Myzus persicae* (Sulzer), infesting burley tobacco, *J. Econ. Entomol.*, 58, 1127, 1965.

905. Lanham, W. M., Bicyclo heterocyclic phosphorus compounds, U.S. Patent 2,910,499; *C.A.*, 54, 3465, 1960.

906. Coffin, D. E., Oxidative metabolism and persistence of trithion on field sprayed lettuce, *J. Assoc. Off. Agric. Chem.*, 47, 662, 1964.

907. Sherman, M., Komatsu, G. H., and Ikeda, J., Larvicidal activity to flies of manure from chicks administered insecticide-treated feed, *J. Econ. Entomol.*, 60, 1395, 1967.

908. Bassand, D., O,O,O,O-Tetraalkyltetrathiopyrophosphates as additives in insecticidal and acaricidal formulations, Ger. Offen., 2,001,817; *C. A.*, 73, 86915, 1970.

909. Menn, J. J., McBain, J. B., Adelson, B. J., and Patchett, G. G., Degradation of N-(mercaptomethyl) phthalimide-S-(O,O-dimethylphosphorodithioate) (Imidan) in soils, *J. Econ. Entomol.*, 58, 875, 1965.

910. Loomis, E. C., Grenshaw, G. L., Bushnell, R. B., and Dunning, L. L., Systemic insecticide study on livestock in California, 1965-67, 1. Cattle grub control, *J. Econ. Entomol.*, 63, 1237, 1970.

911. Ford, J. M., Menn, J. J., and Meyding, G. D., Metabolism of N-(mercaptomethyl) phthalimide-carbonyl-C^{14}-S-(O,O-dimethylphosphorodithioate) (Imidan-C^{14}): balance study in the rat, *J. Agric. Food Chem.*, 14, 83, 1966.

912. Cothran, W. R., Arumbrust, E. J., Horn, D. J., and Gyrisco, G. G., Field evaluation of experimental and recommended insecticides for control of alfalfa weevil in New York, *J. Econ. Entomol.*, 60, 1151, 1967.

913. Colinese, D. L. and Terry, H. J., Phosalone — a wide spectrum organo-phosphorus insecticide, *Chem. Ind.* (London), 44, 1507, 1968.

914. Brair, E. M., Kauer, K. C., and Kenaga, E. E., Synthesis and insecticidal activity of O-methyl O-(2,4,5-trichloro-phenyl) phosphoramidothionates and related compounds, *J. Agric. Food Chem.*, 11, 237, 1963.

915. Founk, J. and McClanahan, R. J., Laboratory studies on the toxicity of insecticides to larvae of the Colorado potato beetle, *J. Econ. Entomol.*, 63, 2006, 1970.

916. Hammann, I., Tamaron, a new insecticide and acaricide, *Pflanzenschutz-Nachr.*, 23, 133, 1970.

917. Eto, M. and Ishida, S., unpublished data, 1973.

918. Wolfenbarger, D. A., Toxicity of certain insecticides to three Lepidoperan cotton insects, *J. Econ. Entomol.*, 63, 463, 1970.

919. Toy, A. D. F., Tetraethyl dithionopyrophosphate and related tetraalkyl dithionopyrophosphates, *J. Am. Chem. Soc.*, 73, 4670, 1951.

920. Menn, J. J., Terminal residues of phosphonate insecticides, in *Pesticide Terminal Residues*, Tahori, A. S., Ed., Butterworths, London, 1971, 57.

921. Robbins, W. E., Hopkins, T. I., and Eddy, G. W., The metabolism of P^{32}-labeled Bayer L 13/59 in a cow, *J. Econ. Entomol.*, 49, 801, 1956.

922. Dedek, W. and Schwarz, H., Untersuchungen zur Ausscheidung von ^{32}P-markiertem Trichlorphon und seinen Abbauprodukten in der Milch nach unterschiedlicher Applikation am Rind, *Arch. Exp. Veterinarmed.*, 20, 849, 1966.

923. Hassan, A., Zayed, S. M. A. D., and Abdel-Hamid, F. M., Metabolism of O,O-dimethyl 2,2,2-trichloro-1-hydroxy-ethylphosphonate (dipterex) in mammalian nervous tissue and kinetics involved in its reaction with acetyl-cholinesterase, *Can. J. Biochem.*, 43, 1263, 1965.

924. **Dedek, W.,** Abbau und Rückstande von ^{32}P-Butonat in Früchten, *Z. Naturforsch.,* 23b, 504, 1968.

925. **Otsubo, I., Ura, Y., Sato, S., Hayakawa, M., and Sakata, K.,** Pesticidal phosphonates, U.S. Patent 3,318,764; *C. A.,* 67, 31901g, 1967.

926. **Ahronson, N. and Resnick, Ch.,** The dissipation of some new organo phosphorus insecticides, in *Pesticide Chemistry, Proceedings 2nd International IUPAC Congress,* VI., Tahori, A. S., Ed., Gordon & Breach, London, 1972, 333.

926b.**Holmstead, R. L., Fukuto, T. R., and March, R. B.,** The metabolism of *O*-(4-bromo-2,5-dichlorophenyl) *O*-methyl phenylphosphonothioate (leptophos) in white mice and on cotton plants, *Arch. Environ. Contam. Toxicol.,* 1, 133, 1973.

927. **Marco, G. J. and Jaworski, E. G.,** Metabolism of *O*-phenyl *O'*-(4-nitrophenyl) methylphosphonothionate (Colep) in plants and animals, *J. Agric. Food Chem.,* 12, 305, 1964.

928. **Fearing, R. B., Walsh, E. N., Menn, J. J., and Freiberg, A. H.,** Synthesis and insecticidal properties of *O*-alkyl S-aryl chloromethylphosphonodithioate, *J. Agric. Food Chem.,* 17, 1261, 1969.

929. **Schrader, G., Lorenz, W., Coelln, R., and Schloer, H.,** Phosphinic acid esters, U.S. Patent 3,232,830; *C. A.,* 64, 15923, 1966.

930. **Ripper, W. E.,** The status of systemic insecticides in pest control practices, *Advances in Pest Control Research,* Vol. 1, Metcalf, R. L., Ed., Interscience, New York, 1957, 305.

931. **Crisp, C. E.,** The molecular design of systemic insecticides and organic functional groups in translocation, in *Pesticide Chemistry, Proceedings 2nd International IUPAC Congress,* Vol. I, Tahori, A. S., Ed., Gordon & Breach, 1972, 211.

931a.**Hussain, M., Fukuto, T. R., and Reynolds, H. T.,** Physical and chemical basis for systemic movement of organophosphorus esters in the cotton plant, *J. Agric. Food Chem.,* 22, 225, 1974.

932. **Stopkan, V. V.,** Systemic insecticide demuphos, *Khim. Prom. Ukr.,* p. 18, 1968; *C. A.,* 69, 85698, 1968.

933. **Hacskaylo, J. and Bull, D. L.,** Metabolism of dimethoate in cotton leaves, *J. Agric. Food Chem.,* 11, 464, 1963.

934. **Rose, J. A. and Voss, G.,** Anticholinesterase activity and enzymic degradation of phosphamidon and γ-chlorophosphamidon. Comparative study, *Bull. Environ. Contam. Toxicol.,* 6, 205, 1971.

935. **Gunther, F. A., Ed.,** Phosphamidon, *Residue Rev.,* 37, 1971.

936. **Gaher, S., Drabek, J., Sirota, T., and Stanova, A.,** New insecticidal drench for hop protection, *Agrochemia* (Bratislava), 7, 362, 1967; *C. A.,* 71, 2473, 1969.

937. **Redfern, R. E., Walker, R. L., and Cantu, E.,** Promising candidate insecticides and acaricides evaluated at Brownsville, Tex., July 1963 to July 1966, *U.S. Dep. Agric., Agric. Res. Serv.,* No. 33-122, 11, 1967; *C. A.,* 67, 99160, 1967.

937a.**Braunholtz, J. T.,** Organo-phosphorus insecticides, *PANS (Pest Artic. News Summ.),* Sec. A, 14, 467, 1968.

938. **Rogoff, W. H.,** Chemical control of insect pests of domestic animals, in *Advances in Pest Control Research,* Vol. 4, Metcalf, R. L., Ed., Interscience, New York, 1961, 153.

939. **Bushland, R. C., Radeleff, R. D., and Drummond, R. O.,** Development of systemic insecticides for pests of animals in United States, *Annu. Rev. Entomol.,* 8, 215, 1963.

940. **Knipling, E. F. and Westlake, W. E.,** Insecticide use in livestock production, *Residue Rev.,* 13, 1, 1966.

941. **Khan, M. A.,** Systemic pesticides for use on animals, *Annu. Rev. Entomol.,* 14, 369, 1969.

942. **Weidenbah, C. P. and Younger, R. L.,** The toxicity of dimethyl 2-(alpha-methylbenzyloxycarbonyl)-1-methylvinyl phosphate (Shell Compound 4294) to livestock, *J. Econ. Entomol.,* 55, 793, 1962.

943. **Kingsbury, P. A.,** Organo-phosphorus esters and phenothiazine acting synergistically as anthelmintics, *Res. Vet. Sci.,* 2, 265, 1961.

944. **O'Brien, R. D. and Wolfe, L. S.,** The metabolism of co-ral (Bayer 21/199) by tissues of the house fly, cattle grub, ox, rat, and mouse, *J. Econ. Entomol.,* 53, 692, 1959.

945. **Roth, A. R. and Eddy, G. W.,** Tests with Dow ET-57 against cattle grubs in Oregon, *J. Econ. Entomol.,* 50, 244, 1957.

946. **Gatterdam, P. E., Wozniak, L. A., Bullock, M. W., Parks, G. L., and Boyd, J. E.,** Absorption, metabolism, and excretion of tritium-labeled famphur in the sheep and calf, *J. Agric. Food Chem.,* 15, 845, 1967.

947. **Smith, H. G., Jr. and Goulding, R. L.,** Toxicological aspects of three organophosphorus compounds (cythioate, famphur, and fenthion) in the host-ectoparasite system, *J. Econ. Entomol.,* 63, 1640, 1970.

948. **Cox, D. D., Mullee, M. T., and Allen, A. D.,** Cattle grub control with feed additives (coumaphos and fenthion) and pour-ons (fenthion and trichlorfon), *J. Econ. Entomol.,* 60, 522, 1967.

949. **Drummond, R. O.,** Control of larvae of *Oestrus ovis* in sheep with systemic insecticides, *J. Parasitol.,* 48, 211, 1962.

949a.**Casapieri, P. and Sear, R. W.,** Pesticidal phosphorothionates against sheep parasites, British Patent 1,144,003; *C. A.,* 70, 114166j, 1969.

950. **Chamberlain, W. F. and Hopkins, D. E.,** Metabolism of ^{14}C-labeled Stauffer R-3828 by a steer, *J. Econ. Entomol.,* 66, 119, 1973.

951. **Rogoff, W. M., Brody, G., Roth, A. R., Batchelder, G. H., Meyding, G. D., Bigley, W. S., Gretz, G. H., and Orchard, R.,** Efficacy, cholinesterase inhibition, and residues persistence of Imidan® for the control of cattle grubs, *J. Econ. Entomol.,* 60, 640, 1967.

952. **Landram, J. F.,** Anthelmintic activity of a new organic phosphate in cattle and sheep, *J. Parasitol.* (Suppl.), 45, 55, 1959.

953. **Brady, U. E., Jr. and Arthur, B. W.,** Absorption and metabolism of ruelene by arthropods, *J. Econ. Entomol.,* 55, 833, 1962.

954. **Drummond, R. O., Darrow, D. I., and Gladney, W. J.,** Further evaluation of animal systemic insecticides, 1969, *J. Econ. Entomol.,* 63, 1103, 1970.

955. **Suzuki, S., Miyamoto, J., Fujimoto, K., Sakamoto, H., and Nishizawa, Y.,** Organophosphorus insecticides, XI. Preparation and biological activity of sulfamoylphenyl phosphorothioates, *Agric. Biol. Chem.* (Tokyo), 34, 1967, 1970.

956. **Igbal, Z. M. and Menzer, R. E.,** Metabolism of *O*-ethyl *S,S*-dipropyl phosphorodithioate in rats and liver microsomal systems, *Biochem. Pharmacol.,* 21, 1569, 1972.

957. **Homeyer, B.,** Nemacur, a highly effective nematocide for protective and curative application, *Pflanzenschutz-Nachr.,* 24, 48, 1971.

958. **Meikle, R. W.,** Metabolism of Nellite nematocide (phenyl *N,N'*-dimethylphosphorodiamidate) in cucumber plants, *J. Agric. Food Chem.,* 16, 928, 1968.

959. **Baumhover, A. H., Graham, A. J., Bitter, B. A., Hopkins, D. E., New, W. D., Dudley, F. H., and Bushland, R. C.,** Screw-worm control through release of sterilized flies, *J. Econ. Entomol.,* 48, 462, 1955.

960. **Knipling, E. F.,** The eradication of the screw-worm, *Sci. Am.,* 203, 54, 1960.

961. **LaBrecque, G. C.,** Studies with three alkylating agents as housefly sterilants, *J. Econ. Entomol.,* 54, 684, 1961.

962. **Nagasawa, S.,** Kagakufuninzai (Chemosterilants), in *Shin-Noyaku-Soseiho,* Yamamoto, R. and Noguchi, T., Eds., Nankodo, Tokyo, 1965, 288.

963. **Borkovec, A. B.,** Chemosterilants for male insects, in *Pesticide Chemistry, Proceedings 2nd International IUPAC Congress, Insecticides,* Vol. I, Tahori, A. S., Ed., Gordon & Breach, London, 1972, 469.

964. **Turner, R. B.,** Chemistry of insect chemosterilants, in *Principles of Insect Chemosterilization,* LaBrecque, G. C. and Smith, C. N., Eds., Appleton-Century-Crofts, New York, 1968, 159.

965. **Beroza, M. and Borkovec, A. B.,** The stability of tepa and other aziridine chemosterilants, *J. Med. Chem.,* 7, 44, 1964.

966. **Borkovec, A. B.,** Alkylating agents as insect chemosterilants, *Ann. N.Y. Acad. Sci.,* 163, 860, 1969.

967. **Gadallah, A. I., Kilgore, W. W., and Painter, R. R.,** Effect of chemosterilant *P,P'*-bis(1-aziridinyl)-*N*-(3-methoxypropyl) phosphinothioic amide on same dehydrogenases during oögenesis and embryogenesis of the house fly, *J. Econ. Entomol.,* 65, 36, 1972.

968. **Kilgore, W. W., Painter, R. R., and Gadallah, A. I.,** Characteristics of the nucleic acids of eggs from house flies fed aziridinyl compounds, *J. Econ. Entomol.,* 64, 30, 1971.

969. **Kilgore, W. W. and Painter, R. R.,** Effect of the chemosterilant apholate on the synthesis of cellular components in developing house fly eggs, *Biochem. J.,* 92, 353, 1964.

970. **Painter, R. R. and Kilgore, W. W.,** The effect of apholate and thiotepa on nucleic acid synthesis and nucleotide ratios in house fly eggs, *J. Insect Physiol.,* 13, 1105, 1967.

971. **Cline, R. E.,** Evaluation of chemosterilant damage to the testes of the housefly, *Musca domestica,* by microscopic observation and by measurement of the uptake of ^{14}C-compounds, *J. Insect Physiol.,* 14, 945, 1968.

972. **Turner, R. B.,** Design of insect chemosterilants, in *Drug Design,* Vol. III, Ariëns, E. J., Ed., Academic Press, New York, 1972, 393.

973. **Chamberlain, W. F. and Barrett, C. C.,** Incorporation of tritiated thymidine into the ovarian DNA of stable flies: effects of treatment with apholate, *Nature,* 218, 471, 1968.

974. **Gadallah, A. I., Kilgore, W. W., and Painter, R. R.,** Effect of the chemosterilant *P,P'*-bis(1-aziridinyl)-*N*-(3-methoxypropyl)phosphinothioic acid on ribosomal RNA and ribosomal protein in the eggs of house flies, *J. Econ. Entomol.,* 64, 371, 1971.

975. **Gadallah, A. I., Kilgore, W. W., and Painter, R. R.,** Protein synthesis by ribosomes from eggs of normal and thiotepa-chemosterilized house flies, *J. Econ. Entomol.,* 64, 819, 1971.

976. **Gadallah, A. I., Kilgore, W. W., and Painter, R. R.,** Characteristics of DNA-RNA hybrids prepared from the eggs of normal and thiotepa chemosterilized houseflies, *Musca domestica* L., *Pest. Biochem. Physiol.,* 1, 166, 1971.

977. **Hayes, W. J., Jr.,** Toxicological aspects of chemosterilants, in *Principles of Insect Chemosterilization,* LaBrecque, G. C. and Smith, C. N., Eds., Appleton-Century-Crofts, New York, 1968, 315.

978. **Chang, S. C., Terry, P. H., and Borkovec, A. B.,** Insect chemosterilants with low toxicity for mammals, *Science,* 144, 57, 1964.

979. **Terry, P. H. and Borkovec, A. B.,** Insect chemosterilants. IV. Phosphoramides, *J. Med. Chem.,* 10, 118, 1967.

980. **Terry, P. H. and Borkovec, A. B.,** Insect chemosterilants. XII. Phosphorus amides, *J. Agric. Food Chem.,* 21, 500, 1973.

981. **Osborne, G. O., Penman, D. R., and Hoyt, C. P.,** Chemosterilant activity of trialkyl phosphates on *Tetranychus urticae, J. Sci.,* 12, 564, 1969; *C. A.,* 71, 120989s, 1969.

982. **Chang, S. C., Terry, P. H., Woods, C. W., and Borkovec, A. B.,** Metabolism of hempa uniformly labeled with C^{14} in male house flies, *J. Econ. Entomol.,* 60, 1623, 1967.

983. **Akov, S., Oliver, J. E., and Borkovec, A. B.,** *N*-Demethylation of the chemosterilant hempa by housefly microsomes, *Life Sci.,* 7, 1207, 1968.

984. **Terry, P. H. and Borkovec, A. B.,** Insect chemosterilants. IX. *N*-(hydroxymethyl)-*N,N',N',N'',N''*-pentamethylphosphoric triamide, *J. Med. Chem.,* 13, 782, 1970.

985. Palinquist, J. and LaChance, L. E., Comparative mutagenicity of two chemosterilants, tepa and hempa, in sperm of *Bracon hebetor, Science,* 154, 915, 1966.

986. Borkovec, A. B., Sexual sterilization of insects by chemicals, *Science,* 137, 1034, 1962.

987. Borkovec, A. B., Insect chemosterilants: their chemistry and application, *Residue Rev.,* 6, 87, 1964.

988. Smith, C. N., LaBrecque, G. C., and Borkovec, A. B., Insect chemosterilants, *Annu. Rev. Entomol.,* 9, 269, 1964.

989. Proverbs, M. D., Induced sterilization and control of insects, *Annu. Rev. Entomol.,* 14, 81, 1969.

990. LaBrecque, G. C. and Smith, C. N., Eds., *Principles of Insect Chemosterilization,* Appleton-Century-Crofts, New York, 1968.

991. Chang, S. C., Woods, C. W., and Borkovec, A. B., Sterilizing activity of bis(1-aziridinyl)phosphine oxides and sulfides in male houseflies, *J. Econ. Entomol.,* 63, 1744, 1970.

992. Seawright, J. A., Bowman, M. C., and Lofgren, C. S., Insect chemosterilants: gas chromatography, p-values, and relationship of p-values to sterilant activity in pupae of *Anopheles albimanus, J. Econ. Entomol.,* 66, 613, 1973.

993. Fye, R. L. and LaBrecque, G. C., Sterility in house flies offered a choice of untreated diets and diets treated with chemosterilants, *J. Econ. Entomol.,* 60, 1284, 1967.

994. Woods, C. W. and Beroza, M., Antifertility composition for insects, U.S. Patent 3,126,315.

995. Hays, S. B., Chemosterilizing activity and toxicity of P,P-bis(1-aziridinyl)-N-methylphosphinic amide and P,P-bis(1-aziridinyl)-N-(3-methoxypropyl)phosphinothioic amide against the housefly, *J. Econ. Entomol.,* 61, 800, 1968.

996. Fye, R. L., Screening of chemosterilants against houseflies, *J. Econ. Entomol.,* 60, 605, 1967.

997. Schuster, M. F. and Boling, J. C., Insect sterilant experiments with apholate and five bifunctional aziridine chemicals in outdoor cages against the boll weevil, *J. Econ. Entomol.,* 62, 1372, 1969.

998. Kropacheva, A. A., Sazonov, N. V., and Vronskikh, M. D., Sterilization of insects, especially of the colorado potato beetle, U.S.S.R. Patent 273,577; *C.A.,* 74, 31084, 1971.

999. Vashkov, V. I., Nifant'ev, E. E., Sidorova, M. V., Zavalishina, A. I., and Volkov, Yu. P., Chemical sterilization of houseflies, U.S.S.R. Patent 217,138; *C.A.,* 69, 66480, 1968.

1000. Zinov'eva, C. A., Osmolovskaya, R. E., Khokhlov, P. S., Klimov, O. V., and Bliznyuk, N. K., Control of fly reproduction, U.S.S.R. Patent 190,142; *C.A.,* 67, 116138, 1967.

1001. Klassen, W., Norland, J. F., and Borkovec, A. B., Potential chemosterilants for boll weevils, *J. Econ. Entomol.,* 61, 401, 1968.

1002. Hamm, P. C., Phosphorodiamidate insect chemosterilants, U.S. Patent 3,397,270; *C.A.,* 69, 105339, 1968.

1003. Hamm, P. C., Organophosphate insect chemosterilants, U.S. Patent 3,492,405; *C.A.,* 72, 99509, 1970.

1004. Hamm, P. C., Haloethyl ethyleneglycol phosphites as insect chemosterilant, U.S. Patent 3,562,390; *C.A.,* 74, 98864, 1971.

1005. Ihara Noyaku Co., Imochi-byo bojo-hoho (Control method for rice blast disease), Japanese Patent 536,423.

1006. Scheinpflug, H. and Jung, H. F., Use of organophosphates for the control of fungal diseases of crops, *Pflanzenschutz-Nachr.,* 21, 79, 1968.

1007. Mel'nikov, N. N., Phosphorsäurederivate als Fungizide, *Arch. Pflanzenschutz.,* 5, 3, 1969.

1008. Uesugi, Y., Pharmacology of organophosphorus fungicides, *Shokubutsu-Boeki,* 26, 103, 1972.

1009. Hacskaylo, J. and Stewart, R. B., Efficacy of phorate as a fungicide, *Phytopathology,* 52, 371, 1962.

1010. Eto, M., Ohkawa, H., Kobayashi, K., and Hosoi, T., Saligenin cyclic phosphorus esters as biological alkylating agents and fungicides, *Agric. Biol. Chem.* (Tokyo), 32, 1056, 1968.

1011. El-Helaly, A. F., Abo-el-Dahab, M. K., and Zeitoum, F. M., Effect of organic phosphorus pesticides on certain phytopathogenic bacteria, *Phytopathology,* 53, 762, 1963.

1012. Maeda, T., Abe, H., Kakiki, K., and Misato, T., Studies on the mode of action of organophosphorus fungicide, Kitazin. II. Accumulation of an amino sugar derivative on Kitazin-treated mycelia of *Piricularia oryzae. Agric. Biol. Chem.* (Tokyo), 34, 700, 1970.

1013. Ohta, N., Kakiki, K., and Misato, T., Studies on the mode of action of polyoxin D. II. Effect of polyoxin D on the synthesis of fungal cell wall chitin, *Agric. Biol. Chem.* (Tokyo), 34, 1224, 1970.

1014. Hori, M., Kakiki, K., Suzuki, S., and Misato, T., Studies on the mode of action of polyoxins. III. Relation of polyoxin structure to chitin synthetase inhibition, *Agric. Biol. Chem.* (Tokyo), 35, 1280, 1971.

1015. Misato, T., Mode of action of agricultural antibiotics developed in Japan, *Residue Rev.,* 25, 93, 1969.

1016. Hendlin, D., Stapley, E. O., Jackson, M., Wallick, H., Miller, A. K., Wolf, F. J., Miller, T. W., Chaiet, L., Kahan, F. M., Foltz, E. L., Woodruff, H. B., Mata, J. M., Hernandz, S., and Mochales, S., Phosphonomycin, a new antibiotic produced by strains of *Streptomyces, Science,* 166, 122, 1969.

1017. Rogers, H. J. and Perkins, H. R., *Cell Walls and Membranes,* Spon, London, 1968.

1018. Ishaya, I. and Casida, J. E., Dietary TH-6040 alters compositions and enzyme activity of housefly larval cuticle, *Pest. Biochem. Physiol.,* in press, 1974.

1019. Yamamoto, H., Tomizawa, C., Uesugi, Y., and Murai, T., Absorption, translocation and metabolism of O,O-diisopropyl S-benzyl phosphorothiolate (Kitazin®) in rice plant, *Agric. Biol. Chem.,* 37, 1553, 1973.

1020. Uesugi, Y., Katagiri, M., and Fukunaga, K., Resistance in *Piricularia oryzae* to antibiotics and organophosphorus fungicides, *Bull. Natl. Inst. Agric. Sci. (Japan) Ser. C.,* 23, 93, 1969.

1021. Uesugi, Y., Katagiri, M., and Noda, M., Snyergistic effect and negatively correlated cross resistance among organophosphorus fungicides, *Ann. Meeting Agric. Chem. Soc. Jap.,* abstract of papers, p. 76, 1973.

1022. **Ishizuka, K., Takase, I., Tan, K. E., and Mitsui, S.,** Absorption and translocation of O-ethyl S,S-diphenyl phosphorodithiolate (Hinosan®) in rice plants, *Agric. Biol. Chem.* (Tokyo), 37, 1307, 1973.

1023. **Ueyama, I., Uesugi, Y., Tomizawa, C., and Murai, T.,** Metabolic fate of O-ethyl S,S-diphenyl phosphorodithiolate (Hinosan®) in rice plant, *Agric. Biol. Chem.* (Tokyo), 37, 1543, 1973.

1024. **Uesugi, Y. and Tomizawa,C.,** Metabolism of O-ethyl S,S-diphenyl phosphorodithioate (Hinosan®) by mycelial cells of *Piricularia oryzae, Agric. Biol. Chem.* (Tokyo), 35, 941, 1971.

1025. **Takase, I., Tan, K. E., and Ishizuka, K.,** Metabolic transformation and accumulation of O-ethyl S,S-diphenyl phosphorodithiolate (Hinosan®) in rice plants, *Agric. Biol. Chem.* (Tokyo), 37, 1563, 1973.

1026. **Smit, F. M.,** Diethyl 6-carbethoxy-5-methylpyrazolo-[1,5-a] pyrimidin-2-yl phosphorothioate, a new systemic fungicide against mildew, *Meded. Rijksfac. Landbouwwet. Gent.,* 34, 763, 1969; *C.A.,* 73, 76038, 1970.

1027. **Fearing, R. B. and McClelland, M. C.,** Control of fungi with O,O-diethyl S-trichloromethyl phosphorodithioate, U.S. Patent 3,439,092; *C.A.,* 71, 37895, 1969.

1028. **Arbuzov, B. A., Zoroastrova, V. M., and Ibragimova, N. D.,** Derivatives of phosphorus acids and α-chloroallyl alcohol, *Izv. Akad. Nauk SSSR, Ser. Khim.,* p. 1278, 1967; *C.A.,* 68, 48957, 1968.

1029. **Wadsworth, W. and Emmans, N.,** Bicyclic thiophosphates, U.S. Patent 3,038,001; *C.A.,* 57, 12322, 1962.

1030. **Gilbert, E. E.,** Thiophosphate ester fungicides, U.S. Patent 2,911,335; *C.A.,* 54, 3839, 1960.

1031. **Stern, C. J.,** Organotin-organophosphorus compounds and their preparation, U.S. Patent 3,179,676; *C.A.,* 63, 2999, 1965.

1032. **Mandelbaum, Ya. A., Soifer, R. S., Melnikov, N. N., and Fedoseenko, L. G.,** Acaricidal and fungicidal activity of dialkylamides of O-alkyl S-(N-alkylcarbamoylmethyl) dithiophosphates, *Khim. Sel'sk. Khoz.,* 6, 107, 1968; *C.A.,* 69, 2087, 1968.

1033. **Mandelbaum, Ya. A., Abramova, G. L., Melnikov, N. N., and Fedoseenko, L. G.,** Fungicidal and insecticidal properties of O-alkyl S-alkyl dithiophosphoric acid amides, *Khim. Sel'sk. Khoz.,* 6, 596, 1968; *C.A.,* 70, 27967, 1969.

1034. **Kishino, S., Yamada, Y., and Uchihira, S.,** Organic phosphate esters as agricultural and horticultural fungicides, Ger. Offen., 2, 056, 176; *C.A.,* 75, 63384, 1971.

1035. **Demecko, J. and Konecny, V.,** Fungicidal and acaricidal properties of a new organophosphorus compound, bis-(N,N-dimethylamido)-O-(2,4,5-trichloro-6-nitrophenyl) phosphate, *Agrochemia,* 10, 127, 1970; *C.A.,* 74, 41374, 1971.

1036. **Tolkmith, H. and Mussell, D. R.,** Novel N-heterocyclic fungicides, *World Rev. Pest Control,* 6, 74, 1967.

1037. **Spencer, E. Y.,** *Guide to the Chemicals Used in Crop Protection,* 6th ed., Information Canada, 1973, 266.

1038. **Tolkmith, H., Buddie, P. B., Mussell, D. R., and Nyquist, R. A.,** Imidazolylphosphinamidothionates, *J. Med. Chem.,* 10, 1074, 1967.

1039. **Van den Bos, B. G., Koopmans, M. J., and Hoisman, H. O.,** Investigations on pesticidal phosphorus compounds. I. Fungicides, insecticides and acaricides derived from 3-amino-1,2,4-triazole, *Rec. Trav. Chim.,* 79, 807, 1960.

1040. **Demozay, D.,** Organophosphorus pesticides, S. African Patent 68 00,526; *C.A.,* 72, 2551, 1970.

1041. **Bliss, A. D. and Raetz, R. F. W.,** Biscarbamoyl spirophosphonthioates, U.S. Patent 3,355,523; *C.A.,* 68, 29656, 1968.

1042. **Leber, J. P.,** Imidazolidine-2-thione-N-phosphoramides, Swiss Patent 439,306; *C.A.,* 69, 43913, 1968.

1043. **Schmulbach, C. D.,** Fungicidal cyclic boraphosphonitriles, U.S. Patent 3,538,155; *C.A.,* 74, 23006, 1971.

1044. **Wagner, F. S., Jr.,** Phosphite esters for preserving wood, S. African Patent 67 02,846; *C.A.,* 71, 37,904, 1969.

1045. **Nihon Tokushu Noyaka Seizo K. K.,** Fungicidal compounds containing organic esters of phosphorodithious acid, French patent 1,583,740; *C.A.,* 73, 76868, 1970.

1046. **Nikonorov, K. V., Gurylev, E. A., Urazaeva, L. G., Nazypov, M. N., Asadov, R. A., and Anisin, S. D.,** Synthesis and biological properties of some chloroorganophosphorus compounds, *Izv. Akad. Nauk SSSR Ser. Khim.,* 10, 2241, 1966; *C.A.,* 72, 31914, 1970.

1047. **Shindo, N., Ura, Y., Takeuchi, H., Yamashita, Y., and Ono, K.,** EPN derivatives. VI. *Noyaku Seisan Gijutsu,* 19, 15, 1968.

1048. **Birum, G. H. and Mattews, C. N.,** Fungicidal compositions and methods using mesomeric phosphonium compounds, U.S. Patent 3,445,570; *C.A.,* 71, 37898, 1969.

1049. **Pappas, J. J. and Gancher, E.,** (Dithiocarboxy)cyanomethylene triphenylphosphines, U.S. Patent 3,394,166; *C.A.,* 69, 77484, 1968.

1050. **Ashton, F. M. and Crafts, A. S.,** *Mode of Action of Herbicides,* Wiley Interscience, New York, 1973.

1051. Weed Science Society of America, *Herbicide Handbook,* 2nd ed., 1970.

1052. **Riov, J. and Jaffe, M. J.,** Cholinesterases from plant tissues. I. Purification and characterization of a cholinesterase from mung bean roots, *Plant Physiol.,* 51, 520, 1973.

1053. **Jaffe, M. J.,** Evidence for the regulation of phytochrome-mediated processes in bean roots by the neurohumor, acetycholine, *Plant Physiol.,* 46, 768, 1970.

1054. **Wegler, R. and Eue, L.,** Herbizide, in *Chemie der Pflanzenschutz- und Schädlings-bekämpfungsmittel,* II. Wegler, R., Ed., Springer-Verlag, Berlin, 1970, 384.

1055. **Buchner, B. and Jacoves, E.,** O-(2-Naphthyl) phosphorothionates, U.S. Patent 3,328,494; *C.A.,* 67, 73440, 1967.

1056. **Young, S. T.,** m-Ureidophenyl esters of O,O-dimethyl phosphorothioic acid as herbicides, U.S. Patent 3,384,194; *C.A.,* 69, 86617, 1968.

1057. **Gutman, A. D.,** (O-Carbamoyl oxime)-phosphate, phosphonate, and phosphinate compositions and their utility as herbicides and pesticides, S. African Patent 68 03,662; *C.A.,* 71, 30236, 1969.

1058. **Konecny, V., Truchlik, S., Drabek, J., and Patorek, I.,** Herbicidal phosphate esters, Czechoslovakian Patent 132,843; *C.A.,* 73, 120319, 1970.

1059. **Gubler, K. and Kristiansen, O.,** Herbicidal thiazolinylpyridyl phosphorothioates, Ger. Offen., 2,014,886; *C.A.,* 74, 53766, 1971.

1060. **Price, G. R., Walsh, E. N., Dewald, C. L., and Soong, S. Y. C.,** Herbicidal thiocarbamidates, phosphorothiocarbimidates, and phosphorothioimides, U.S. Patent 3,161,498; *C.A.,* 63, 14762, 1965.

1060a. **Richter, S. B.,** Herbicidal phosphorodithioates, U.S. Patent 3,450,520; *C.A.,* 71, 60996, 1969.

1061. **Baker, D. R., Brokke, M. E., and Arneklev, D. R.,** 3-Dithiophosphorylacetyl-3-azabicyclo[3.3.2]nonane compounds, U.S. Patent 3,347,850; *C.A.,* 68, 95702, 1968.

1062. **Melnikov, N. N., Kaskin, B. A., and Sablian, I. V.,** Herbicidal and plant growth regulating amines via esters of thio- and dithiophosphorus acids, *Zh. Obshch. Khim.,* 38, 1558, 1968; *C.A.,* 70, 3803, 1969.

1063. **Below, J. F., Jr.,** Stabilized herbicides containing N-[2-(dialkoxyphosphinothioylthio)ethyl] arene sulfonamides and a Lewis base, S. African Patent 68 03,592; *C.A.,* 71, 100732, 1969.

1064. **Ashton, F. M., Penner, D., and Hoffman, S.,** Effect of several herbicides on proteolytic activity of squash seedlings, *Weed Sci.,* 16, 169, 1968.

1065. **Aya, M., Fukazawa, A., Kishino, S., and Kume, T.,** Tokunol – on the structure and herbicidal activity, *Noyaku-Kenkyu,* 17, 62, 1971.

1066. **Smith, T. D.,** Controlling growth of plants with phosphorodiamidate compositions, U.S. Patent 3,539,331; *C.A.,* 75, 34391, 1971.

1067. **Bayer, H. O. and Unger, V. H.,** Herbicidal phosphoric triamides, U.S. Patent 3,433,623; *C.A.,* 70, 105939, 1969.

1068. **Pitina, M. R., Stonov, L. D., Bakumenko, L. A., Kol'tsova, S. S., and Shvetsov, N. I.,** Methods for combatting weeds, USSR Patent 204,054; *C.A.,* 72, 30540, 1970.

1069. Deutsche Gold- und Silber-Scheidenanstalt vorm. Roessler, 2-(N-Phosphinyl-N-ethylamino)-4-alkylamino-6-halo-(or trihalomethyl)-s-triazines, Belgian Patent 635,690; *C.A.,* 61, 10693, 1964.

1070. **Olah, G. A. and Kuhn, S. J.,** Phosphorisocyanatidic difluoride, U.S. Patent 3,096,371; *C.A.,* 59, 12435, 1963.

1071. **Watson, A. J. and Leasure, J. K.,** Zytron – for crabgrass control, *Down to Earth* (winter issue), 1, 1959.

1072. **Sumida, S. and Ueda, M.,** Chlorella as a model system to study herbicidal mode of action and its application to a new herbicide, O-ethyl O-(3-methyl-6-nitrophenyl) N-sec-butyl phosphorothioamidate, Ann. Meeting Am. Chem. Soc., Los Angeles, 1974.

1073. **Boswell, S. B., Burns, R. M., and McCarty, C. D.,** Chemical inhibition of avocado top growth, *Citrograph,* 56, 386, 1971.

1074. **Regel, F. K. and Motts, M. F.,** Vinyl phosphonodithioates having selective preemergent and postemergent herbicidal activity, U.S. Patent 3,416,912; *C.A.,* 71, 22185, 1969.

1075. **Vasilev, G. S., Prilezhaceva, E. N., Shcheglov, Y. V., and Shostakovskii, M. F.,** Structure and biological activity of reaction products of 1-alkoxy- or 1-(alkylthio)butenynes with PCl_5, *Khim Atsetilena,* 288, 1968; *C.A.,* 71, 30552, 1969.

1076. Chemagro Corp., Herbicidal vinylphosphonodithioates, British Patent 1,183,130; *C.A.,* 72, 121703, 1970.

1077. **Schräder, H.,** Phosphonylchloromethyl triazines and thiadiazoles, U.S. Patent 3,299,061; *C.A.,* 66, 65630, 1967.

1078. **Grapov, A. F., Lebedeva, N. V., and Melnikov, N. N.,** Synthesis and herbicidal activity of some O-arylmethyl-, dichloromethyl-, and trichloromethylphosphonic acid amides, *Zh. Obshch. Khim.,* 38, 1751, 1968; *C.A.,* 70, 4227, 1969.

1079. **Azerbaev, I. N., Sarbaev, T. G., Abiyurov, B. D., and Gorelova, L. G.,** Dialkyl 2,2-dimethyl-4-hydroxytetrahydro-4-pyranophosphonate plant-growth regulators, USSR Patent 257,219; *C.A.,* 72, 132969, 1970.

1080. **Gier, D. W.,** Dialkyl and diphenyl esters of aryloxyacetylphosphonic acids as herbicides, U.S. Patent 3,382,060; *C.A.,* 70, 4273, 1969.

1081. **Bucha, H. C., Langsdort, P., Jr., and Jelinek, A. G.,** Dialkyl carbamoylphosphonates for retardation of plant growth, Ger. Offen., 2,040,367; *C.A.,* 74, 112211, 1971.

1082. **Kohler, J. J.,** N,N'-Sulfonylbis(iminocarbonyl)diphosphonic acid esters, U.S. Patent 3,401, 214; *C.A.,* 69, 106869, 1968.

1083. **Hagen, H. and Becke, F.,** Herbicidal and acaricidal thiocarbamoylphosphonic diamides, Ger. Offen., 1,947,192; *C.A.,* 74, 125846, 1971.

1084. **Weil, E. D.,** Cyclic phosphine oxides or sulfides as herbicides, East German Patent 54,836; *C.A.,* 67, 107587, 1965.

1085. **Kabachnik, M. I. and Rossiiskaya, P. A.,** Organophosphorus compounds, I. and II., *Izv. Akad. Nauk SSSR Otd. Khim. Nauk,* 406, 295 and 403, 1946; *C.A.,* 42, 7241 and 7242, 1948.

1086. **DeWilde, R. C.,** Ethephon, practical application of (2-chloroethyl)phosphonic acid in agricultural production, *Hortic. Sci.,* 6, 364, 1971.

1087. **Caseley, J.,** Effect of 2-chloroethylphosphonic acid on the morphology and apical dominance of *Agropyron repens, Pest. Sci.,* 1, 114, 1970.

1088. CIBA-Geigy AG, Esters of phosphonodithioic acid, DOS 2144926, *Angew. Chem. Int. Ed.,* 12, 344, 1973.

1089. **Tompkins, D. R., Sistrunk, W. A., and Fleming, J. W.,** Yield of snap beans (*Phaseolus vulgaris*) as influenced by 5-chloro-2-thenyl-tributylphosphonium chloride, *Hortic. Sci.,* 6, 393, 1971.

1090. **Lang, A.,** Gibberellins: structure and metabolism, *Annu. Rev. Plant Physiol.,* 21, 537, 1970.

1091. **Riov, J. and Jaffe, M. J.,** A cholinesterase from bean roots and its inhibition by plant growth retardants, *Experientia,* 29, 264, 1973.

1092. **Dubois, K. P., Kinoshita, F., and Jackson, P.,** Acute toxicity and mechanism of action of cholinergic rodenticide, *Arch. Int. Pharmacodyn. Ther.,* 169, 108, 1967.

1093. **Bowers, W. S.,** Juvenile hormone: activity of natural and synthetic synergists, *Science,* 161, 895, 1968.

1094. **Bowers, W. S.,** Insect hormones and their derivatives as insecticides, *Bull. W. H. O.,* 44, 381, 1971.

1095. **Montogomery, R. E. and Incho, H. H.,** Alken-ω-ynyl phosphinate synergists for cyclopropanecarboxylate insecticides, Def. Publ., U.S. Pat. Off., 869,008; *C.A.,* 72, 30578, 1970.

1096. **Kashafutdinov, G. A. and Il'ina, N. A.,** Repellent activity of cyclohexenyl esters of amidophosphoric acid, *Med. Parazitol. Parazit. Bolez.,* 40, 36, 1971; *C.A.,* 75, 75290, 1971.

A

Coumaphos, 20, 108, 172, 186, 297
  fluorimetry of, 108
  metabolic breakdown of, 186
  reductive dechlorination of, 172
Coumithoate, 297
Cremart, 329
Cross resistance, 198
  in strain Fc housefly, 198
Crotoxyphos, 295
Crufomate, 14, 41, 103, 184, 300–301
  fluorogenic reaction of, 103
  main metabolite of, 184
  synthesis of, 41
CTBP, 332
Cyanofenphos, 270
Cyanogen bromide, 106
p-Cyanophenyl dimethyl phosphorothionate see
    Cyanophos
p-Cyanophenyl ethyl phenylphosphonothionate see
    Cyanofenphos
Cyanophos, 190, 240, 244
  main metabolite of, 190
5-Cyano-2-pyridyl diethyl phosphorothionate, 248
Cyanox® see Cyanophos
Cyanthoate, 262
S-[N-(1-Cyano-1-methylethyl)carbamoylmethyl] diethyl
    phosphorothiolate see Cyanthoate
CYAP see Cyanophos
Cyclic ethylenesulfonium, 85
Cyclic phosphate esters, 72–75, 152, 153
  anticholinesterase activity of, 152
  electronic structure of, 74
  hydrolysis of, 72–75
  insecticidal activities of, 153
Cyclic phosphoramidates, 38–39, 152
  anticholinesterase activity of, 152
  reaction with disulfides, 38–39
Cyclophosphamide, 42–43, 75, 166, 233, 308
  hydrolysis of, 75
  metabolic pathway of, 166
  synthesis of, 42–43
Cygon® see Dimethoate
Cymetox® see Demephion
Cynem® see Thionazin
Cyolane® see Phosfolan
CYP see Cyanofenphos
Cythioate, 167
Cythion® see Malathion
Cytochrome b$_5$, 160
Cytochrome c reductase, 160
Cytochrome P-450, 160
Cytrolane® see Mephosfolan

D

2,4-D, 331
DAEP see Amiphos
Danifos® see PTMD
Dasanit® see Fensulfothion
DCQ, 102, 105
DDT, 162

DDVP see Dichlorvos
Dealkylation, 88, 142–144
  in dialkoxyphosphinyl enzyme, 142
Dealkylation reactions, 63
DEF®, 40, 105, 207, 228, 327
  determination of, 105
  synthesis of, 40
  teratogenic effect of, 228
Defoliant, 327
Dehydrogenase, 228
  inhibition by organophosphorus esters, 228
Delnav® see Dioxathion
Demephion, 289
Demethylation, 179
Demeton, 25, 26, 64, 288
  anticholinesterase activity of metabolites of, 288
  isomerization of, 26
  metabolic routes of, 288
  phosphate analog of, 64
  synthesis of, 25
  synthesis of thiono isomer, 26
Demeton-methyl, 82, 289
Demeton-O, 36, 90, 94, 288
  isomerization of, 90
  oxidation of, 94
  oxidation with bromine, 36
  sulfone derivatives of, 90
Demeton-O-methyl, 85
  formation of S-(ethylthioethyl) demeton-S-methyl, 85
Demeton-S, 33, 38, 58, 64, 94, 288
  hydrolysis rates, 58
  oxidation of, 94
  synthesis of, 33, 38
Demeton-S-methyl, 35, 40, 58, 65, 84
  half-life, 65
  hydrolysis rates, 58
  synthesis from trimethyl phosphorothionate, 40
  synthesis of, 35
  transmethylation, 84
Demeton-S-methyl-sulfoxide see Oxydemeton-methyl
Demeton-S-methyl-sulphone see Dioxydemeton-S-methyl
Demuphos, 44, 200, 277
  synthesis of, 44
Demyelination, 219
2,4-DEP, 331
Desmethyl dichlorvos, 85
Detoxication, 158
DFP, 46, 63, 67, 131
  bimolecular hydrolysis rates, 67
  formal positive charges, 63
  synthesis of, 46
Dialifor, 259–260
bis-(Dialkoxyphosphinothioyl) disulfides, 48
  synthesis of, 48
Dialkyl N-acetyl-N-phenylphosphoramidates, 150
O,O-Dialkyl hydrogen phosphorothioates, 34, 36
  reaction wtih alkyl halides, 34
  synthesis of, 36
  tautomerism of, 34
Dialkyl phosphites, 8, 19, 36, 38, 39, 43, 52, 53, 54
  addition to carbonyl compounds, 53
  addition of elemental sulfur, 36

O,O-Diethyl 2-ethylthioethyl phosphorothioate *see* Demeton

Diethyl S-(2-ethylthioethyl) phosphorothiolothionate *see* Disulfoton

Diethyl S-(ethylthiomethyl) phosphorothiolothionate *see* Phorate

Diethyl hydrogen phosphorodithioate, 31
abnormal addition to olefins, 31

O,O-Diethyl hydrogen phosphorodithioate, 31
normal addition to olefins, 31

Diethyl S-(N-isopropylcarbamoylmethyl) phosphoro-thiolothionate *see* Prothoate

Diethyl 2-isopropyl-6-methyl-4-pyrimidinyl phosphoro-thionate *see* Diazinon

Diethyl 4-methyl-7-coumarinyl phosphorothionate *see* Potasan

Diethyl 3-methyl-4-methylthiophenyl phosphorothionate *see* Lucijet

Diethyl methyl phosphate, 117
fragmentation scheme of, 117

Diethyl S-methyl phosphorothiolate, 117–118
fragmentation scheme of, 117–118

Diethyl S-methyl phosphorothiolothionate, 314
as fungicide, 314

Diethyl methyl phosphorothionate, 117–118
fragmentation scheme of, 117–118

Diethyl 3-methylpyrazol-5-yl phosphorothionate *see* Pyrazothion

Diethyl 2-methylquinolin-4-yl phosphorothionate *see* Quinothion

Diethyl p-methylsulfinylphenyl phosphorothionate *see* Fensulfothion

O,S-Diethyl O-p-nitrophenyl phosphorothiolate, 64

Diethyl p-nitrophenyl phosphorothionate *see* Parathion

Diethyl S-(4-oxobenzotriazino-3-methyl) phosphoro-thiolothionate *see* Azinphosethyl

Diethyl 3-oxo-2-phenyl-2H-pyridazin-6-yl phosphoro-thionate *see* Pyridafenthion

Diethyl 5-phenyl-3-isoxazolyl phosphorothionate *see* Isoxathion

Diethyl phenyl phosphate, 147
anticholinesterase activity in a series of, 147

Diethyl phenyl phosphorothionate *see also* SV₁, 253
effect of para-substitution on insecticidal activity of, 253

Diethyl phthalimidophosphonothionate *see* Dowco 199

Diethyl 2-quinoxalyl phosphorothionate *see* Diethquinalphion

Diethyl substituted phenyl phosphates, 146
biological activities of, 146

Diethyl trichloroacetylphosphonate, 52

Diethyl 3,5,6-trichloro-2-pyridyl phosphorothionate *see* Chloropyrifos

Diethyl vinyl phosphates, 71
hydrolysis of, 71

N-[2-(Diisopropoxyphosphinothioylthio) ethyl] benzenesulfonamide *see* Bensulide

Diisopropyl S-(ethylsulfinylmethyl) phosphorothiolo-thionate *see* Aphidan

N,N'-Diisopropylphosphorodiamidic fluoride *see* Mipafox

Diisopropyl phosphorofluoridate *see* DFP

Dimecron® *see* Phosphamidon

Dimefox, 46, 60, 277
hydrolysis rates, 60
synthesis of, 46

Dimephenthoate *see* Phenthoate

Dimethoate, 28–29, 32, 59, 86, 88, 103, 119, 182, 184, 187, 199, 204, 207, 280–281
carboxylic acid, 187
cleavage sites of, 184
color reaction of, 103
degradation of, 182
demethylation of, 86
deterioration by methyl "cellosolve", 204
enzymic carboxyamide cleavage of, 187
fragmentation of, 119
fluorogenic reaction of, 103
hydrolysis rates, 59
reaction with triaminophosphines, 88
species specific toxicity of, 199
synthesis of, 28–29, 32

Dimethoxon, 120
rearrangement ion peak of, 120

1,3-Di(methoxycarbonyl)-1-propen-2-yl dimethyl phosphate *see* Bomyl

3-(Dimethoxyphosphinyloxy)-N,N-dimethyl-*cis*-crotonamide *see* Dicrotophos

3-(Dimethoxyphosphinyloxy)-N-methyl-*cis*-crotonamide *see* Monocrotophos

O,S-Dimethyl N-acetylphosphoramidothiolate *see* Acephate

Dimethyl 1-n-butyryloxy-2,2,2-trichloroethylphosphonate *see* Butonate

Dimethyl 3,5-dimethyl-4-methylthiophenyl phosphoro-thionate *see* Bay 37342

Dimethyl p-(dimethylsulfamoyl)phenyl phosphoro-thionate *see* Famphur

Dimethyl S-[(α-ethoxycarbonyl)benzyl] phosphoro-thiolothionate *see* Phenthoate

Dimethyl S-(N-ethylcarbamyolmethyl) phosphoro-thiolothionate *see* Ethoate-methyl

Dimethyl S-(2-ethylsulfinylisopropyl) phosphorothiolate *see* Metasystox S

Dimethyl S-(2-ethylsulfonylethyl) phosphorothiolate *see* Dioxydemeton-S-methyl

O,O-Dimethyl 2-ethylthioethyl phosphorothioate *see* Demeton-methyl

Dimethyl S-(2-ethylthioethyl) phosphorothiolothionate *see* Thiometon

Dimethyl S-(N-formyl-N-methylcarbamoylemethyl) phosphorothiolothionate *see* Formothion

Dimethyl 1-hydroxy-2,2,2-trichloroethylphosphonate *see* Trichlorfon

Dimethyl N-(isopropoxycarbonyl) phosphoramidate *see* Avenin

Dimethyl S-(2-isopropylthioethyl) phosphorothiolo-thionate *see* Isothioate

Dimethyl isoxazolyl phosphorothionate, 91
thermal decomposition of, 90

Dimethyl 1-methoxycarbonyl-1-propen-2-yl phosphate *see* Mevinphos

Dimethyl S-(N-methoxyethylcarbamoylmethyl) phosphorothiolothionate *see* Amidithion

Dimethyl S-(5-methoxy-4-oxo-4H-pyran-2-ylmethyl) phosphorothiolate see Endothion
Dimethyl S-(2-methoxy-1,3,4-thiadiazol-5-(4H)-onyl-4-methyl) phosphorothiolothionate see Methidathion
Dimethyl S-[2-(methylcarbamoylethylthio)ethyl] phosphorothiolate see Vamidothion
Dimethyl S-(N-methylcarbamoylmethyl) phosphorothiolate see Omethoate
Dimethyl S-(N-methylcarbamoylmethyl) phosphorothiolothionate see Dimethoate
Dimethyl 3-methyl-4-methylthiophenyl phosphorothionate see Fenthion
Dimethyl 3-methyl-4-nitrophenyl phosphorothionate see Fenitrothion
Dimethyl cis-1-methyl-2-(1-phenylethoxycarbonyl) vinyl phosphate see Crotoxyphos
O,O-Dimethyl 2-methylthioethyl phosphorothioate see Demephion
Dimethyl S-(morpholinocarbonylmethyl) phosphorothiolothionate see Morphothion
Dimethyl p-nitrophenyl phosphorothionate see Parathion-methyl
Dimethyl S-(4-oxobenzotriazino-3-methyl) phosphorothiolothionate see Azinphosmethyl
Dimethyl-p-phenylenediamine, 102
O,S-Dimethyl phosphoramidothiolate see Methamidophos
Dimethyl phosphorochloridothionate, 24
  reaction with tertiary amines, 24
Dimethyl S-phthalimidomethyl phosphorothiolothionate see Phosmet
Dimethyl sulfoxide, 205
  as a "penetrant carrier", 205
Dimethyl 2,4,5-trichlorophenyl phosphorothionate see Ronnel
Dimex®, 251
Dinitrogentetroxide, 20
1,4-Dioxan-2,3-ylidene bis(O,O-diethyl phosphorothiolothionate) see Dioxathion
Dioxathion, 31–32, 87, 100, 107, 182, 258–259
  colorimetry of, 107
  degradation of, 182
  methyl ester homolog of, 259
  pyrolysis of, 100
  reaction with mercuric chloride, 87
  synthesis of, 31–32
Dioxydemeton-S-methyl, 58, 65, 293
  half-life, 65
  hydrolysis rates, 58
S,S-Diphenyl ethyl phosphorodithiolate see Edinfenphos
Diphosphoramide, 44
  tautomerism of, 44
S,S-Dipropyl ethyl phosphorodithiolate see Prophos
S,S-Dipropyl methyl phosphorothionodithiolate see VC 3-668
Di-n-propyl p-methylthiophenyl phosphate see Propaphos
Dipterex® see Trichlorfon
Disulfoton, 58, 119, 165, 195, 290
  base peak of, 119
  hydrolysis rates, 58
Disulfoton sulfoxide, 188
Disyston® see Disulfoton
Disyston S® see Oxydisulfoton

Disyston-sulfoxide® see Oxydisulfoton
Dithiometasystox® see Thiometon
Dition® see Coumithoate
Dithiosystox® see Disulfoton
Diuretic factor, 230
  liberation of, 230
DMCP, 262–263
DMPA, 15, 42, 69–70, 103, 328
  fluorogenic reaction of, 103
  oxo-analog of, 69
    acid catalyzed hydrolysis of, 69–70
  synthesis of, 42
DNA, 307
  in chemosterilized fly eggs, 307
DNA biosynthesis, 307
  inhibition of, 307
Dopamine, 126
Double hydrogen rearrangement, 114
Dowco 109® see Narlene
Dowco 169 see Nellite
Dowco 199, 43, 44, 322
  synthesis of, 43, 44
Dowco® 214, 248
Dowco® 217, 248
Dow ET-15, 263
DSP, 302
Dursban® see Chloropyrifos
Dyfonate® see Fonofos

E

E605® see Parathion
Ecological selectivity, 196
EDDP see Edinfenphos
Edifenphos, 34, 37, 98, 99, 313, 319
  photolysis of, 99
  synthesis of, 34, 37
Ekatin® see Thiometon
Ekatin F® see Morphothion
Ekatin M® see Morphothion
Ektafos® see Dicrotophos
Electronegativities, 64
β-Elimination, 65–66
Ellman method, 109
Elsan® see Phenthoate
Endocide® see Endothion
Endothion, 295
End-plate potential, 126
Enol phosphates, 70–71, 150–151
  anticholinesterase activity of, 150–151
  hydrolysis rate of, 70–71
Enzyme activity, 134–136
  inhibition of, 134–136
Enzyme-inducing agents, 218
  effect on parathion metabolism, 218
Enzyme induction, 217–219
  effect of, 217–219
Enzyme inhibitor complex, 135
EPBP, 270
EPN, 14, 20, 49, 75, 205, 209, 270
  alkaline hydrolysis of, 75

effect on the metabolism of malathion, 205
synthesis of, 49
EPN-type compounds, 266, 267, 268
structure-activity relationship of, 266, 267, 268
(-)(1R,2S)1,2-Epoxypropylphosphonic acid *see*
Phosphonomycin
ESBP *see* Inezin
Esterases, 110, 128–129, 205–210
chromogenic substrate of, 110
classification of, 128–129
fluorogenic substrate of, 110
inhibitors of, 205
as synergists, 205–210
Esteratic site, 131, 134
phosphorylation of, 134
Estox® *see also* Metasystox S, 16, 65
Ethephon, 77, 330, 329
Ethion, 59, 103, 258
color reaction of, 103
hydrolysis rates, 59
Ethoate-methyl, 281
Ethrel® *see* Ethephon
Ethylene, 231
Ethyleneimides, 305
Ethyl S-(2-ethylthioethyl) methyl phosphorothiolo-
thionate *see* Tetrathion
Ethyl guthion® *see* Azinphosethyl
Ethyl hydrogen propylphosphonate *see* NIA 10637
Ethyl S-(N-methoxy-N-methylcarbamoylemthyl) N-
isopropylphosphoramidothiolate *see* FCS 13
Ethyl 3-methyl-4-methylthiophenyl N-isopropylphos-
phoramidate *see* Phenamiphos
Ethyl 3-methyl-6-nitrophenyl N-sec-butylphosphoramido-
thionate *see* Cremart
Ethyl methyl parathion, 241
O-Ethyl S-methyl phosphoramidothiolate *see* Bay 65258
Ethyl 2-nitro-4-methyl N-isopropylphosphoramido-
thionate *see* Amiprophos
Ethyl p-nitrophenyl alkylphosphonate, 151–152
inhibition rate for housefly-head AChE, 151–152
Ethyl p-nitrophenyl phenylphosphonothionate *see* EPN
Ethyl S-phenyl N-butylphosphoramidothiolothionate *see*
Phosbutyl
Ethyl S-phenyl ethylphosphonothiolothionate *see*
Fonofos
N-Ethylphosphoramidothionate, 42
Ethyl quinolin-8-yl phenylphosphonothionate *see*
Oxinothiophos
Ethyl 2,4,5-trichlorophenyl ethylphosphonothionate *see*
Trichloronate
Excitatory postsynaptic potential, 126
Excretion, 201
Exothion® *see* Endothion

F

Fac 20® *see* Prothoate
Factors, 203
interaction of, 203
Falodin® *see* 2,4-DEP
Falone® *see* 2,4-DEP

Famophos® *see* Famphur
Famphur, 167, 189, 203, 299
N-demethylation of, 167
glucuronide formation from, 189
FCS 13, 304
Fenchlorphos *see also* Ronnel, 58
hydrolysis rates, 58
Fenitrothion, 25, 58, 86, 115, 181, 200, 202, 203, 217,
240, 243–244, 267–268
as an antagonist for organothiocyanates, 217
O-demethylation of, 181
desmethyl, 86
fragment ion, 115
hydrolysis rates, 58
phosphonate and phosphinate analogs of, 267
biological activities of, 267–268
selective toxicity of, 200
synthesis of, 25
Fensulfothion, 93, 172, 188, 302
metabolites of, 188
thiono-thiolo rearrangement of, 172
Fenthion, 16, 58, 93, 97, 240, 245–246, 275, 299
hydrolysis rates, 58
oxidation of, 93
phosphinate analogs of, 275
photooxidation of, 97
Fitios B/77 *see* Ethoate-methyl
Fluorothion®, 240
Folex® *see* Merphos
Folidol® *see* Parathion
Folimat® *see* Omethoate
Folithion® *see* Fenitrothion
Folpet, 323
Fonofos, 14, 50, 93, 119, 163, 191, 272–274
base peak of, 119
metabolic pathway of, 191
oxidation of, 93
oxidative desulfuration, 163
synthesis of, 50
Forestenon, 236
Formanzane, 106
Formothion, 167, 282
dimethoate from, 167
Fostion® *see* Prothoate
Fujithion® *see* DMCP
Fujiwara reaction, 107
Fungicides, 312–325

G

G-30493, 257
GABA, 126
GC-6506, 239
GC-9,879, 256
Gardona® *see* Tetrachlorvinphos
Garrathion® *see* Carbophenothion
Gas chromatography, 110–113
derivatization for, 110–113
followed by alkylation, 112
hydrolysis of, 112
methylation, 111
oxidation, 113

H

I

synthesis of, 30
Methoxide ion, 87
  reaction with *O,O,S*-trimethyl phosphorothiolate, 87
2-Methoxy-4*H*-1,3,2-benzodioxaphosphorin-2-sulfide *see*
    Salithion
S-(*N*-methoxycarbonyl-*N*-methylcarbamoylmethyl)
    methyl methylphosphonothiolothionate *see*
    Mecarphon
2-(Methoxymethylthiophosphinylimino)-3-ethyl-5-
    methyl-1,3-oxazoline *see* Stauffer R-16661
Methylation, 191
Methyl-1-carboisopropoxy-1-propen-2-yl, 42
  synthesis of, 42
    stereospecificity in, 42
Methyl carbophenothion, 257
3-Methylcholanthrene, 162
Methyldemeton *see* Demeton-methyl
Methyldemeton-methyl *see* Demephion
Methylenedioxyphenyl compounds *see* MDP
Methyl esters, 239
3-Methyl Group, 156
  effects on biological activities of dimethyl *p*-nitrophenyl
    phosphate and phosphorothionate, 156
Methyl iodide, 181
  as potentiator for fenitrothion, 181
Methylnitrophos, 243
Methyl paraphonothion, 268
Methyl parathion *see* Parathion-methyl
Methyl phenkapton, 257
Methyl phenyl *N*-methylphosphoramidates, 150
Methyl saligenin cyclic phosphorothionate *see* Salithion
Methyl 2,4,5-trichlorophenyl phosphoramidate *see*
    Dow ET-15
Methyl Trithion® *see* Methyl carbophenothion
Metrifonate *see* Trichlorfon
Mevinphos, 18, 70, 96, 98, 106, 107, 115, 151, 180, 207,
    228, 286
  base peak ion of, 115
  trans-crotonate isomer, 18
  determination of, 106
  α-form, 18
  β-form, 18
  formulation analysis, 107
  geometrical isomers of, 70
    hydrolysis of, 70
  geometric isomers of, 180
  cis-isomer, 18, 151
    as an anticholinesterase agent, 151
  trans-isomer, 151
    as an anticholinesterase agent, 151
  cis-trans isomerization of, 98
  oxidative breakdown of, 96
  synthesis of, 18
  teratogenic effect of, 228
Mevinphos α, 60
  hydrolysis rates, 60
Mevinphos β, 60
  hydrolysis rates, 60
α-Mevinphos, 150–151
  anticholinesterase activity of, 150–151
cis-Mevinphos, 70
trans-Mevinphos, 70

mfo, 160, 162
  inhibitors of, 162
mfo inhibitors, 219
  as enzyme inducers, 219
MGK 264, 215
  structure, 215
Michaelis-Arbuzov reaction, 10, 21, 51–52, 56
Michaelis-Becker reaction, 52
Michalski reaction, 41
Microsomal hydroxylation, 161
  schematic representation for, 161
Microsomes, 160
Mintacol, 233
Mipafox, 63, 67, 142, 154, 277
  bimolecular hydrolysis rates, 67
  formal positive charges, 63
  selectivity in esterase inhibition, 154
Mitemate® *see* Amidothioate
Mites, 209
  organophosphate resistance, 209
Mixed-function oxidase, 161, 210–217
  inhibitors of, 210–217
  reactions catalyzed by, 161
Mixed-function oxidase inhibitors, 211, 214–216
  effect on parathion metabolism, 211
  mode of action of, 214–216
Mixed-function oxidase *see also* mfo, 160–162
MOCAP, 91
  ethyl propyl sulfide from, 91
Mocap® *see* Prophos
Monitor® *see* Methamidophos
Monoalkyl phosphonites, 56
  addition to carbonyl, 56
  addition to unsaturated hydrocarbons, 56
Monocrotophos, 167, 228, 286
  teratogenic effect of, 228
Monomeric metaphosphate, 61, 68, 77
Mononitrosoacetone, 79
Monooxygenases, 160
Morphothion, 59, 282
  hydrolysis rates, 59
Morphotox® *see* Morphothion
Murfotox® *see* Mecarbam
Muscatox® *see* Coumaphos
Mutant aliesterase, 177, 200

N

N-2596, 273
N-3727, 273
N-3794, 273
N-4543, 274
NADPH, 160
Naled, 96, 171, 236
  debromination of, 171
Nankor® *see* Ronnel
Naphthalophos *see* Maretin
Naphthyl acetates, 109
Narlene®, 301
Navadel® *see* Dioxathion
Neguvon® *see* Trichlorfon

P

Palladium chloride, 102
2-PAM, 140
PAPS, 190
Papthion® *see* Phenthoate
Paraoxon, 17, 20, 58, 63, 73, 79, 81, 115, 181, 241
  S-aryl thiolate analog, 64
  formal positive charges, 63
  hydrolysis of, 73, 79
    by several bases, 79
  hydrolysis rates, 58
  mass spectrum of, 115
  synthesis of, 17
Parathion, 20, 37, 58, 63, 81, 104, 115, 162, 173, 189,
    193, 228, 229, 240, 241–242, 267–268
  colorimetric determination of, 104
  dealkylation, 63
  dearylation (detoxication) of, 173
  depletion of myocardial phosphorylase activity, 229
  desulfuration (activation) of, 173
  formal positive charges, 63
  glucuronide formation from, 189
  hydrolysis rates, 58
  mass spectrum, 115
  metabolic pathways of, 241
  oxidative desulfuration, 162
  phosphonate and phosphinate analogs of, 267
    biological activities of, 267–268
  teratogenic effect of, 228
  thio-isomer, 37
  toxic products in the environment, 193
Parathion-methyl, 15, 58, 63, 68, 84, 88, 115, 179, 240,
    242
  amido analog of, 68
  dealkylation, 63
  decomposition by pyridine, 84
  demethylation of, 179
  fragment ion, 115
  hydrolysis rates, 58
  isomerization of, 88, 89
    by heat, 89
    dimethylformamide, 88
  S-methyl isomer of, 15
  reaction with diethylphenylphosphine, 88
Partition coefficient, 146
Pentacoordinate intermediate, 61
Pentavalent phosphorus compounds, 9
Perkow reaction, 18, 21–23
  possible mechanisms, 21
Permeability factor P, 197
Peroxidases, 162
Peroxides, 79
Perphosphoric acid, 101
Persistency, 195
  in soils, 195
Pestan® *see* Mecarbam
Phenamiphos, 304
Phencapton® *see* Phenkapton
Phenkapton, 59, 94, 105, 106, 257
  colorimetric determination of, 106

determination of, 105
hydrolysis rates, 59
reaction with bromine, 94
Phenobarbital, 162
Phenobarbital pretreatment, 217
  effect on the toxicity of organophosphorus insecticides,
    217
Phenthoate, 181, 207, 231, 255
  analogs, 255
    mammalian toxicity of, 255
  as inhibitor of anthocyanin development, 231
  degradation of, 181
Phenyl benzylcarbamate, 227
  protective effect of, 227
    from the delayed neurotoxicity, 227
1-Phenyl-3-(diethoxyphosphinothioyloxy)-1,2,4-triazole
    see Triazophos
Phenyl N,N'-dimethylphosphorodiamidate *see* Nellite
Phenylmethanesulfonyl fluoride, 227
  protective effect of, 227
    from the delayed neurotoxicity, 227
Phenylphosphonothionates, 267
  alkyl phenyl, 267
  biological activities of, 267
Phorate, 30, 58, 65, 67, 94, 107, 164, 182, 291, 314
  colorimetric determination of, 107
  fungicidal effect of, 314
  hydrolysis of, 67
  hydrolysis rates, 58
  metabolic degradation of, 182
  metabolic pathways of, 164
  oxidation of, 94
  sulfoxide and sulfone derivatives of, 65
    hydrolysis of, 65
  synthesis of, 30
Phorate-O-analog, 58
  hydrolysis rates, 58
Phosalone, 67, 182–183, 260
  hydrolysis of, 67
  metabolic pathways of, 182–183
Phosbutyl, 321
Phosdiphen, 313
Phosdrin acid, 70
Phosdrin® *see* Mevinphos
Phosfolan, 279
Phosfon®, 332
Phosmet, 29, 50, 104, 119, 182, 259, 300
  alkylphosphonate analogs of, 50
  as an animal systemic, 300
  base peaks of, 119
  colorimetric determination of, 104
  metabolism of, 182
  synthesis of, 29
Phosnichlor, 243
Phosphamidon, 86, 106, 115, 167, 172, 187, 228,
    284–285
  base peak ion of, 115
  colorimetric determination of, 106
  demethylation of, 86
  metabolic pathways of, 284
  oxidative N-dealkylation of, 167
  teratogenic effect of, 228

Vinyl phosphates, 21
  preparation of, 21